Biogenic Wastes-Enabled Nanomaterial Synthesis

Abhishek Kumar Bhardwaj
Arun Lal Srivastav • Swapnil Rai
Editors

Biogenic Wastes-Enabled Nanomaterial Synthesis

Applications in Environmental Sustainability

Springer

Editors
Abhishek Kumar Bhardwaj
Department of Environmental Science
Amity School of Life Sciences (ASLS)
Amity University Madhya Pradesh (AUMP)
Gwalior, Madhya Pradesh, India

Arun Lal Srivastav
School of Engineering and Technology
Chitkara University
Shimla, Himachal Pradesh, India

Swapnil Rai
Department of Environmental Science
Amity School of Life Sciences (ASLS)
Amity University Madhya Pradesh (AUMP)
Gwalior, Madhya Pradesh, India

ISBN 978-3-031-59082-5 ISBN 978-3-031-59083-2 (eBook)
https://doi.org/10.1007/978-3-031-59083-2

This Springer imprint is published by the registered company Springer Nature Switzerland AG
The registered company address is: Gewerbestrasse 11, 6330 Cham, Switzerland

Paper in this product is recyclable.

Contents

A Review on Agricultural Wastes–Based Green Metal and Metal Oxide Nanoparticles

Sakshi Kabra Malpani, Renu Hada, and Deepti Goyal

1 Introduction

Nanoparticles are small-sized particles with at least one dimension in the nano range, that is, less than 100 nm. These particles have high specific surface area and exhibit size-dependent features such as high reactivity; strong sorption capacity; superparamagnetism; excellent mechanical, optical, and electrical properties; uniform porosity; surface functionalization; and superior attachment with functional groups (Khan et al., 2019). Owing to these unique properties, nanoparticles are extensively used in diverse fields of applications such as catalysis (Astruc, 2020), environmental remediation (Del Prado-Audelo et al., 2021), fuel cells, adsorption (Osagie et al., 2021), coatings, sensing, medicinal chemistry, drug delivery (Mitchell et al., 2020), and energy storage (Awad et al., 2021). These days nano metals, nano metal oxides, nano silica, and carbon nanoparticles are the most usual types of nanoparticles, which have been extensively synthesized (Joudeh & Linke, 2022). Especially nanoscale metals and metal oxides are potential candidates in the field of adsorption, catalysis, and pollution control and can easily substitute traditional adsorbents, catalysts such as activated carbon, silica, alumina, and polymers (Chavali & Nikolova, 2019). Nano metals and nano metal oxides are used as such in the form of powder or compressed as pellets for their larger scale use in many industrial applications. Depending upon the application of nanoparticles, their methods

S. K. Malpani
Save the Water, Inc., Fort Lauderdale, FL, USA

R. Hada
Department of Chemistry, Shri G. S. Institute of Technology & Science, Indore, Madhya Pradesh, India

D. Goyal (✉)
Department of Applied Chemistry, University School of Vocational Studies & Applied Sciences, Gautam Buddha University, Greater Noida, Uttar Pradesh, India

A. K. Bhardwaj et al. (eds.), *Biogenic Wastes-Enabled Nanomaterial Synthesis*,
https://doi.org/10.1007/978-3-031-59083-2_1

of synthesis may vary. Two approaches—top-down and bottom-up—are majorly used to synthesize nanoparticles. Many physical, mechanical, chemical, and biological methods are used within these two approaches to produce such nanoparticles. Ball milling, etching, mechanical alloying, reactive milling, and micromachining are common top-down synthesis methods. While, molecular self-assembly, sol-gel, electron/ion beams, chemical vapor deposition, inert gas condensation, electrodeposition, ultrasonic dispersion, vacuum arc deposition, etc., are conventional bottom-up synthesis procedures (Abid et al., 2022). Nanoparticles extracted from agricultural wastes possess some lucrative features such as smaller size, more stability, reduced toxicity levels, decreased environmental and health issues associated with their preparation, and, above all, they are highly cost-effective. Nowadays, metal and metal oxide nanoparticles synthesized by using these approaches are of tremendous therapeutic, chemical, eco-friendly, sustainable importance and can be efficiently used in real-life applications (Zamani et al., 2019). For example, silver and gold nanoparticles can be produced from biomolecules available in agricultural wastes (Thangadurai et al., 2021). Apart from metal nanoparticles, production of metal oxide nanoparticles has also gained a lot of attention. For example, silicon oxide or silica nanoparticles are extensively prepared from agricultural waste materials such as rice husk ash, sugarcane bagasse, and corncobs (Malpani & Goyal, 2022). These nanoparticles are chemically inert, stable at higher temperatures, can be modified easily, and thus they have numerous industrial applications such as drug delivery, furnace lining, catalysis, adsorption, and energy storage. Several types of important carbon nanoparticles can also be derived from agricultural and biomass waste materials (Thangadurai et al., 2021).

About 2.6 billion tons of solid waste will be generated across the globe by the end of 2030 (Peng et al., 2023). Out of this total solid waste, about 1300 million tons of waste has been generated from the agricultural sector (Amran et al., 2021). Agricultural or agro-wastes are results of agricultural practices right from crop cultivation, dairy farming, livestock rearing to the first processing of raw agricultural products. Approximately 30% of agricultural goods produced globally are discarded as agricultural wastes (El-Ramady et al., 2022). These agricultural wastes are rich sources of nutrients, organic material, and metal and metal oxide–based nanoparticles. They comprise husks from cereal and oil industries, fruit peels from beverage and juice industries, and tea and coffee wastes from their respective manufacturing units. In low- or middle-income countries, more than 50% of agricultural waste is either burnt or dumped as such. Such careless waste management practices are considerable reasons for environmental pollution and harmful diseases. Thus, there is a pressing need to develop useful and sustainable methods to utilize such wastes in a fruitful manner. Production of nanoparticles is one of the value-added methods to utilize or recycle agricultural wastes. It not only helps in attaining goals of circular economy, green chemistry, and sustainable development but also minimizes the negative footprint of ill management or disposal of agricultural wastes. It also adds up to the total income of local people and helps with their welfare.

Nanoparticles synthesized by conventional methods are highly expensive and could turn out to be toxic also, thus can pose health, legal, and environmental threats. If agricultural wastes are used as raw materials for preparing nanoparticles, then by using green or clean manufacturing processes such wastes can be converted into value-added materials. Such solid wastes can be practically recycled, embark on the economic value of the final products, and deploy mass-scale manufacturing. This chapter provides a perception about various solid agricultural wastes that are generally used for bulk-scale production of metal and metal oxide nanoparticles. Results of characterization techniques, which give a clear vision about features of nanoparticles, are also summarized here. Then, several research works about synthesis of metal and metal oxide nanoparticles from these agricultural wastes are discussed. In the end, general applications of these nanoparticles are discussed briefly. Prospects and concerns are elucidated. This chapter also sketches that production of clean, safe metal and metal-oxide nanoparticles from solid agricultural wastes can become a golden chance in terms of research in areas of nanotechnology, solid waste management, mass-scale production of green nanoparticles, and concurrently also divulged in solving the issues of circular economy and sustainable growth.

2 Agricultural Wastes and Their Types

The agriculture sector plays a vital role in the social and economic development of India, but simultaneously, a huge amount (approximately 350 MTons) of agricultural wastes is generated every year. As per the data released by Ministry of New and Renewable Energy, these wastes can be utilized to generate about 18,000 MW of power every year (Mohite et al., 2022). Agricultural wastes, such as rice husk, wheat husk, rice straw, sugarcane bagasse, bamboo leaves, empty seed shells, and fruit and vegetable wastes, are the most common examples which have been produced worldwide (FAO, 2000; Mohite et al., 2022). If these wastes are released into the environment without applying proper disposal procedures, it may cause environment pollution and create threat to the mankind and animals as well (da Luz Corrêa et al., 2020; Lopes et al., 2021).

The agricultural waste materials mainly contain organic and inorganic components. The organic components can be oxidized easily, and the inorganic component consists of various minerals applicable for manufacturing various metal nanoparticles such as silver and gold nanoparticles and metal oxide nanoparticles such as silica nanoparticles. Agricultural wastes are classified into two different types, that is, agro-wastes and agro-industrial wastes. Agro-wastes can be further divided into field wastes and process wastes. Field wastes are wastes that are present in the field after the process of crop harvesting. These wastes consist of stalks, stems, leaves, and seed pods, whereas the process wastes are wastes present after the produced crop is being processed. These wastes consist of husk, bagasse, seeds, peels, roots, etc. The process waste and the field waste are commonly used for animal food, fertilizing soil, manufacturing, and other processes (Sadh et al., 2018). The agro-industrial wastes involve farm products and the wastes generated every year through the food processing industries such as juice, fruits, chips, and meat industries (Rudra et al., 2015). These wastes are schematically presented in Fig. 1.

2.1 Straw

Straw falls in the category of agricultural waste, and it contains dry stalks of cereal plants without grain. It consists about half of the yield of cereal crops such as wheat, barley, and rice (FAO, 2000). They can be used as energy sources, additives of concrete materials, and nanoparticles production. They mainly consist of cellulose, hemicellulose, lignin, lipids, and protein. Wheat straw and rice straw are rich in mineral content and poor in nitrogen. They show high carbon to nitrogen ratio, which leads to lower biodegradability as compared to other agricultural residues. Most of the quantity of straws are being utilized traditionally in major developing countries where they are produced largely (Álvarez et al., 2021; Belewu & Babalola, 2009).

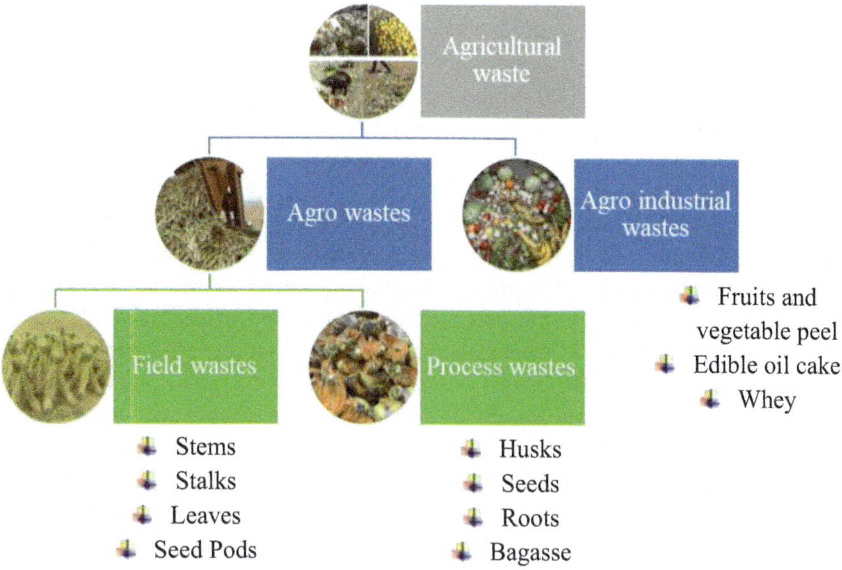

Fig. 1 Schematic classification and major sources of agricultural wastes

2.2 *Rice Husk and Rice Husk Ash*

Rice husks are the hard protecting covers of rice grains, which are considered as a waste during rice grain processing. They consist of mainly cellulose, lignin, and minerals (Das et al., 2020; Esra et al., 2013). The mineral quantity in rice husks varies from 15% to 20% by weight of rice husks. Due to high silica content in rice husks, it burns with difficultly under natural conditions, and biodegradation of rice husk is quite difficult and can also create environmental and ecological issues. Rice husk has high calorific value, approximately 17,000 KJ/Kg, due which it is mainly used in energy production through direct burning. This inappropriate combustion of rice husk generates another waste known as rice husk ash, which again causes environmental and disposal problem (Hossain et al., 2018). Being rich in silica, it also acts as a source for producing silica nanoparticles.

2.3 *Sugarcane Bagasse*

Sugarcane bagasse is an abundantly available biomass, obtained after sugarcane processing. It is majorly found in tropical countries (Ajala et al., 2021; Vaibhav et al., 2015). Due to its abundance in nature, it is harnessed for numerous applications such as energy and environment sustainability; paper industries; and feedstock for various products such as biopolymers, biochemicals, and biofuels (Hernández-Salas et al., 2009; Loh et al., 2013; Mahmud & Anannya, 2021; Toscano Miranda et al., 2021). It is typically rich in cellulose (44%) (Karp et al., 2013), which can be easily extracted and further utilized in paper and fiber production. The fibrous part may also be used in the textile and civil engineering sectors. Other than that, the bagasse can also be a feedstock for producing nanoparticles, for example, cellulose nanocrystals and starch-fiber nanocomposites (Gilfillan et al., 2012; Slavutsky & Bertuzzi, 2014).

2.4 *Corncob*

Corn is one of the most commonly planted and consumed crops round the globe. At the time of corn processing stems, leaves, husks, and corncobs remain as waste. Corncob is essential inedible residue generated in large amount after the removal of corn kernels and considered as an agricultural waste. Approximately 18 kg corncob waste is generated from 100 Kg of corn grain production (Tsai et al., 2001). Corncob has low density; fibrous, porous sponge like structures; and excellent adsorbing capacity. It is rich biomass resource that contains cellulose, hemicelluloses, and lignin. Some agricultural wastes can be treated as feed or compost, but it is difficult to treat corncob in the similar manner; hence, a large amount of corncob remains untreated. Since corncob contains wood like fibrous components, it is suitable as a building material, and it can also be used as mild abrasives to clean building surfaces. Most of their properties such as macrostructure, microstructure, chemical composition, water absorption, fire resistance, and thermal insulation capacities are found to be comparable with extruded polystyrene and expanded polystyrene (Berber-villamar et al., 2018; Choi et al., 2022). It also acts as an excellent source of producing silica and carbon nanoparticles.

2.5 *Plant Leaves*

Leaves of many plants, for example, papaya, mango, cereal, and bamboo find application in nanoparticle synthesis. Bamboo is one of the abundant and fast-growing renewable bio-resources. Bamboo leaves obtained from bamboo are commonly used (Silviana & Bayu, 2018) for making papers, handicrafts, and medicines. Apart from cellulose, hemicelluloses, and lignin, bamboo leaves are also rich in a variety of bioactive components such as flavonoids, polysaccharides, phenolic acids, and amino acids, thereby exhibiting multiple pharmacological activities (Cheng et al., 2023; Shen et al., 2022; Wang et al., 2015b). Bamboo leaf extracts containing bioactive components have been found to have significant effects as anti-virus, antioxidants, anti-fatigue, and anti-inflammatory based on modern nano pharmacological research (Kimura et al., 2022). Bamboo leaves are also used as a source for nanoparticle synthesis such as spherical silica nanoparticles with average size of 25 nm that are synthesized by thermal combustion and alkaline extraction method (Rangaraj & Venkatachalam, 2017).

2.6 *Fruit and Vegetable Peels*

Peel is an outer protective layer of a fruit or vegetable, which can be peeled and separated out. Their chemical composition shows the presence of proteins, lipids, fibers, and carbohydrates as well as minerals of zinc, iron, calcium, and manganese (Dibanda Romelle et al., 2016). Peel is renewable, green, nontoxic material like most of other agricultural waste materials. Potato peels are used for the synthesis of various nanoparticles such as silver, copper (Idhayadhulla et al., 2021), and zinc oxide (Bhuvaneswari, 2017). They contain starch, non-starch polysaccharide, vitamins, protein, lipids, acid soluble/insoluble lignin, and ash (Bhuvaneswari, 2017). Lemon peel and orange peel are also used for green synthesis of silver nanoparticles, and the method used was clean, safe, and eco-friendly; thus, obtained silver nanoparticles showed efficient antibacterial activity toward *Escherichia coli* (Niluxsshun et al., 2021).

3 Synthesis of Metal/Metal Oxide Nanoparticles

Metal and metal oxide nanoparticles possess distinctive properties due to their high surface-to-volume ratio. Owing to their unique properties resulting from the exceptionally high surface-to-volume ratio, metal/metal oxide nanoparticles (NPs) have attracted much attention in the field of science as well as in different technologies. The conventional chemical methods for the synthesis of these metal/metal oxide nanoparticles consume hazardous chemicals as precursors or as reducing agents, which can lead to severe environmental issues (Gutiérrez-Wing et al., 2012). In addition, these methods are quite expensive and time-consuming. To overcome these issues, scientists are exploring the greener route for the synthesis of these nanoparticles. For this purpose, several plant-based materials have been successively utilized as a raw material (Kulkarni & Muddapur, 2014). Different parts of a plant or crop such as leaves, stem, fruit, and roots can be used for this purpose (Suman et al., 2013). The size and shape of the synthesized nanoparticles can be altered by optimizing the reaction conditions such as time, temperature, and pH of the solution (Kharissova et al., 2013). In the recent scenario, for the further improvement and cost-cutting of synthesis process of nanoparticles, agricultural wastes have been utilized in place of plant-based materials (Saha & Kim, 2022).

Below are some green approaches adopted for synthesis of some metal and metal oxide nanoparticles from agricultural wastes.

3.1 Synthesis of Silver Nanoparticles

Silver nanoparticles are the most widely utilized nanoparticles in the area of food, medicine, and health care due to their extraordinary antibacterial and antifungal activities. Every year an estimated amount of 500 tons of silver nanoparticles are produced globally. Nowadays, several environmental benign methods are being developed to produce these silver nanoparticles by the use of agricultural wastes.

Wheat straw extracted lignin has been utilized to synthesize silver nanoparticles under optimized reaction conditions. Lignin was successfully used as reducing, capping, and stabilizing agent. Thus, prepared silver nanoparticles possessed face cubic centered (FCC) crystal structure with the average particle size of 15–20 nm. Silver nanoparticles showed excellent antimicrobial and antioxidant activities (Saratale et al., 2019).

Francisco et al. have reported the synthesis of silver nanoparticles from safflower waste (*Carthamus tinctorius* L.) (Figs. 2 and 3). As-synthesized silver nanoparticles were characterized using SEM and TEM techniques, and the results showed that the synthesized nanoparticles were uniform and spherical with the average particle diameter of 8.67 ± 4.7 nm. The prepared silver nanoparticles showed excellent antibacterial activity toward *Staphylococcus aureus* (Gram positive) and *Pseudomonas fluorescens* (Gram negative) bacteria (Rodríguez-Félix et al., 2021).

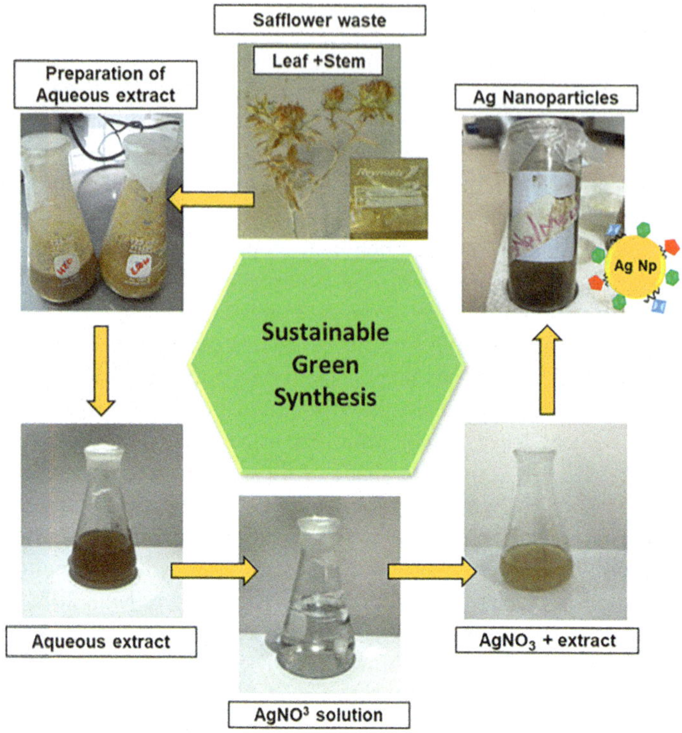

Fig. 2 Schematic representation of obtaining silver nanoparticles from aqueous extract of safflower waste. (Rodríguez-Félix et al., 2021)

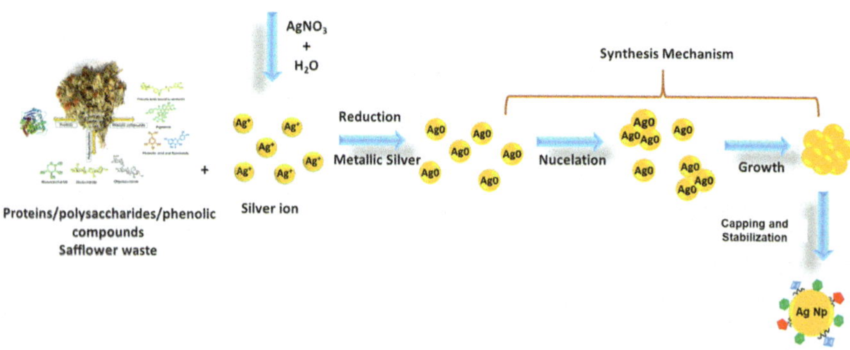

Fig. 3 Possible macromolecules and phenolic compounds present in the safflower waste aqueous extract

In two separate studies, silver nanoparticles were synthesized using the same material, that is, pomegranate fruit peel extracts. In the first study, authors reported that the use of ellagic acid present in the peel extract as a reducing agent to prepare silver nanoparticles (Ahmad et al., 2012). Thus resulting nanoparticles were spherical, which absorbed at UV-Vis wavelength of 427 nm and possessed FCC crystal structure with 5–10 nm average particle size. In the other study performed by Shanmugavadivu et al. (Shanmugavadivu et al., 2014), the prepared silver nanoparticles were in the range of 5–50 nm and showed absorption at 371 nm. The prepared particles showed great antibacterial activity against *Staphylococcus aureus*. In this study, primary and secondary amines groups were proposed as major functional groups, and other aromatic groups as implementers of silver nanoparticles synthesis.

In a similar study, a group of researchers have demonstrated the synthesis of silver nanoparticles using *Punica granatum* peel extract. The presence of silver nanoparticles was confirmed by the brownish yellow solution along with the presence surface plasmon peak at 432 nm. The particle size was observed as 30 nm with distorted spherical shape (Edison & Sethuraman, 2013). The prepared silver nanoparticles were successfully utilized in catalytic reduction of 4-nitophenol.

Synthesis of silver nanoparticles has also been performed using oak fruit hull extract under optimized reaction conditions ($AgNO_3$ concentration = 40 g/L; pH = 9; and temperature 45 °C) (Heydari & Rashidipour, 2015). The characterization results showed that the prepared silver nanoparticles were spherical and of 40 nm particle size. The cytotoxic activity of the particles was checked against human breast cancer cell. Another study demonstrated the use of defatted cashew nut shell starch to produce silver nanoparticles. The prepared silver nanoparticles (particle size 10–50 nm) were found structurally identical with the commercial silver nanoparticles (Velmurugan et al., 2015b).

The synthesis of silver nanoparticles was also reported using coconut shell extract as raw material. UV-visible spectroscopy (absorption peak at 432 nm) and TEM (particle size approximately 10–25 nm) techniques were used to confirm the production of silver nanoparticles. The prepared particles were found efficient toward antibacterial activity against human pathogens *Staphylococcus aureus*, *Listeria monocytogenes*, *E. coli*, and *Salmonella typhimurium* (Sinsinwar et al., 2018).

In a similar study, bilberry waste and spend coffee grounds were chosen as raw materials for the synthesis of silver nanoparticles. The phenolic extract of bilberry waste and spend coffee grounds was prepared using ethanol solvent and mixed with $AgNO_3$ solution at 25 °C. The size and shape (10–20 nm and spherical) of silver nanoparticles were observed via XRD, SEM, and DLS techniques (Chianese et al., 2016).

3.2 *Synthesis of Gold Nanoparticles*

Gold nanoparticles have potential applications in biomedical fields due to their great biocompatibility and nontoxic nature. Also gold nanoparticles are widely utilized in several applications such as catalytic reactions, antioxidant and antimicrobial activities, and in bio-sensing and bio-imaging purposes (Oladipo et al., 2017). In the previous studies, gold nanoparticles have been synthesized via various conventional physical and chemical methods, but these methods are costly and produce hazardous by-products. Thus, in the recent studies, the synthesis of gold nanoparticles via green methods using plant extracts and agricultural wastes such as banana peels and custard apple peels are being explored. The presence of phenolic compounds in these agricultural wastes is responsible for the production of nanoparticles (Mojumdar & Deka, 2019; Patra & Baek, 2015).

A recent study provides the green synthesis of biocompatible gold nanoparticles using waste fruit peel extracts of *Ananas comosus* (pineapple) and *Passiflora edulis* (passion fruit) for reduction of gold (III) chloride trihydrate ($HAuCl_4$) (Fig. 4a). The chemicals present in the fruit extracts, including proteins, minerals, lipids, vitamin, phenolic compounds, flavonoids, and carotenoid, act as reductants and nanoparticle stabilizers. The prepared gold nanoparticles exhibit a characteristic red coloration and UV-visible spectral absorption maximum around 545.5 nm and 545 nm using pineapple and passion fruit extracts, respectively (Fig. 4b). The resulting gold nanoparticles showed non-cytotoxic effects toward normal and cancer cells, which indicate their biocompatibility and thus their wide applications in medical field (Chiravoot et al., 2021).

Some researchers have synthesized gold nanoparticles using watermelon rind extract and evaluated their antibacterial activity against foodborne pathogens. The results indicated that the prepared gold nanoparticles possessed high antioxidant and anti-proteasome inhibitory potential, showing potential as an anticancer agent. The particles were spherical with 20–140 nm size and showed absorption maxima at 560 nm (Patra & Baek, 2015).

Another agricultural waste reported in the literature for the production of gold nanoparticles is mango peel extract. The particles were obtained with approximately

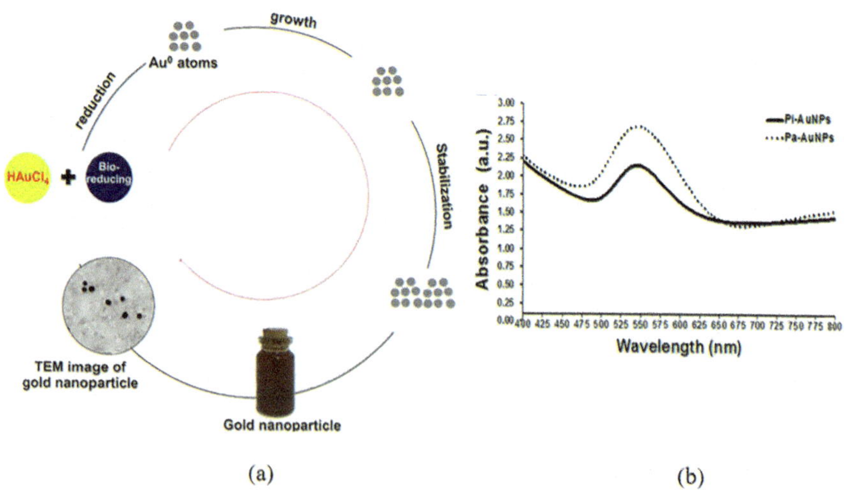

Fig. 4 (**a**) Synthesis mechanism of gold nanoparticles using fruit peel extract. (**b**) UV-Vis absorption spectra of gold nanoparticles using pineapple and passion fruit extracts, respectively

20 nm size and were found biocompatible with African green monkey kidney normal cells (CV-1) and normal human fetal lung fibroblast cells (wI-38) at a high concentration of 160 μg/mL (Yang et al., 2014).

In a similar study, grape waste was utilized to produce gold nanoparticles with the particle size in the range of 18–25 nm. The particles showed maximum absorbance at 537 nm in UV-visible spectrophotometer. The researchers proposed that the gold nanoparticles were formed due to the presence of some phenolic compounds in the grape waste (Krishnaswamy et al., 2014).

Malhotra et al. have reported the synthesis of gold nanoparticles using rice bran extract. The characterization results revealed that the ferulic acid present in the extract is responsible for the reduction of Au^{+3} to Au^0 (Malhotra et al., 2014).

Furthermore, some researchers have utilized aqueous extract of de-oiled *Jatropha* waste to produce gold nanoparticles under optimized reaction conditions. The produced nanoparticles were of different shapes (triangle, hexagonal, and spherical) with the particle size in the range of 10–15 nm. The surface plasmon resonance peak was obtained at 520 nm, which confirms its high compatibility in several biomedical applications such as tissue imaging, drug delivery, and photothermal therapy (Kanchi et al., 2018).

In a similar study, seed shells and detoxified-defatted seed meal aqueous extracts of *Jatropha curcas* were utilized to produce gold nanoparticles with the particle size in the range of 5–20 nm under optimized reaction conditions (temperature = 60 °C and seed meal/shell:$HAuCl_4$ = 1:1). The characterization results showed that on increasing concentration of reducing agent, time, and temperature, the shape of the particles also changes. Thus, produced gold nanoparticles were found biocompatible with brain cancer glioma GI-1 and neuronal HCN-1A cell lines. Along with this, the prepared particles possessed significant bio-imaging and photoluminescent properties (Sheikh Mohamed et al., 2013).

Moringa oleifera flower extract is also reported to produce gold nanoparticles. The presence of proteins, carotenoids, phenolics, flavonoids, and alkaloids were responsible for the conversion of Au^{+3} to Au^0. The prepared particles showed absorption maxima at 540 nm in UV-visible spectroscopy. The particles were of different shapes and sizes (triangular, hexagonal, and spherical and approximately 5 nm) showing efficient catalytic activity toward reduction of 4-nitrophenol and 4-nitroaniline (Anand et al., 2015).

In a separate study, palm oil mill waste was utilized to produce spherical gold nanoparticles with average particle size of 18.75–5.95 nm (Gan et al., 2012). In addition, banana peel has been utilized to produce gold nanoparticles. The extracted banana peels were first crushed, boiled and precipitated using acetone, and mixed with $HAuCl_4$ and stirred under optimized reaction conditions (Bankar et al., 2010). Thus, produced nanoparticles were characterized using UV-visible spectroscopy, X-ray diffraction pattern, SEM, and FTIR analysis. The particles showed significant antimicrobial activities.

3.3 *Synthesis of Silica Nanoparticles*

Silica nanoparticles are gaining incredible attention of researchers due to their potential applications in different industries and medical fields (Briceño-Ahumada et al., 2021; Gu et al., 2013). Conventional synthesis of silica nanoparticles uses toxic and costly precursors. Therefore, green alternatives of these precursors such as agricultural wastes (rice husk, wheat husk, rice straw, corncob, sugarcane bagasse, and bamboo leaves) (Shim et al., 2015; Velmurugan et al., 2015a) are widely utilized in the recent scenario. Synthesis of silica nanoparticles using agricultural wastes makes the process green and environmentally friendly.

The use of rice husk for the synthesis of silica nanoparticles is well documented in the literature. A group of researchers have reported the use of different acids for extraction of silica nanoparticles from rice husk. Among all acids, HCl was found efficient to produce silica nanoparticles with the highest surface area (Liou & Yang, 2011). Similarly, alkali fusion and acid precipitation method are also reported to extract silica nanoparticles from rice husk. The produced silica nanoparticles were in the range of 20–25 nm and 274 m^2/g surface area (Yuvakkumar et al., 2014a). Furthermore, a separate study shows the extraction of silica nanoparticles from rice husk via pyrolysis method. The process involves the mixing of H_3PO_4 with sodium silicate solution (between pH 3 and 6) and further addition of PEG (polyethylene glycol), which helps to generate porosity on the silica surface. Thus produced silica nanoparticles possessed high surface area with 2.30 nm diameter (Li et al., 2011). Similar investigation was done at different temperatures ranging from 500 to 650 °C. Maximum yield of silica nanoparticles were obtained in temperature from 500 to 610 °C, which decreased on further increasing temperature (Ghaferi et al., 2021). Rice husk–derived silica nanocomposites with good tensile strength and uniform distribution are also well documented in the literature (Ban & Saddam, 2017).

Chen et al. have demonstrated the synthesis of highly porous silica nanoparticles from rice husk via extraction with ionic liquids. For this purpose, firstly rice husk was dissolved in ionic liquids and filtered (Fig. 5). The obtained rice husk residue

Fig. 5 Conversion of rice husk into nanosilica

(contains high silica content) was then thermally activated to get highly pure and amorphous nanosilica (Chen et al., 2013).

Apart from the use of rice husk, bamboo leaves, high silica–containing material have also been utilized for the production of silica nanoparticles. Wang et al. have demonstrated the use of bamboo leaves in the production of silica nanoparticles (5–8 nm size and 300 m^2/g surface area) via thermal decomposition and magnesiothermic reduction (Wang et al., 2015a). In a separate study, silica nanoparticles were produced from bamboo leaves via ultrasonic method. Thus, the produced nanoparticles were comparatively of smaller size (Dirna et al., 2020). Furthermore, bamboo leaves derived silica nanoparticles have been utilized in producing silica supported Fe_2O_3 photocatalyst with crystallite size 20–70 nm. The photocatalytic activity of the catalyst was checked in degradation of Rhodamine B dye (Fatimah et al., 2019a). Similarly, silica supported TiO_2 and silica supported ZnO photocatalysts have also been synthesized using bamboo leaves, and the photocatalytic activity was evaluated in degradation of methylene blue and Rhodamine B dye, respectively. The degradation study suggested that the efficiency of the catalyst increased with increase in TiO_2 content (Fatimah et al., 2019b, 2021).

The use of corncob for the synthesis of silica nanoparticles is also well documented in the previous studies. The high silica content in corncob makes it suitable for the production of silica nanoparticles (Duque-Acevedo et al., 2020; Okoronkwo et al., 2013; Velmurugan et al., 2015a). A study reports the synthesis of silica nanoparticles via sol-gel method followed by alkali treatment of corncob. The produced silica nanoparticles were found in the range of 50 nm particle size (Velmurugan, Shim et al., 2015; Duque-Acevedo et al., 2020) Corncob ash–derived silica nanoparticles were utilized to prepare faujasite nanosheets under optimized conditions (temperature 85 °C and 0.030 M TPOAC) (Salakhum et al., 2018). In a separate study, corncob was converted into carbon nanocomposite material. The prepared nanocomposite was found efficient for the removal of methylene blue dye and U(VI)/Cr(VI) ions from the aqueous solution.

Sugarcane bagasse, a by-product of ethanol industries, contains high amount of silica. A group of researchers have synthesized silica nanoparticles from sugar cane bagasse via alkali fusion followed by acid precipitation method (Falk et al., 2019; Vaibhav et al., 2015). Similarly, silica nanoparticles with high surface area and pore volume have been produced via acid treatment of sugarcane bagasse. A separate study demonstrated the use of sugar cane bagasse in the synthesis of silica supported TiO_2 photocatalyst with 5–20 nm particle size. The catalytic activity was tested on degradation of methyl orange dye. Falk et al. have synthesized high purity silica nanoparticles with 10 nm particle size via sol-gel method (Falk et al., 2019). Similarly, the synthesis of silica nanoparticles (80–100 nm particle size) via sol-gel method followed by freeze drying and heat drying was also investigated. The study revealed that the nanoparticles formed via freeze drying method were possessed high surface area and porosity as compared to the particles formed via heat drying method (Boonmee & Jarukumjorn, 2020). In a recent study, biogenic nanosilica has been synthesized using different agricultural wastes such as groundnut shell, banana peel, coconut husk, orange peel, and walnut shell via alkali treatment. The SEM and

TEM analysis of the obtained nanosilica confirmed the presence of agglomerated nanosized particles. The results indicated that nanosilica derived from groundnut shell possessed high surface area (0.829 μm) as compared to the nanosilica derived from other agricultural wastes. The particle size distribution for walnut shell–derived nanosilica was found 12.18 μm to 912 μm for nanosilica derived from banana peel (Peerzada & Chidambaram, 2021).

In a separate study, magnetic silica nanoparticles were synthesized using waste barley husk via a two-step green route (Fig. 6). Barley husk was treated with acid and then thermally activated to prepare silica nanoparticles, which then treated with Fe_3O_4 to synthesize magnetic silica nanoparticles. The characterization results of thus obtained silica nanoparticles confirmed the average diameter of 162 nm with a large surface area ~ 120 m^2/g. The prepared magnetic silica nanoparticles were found to be an efficient adsorbent for removing petrol contaminants (with removal efficiency of more than 80%) from waste (Akhayere et al., 2020).

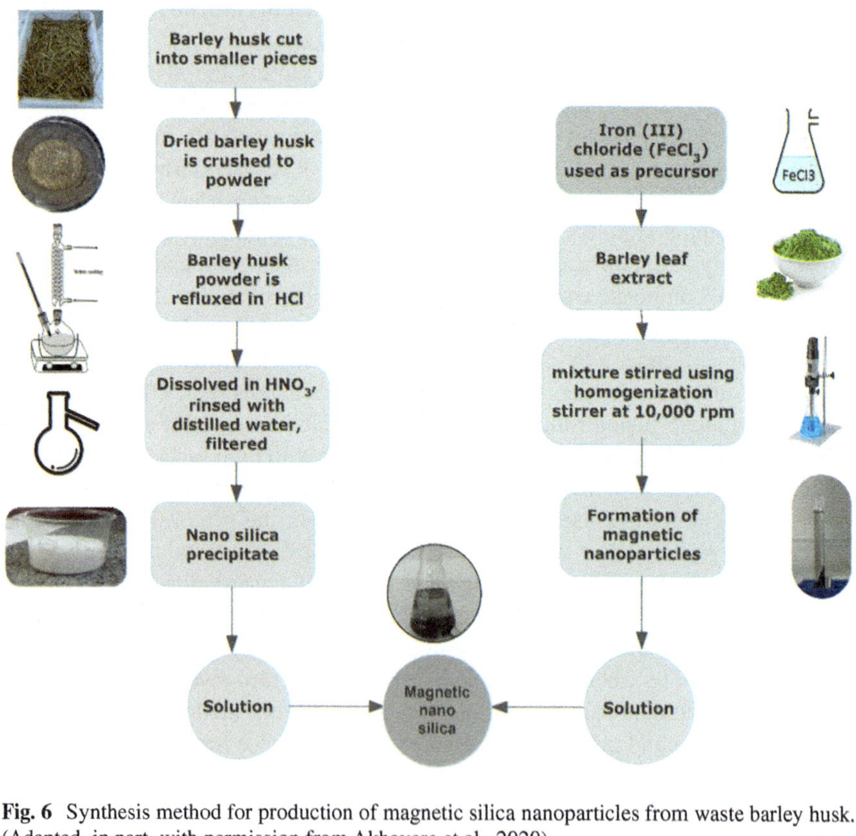

Fig. 6 Synthesis method for production of magnetic silica nanoparticles from waste barley husk. (Adapted, in part, with permission from Akhayere et al., 2020)

3.4 Synthesis of Other Metal/Metal Oxide Nanoparticles

The synthesis process of metal and metal oxide nanoparticles from different waste materials is less expensive and environmental friendly. Papaya leaves have been utilized to prepare iron oxide nanocomposites via thermal decomposition method (Ahmed & Ahmaruzzaman, 2015). These agricultural wastes well preserve several minerals and food reserves and also contain some organic compounds such as phenols, alkaloids, flavonoids, terpenoids, proteins, enzymes, and carbohydrates. These organic compounds act as reducing or stabilizing agents to produce nanoparticles (Bachheti et al., 2020). A study revealed the synthesis of Mn_3O_4 nanoparticles from banana peel extract, which reduce $KMnO_4$ to Mn_3O_4 (Yan et al., 2014). Another study demonstrated the use of tea wastes in producing hydrated aluminum nanoparticles via co-precipitation method. In this process, $Al_2(SO_4)_3$ and NaOH were mixed with tea waste and polyacrylamide, and the obtained precipitate of porous aluminum was washed and dried (Cai et al., 2015).

Another study reports the synthesis of biogenic ZnO nanoparticles from rambutan peels. The process includes the mixing of zinc nitrate hexahydrate with rambutan peel extract at 80 °C for 2 h. The obtained zincellagate complex was dried at 40 °C and further calcined at 450 °C. The resulting ZnO nanoparticles fabricated on cotton were found efficient antibacterial agent toward *E. coli* and *S. aureus* bacteria (Yuvakkumar et al., 2014b). A similar study has been demonstrated by Yuvakkumar et al., showing the synthesis of NiO nanocrystals from rambutan peels (Yuvakkumar et al., 2014c). The presence of polyphenols in the extract was responsible for the reduction of $Ni(NO_3)_2$ to NiO nanocrystals. The obtained NiO crystals fabricated on cotton were shown antibacterial properties against *E. coli* and *S. aureus*, respectively.

In a recent study, crystalline platinum nanoparticles with 15–25 nm size have been synthesized using *Punica granatum* peel extract. The negative zeta potential shows the high stability of the prepared platinum nanoparticles. The study revealed that the presence of phenolic compounds in the peel extract was responsible for the capping and stabilization of prepared particles. The prepared platinum nanoparticles were found efficient toward the reduction of 3-nitrophenol (Dauthal & Mukhopadhyay, 2015).

From the above discussion, it is apparent that agricultural wastes have a wide scope for the synthesis of various metal and metal oxide nanoparticles due to the presence of various biomolecules, which can act as reducing or capping agents. The richness and easy availability of agricultural wastes can be successfully exploited on a large scale for the biogenic synthesis of nanoparticles. The system is cost-effective and can create wealth and useful products through nanotechnology.

A summary of nanoparticles synthesized from agricultural wastes is given in Table 1.

Table 1 Nanoparticles synthesized from various agricultural wastes

Agriculturalwastes	Name of nanoparticles	Synthesis method	Properties	References
Plant extract	Au, Ag, Cu, Pt, Pd, ZnO, TiO_2	Biosynthesis	Show potential applications in environmental and biomedical fields	Jadoun et al. (2021), Barnawi et al. (2022), and Chaudhary et al. (2023)
Fruit and vegetable peels	Ag and Au, TiO_2, ZnO	Green method using fruit and vegetable peel extracts	Silver and gold nanoparticles have average particle size range of 5–20 nm, exhibited potential antioxidant and antibacterial activity	Mojumdar and Deka (2019) and Oyekanmi et al. (2022)
Plant roots	Au	Green synthesis	Show antioxidant, antimicrobial, anti-inflammatory, anticancer activities, catalytic action	Varghese et al. (2021)
Safflower waste	Ag	Green synthesis using safflower waste aqueous extract	Show potential application as antibacterial agents in food and pharma industries	Rodríguez-Félix et al. (2021)
Bamboo leaves	Si and SiO_2	Magnesiothermal reduction, ultrasonic method	Less than 10 nm size and high lithium storage capacity	Wang et al. (2015b) and Dirna et al. (2020)
Wheat straw	Ag	Reduction of silver salt using straw extract	Antimicrobial and antioxidant silver nanoparticles showed FCC crystal structure with the average particle size of 15–20 nm	Saratale et al. (2019)
Watermelon rind extract	Au	Green synthesis using rind extract	Particles had spherical shape and have a size distribution of 20–140 nm, showed bio potential applications	Patra and Baek (2015)
Grape waste	Au	Green reduction using phenolic compounds present in grape fruit waste	Mean particle size range from 20 to 25 nm	Krishnaswamy et al. (2014)

(continued)

Table 1 (continued)

Agriculturalwastes	Name of nanoparticles	Synthesis method	Properties	References
Rice bran	Au	Bio mineralization of gold using ferulic acid present in rice bran extract	Spherical gold nanoparticles with average particle size of 50–100 nm	Malhotra et al. (2014)
Rice husk	SiO_2	Extraction using mineral acids, alkali fusion, pyrolysis, and acid precipitation method	Higher surface area with particle size range 20–25 nm, good tensile strength	Liou and Yang (2011) and Yuvakkumar et al. (2014b)
Seed shell	Au	Reduction method	5–10 nm sized gold nanoparticles were found biocompatible with brain cancer cells and possessed bio-imaging and photoluminescent properties	Sheikh Mohamed et al. (2012)

4 General Applications

Metal and metal oxide nanoparticles produced from solid agricultural waste materials have wide applications in many fields such as catalysis, drug-delivery, industrial, environmental, and energy sectors (Fig. 7). Silver and gold nanoparticles show antimicrobial and antioxidant properties; thus, these can be utilized in drug-delivery and food industry applications (Rodríguez-Félix et al., 2021). Silver nanoparticles extracted from coconut shell waste have been used as an antibacterial agent against severe human pathogens such as *E.coli* and *S. typhimurium* (Sinsinwar et al., 2018). Platinum, iron, and palladium nanoparticles extracted from different fruit peel waste have been utilized as catalysts, while ZnO and NiO nanocrystals derived from plant waste have also shown antibacterial activities (Adelere & Lateef, 2016). Bio-sensing, bio-imaging, and bio-labeling are other applications of metal and metal oxide-based nanoparticles. Such nanoparticles are also used in degradation of dyes, additives in paints, and removal of anthropogenic pollutants. Magnetic iron oxide nanoparticles can be produced from fruit peel waste and can be used as catalysts in different organic and environmental reactions (Adelere & Lateef, 2016). Thermal insulation, semi conductivity, and energy storage are other applications where such nanoparticles can also be utilized. Silica nanoparticles have been also used in multifaceted applications such as catalysts in Suzuki coupling reactions, oxidation reactions, and multicomponent organic reactions (Malpani & Goyal, 2022). They are also used in adsorption of toxic contaminants such as nitrate ions, carcinogenic polycyclic aromatic hydrocarbons, and dyes such as methylene blue, Rhodamine B, and methyl orange. They also behaved as nanocarriers in various drug-delivery

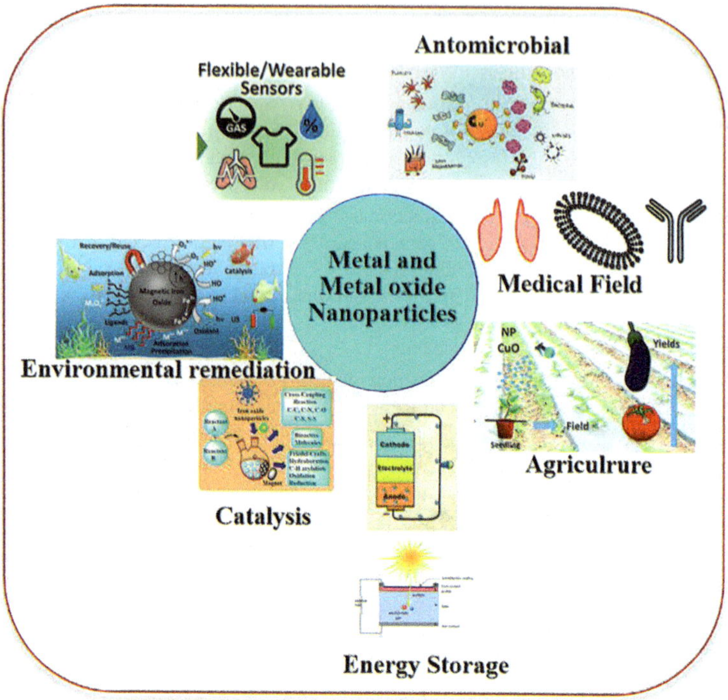

Fig. 7 Applications of metal and metal oxide nanoparticles

applications. They have been also utilized as anodes in Li-ion batteries, solar cells, and energy storage batteries.

5 Future Perspectives

Nanoparticles, particularly metal and metal oxide nanoparticles, have garnered substantial attention over the past few decades due to their unique physicochemical properties that differ significantly from bulk materials. The reduced size and high surface-to-volume ratio provide these nanoparticles with remarkable reactivity, strength, electrical conductivity, and optical properties. Some future perspectives of metal and metal oxide nanoparticles are summarized below:

1. *Medicine:*
   - *Targeted Drug Delivery:* Metal nanoparticles can be functionalized with specific ligands or antibodies to target specific cells or tissues, enabling localized treatment and minimizing side effects.

- *Thermal Ablation and Photothermal Therapy:* Gold nanoparticles, for instance, can absorb near-infrared light and convert it into heat, which can be used to destroy cancer cells.
- *Diagnostics:* Metal nanoparticles can be used as contrast agents in various imaging techniques, such as MRI and CT scans.

2. *Catalysis:*

- Due to their high surface area and enhanced reactivity, metal and metal oxide nanoparticles serve as efficient catalysts for various chemical reactions, potentially reducing the energy input required and increasing yield.

3. *Energy:*

- *Hydrogen Storage:* Some metal nanoparticles can absorb hydrogen and thus can be utilized in fuel cells.
- *Solar Cells:* Metal nanoparticles can enhance light absorption, leading to increased efficiency in photovoltaic devices.
- *Batteries:* Metal oxide nanoparticles can enhance the charge storage capacity and rate capabilities of batteries.

4. *Environmental Applications:*

- *Water Treatment:* Metal oxide nanoparticles can degrade organic pollutants in water.
- *Air Purification:* Metal/metal oxide nanoparticles can help remove harmful pollutants from air, especially when incorporated into filters or coatings.

5. *Electronics:*

- Metal nanoparticles can be used in memory storage devices, sensors, and other electronic components due to their unique electrical properties.

6. *Agriculture*:

- Metal nanoparticles can be used for the controlled release of pesticides or fertilizers, reducing the amount needed and minimizing environmental impact.

7. *Antimicrobial Applications:*

- Silver nanoparticles, for instance, exhibit strong antibacterial properties and can be incorporated into textiles, coatings, and medical devices to prevent microbial growth.

6 Concerns

We have tried to point out the most familiar types of solid agricultural wastes, metal and metal oxide nanoparticles produced from these wastes, and a few common applications of such nanoparticles in this chapter. However, many possibilities exist which can be unveiled by exerting more serious attempts from researchers, environmentalists, industrialists and government officials. The conception of generating value-added materials from solid wastes is quite tempting and satisfies the notion of principles of green chemistry and circular economy. But this area is lacking from gaps in knowledge, information, and implementation. Many engineered nanoparticles developed in research labs fail to compete in the real world/market due to lack of sufficient technologies for their mass production, generation of by-products or secondary wastes, easy degradation, costlier handling, transportation, etc. Prior to transition from lab to practical world, life-cycle assessment, risk analysis, market research, cost-benefit analysis, management of toxic by-products should have been conducted thoroughly. Data available in this field is deprived of these factors. Thus, it has become a limiting factor and should be addressed primarily. Precautions should have been applied throughout all phases of the life cycle, that is, production, usage, and then final disposal of these waste-derived nanoparticles. Even though the green chemistry and sustainability principles have been completely followed during the development phase, it is not necessary that nanoparticles produced are safe to use and do not cause ill effects on human beings and ecological systems. For example, copper, nickel nanoparticles, carbon nanotubes, and nano fullerenes could be carcinogenic. Some of the major concerns associated with the use of metal/metal oxide nanoparticles in our day-to-day life are outlined as follows:

1. *Toxicity and Environmental Impact:* The potential toxicological effects of metal and metal oxide nanoparticles on human health and the environment are still not fully understood. More comprehensive studies are needed to ascertain their safety.
2. *Scalability:* While lab-scale synthesis of nanoparticles is well established, large-scale production maintaining the same quality, cost, and properties remains a challenge.
3. *Stability:* Some nanoparticles tend to agglomerate over time, losing their unique properties. Methods to ensure their stability and shelf life during storage and use are essential.
4. *Regulations:* As the use of nanoparticles becomes more widespread, appropriate regulatory frameworks will need to be developed to ensure their safe application.

Given the rapid advancements in nanotechnology, it is expected that new applications and improvements in existing technologies will continue to emerge. Collaborative efforts between scientists from various disciplines, industries, and policymakers will be crucial in harnessing the full potential of metal and metal oxide nanoparticles while ensuring their responsible and sustainable use.

7 Conclusion

In conclusion, this chapter summarizes the concept of solid agricultural wastes, their types with some contemporary examples, and how most popular metal and metal oxide nanoparticles can be produced from such wastes. Some researches related to how these nanoparticles are derived from solid agricultural wastes are also summarized with their characteristic results. It also outlines the common applications of these nanoparticles in the field of catalysis, environmental remediation, energy storage, drug delivery, and other biomedical sectors. This work also suggests that an overall target-oriented system should be designed of both government and private organizations, which can help in mass-scale production of metal and metal oxide–based nanoparticles from agri-wastes. The process parameters should be optimized to achieve these goals. Researchers and their labs should be equipped with sufficient data and market study to make these products industry fit. Future prospects of use of metal and metal oxide nanoparticles in various sectors are also discussed. This chapter also pointed out a few concerns correlated with extensive usage of these nanoparticles.

References

Abid, N., et al. (2022). Synthesis of nanomaterials using various top-down and bottom-up approaches, influencing factors, advantages, and disadvantages: A review. *Advances in Colloid and Interface Science, 300*, 102597. https://doi.org/10.1016/J.CIS.2021.102597

Adelere, I. A., & Lateef, A. (2016). A novel approach to the green synthesis of metallic nanoparticles: The use of agro-wastes, enzymes, and pigments. *Nanotechnology Reviews, 5*(6), 567–587. https://doi.org/10.1515/ntrev-2016-0024

Ahmad, N., Sharma, S., & Rai, R. (2012). Rapid green synthesis of silver and gold nanoparticles using peels of Punica Granatum. *Advanced Materials Letters, 3*(5), 376–380. https://doi.org/10.5185/AMLETT.2012.5357

Ahmed, M. J. K., & Ahmaruzzaman, M. (2015). Fabrication and characterization of novel lignocellulosic biomass tailored Fe3O4 nanocomposites: Influence of annealing temperature and chlorazol black E sequestration. *RSC Advances, 5*(130), 107466–107473. https://doi.org/10.1039/C5RA20605G

Ajala, E. O., et al. (2021). Sugarcane bagasse: A biomass sufficiently applied for improving global energy, environment and economic sustainability. *Bioresources and Bioprocessing, 8*(1), 87. https://doi.org/10.1186/s40643-021-00440-z

Akhayere, E., Vaseashta, A., & Kavaz, D. (2020). Novel magnetic nano silica synthesis using barley husk waste for removing petroleum from polluted water for environmental sustainability. *Sustainability (Switzerland), 12*(24), 1–16. https://doi.org/10.3390/su122410646

Álvarez, C., et al. (2021). Classification and target compounds. In *Food waste recovery: Processing technologies, industrial techniques, and applications* (pp. 21–49). https://doi.org/10.1016/B978-0-12-820563-1.00024-X

Amran, M. A., et al. (2021). Value-added metabolites from agricultural waste and application of green extraction techniques. *Sustainability (Switzerland)*. https://doi.org/10.3390/su132011432

Anand, K., et al. (2015). Agroforestry waste Moringa oleifera petals mediated green synthesis of gold nanoparticles and their anti-cancer and catalytic activity. *Journal of Industrial and Engineering Chemistry, 21*, 1105–1111. https://doi.org/10.1016/J.JIEC.2014.05.021

Astruc, D. (2020). Introduction: Nanoparticles in catalysis. *Chemical Reviews, 120*(2), 461–463. https://doi.org/10.1021/ACS.CHEMREV.8B00696

Awad, A., Ahmed, W., & Waleed, M. (2021). Nanotechnology for energy storage. In *Emerging nanotechnologies for renewable energy* (pp. 495–516). https://doi.org/10.1016/B978-0-12-821346-9.00009-2

Bachheti, R. K., et al. (2020). Root-based fabrication of metal/metal-oxide nanomaterials and their various applications. In *Nanomaterials for agriculture and forestry applications* (pp. 135–166). https://doi.org/10.1016/B978-0-12-817852-2.00006-8

Ban, F., & Saddam, A. M. (2017). Synthesis and characterization of Nanosilica from Rice husk with applications to polymer composites. *Americal Journal of Material Science, 7*(6), 223–231. Available at: http://article.sapub.org/10.5923.j.materials.20170706.01.html. Accessed 11 Mar 2023.

Bankar, A., et al. (2010). Banana peel extract mediated synthesis of gold nanoparticles. *Colloids and Surfaces B: Biointerfaces, 80*(1), 45–50. https://doi.org/10.1016/J.COLSURFB.2010.05.029

Barnawi, N., Allehyani, S., & Seoudi, R. (2022). Biosynthesis and characterization of gold nanoparticles and its application in eliminating nickel from water. *Journal of Materials Research and Technology, 17*, 537–545. https://doi.org/10.1016/J.JMRT.2021.12.013

Belewu, M. A., & Babalola, F. (2009). Nutrient enrichment of some waste agricultural residues after solid state fermentation using Rhizopus oligosporus. *Journal of Applied Biosciences, 13*, 695–699.

Berber-villamar, N. K., et al. (2018). Corncob as an effective, eco-friendly, and economic biosorbent for removing the azo dye direct yellow 27 from aqueous solutions. *PLoS One, 13*, 1–30.

Bhuvaneswari, S. (2017). Green synthesis of zinc oxide nanoparticles using potato Peel and degradation of textile mill effluent by photocatalytic activity. *World Journal of Pharmaceutical Research, 6*(6), 774–785. https://doi.org/10.20959/wjpr20176-8496

Boonmee, A., & Jarukumjorn, K. (2020). Preparation and characterization of silica nanoparticles from sugarcane bagasse ash for using as a filler in natural rubber composites. *Polymer Bulletin, 77*(7), 3457–3472. https://doi.org/10.1007/S00289-019-02925-6/FIGURES/14

Briceño-Ahumada, Z., Soltero-Martínez, J. F. A., & Castillo, R. (2021). Aqueous foams and emulsions stabilized by mixtures of silica nanoparticles and surfactants: A state-of-the-art review. *Chemical Engineering Journal Advances, 7*, 100116. https://doi.org/10.1016/J.CEJA.2021.100116

Cai, H., et al. (2015). Enhanced removal of fluoride by tea waste supported hydrous aluminium oxide nanoparticles: Anionic polyacrylamide mediated aluminium assembly and adsorption mechanism. *RSC Advances, 5*(37), 29266–29275. https://doi.org/10.1039/C5RA01560J

Chaudhary, J., et al. (2023). An overview of biosynthesized metal nanoparticles via medicinal plant extracts of the Moraceae family. *Biocatalysis and Agricultural Biotechnology, 52*, 102812. https://doi.org/10.1016/J.BCAB.2023.102812

Chavali, M. S., & Nikolova, M. P. (2019). Metal oxide nanoparticles and their applications in nanotechnology. *SN Applied Sciences, 1*(6), 1–30. https://doi.org/10.1007/S42452-019-0592-3

Chen, H., et al. (2013). Extraction of lignocellulose and synthesis of porous silica nanoparticles from rice husks: A comprehensive utilization of rice husk biomass. *ACS Sustainable Chemistry and Engineering, 1*(2), 254–259. https://doi.org/10.1021/SC300115R/ASSET/IMAGES/MEDIUM/SC-2012-00115R_0005.GIF

Cheng, Y., et al. (2023). Bamboo leaf: A review of traditional medicinal property, phytochemistry, pharmacology, and purification technology. *Journal of Ethnopharmacology, 306*, 116166. https://doi.org/10.1016/j.jep.2023.116166

Chianese, A., et al. (2016). Production of metal nanoparticles by agro-industrial wastes: A green opportunity for nanotechnology. *Chemical Engineering Transactions, 47*, 67–72. https://doi.org/10.3303/CET1647012

Chiravoot, P., Khanittha, P., & Benchamaporn Tangnorawich, N. N. (2021). Fruit peel – Mediated green synthesis of biocompatible gold nanoparticles. *Journal of Materials Research and Technology, 14*, 2982–2991. https://doi.org/10.1016/j.jmrt.2021.08.111

Choi, J. Y., et al. (2022). Utilization of corn cob, an essential agricultural residue difficult to disposal: Composite board manufactured improved thermal performance using microencapsulated PCM. *Industrial Crops and Products, 183*, 114931. https://doi.org/10.1016/J.INDCROP.2022.114931

da Luz Corrêa, A. P., et al. (2020). Preparation of sulfonated carbon-based catalysts from murumuru kernel shell and their performance in the esterification reaction. *RSC Advances, 10*(34), 20245–20256. https://doi.org/10.1039/d0ra03217d

Das, S. K., et al. (2020). Characterization and utilization of rice husk ash (RHA) in fly ash – Blast furnace slag based geopolymer concrete for sustainable future. *Materials Today: Proceedings, 33*, 5162–5167. https://doi.org/10.1016/j.matpr.2020.02.870

Dauthal, P., & Mukhopadhyay, M. (2015). Biofabrication, characterization, and possible bio-reduction mechanism of platinum nanoparticles mediated by agro-industrial waste and their catalytic activity. *Journal of Industrial and Engineering Chemistry, 22*, 185–191. https://doi.org/10.1016/J.JIEC.2014.07.009

Del Prado-Audelo, M. L., et al. (2021). Nanoremediation: Nanomaterials and nanotechnologies for environmental cleanup. *Frontiers in Environmental Science, 9*, 645. https://doi.org/10.3389/FENVS.2021.793765/BIBTEX

Dibanda Romelle, F., Ashwini, R. P., & Manohar, R. S. (2016). Chemical composition of some selected fruit peels. *European Journal of Food Science and Technology, 4*(4), 12–21. Available at: www.eajournals.org

Dirna, F. C., et al. (2020). Nanosilica synthesis from betung bamboo sticks and leaves by ultrasonication. *Nanotechnology, Science and Applications, 13*. https://doi.org/10.2147/NSA.S282357

Duque-Acevedo, M., et al. (2020). Agricultural waste: Review of the evolution, approaches and perspectives on alternative uses. *Global Ecology and Conservation, 22*, e00902. https://doi.org/10.1016/J.GECCO.2020.E00902

Edison, T. J. I., & Sethuraman, M. G. (2013). Biogenic robust synthesis of silver nanoparticles using Punica granatum peel and its application as a green catalyst for the reduction of an anthropogenic pollutant 4-nitrophenol. *Spectrochimica Acta Part A: Molecular and Biomolecular Spectroscopy, 104*, 262–264. https://doi.org/10.1016/J.SAA.2012.11.084

El-Ramady, H., et al. (2022). An overview of agro-waste management in light of the water-energy-waste nexus. *Sustainability (Switzerland)*. https://doi.org/10.3390/su142315717

Esra, A., Seyhan, B., & Gulnur, A. (2013). Preparation and antibacterial activity of silver coated carbon from rice husk. *Research Journal of Biotechnology, 8*(9), 17–21.

Falk, G., et al. (2019). Synthesis of silica nanoparticles from sugarcane bagasse ash and nano-silicon via magnesiothermic reactions. *Ceramics International, 45*(17), 21618–21624. https://doi.org/10.1016/J.CERAMINT.2019.07.157

FAO. (2000). *Case study 13. Dry-season feeding: a case study from Nicaragua. Hay and straw conservation: For small-scale farming and pastoral conditions* (J. M. Suttie, Eds.). FAO. Available at: http://www.fao.org/docrep/005/x7660e/x7660e0s.htm#bm28. Accessed 26 Feb 2023.

Fatimah, I., Amaliah, S. N., et al. (2019a). Iron oxide nanoparticles supported on biogenic silica derived from bamboo leaf ash for rhodamine B photodegradation. *Sustainable Chemistry and Pharmacy, 13*, 100149. https://doi.org/10.1016/J.SCP.2019.100149

Fatimah, I., Prakoso, N. I., et al. (2019b). Physicochemical characteristics and photocatalytic performance of TiO2/SiO2 catalyst synthesized using biogenic silica from bamboo leaves. *Heliyon, 5*(11), e02766. https://doi.org/10.1016/j.heliyon.2019.e02766

Fatimah, I., et al. (2021). Nanoflower-like composites of ZnO/SiO2 synthesized using bamboo leaves ash as reusable photocatalyst. *Arabian Journal of Chemistry, 14*(3), 102973. https://doi.org/10.1016/J.ARABJC.2020.102973

Gan, P. P., et al. (2012). Green synthesis of gold nanoparticles using palm oil mill effluent (POME): A low-cost and eco-friendly viable approach. *Bioresource Technology, 113*, 132–135. https://doi.org/10.1016/J.BIORTECH.2012.01.015

Ghaferi, M., et al. (2021). Mesoporous silica nanoparticles: Synthesis methods and their therapeutic use-recent advances. *Journal of Drug Targeting*. Taylor & Francis. https://doi.org/10.1080/1061186X.2020.1812614

Gilfillan, W. N., et al. (2012). Preparation and characterisation of composites from starch and sugar cane fibre. *Industrial Crops and Products, 40*(1), 45–54. https://doi.org/10.1016/j.indcrop.2012.02.036

Gu, S., et al. (2013). A detailed study of the effects of pyrolysis temperature and feedstock particle size on the preparation of nanosilica from rice husk. *Industrial Crops and Products, 50*, 540–549. https://doi.org/10.1016/J.INDCROP.2013.08.004

Gutiérrez-Wing, C., Velázquez-Salazar, J. J., & José-Yacamán, M. (2012). Procedures for the synthesis and capping of metal nanoparticles. *Methods in Molecular Biology, 906*, 3–19. https://doi.org/10.1007/978-1-61779-953-2_1/COVER

Hernández-Salas, J. M., et al. (2009). Comparative hydrolysis and fermentation of sugarcane and agave bagasse. *Bioresource Technology, 100*(3), 1238–1245. https://doi.org/10.1016/j.biortech.2006.09.062

Heydari, R., & Rashidipour, M. (2015). Green synthesis of silver nanoparticles using extract of oak fruit hull (Jaft): Synthesis and in vitro cytotoxic effect on MCF-7 cells. *International Journal of Breast Cancer, 2015*. https://doi.org/10.1155/2015/846743

Hossain, S. K. S., Mathur, L., & Roy, P. K. (2018). Rice husk/rice husk ash as an alternative source of silica in ceramics: A review. *Journal of Asian Ceramic Societies, 6*(4), 299–313. https://doi.org/10.1080/21870764.2018.1539210

Idhayadhulla, A., et al. (2021). Potato peels mediated synthesis of cu(Ii)-nanoparticles from tyrosinase reacted with bis-(n-aminoethylethanolamine) (tyr-cu(ii)-aeea nps) and their cytotoxicity against Michigan cancer foundation-7 breast cancer cell line. *Molecules, 26*(21). https://doi.org/10.3390/molecules26216665

Jadoun, S., et al. (2021). Green synthesis of nanoparticles using plant extracts: A review. *Environmental Chemistry Letters, 19*(1), 355–374. https://doi.org/10.1007/s10311-020-01074-x

Joudeh, N., & Linke, D. (2022). Nanoparticle classification, physicochemical properties, characterization, and applications: A comprehensive review for biologists. *Journal of Nanobiotechnology, 20*(1), 1–29. https://doi.org/10.1186/S12951-022-01477-8

Kanchi, S., et al. (2018). Exploitation of de-oiled jatropha waste for gold nanoparticles synthesis: A green approach. *Arabian Journal of Chemistry, 11*(2), 247–255. https://doi.org/10.1016/J.ARABJC.2014.08.006

Karp, S. G., et al. (2013). Pretreatment strategies for delignification of sugarcane bagasse: A review. *Brazilian Archives of Biology and Technology, 56*, 679–689.

Khan, T. A., et al. (2019). Hydrothermal carbonization of lignocellulosic biomass for carbon rich material preparation: A review. *Biomass and Bioenergy, 130*(September), 105384. https://doi.org/10.1016/j.biombioe.2019.105384

Kharissova, O. V., et al. (2013). The greener synthesis of nanoparticles. *Trends in Biotechnology, 31*(4), 240–248. https://doi.org/10.1016/J.TIBTECH.2013.01.003

Kimura, I., et al. (2022). Multitasking bamboo leaf-derived compounds in prevention of infectious, inflammatory, atherosclerotic, metabolic, and neuropsychiatric diseases. *Pharmacology & Therapeutics, 235*, 108159. https://doi.org/10.1016/j.pharmthera.2022.108159

Krishnaswamy, K., Vali, H., & Orsat, V. (2014). Value-adding to grape waste: Green synthesis of gold nanoparticles. *Journal of Food Engineering, 142*, 210–220. https://doi.org/10.1016/J.JFOODENG.2014.06.014

Kulkarni, N., & Muddapur, U. (2014). Biosynthesis of metal nanoparticles: A review. *Journal of Nanotechnology, 2014*. https://doi.org/10.1155/2014/510246

Li, D., Chen, D., & Zhu, X. (2011). Reduction in time required for synthesis of high specific surface area silica from pyrolyzed rice husk by precipitation at low pH. *Bioresource Technology, 102*(13), 7001–7003. https://doi.org/10.1016/J.BIORTECH.2011.04.020

Liou, T. H., & Yang, C. C. (2011). Synthesis and surface characteristics of nanosilica produced from alkali-extracted rice husk ash. *Materials Science and Engineering: B, 176*(7), 521–529. https://doi.org/10.1016/J.MSEB.2011.01.007

Loh, Y. R., et al. (2013). Sugarcane bagasse—The future composite material: A literature review. *Resources, Conservation and Recycling, 75*, 14–22. https://doi.org/10.1016/j.resconrec.2013.03.002

Lopes, F. C., Ligabue-braun, R., & Bernardo, S. (2021). Agro-industrial residues: Eco-friendly and inexpensive substrates for microbial pigments production. *Fronties in Sustainable Food Systems, 5*, 589414. https://doi.org/10.3389/fsufs.2021.589414

Mahmud, M. A., & Anannya, F. R. (2021). Sugarcane bagasse – A source of cellulosic fiber for diverse applications. *Heliyon, 7*(8), e07771. https://doi.org/10.1016/J.HELIYON.2021.E07771

Malhotra, A., et al. (2014). Multi-analytical approach to understand biomineralization of gold using rice bran: A novel and economical route. *RSC Advances, 4*(74), 39484–39490. https://doi.org/10.1039/C4RA05404K

Malpani, S. K., & Goyal, D. (2022). Synthesis, analysis, and multi-faceted applications of solid wastes-derived silica nanoparticles: A comprehensive review (2010–2022). *Environmental Science and Pollution Research*. Springer Science and Business Media Deutschland GmbH. https://doi.org/10.1007/s11356-022-23873-1

Mitchell, M. J., et al. (2020). Engineering precision nanoparticles for drug delivery. *Nature Reviews Drug Discovery, 20*(2), 101–124. https://doi.org/10.1038/s41573-020-0090-8

Mohite, A. S., et al. (2022). Recycling of major agriculture crop residues and its application in polymer industry: A review in the context of waste to energy nexus. *Energy Nexus, 7*, 100134. https://doi.org/10.1016/j.nexus.2022.100134

Mojumdar, A., & Deka, J. (2019). Recycling agro-industrial waste to produce amylase and characterizing amylase–gold nanoparticle composite. *International Journal of Recycling of Organic Waste in Agriculture, 8*(1), 263–269. https://doi.org/10.1007/S40093-019-00298-4/FIGURES/3

Niluxsshun, M. C. D., Masilamani, K., & Mathiventhan, U. (2021). Green synthesis of silver nanoparticles from the extracts of fruit peel of *Citrus tangerina*, *Citrus sinensis*, and *Citrus limon* for antibacterial activities. *Bioinorganic Chemistry and Applications*. Edited by G. Ciccarella, *2021*, 6695734. https://doi.org/10.1155/2021/6695734

Okoronkwo, E. A., Imoisili, P. E., & Olusunle, S. O. O. (2013). Extraction and characterization of Amorphous Silica from Corn Cob Ash by Sol-Gel Method. *Chemistry and Materials Research, 3*(4), 68–72. Available at: https://www.iiste.org/Journals/index.php/CMR/article/view/5059. Accessed 11 Mar 2023.

Oladipo, I. C., et al. (2017). Enterococcus species for the one-pot biofabrication of gold nanoparticles: Characterization and nanobiotechnological applications. *Journal of Photochemistry and Photobiology B: Biology, 173*, 250–257. https://doi.org/10.1016/J.JPHOTOBIOL.2017.06.003

Osagie, C., et al. (2021). Dyes adsorption from aqueous media through the nanotechnology: A review. *Journal of Materials Research and Technology, 14*, 2195–2218. https://doi.org/10.1016/J.JMRT.2021.07.085

Oyekanmi, A. A., et al. (2022). Development of nanoparticles loaded composites from agricultural wastes for cationic dye removal from aqueous solution – A review. *Journal of Environmental Chemical Engineering, 10*(5), 108263. https://doi.org/10.1016/j.jece.2022.108263

Patra, J. K., & Baek, K.-H. (2015). Novel green synthesis of gold nanoparticles using Citrullus lanatus rind and investigation of proteasome inhibitory activity, antibacterial, and antioxidant potential. *International Journal of Nanomedicine, 10*, 7253–7264. https://doi.org/10.2147/IJN.S95483

Peerzada, J. G., & Chidambaram, R. (2021). A statistical approach for biogenic synthesis of Nano-silica from different agro-wastes. *Silicon, 13*(7), 2089–2101. https://doi.org/10.1007/S12633-020-00629-5/METRICS

Peng, X., et al. (2023). Recycling municipal, agricultural and industrial waste into energy, fertilizers, food and construction materials, and economic feasibility: A review. *Environmental Chemistry Letters*. Springer Science and Business Media Deutschland GmbH. https://doi.org/10.1007/s10311-022-01551-5

Rangaraj, S., & Venkatachalam, R. (2017). A lucrative chemical processing of bamboo leaf biomass to synthesize biocompatible amorphous silica nanoparticles of biomedical importance. *Applied Nanoscience, 7*(5), 145–153. https://doi.org/10.1007/s13204-017-0557-z

Rodríguez-Félix, F., et al. (2021). Sustainable-green synthesis of silver nanoparticles using safflower (Carthamus tinctorius L.) waste extract and its antibacterial activity. *Heliyon, 7*(4). https://doi.org/10.1016/j.heliyon.2021.e06923

Rudra, S. G., et al. (2015). Food industry waste. *Mine of Nutraceuticals, 4*(1), 205–229.

Sadh, P. K., Duhan, S., & Duhan, J. S. (2018). Agro-industrial wastes and their utilization using solid state fermentation: A review. *Bioresources and Bioprocessing, 5*(1), 1. https://doi.org/10.1186/s40643-017-0187-z

Saha, P., & Kim, B. S. (2022). *Plant extract and agricultural waste-mediated synthesis of silver nanoparticles and their biochemical activities* (pp. 285–315). Green Synthesis of Silver Nanomaterials. https://doi.org/10.1016/B978-0-12-824508-8.00010-1

Salakhum, S., et al. (2018). Synthesis of hierarchical faujasite nanosheets from corn cob ash-derived nanosilica as efficient catalysts for hydrogenation of lignin-derived alkylphenols. *Microporous and Mesoporous Materials, 258*, 141–150. https://doi.org/10.1016/J.MICROMESO.2017.09.009

Saratale, R. G., et al. (2019). Wheat straw extracted lignin in silver nanoparticles synthesis: Expanding its prophecy towards antineoplastic potency and hydrogen peroxide sensing ability. *International Journal of Biological Macromolecules, 128*, 391–400. https://doi.org/10.1016/J.IJBIOMAC.2019.01.120

Shanmugavadivu, M., Ranjithkumar, S. K., & Ranjithkumar, R. (2014). Synthesis of pomegranate peel extract mediated silver nanoparticles and its antibacterial activity. *American Journal of Advanced Drug Delivery, 2*, 174–182.

Sheikh Mohamed, M., et al. (2012). Non-destructive harvesting of biogenic gold nanoparticles from Jatropha curcas Seed meal and shell extracts and their application as bio-diagnostic photothermal ablaters-lending shine to the biodiesel byproducts. *Nanomaterials and the Environment, 1*. Available at: https://www.academia.edu/30093144/Non_Destructive_Harvesting_of_Biogenic_Gold_Nanoparticles_from_Jatropha_curcas_Seed_Meal_and_Shell_Extracts_and_their_Application_as_Bio_Diagnostic_Photothermal_Ablaters_Lending_Shine_to_the_Biodiesel_Byproducts. Accessed 13 Mar 2023.

Sheikh Mohamed, M., et al. (2013). Non-destructive harvesting of biogenic gold nanoparticles from Jatropha curcas seed meal and shell extracts and their application as bio-diagnostic Photothermal Ablaters-lending Shine to the biodiesel byproducts. *Nanomaterials and the Environment, 1*. https://doi.org/10.2478/nanome-2012-0002

Shen, D.-Y., et al. (2022). Characterization of odor-active compounds in moso bamboo (Phyllostachys pubescens Mazel) leaf via gas chromatography-ion mobility spectrometry, one- and two-dimensional gas chromatography-olfactory-mass spectrometry, and electronic nose. *Food Research International, 152*, 110916. https://doi.org/10.1016/j.foodres.2021.110916

Shim, J., Velmurugan, P., & Oh, B.-T. (2015). Extraction and physical characterization of amorphous silica made from corn cob ash at variable pH conditions via sol gel processing. *Journal of Industrial and Engineering Chemistry, 262*(4). https://doi.org/10.1016/j.jiec.2015.05.029

Silviana, S., & Bayu, W. J. (2018). Silicon conversion from bamboo leaf silica by magnesiothermic reduction for development of Li-ion baterry anode. *MATEC Web of Conferences, 156*, 05021.

Sinsinwar, S., et al. (2018). Use of agricultural waste (coconut shell) for the synthesis of silver nanoparticles and evaluation of their antibacterial activity against selected human pathogens. *Microbial Pathogenesis, 124*, 30–37. https://doi.org/10.1016/J.MICPATH.2018.08.025

Slavutsky, A. M., & Bertuzzi, M. A. (2014). Water barrier properties of starch films reinforced with cellulose nanocrystals obtained from sugarcane bagasse. *Carbohydrate Polymers, 110*, 53–61. https://doi.org/10.1016/j.carbpol.2014.03.049

Suman, T. Y., et al. (2013). Biosynthesis, characterization and cytotoxic effect of plant mediated silver nanoparticles using Morinda citrifolia root extract. *Colloids and Surfaces B: Biointerfaces, 106*, 74–78. https://doi.org/10.1016/J.COLSURFB.2013.01.037

Thangadurai, D., et al. (2021). Nanomaterials from Agrowastes: Past, present, and the future. In *Handbook of nanomaterials and nanocomposites for energy and environmental applications* (pp. 1–17). Springer International Publishing. https://doi.org/10.1007/978-3-030-11155-7_43-1

Toscano Miranda, N., et al. (2021). Sugarcane bagasse pyrolysis: A review of operating conditions and products properties. *Renewable and Sustainable Energy Reviews, 149*, 111394. https://doi.org/10.1016/j.rser.2021.111394

Tsai, W. T., et al. (2001). Adsorption of acid dye onto activated carbons prepared from agricultural waste bagasse by ZnCl2 activation. *Chemosphere, 45*(1), 51–58. https://doi.org/10.1016/S0045-6535(01)00016-9

Vaibhav, V., Vijayalakshmi, U., & Roopan, S. M. (2015). Agricultural waste as a source for the production of silica nanoparticles. *Spectrochimica Acta – Part A: Molecular and Biomolecular Spectroscopy, 139*, 515–520. https://doi.org/10.1016/j.saa.2014.12.083

Varghese, B. A., et al. (2021). Green synthesis of gold nanoparticles using Kaempferia parviflora rhizome extract and their characterization and application as an antimicrobial, antioxidant and catalytic degradation agent. *Journal of the Taiwan Institute of Chemical Engineers, 126*, 166–172. https://doi.org/10.1016/J.JTICE.2021.07.016

Velmurugan, P., Shim, J., et al. (2015a). Extraction, characterization, and catalytic potential of amorphous silica from corn cobs by sol-gel method. *Journal of Industrial and Engineering Chemistry, 29*, 298–303. https://doi.org/10.1016/J.JIEC.2015.04.009

Velmurugan, P., Park, J. H., et al. (2015b). Reduction of silver (I) using defatted cashew nut shell starch and its structural comparison with commercial product. *Carbohydrate Polymers, 133*, 39–45. https://doi.org/10.1016/J.CARBPOL.2015.06.097

Wang, L., et al. (2015a). Bamboo leaf derived ultrafine Si nanoparticles and Si/C nanocomposites for high-performance Li-ion battery anodes. *Nanoscale, 7*(33), 13840–13847. https://doi.org/10.1039/C5NR02578H

Wang, Y., et al. (2015b). Vitexin protects brain against ischemia/reperfusion injury via modulating mitogen-activated protein kinase and apoptosis signaling in mice. *Phytomedicine, 22*(3), 379–384. https://doi.org/10.1016/j.phymed.2015.01.009

Yan, D., et al. (2014). Supercapacitive properties of Mn3O4 nanoparticles bio-synthesized from banana peel extract. *RSC Advances, 4*(45), 23649–23652. https://doi.org/10.1039/C4RA02603A

Yang, N., Weihong, L., & Hao, L. (2014). Biosynthesis of Au nanoparticles using agricultural waste mango peel extract and its in vitro cytotoxic effect on two normal cells. *Materials Letters, 134*, 67–70. https://doi.org/10.1016/J.MATLET.2014.07.025

Yuvakkumar, R., Elango, V., et al. (2014a). High-purity nano silica powder from rice husk using a simple chemical method. *Journal of Experimental Nanoscience, 9*(3), 272–281. https://doi.org/10.1080/17458080.2012.656709

Yuvakkumar, R., Suresh, J., et al. (2014b). Novel green synthetic strategy to prepare ZnO nanocrystals using rambutan (Nephelium lappaceum L.) peel extract and its antibacterial applications. *Materials Science and Engineering C, 41*, 17–27. https://doi.org/10.1016/J.MSEC.2014.04.025

Yuvakkumar, R., Suresh, J., et al. (2014c). Rambutan (Nephelium lappaceum L.) peel extract assisted biomimetic synthesis of nickel oxide nanocrystals. *Materials Letters, 128*, 170–174. https://doi.org/10.1016/J.MATLET.2014.04.112

Zamani, A., Marjani, A. P., & Mousavi, Z. (2019). Agricultural waste biomass-assisted nanostructures: Synthesis and application. *Green Processing and Synthesis, 8*(1), 421–429. https://doi.org/10.1515/gps-2019-0010

Fabrication of Metal NPS from Plant Root and Tuber

Milan Hait, Nand Kumar Kashyap, Pradeep Sahu, and Utpal Jana

Abbreviations

AA	Ascorbic acid
AgNPs	Silver nanoparticles
AIDS	Acquired immunodeficiency syndrome
AuNPs	Gold nanoparticles
CuNPs	Copper nanoparticles
DM	Diabetic mellitus
EDX	Energy-dispersive X-ray
FA	Fatty acid
FCC	Face-centered cubic
FDA	Food and Drug Administration
FRS	Free radical scavenging
FTIR	Fourier transform infrared spectroscopy
GLC	Gas liquid chromatography
GRAS	Generally recognized as safe
HbA	Glycosylated hemoglobin
HCPDEC	Health and Consumer Protection Directorate of the European Commission
HFR	Hydrogen free radical
HPx	Hydrogen peroxidase
IOSOECD	International Organization for Standardization and Organisation for Economic Co-operation and Development
IV	Intravenously

M. Hait · N. K. Kashyap · P. Sahu
Department of Chemistry, Dr. C. V. Raman University, Bilaspur, Chhattisgarh, India

U. Jana (✉)
Department of Pharmaceutics, Shiva Institute of Pharmacy, Bilaspur, HP, India

A. K. Bhardwaj et al. (eds.), *Biogenic Wastes-Enabled Nanomaterial Synthesis*,
https://doi.org/10.1007/978-3-031-59083-2_2

LC-MS	Liquid chromatography-mass spectrometry
MB	Methylene blue
MBR	Membrane bioreactor
NIOSH	National Institute for Occupational Safety and Health
NPs	Nanoparticles
NPSS	Nanoporous stainless steel
PdNPs	Palladium nanoparticles
PtNPs	Platinum nanoparticles
ROS	Reactive oxygen species
SEM	Scanning electron microscopy
SOD	Superoxide dismutase
SPR	Surface plasma resonance
TEM	Transmission electron microscopy
TiNPs	Titanium nanoparticles
TiO2NPs	Titanium dioxide nanoparticles
USEPA	United States Environmental Protection Agency
XRD	X-ray diffraction
ZnONPs	Zinc oxide nanoparticles

1 Introduction

The production of MNPs via root and tuber extracts is the easiest, most practical, cost-effective, and ecologically appropriate approach available for reducing the number of unfavorable components used in the procedure. Therefore, in current times, various environmentally safe procedures have been discovered for the fast manufacturing of MNPs utilizing roots and tubers. These processes may be found in a number of different scientific publications (Vanlalveni et al., 2021). In the field of materials science, "green" synthesis has attracted greater interest as a trustworthy, affordable, acceptable, and environmentally based technique for creating a broad variety of nanosized materials and crystals. These materials include MNPs and MONPs, hybrids, and biocrystals. Green synthesis is recognized as a key instrument for reducing the damaging consequences connected with the conventional techniques of synthesis for MNPs, which are frequently employed in laboratory and industrial settings (Aromal & Philip, 2012). The creation of curative nanocrystals for use in healthcare and pharmacological administration is the primary focus of the interdisciplinary area of nanotechnology. There are a wide variety of macro- and microorganisms that contribute to the biological production of MNPs. These organisms are plants, bacteria, fungi, seaweeds, and microalgae (Hait et al., 2020). The reduction of metal ions to MNPs is accomplished with a green manufacturing step by using biomolecules that are prevalent in root and tuber extracts. Rapid, room-temperature, and pressure-free reduction of metal ions to metals is simple to implement and scale up. Plant extracts can be used as catalysts for a greener synthesis.

Plant-natural bioactive components and enzymes serve as the reducing agents here. Plant-based syntheses have focused mostly on silver (Ag) and gold (Au) nanoparticles. Nanoparticles (Ag, Au, Zn, Ti, and Pd) have been effectively fabricated using extracts from various plant species (Chandran et al., 2006). The synthesis can be done using either plant extracts or live plants. The biosynthesized nanomaterials have reduced side effects while efficiently combating multiple endemic diseases. Alkaloids, flavonoids, saponins, steroids, tannins, and other nutritious substances are plentiful in the plant (Rathore et al., 2023). Leaves, stems, roots, shoots, flowers, tubers, barks, and seeds are all sources for these natural compounds (Bhardwaj et al., 2023a, b). Recent research has shown that plant extracts serve as a safe source of nanomaterials. A root is a compact, usually expanded storage organ that grows from root tissue and has hairy stems. Tubers are a type of root. It is essentially a big storage organ that grows out of a rhizome or other forms of elongated stem tissue (Hait et al., 2018). The production of AgNPs from extracts of the medicinal plant's root bark and tuber was shown to be a straightforward, trouble-free, and environmentally benign process. Root tubers are enlarged roots that have developed for the purpose of storing water and nutrients. They are also referred to as "root crops." Root tubers act as substitutes for true roots, allowing the plant to take up nutrients and water from the soil and hold itself firmly in place (Shafey, 2020).

Roots and tubers include dahlias, daikon radishes, and chives. The bioreduction procedure used to synthesize the unique metallic nanoparticles relies on the root and tuber extracts, which have numerous active molecules that act as reducing and stabilizing agents. Metal nanoparticles (MNPs) have the potential to revolutionize various sectors, such as the medical field, the food industry, and energy production (Cruz et al., 2010). In the field of biomaterial science, synthesizing monodispersed nanoparticles of the correct size and shape through biogenetic synthesis has proved challenging. The pharmaceutical industry has reaped incredible profits from its use as a result of its success in treating a wide variety of bacterial and viral illnesses (Song & Kim, 2009). When compared to more conventional methods of synthesis, biosynthetic processes provide more compensation due to the greater accessibility of biogenic establishments, which are less harmful to the environment. The synthesis of metal nanomaterials from plant entities has attracted a significant amount of interest as of late due to the immense variety and accessible availability of plant entities. According to many reports, metal nanoparticles (MNPs) can be biosynthesized from flowers, fruits, leaves, roots, and tubers (Bar et al., 2009). These biologically synthesized metal nanoparticles have a variety of potential applications, including medical recognition and therapy, the development of nanoscale tools for surgical procedures, and even the manufacturing of consumer goods. Nanomedicine has already made a significant contribution to the treatment of a broad variety of conditions that may severely limit one's quality of life. For future generations to be successful in treating a broad variety of diseases, they will need to learn how to manufacture metal nanoparticles in a safe manner for the environment (Monda et al., 2011). Root and tuber extracts contain phenolic compounds, flavonoids, alkaloids, and terpenoids, which are unique secondary metabolites that are needed to turn ions into massive MNPs. These natural bioactive components are involved in the redox process, which creates MNPs that are good for the environment (Mittal et al., 2013).

In a great number of prior studies, it was seen that biologically synthesized MNPs were able to significantly lower oxidative cell damage and mutagenic- and carcinogenic-related changes. Additionally, MNPs have great administration skills in plant science and agriculture. MNPs may be able to recycle food and agricultural waste into usable fuel and byproducts by using a technology that is associated with bioprocessing. Currently, research is focused on MNPs that are biologically derived from plants (roots, tubers) and their application, especially in healthcare, Medicare, wastewater, cosmetics, and the food sector (Kim et al., 2007). Root and tuber MNPs are synthesized via chemical and physical processes that are not biological, posing a high risk of harm to living organisms. The biological synthesis of MNPs is simple, cheap, and environmentally friendly. Several environmentally preferable MNPs, including Co, Cu, Ag, Au, Pd, Pt, ZnO, and Fe_3O_4, have been synthesized using roots and tubers. Cosmetics, food, and the pharmaceutical industries could all benefit from the root- and tuber-mediated MNPs' various therapeutic properties, such as antimicrobial, antifungal, anti-inflammatory, anticancer, antiviral, and antioxidant properties (Khanal et al., 2022).

2 Biosynthesis of MNPs Using Plant Extract

There are several different techniques to synthesize MNPs via synthetic and biological approaches. Many different synthetic approaches exist for synthesizing MNPs, such as plasma arcing, ball milling, thermal evaporation, ultrathin film method, pulsed laser desorption, lithography, sputter deposition, molecular beam epistaxis, flame dispersion, electron deposition, sol-gel technique, chemical deposition, chemical vapor deposition, soft chemical procedure, Langmuir-Blodgett technique, catalytic process, and coprecipitation (Joerger et al., 2000). Synthetic approaches make use of greater amounts of MNPs, which are highly concentrated reductants and stabilizing agents (Gan et al., 2012). These types of substances give rise to the health of both humans and environments (Sathishkumar et al., 2009). It has been shown that a lower amount of energy is required for the synthesis of MNPs, which is ecologically acceptable when MNPs are synthesized via biological processes. In biological processes favorable to the environment, these include plant extracts, microorganisms (bacteria, fungi), algae, diatoms, weeds, and enzymes (Iravani, 2011).

The potential of secondary metabolites present in plants, which include pharmaceuticals, agrochemicals, flavors, and aromas, is mostly unexplored in plant roots. Ipecac (*Cephaelis ipecacuanha*) is a plant that is known for its expectorant and emetic properties, whereas *Glycyrrhiza glabra, C. amada* and *C. arborea* Roxb. (Kashyap et al. 2022a, b) are known for their sweet-tasting glycyrrhizic acid. Numerous plant families are accountable for the production of a variety of compounds in their roots (Flores et al., 1987). The lack of appropriate experimental conditions has significantly impeded research into the phytochemicals of roots, impeding a thorough knowledge of the biosynthesis and processes of accumulation for the majority of these compounds. Khanal et al. (2020) discovered that the root extract of the plant has potent antibacterial and antioxidant properties. Researchers employed root extract from the *Morinda citrifolia* plant in their study to reveal the creation, characterization, and anticancer effectiveness of AgNPs mediated by plants (Suman et al., 2013). It is more effective and less damaging to the environment of AgNPs using the root barks of *R. ellipticus* Sm. There were noticeable changes in the peaks of the FTIR spectra, which might be an indication of capping and stabilizing agents in roots and tubers. The analysis of the MNPs using XRD and SEM both verified that the MNPs were FCC crystals with a spherical form. The EDX examination uncovered information about the levels of abundance of Ag, O_2, C, Ca, and Cl_2 (Olivera et al. 2005). The AgNPs were known to have greater effectiveness as antimicrobials and antioxidants compared to other MNPs. Further investigation is required to discover the various parameters such as size, shape, and phytochemicals that may alter the characterization of the AgNPs, and the biological action of the AgNPs may be better optimized. This leads one to believe the plant might be used in the synthesis of a number of substances that have the potential to be employed in biological and therapeutic applications, and a biosynthetic approach could replace more conventional procedures (Khanal et al., 2022).

Tubers are generally thin and small in length, and they form underground. The majority of the dormant stage of many plant species is made of parenchyma tissue, which is responsible for the storage of starch. This allows plants to endure the harsh conditions of winter. Ghosh et al. (2012) described AgNPs synthesized from *Dioscorea bulbifera* tuber extract, and their synergistic capacity increased the antimicrobial action of medicines, which was subsequently estimated. Biogenic AgNPs were synthesized and characterized by UV-visible procedures, XRD, FT-IR, SEM, TEM, and EDX. The XRD spectroscopy revealed that AgNPs have a FCC crystal composition. The FTIR spectra revealed that a number of different polyphenolic compounds had been used to coat the AgNPs. The SPR (400–430 nm) was discovered in UV-visible spectroscopy tests and provided conclusive evidence for the presence of silver. The SEM analysis exhibited that the newly synthesized AgNPs were spherical in shape, and their sizes were 12–20 nanometers (Panigrahi et al., 2004). The EDX analysis demonstrated the high signals presented in AgNPs. MDR bacteria were used to test the antimicrobial properties of AgNPs made from extracts of roots and tubers (Pileni, 1997). The AgNPs possess significant antimicrobial effectiveness against MDR bacterial strains, most notably *E. coli*. AgNPs synthesized from *Arisaema flavum* extract showed consequential photocatalytic potential to destroy MB because of their diminutive size, spherical form, and high dispersion. AgNPs exhibit very high levels of both biological and photocatalytic actions. This is the first evidence that the biological AgNPs have a potent antibacterial impact against transgenic *E. coli*, which is resistant to various drugs (Rahman et al., 2019). Plant extracts have seen widespread use in the synthesis of eco-friendly nanoparticles due to their powerful metal-ion reduction potential and the bioactive compounds that are helpful to the surrounding ecosystem. Both Hemalatha et al. (2019)

and Bhardwaj et al. (2023b) investigated the antibacterial and cytotoxic capabilities of AgNPs. Moreover, eco-friendly AgNPs were evaluated from *Momordica cymbalaria* and *C. caesia* rhizome (Bhardwaj et al. 2023b), tuber, and fruit extracts. After seeing a shift from green to red coloration, AgNP formation is confirmed and the SPR peak is recoded at a constant wavelength. AgNPs were characterized using XRD, SEM, TEM, and elemental analysis to establish their sizes, elemental compositions, phases, and crystallite sizes. The increased stability of the AgNPs is due to the fact that they have layered phenolic and protein moieties, as shown by FTIR. When compared to *S. pyogenes* and *P. aeruginosa*, the antibacterial efficacy of AgNPs made from tuber and fruit extracts was higher. The cytotoxicity capacity of AgNPs made from the tuber extract was shown to be higher in vitro than fruit extract AgNPs. This was the case for both prostate (PC-3) and breast adenocarcinoma (MDA-MB 231) cells. AgNPs with significant antibacterial and cytotoxic potential have been successfully synthesized using extracts of *Momordica cymbalaria* fruit and tuber (Hemalatha et al., 2019). Tables 1 and 2 display the synthesis of MNPs from biological bacterial and fungal species and their applications.

2.1 *Mechanism of MNP Synthesis*

The physicochemical features of plant-mediated nanomaterials have recently attracted increased attention, expanding their potential applications. Research has focused solely on MNPs synthesized from natural resources, including Au, Ag, Pt, Zn, Cu, Ni, TiO_2, and Fe_3O_4. The synthesis of MNPs that vary in both shape and size is possible via biological processes, and these MNPs may originate from a broad variety of plant parts. Changing the concentration of metals and the quantity of root and tuber extract in the reaction media may influence the biosynthesis process, which affects the MNPs' shape and size (Chandran et al., 2006). The [Ag (*Callicarpa maingayi*)]$^+$ complex was synthesized from *Callicarpa maingayi* root (Shameli et al., 2012). The primary factor in the conversion of AgNPs from Ag^+ is the presence of the aldehyde group in the plant extract. Molecules responsible for encasing ionic compounds into MNPs are known as polypeptides with the name amide I. These molecules are distinguished by the unique functional group C=0, C=N. Molecular biologists do not have a complete understanding of the complicated process of the biosynthesis of silver crystals. This is because the process is so difficult. In contrast, several model interactions of NPs with pathogenic organisms have been hypothesized as a result of previous research. The attachment of biologically synthesized AgNPs to the outer membrane of bacteria, fungi, and viruses results in the destruction of the microorganisms (Dubey et al., 2010). The cell cycle was finally stopped due to the ultimate cell death. Vanaja and Annadurai (2012) showed the biological production of AgNPs using *Cissus quadrangularis* root and tuber extract at 30 °C. The root extract demonstrates the extant variety of functional groups involved in the reduction of Ag^+. These functional groups contain carboxyl, amine, and phenolic compounds. This indicates that the synthetic AgNPs are more

Table 1 Synthesis of MNPs from biological bacterial species and their applications

S. N.	MNPs	Biological species	Application	References
1		*Bacillus pumilus*	Antimicrobial activity against *S. aureus*, *E. coli*, *P. aeruginosa*, and *A. baumannii* ranging from 1.4 to 5.6 µg/ml	Esmail et al. (2022)
2		*Raoultella planticola Pantoea agglomerans*	Antimicrobial activity against *Staphylococcus aureus*, *Streptococcus pyogenes*, *Salmonella*, and *Bacillus amyloliquefaciens*	Ghiuta et al. (2021)
3		*Bacillus thuringiensis*	Used for the biosynthesis of silver nanoparticles with antibacterial and antibiofilm activities	Khaleghi et al. (2019)
4		*Bacillus licheniformis Dahb1*	*Bacillus licheniformis* Dahb1-derived silver nanoparticles for antibiofilm activity and toxicity effects in *Ceriodaphnia cornuta*	Shanthi et al. (2016)
5	AgNPs	*B. cereus*	Antimicrobial activityagainst *E. coli*, *P. aeruginosa*, *S. aureus*, *S. typhi*, and *K. pneumoniae*	Sukumaran and Poulose (2012)
6		*P. proteolytica, B. cecembensis*	Antibacterial action against *A. kerguelensis*, *A.gangotriensis*, *B. indicus*, *P. antarctica*, *P. proteolytica*, and *E. coli*	Shivaji et al. (2011)
7		*L. casei*	Drug administration, anticancer, biofilm	Korbekandi et al. (2012)
8		*K. pneumonia,E. coli, E. cloacae*	Optical receptors, electrical, batteries, antimicrobial	Ahmad et al. (2010)
9		*Bacillus indicus*	Antimicrobial, catalysis,antifungal	Fu et al. (2006)
10	AuNPs	*Plectonema boryanum*	Catalysis, antifungal	Lengke et al. (2006)
11		*Bacillus subtilis*	Antimicrobial, catalysis,antifungal	Southam and Beveridge (1994)
12		*Bacillus megaterium*	Catalysis, bio-sensing	Wang et al. (2016)
13		*Shewanella alga*	Catalysis, bio-sensing	Konishi et al. (2007)
14		*E. coli*	Biochemistry, hemoglobin	Du et al. (2007)
15		*Desulfovibrio desulfuricans*	Enzyme catalysis	Deplanche and Macaskie (2008)
16		*Rhodopseudomonas capsulate*	Anticancer	He et al. (2007)
17	Fe_3O_4 NPs	*Magnetospirillummagneto-tacticum*	Cluster shaping	Philipse and Maas (2002)
18		*Aquaspirillummagneto-tacticum*	Biomolecule cluster	Mann et al. (1984)

(continued)

Table 1 (continued)

S. N.	MNPs	Biological species	Application	References
19	UO_2	*Shewanella oneidensis*	Biomolecule radioactive	Marshall et al. (2006)
20	CdS	*Klebsiella aerogenes*	Biomolecule radioactive	Holmes et al. (1995)
21		*E. coli*	Wurtzite structures	Chen et al. (2009)

Table 2 Synthesis of MNPs from biological fungal species and their applications

S. N.	MNPs	Biological species	Application	References
1		*Penicillium duclauxii*	Inhibit *Bipolaris sorghicola* that cause leaf spot disease on sorghum plants	Almaary et al. (2020)
2		*Penicillium oxalicum*	Catalytic reduction of methylene blue	Du et al. (2015)
3	AgNPs	*Rhizopus nigricans*	Bactericidal, catalytic	Ravindra and Rajasab (2014)
4		*Verticillium*	Catalysis	Mukherjee et al. (2001)
5		*Aspergillus fumigates*	Coating, absorption, batteries	Bhainsa and D'Souza (2006)
6		*Phanerochaete chrysosporium*	Antimicrobial, therapeutic textiles	Vigneshwaran et al. (2007)
7		*Aspergillus flavus*	Catalysis, antifungal	Vigneshwaran et al. (2006)
8		*Aspergillus niger*	Antibacterial agent	Gade et al. (2008)
9		*Fusarium semitectum*	Biolabeling	Basavaraja et al. (2008)
10		*Cladosporium cladosporioides*	Biolabeling	Balaji et al. (2009)
11		*Coriolus versicolor*	Catalysis, biolabeling	Sanghi and Verma (2009)
12		*Fusarium solani*	Biolabeling, sensing, drug delivery	Ingle et al. (2009)
13.		*Penicillium brevicompactum*	Antifungal, antibacterial effectiveness	Shaligram et al. (2009)
14		*Penicillium fellutanum*	Biofilm, coating	Kathiresan et al. (2009)
15	AuNPs	*Helianthus annuus flower*	Reducing agent and a capping agent	Lee et al. (2016)
16	ZnONPs	*Aspergillus niger*	Antibacterial and anticancer activities against human alveoli lung cancer A-549 cell line	Majeed et al. (2018)
17	CuONPs	*Aspergillus terreus*	Antibacterial, anti-Candida, biocompatibility, anticancer, and antioxidant activities	Nassar et al. (2023)

efficient against the potentially harmful bacteria *K. planticola* and *B. subtilis*. The biosynthesized MNPs were successful in their inhibition of bacteria. High concentrations of flavonoids, saponin, lignin, and vitamins are found in *Trigonella foenum-graecum* seeds and roots, which contain potent reducing agents and improve the surfactant properties of chloroauric acid. The extract contains the COOH, C=N, and C=C functional groups. Electrostatically stabilizing AuNPs with flavonoids has been shown to work, and the functional group of metabolites works as a surfactant for AuNPs. *Macrotyloma uniflorum* root extract increased the concentration of silver ions, which reduced bacterial effectiveness because the extract contains a heavy concentration of caffeic acid (Mittal et al., 2013, 2015). Noruzi et al. (2011) synthesized AuNPs from rose petals. The extract contains a significant amount of sugar in addition to the proteins. These bioactive compounds are very important to the process of converting industrial amounts of tetrachloroaurate salt into AuNPs since they play a key role in the production. In a similar manner, the roots and tubers of *Catharanthus roseus* and *Clitoria ternatea* are used in the synthesis of MNPs in the dimensions and configurations of interest. In the synthesis of a medicine out of the extract of AuNPs found in the flowers of the *Nyctanthes arbor-tristis* plant, a green chemistry strategy is used in the formation process (Das et al., 2011; Vankar & Bajpai, 2010). As a reducing agent, an aqueous extract of *Mirabilis jalapa* flowers, roots, and tubers was used to facilitate the synthesis of AuNPs that are favorable to the environment (Mandava, 2017). Figure 1 depicts the relationship between metal ions and their clinical uses.

2.2 AgNPS

AgNPs are one of the nanomaterials that stand out as particularly fascinating and significant in the field of biomedicine. AgNPs are especially valuable in nanomedicine, which is a subfield of drugs that falls under the umbrella of nanoscience and nanotechnology. The promise of AgNPs in cancer detection and therapy has garnered the greatest interest, despite the fact that various other noble metals have been put to application. The anticarcinogenic action of AgNPs and many other bioapplications, such as antimicrobial, antifungal, antiviral, anti-inflammatory, antiangiogenic, antitumor, and cytotoxic medications (Zhang et al., 2016), have been the focus of inquiry as a possible next-generation anticancer treatment agent in both academic and industrial settings. This is due to the typical bad effects that are associated with chemotherapy and radiation therapy. Even though AgNPs play an important part in clinical research, there are great considerations that need to be taken before being utilized on people without risk. Concerns with the product's stability, bio-distribution and controlled release, accumulation in cells, and toxicology are among these issues. Developing AgNPs as antiangiogenic components that may avoid the problems of inefficient transport and therapeutic obstruction is the most promising option for treating cancer and other illnesses linked to angiogenesis. Other angiogenic illnesses, such as obesity, psoriasis, endometriosis, rheumatoid

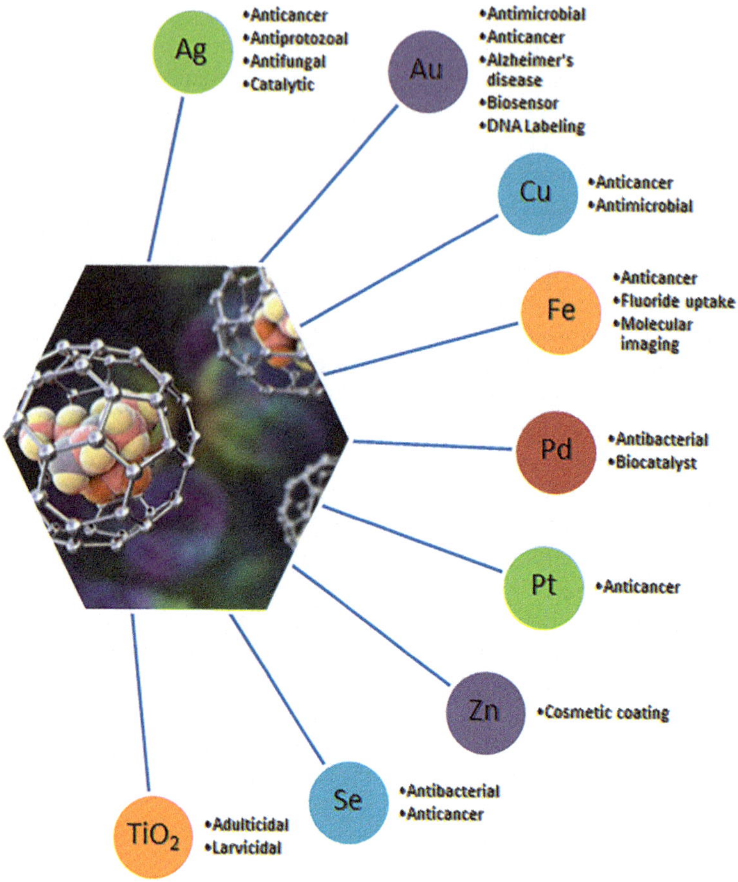

Fig. 1 Metal ions and their medical applications

arthritis, and diabetic retinopathy, may potentially benefit in this manner. The use of plant materials reduces the need for energy, and the possibility of both bacterial and toxic chemical contaminations during AgNP synthesis have far-reaching implications and simplify the process. Rai and Ingle (2012) pointed that one way to make AgNPs in a green way is to use plant extracts that contain a lot of phytochemicals that work to lower the concentration of metal ions. The potential of AgNPs used in the diagnosis of cancer is substantial, and thus, a number of strategies have been developed to capitalize on this possibility. AgNPs and antibiotics, or several therapeutic agents that could operate together to synthesize their reportedly larger benefits, are not yet known, despite the existence of a variety of methods. Therefore, further study is required to explain the mechanism(s) behind the cumulative effect of the two cytotoxic medications simultaneously (Calderon-Jimeneze et al., 2017). The ability to figure out how, why, and how well two or more drugs work together to help each other could lead to the creation of a new system with many parts that work together to treat different types of cancer. Zhang et al. (2010) reported that more research needs to be done in in vivo models to confirm the processes and get a full picture of the biocompatibility versus toxicity of AgNPs for therapeutic purposes. If these results hold up, they will help scientists and engineers in the nanotechnology and nanoscience fields develop cancer and antiangiogenic medicines, such as AgNPs, that are less toxic, are more widely tolerated, and have a greater impact (Mody et al., 2010). Before the use of AgNPs in humans is regarded as safe, biocompatibility tests and research into how AgNPs interact with cells and tissues are unavoidably going to be necessary. The highest significance is that nanotechnology-based therapy is now in the development stage, exceeding the technologies that are already in use and overcoming the constraints of the treatment techniques that are currently in use (Shafey, 2020). In the end, but certainly not least, it must be able to provide patients with treatment that is effective, efficient, and comforting for a wide range of ailments. AgNPs are appealing to a wide variety of biological systems for a variety of reasons, including the possibility of their use in food packaging, medicine, pharmacology, and healthcare (Okaiyeto et al., 2019). Patra and Baek (2016) have explained that AgNPs have features that make them effective as antioxidants, antimicrobial agents, and anti-inflammatory agents. Reduced silver is harmful to microorganisms since it may cause damage to cell walls and interfere with normal activities and growth. This is because the freed Ag ions engage in chemical interactions with the macromolecules of DNA and proteins. They accomplish this goal through the inhibition of protein synthesis, the reduction of membrane permeability, and the induction of cell death. It is possible that AgNPs have a higher level of biological action than ionic Ag because of their very small size and high surface-to-volume ratio.

2.3 AuNPs

The diverse surface chemistry of gold nanoparticles makes them amenable to coating with a wide variety of compounds, including small molecules, biomolecules, and polymers. As a result, you can find uses for them everywhere, from catalysis and sensory probes to medication transport and medicinal agents (Das et al., 2011). The application of AuNPs as a better modifier for electrode surfaces in electrochemical sensor fabrication has recently come to light. Electrochemical sensors frequently use AuNPs due to their many desirable characteristics. Some of these are that they conduct electricity better, are more resistant to oxidation, are easy to make, have a narrow size distribution, are good for living things, can have their surfaces changed, have a larger surface area, and have a good catalytic potential (Nasa-Esfahani & Ensafi, 2022). The surfaces of the various electrodes are good candidates for the simple deposition of AuNPs. They play an essential role in reducing the overpotentials of electrochemical processes, which is one reason why they have evolved into a very attractive kind of surface modification. Another essential quality that contributes to the utility of AuNPs in the formation of electrochemical sensors is their singular form. The appearance of the AuNPs is powdery and reddish brown, and they have a ball shape. In spite of the fact that materials based on AuNPs provide a number of advantages, their high price has precluded their widespread use for the purpose of electrode surface modification. When applied to sensors, AuNPs offer a new vantage point by enabling the miniaturization of electrochemical sensors to the nanoscale. Miniaturizing sensors based on AuNPs using micro- and nanofabrication processes is an exciting concept for developing electrochemical sensor array platforms. The combination of microfluidic instruments with electrochemical sensors allows for the rapid, accurate, and specific detection of analytes. High selectivity, optimum diffusion of electroactive species, a greater signal-to-noise ratio, and improved catalytic potential are advantages of electrodes coated with AuNPs (Zhang et al., 2010).

2.4 ZnNPs

Due to their versatility and potential applications, research into zinc oxide nanoparticles has been extensive. In addition to being cheap, safe, and simple to make, ZnONPs are the second most common metal oxide after Fe metals (Lakshamipriya & Gopinath, 2021). Changing the synthesis process, precursor, or substance used to synthesize the nanomaterial can affect the final product's morphology and hence the nanoparticle's physical and chemical behaviors. For analytical sensing purposes, ZnONPs are useful inorganic chemicals belonging to the semiconductor groups II–IV. They are white, powdery nanoparticles of zinc oxide that do not dissolve in water. The 3.37 eV energy band and the 60 meV bonding energy of ZnONPs provide them with excellent chemical, electrical, and thermal durability. ZnONPs have optical electrical and photocatalytic properties (Zhang et al., 2016). Nanoscale ZnO has found use in solar cells, photocatalysis, and chemical sensors due to its unusual set of properties. ZnONPs show great promise for use in biomedicine due to their superior UV absorption and low toxicity. Because of their high hardness and rigidity, ZnONPs have found use in the ceramics industry and are also a promising surface material with potential biomedical uses. ZnONPs have been recognized for their potent antimicrobial properties. For these reasons, ZnONPs find widespread application in fields as diverse as nanomedicine, drug transport, gene delivery, biological sensing, and drug labeling. Zinc oxide is a substance recognized as safe by the FDA. Because of its solubilization in an acidic environment, zinc oxide has the potential to be identified as a multifunctional nanocarrier that facilitates the delivery and release of drugs (Aritonang et al., 2019).

2.5 *Ti NPs*

Nanoparticles (NPs) made from TiO_2 are produced in huge quantities around the world. TiO_2 NPs differ from their fine particle (FP) analogues in terms of physicochemical properties, which may have an impact on their bioactivity. Inhalation has been highlighted as the key route for TiO_2 NP exposure in the workout by the majority of those cited. Although the rate of translocation of TiO_2 NPs from the lungs and GIT to other systemic organs is modest, it is nonetheless possible. Other forms of human exposure have also been the subject of research. In most cases, people are exposed to TiO_2 NP additives through their diet (Zhang et al., 2010). In the majority of in vivo and in vitro cutaneous exposure research, it was shown that TiO_2 NPs did not penetrate the stratum corneum (SC). TiO_2 NP carriers can be injected directly into a patient's veins for use in nanomedicine. TiO_2 NPs can cause severe lesions in the hepatic, spleen, kidneys, and brain after being administered IV (intravenous). In addition, similar effects may be triggered by ingesting excessive quantities of TiO_2 NPs. Despite its widespread adoption, very little epidemiological information is available on the health effects of TiO_2 NPs. However, lung tumors have been seen in rats subjected to long-term inhalation trials (Shi et al., 2013).

2.6 *PdNPs*

In the field of biological applications, PdNPs have received more attention because of their extensive catalytic and sensor activities. To make simple, selective catalysts and sensors for a wide range of chemical and biological analytes, it is necessary to make PdNPs with carefully controlled size and shape. Because it is so readily available, Pd offers a competitively priced alternative to the use of other noble metals, such as Au and Pt, in electrochemical sensing and biosensing systems. Different PdNPs have been studied recently for the detection of a wide range of biomarkers (Mittal et al., 2015). This is because they have unique electrical properties, better catalytic performance, and the ability to selectively sense. Nanocomposites, bimetallic nanoparticles, carbon nanomaterials with variable compositions, and metal oxide nanomaterials are some examples of these nanomaterials. We were able to develop an electrochemical sensor for *Brucella* by applying PdNPs to an Au surface while maintaining a steady voltage. Gram-negative bacteria from the genus *Brucella* are responsible for the infectious diseases known as brucellosis. This ailment may be transmitted from animals to people through unhygienic food or other products obtained from animals, as well as through direct contact with an infected animal. PdNPs act as transducers, allowing them to successfully immobilize a *Brucella*-specific probe. Voltammetry accomplished the immobilization and hybridization of the material (Mittal et al., 2013). The sensor has demonstrated outstanding stability for 1 month when kept in a cooled place like a freezer at 4 °C. Bimetallic Au-PdNPs have shown excellent uses in drug sensing, making them more prominent alloys in

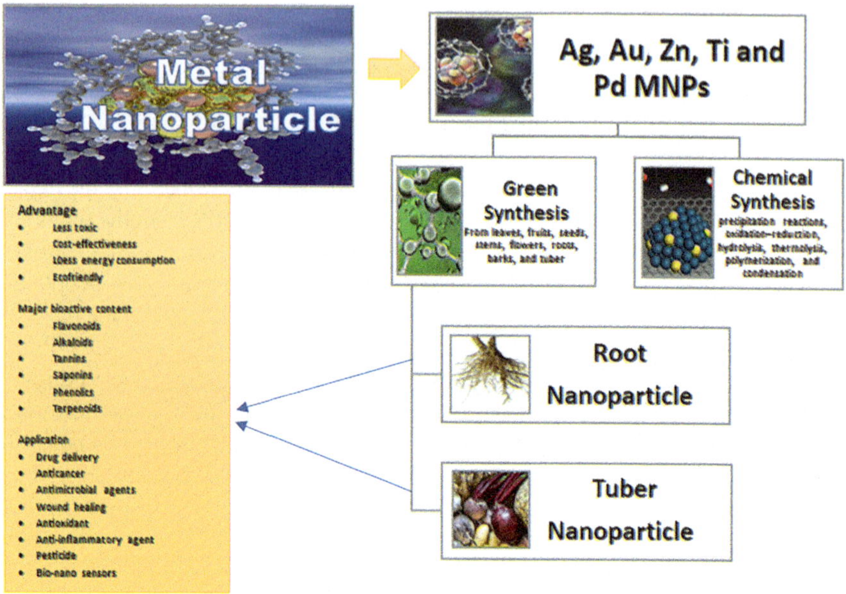

Fig. 2 Different paradigms of root and tuber nanoparticles

research pertaining to catalytic processes and sensitization. The simultaneous detection of UA (uric acid), levodopa in urine, blood serum, and levodopa capsule led to the evaluation of a sensor-based nanoporous stainless steel (NPSS) electrode that had AuNPs and PdNPs added to it. This was necessary for the simultaneous detection of both substances (Madhuraiveeran & Jin, 2020). Different paradigms of root and tuber MNPs are outlined in Fig. 2.

2.7 *Others*

(a) *Iron oxide:* Magnetic image and environment intensification (Schrand et al., 2010).
(b) *Silver oxide:* Antibacterial, antifungal, therapeutic agent, genetic engineering, and tissue culture (Shanmuganathan et al., 2019).
(c) *Cerium oxide:* Biomolecule imagination, surgical instruments (Schrand et al., 2010).
(d) *Silica oxide:* Thermal insulator, electrical insulators, gene therapy, enzyme catalysis therapeutic mediator, natural adsorbents, and filler components (Schrand et al., 2010).
(e) *Zinc oxide:* Dermal protection (Pantic et al., 2019).
(f) *Titanium dioxide: Antifungal, antibacterial,* antimicrobial, coating, and sterilizing agents (Esmaelnejad et al., 2019).

(g) *Nickel (oxide):* Cytotoxic, antitumor, and anticancer agents (Khan et al., 2019).
(h) *Copper oxide:* Antibiotic, antiviral, antibacterial, antifouling, and antifungal agents, inks, coating materials, biocatalysis, lubricants, resistance, and conductance agents (Schrand et al., 2010).
(i) *Gold oxide:* Antimicrobial, antibacterial, antifungal, and therapeutic agents, cell imagination, photodynamic treatment, carcinogenic agent, and surgical instruments (Schrand et al., 2010).
(j) *Aluminum oxide:* Antimicrobial, antifungal, antibacterial, and antiviral agents (Subramaniam et al., 2019).
(k) *Magnetic iron oxide:* Drug administration, tissue culture, cell labeling, and hyperthermal agents (Tong et al., 2019).
(l) *Calcium oxide:* Antimicrobial and antifungal agents and alkalinity (Dizaj et al., 2014).
(m) *Magnesium oxide:* Antimicrobial and antifungal agents (Dizaj et al., 2014).
(n) *Bismuth oxide:* Drug delivery systems (Szostak et al., 2019).
(o) *Chromium oxide:* Collagen stabilizer (Sangeetha et al., 2012).
(p) *Manganese dioxide:* Biocatalyst, fluorescence sensor, drug controller, stimulator, and bio-image (Wu et al., 2019).

3 Secondary Metabolite/Phytochemical Effect in Bioreduction Reaction

Silver The synthesis of AgNPs is the result of a biological reaction involving the interaction of $AgNO_3$ with root or tuber extracts (Tripathy et al., 2010). The following reaction provides the chemical reaction between silver nitrate and root or tuber extracts:

$$\mathrm{AgNO_3} + \mathrm{Root\,/\,Tuber\ extract} \rightarrow \mathrm{AgNPs} + \mathrm{Byproducts}. \tag{1}$$

Gold The extracts of the root or tuber include a variety of active molecules. The bioreduction process relies heavily on the participation of active molecules. In the appearance of active molecules and redox enzymes, the suggested process included the reduction of metallic AuONPs into Au^+ ions (Thakkar et al., 2010). The proposed reaction equation for the synthesis of AuNPs is detailed below:

$$\mathrm{HAuCl_4 \cdot 4H_2O} + \mathrm{Root\,/\,Tuber\ extract} \rightarrow \mathrm{AuONPs} + \mathrm{Byproducts}. \tag{2}$$

Platinum Platinum is necessary for the reduction process as well as the synthesis of PtNPs. The chemical equation for the reaction process is as follows:

$$\mathrm{H_2PtCl_2 \cdot 6H_2O} + \mathrm{Root\,/\,Tuber\ extract} \rightarrow \mathrm{PtONPs} + \mathrm{Byproducts}. \tag{3}$$

Table 3 Ionic liquid MNP synthesis using reduction methods

S. N.	Ionic liquid	MNPs	Metal salt	Reducing agents	References
1	[BMIM][BF_4],[BMIM][PF_6]	AgNPs	$AgBF_4$	H_2	Hulkoti and Taranath (2014)
2	[BMIM][BF_4], [BMpy][TFO]	AgNPs	$AgBF_4$	H_2	Setua et al. (2011)
3	[BMIM][BF_4] and H_2O	AgNPs	$AgBF_4$	[BMIM][BH_4]	Lazarus et al. (2012)
4	[BMIM][PF_6]	AgNPs	$AgNO_3$	Tween 85	Ge et al. (2007)
5	[BMIM] [$C_{12}H_{25}SO_4$] (lauryl sulfate)	AuNPs	$HAuCl_4$	Ascorbic acid	Obliosca et al. (2010)
6	[ShexMIM][cl]	AuNPs	$HAuCl_4$	$NaBH_4$	Itoh et al. (2004)
7	[BMIM][BF_4]	AuNPs	$HAuCl_4$	$NaBH_4$	Lazarus et al. (2010)
8	[EMIM][TfO],[EMIM][$MeSO_3$]	AuNPs	$HAuCl_4$-$3H_2O$	Glycerol	Khare et al. (2010)
9	[$(CH_3)_2NH_2$],[[$(CH_3)_2NCO_2$], DMF	$HAuBr_4$	$HAuCl_4$-$3H_2O$	[$(CH_3)_2NCHO$ (DMF)	Bhatt et al. (2007)
10	[BMIM][BF_4]	CuNPS	$Cu(OAc)_2$-H_2O	H_2NNH_2-H_2O	Raut et al. (2009)

Copper In a reduction method, copper sulfate and extracts of roots or tubers are used to make copper nanoparticles (CuNPs) (Ramanathan et al., 2013). The reaction mechanism is described in more detail below:

$$CuSO_4 \cdot 5H_2O + \text{Root / Tuber extract} \rightarrow \text{CuONPs} + \text{Byproducts}. \quad (4)$$

Zinc oxide The synthesis of ZnONPs followed an appropriate process in which zinc nitrate was dissolved in a root or tuber extract to obtain ZnONPs (Sangeetha et al., 2011). The synthesized process is completed using the below chemical equation:

$$Zn(NO_3)_2 + \text{Root / Tuber extract} \rightarrow \text{ZnONPs} + \text{Byproducts}. \quad (5)$$

Ionic liquid metal nanoparticle synthesis using reduction methods are listed in Table 3.

4 Pharmacological Applications of MNPs

The MNPs‘ antimicrobial properties were extensively researched. It has been observed that both AgNPs and AuNPs have potent antimicrobial action against various human and animal diseases. This is because a pathogenic organism's disruption of the polymer units of the cell membrane effectively disrupts the bacterial system, leading to dysfunction (Sondy & Sondy, 2004). Increasing the quantity of AgNPs leads to greater membrane permeability and cell wall rupture (Kasthuri et al., 2009). AgNPs and AuNPs are shown to interact with cell membranes by attaching to the membrane's active site (Kim et al., 2007). CuNPs and CuONPs were reported to have antibacterial properties (Ren et al., 2009). In a study involving *B. subtilis*, it was discovered that CuNPs acted as a strain-specific antibacterial agent. Many researchers reported the anti-inflammatory effect of AgNPs, mainly due to their ability to reduce matrix metalloproteinase (MMP) and proinflammatory cytokines by inhibiting interferon-alpha and tumor necrosis factor alpha (TNF-α) (Kirsner et al., 2001; Tian et al., 2007). Evidence suggests that biosynthesized MNPs have anti-inflammatory and wound-healing properties (Gurunathan et al., 2009). The fungicidal activities of AgNPs were investigated against *Candida* sp., and the results were comparable to *fluconazole* and amphotericin (Vivek et al., 2011). Damage to fungal intracellular components may be blamed for this action (Logeswari et al., 2012). Due to the problematic nature of currently existing antifungal drugs, research into and development of multifunctional MNPs for use against fungal infections is strongly encouraged. It has been observed that AgNPs have larvicidal effectiveness against malarial vectors (Rajkumar et al., 2011; Santhoshkumar et al., 2011). Malaria can be effectively managed with green-synthesized MNPs of silver, platinum, and palladium. Finding a new method of combating the vectors that spread the most common plasmodial infections is a pressing concern (Gnanadeshigan et al., 2011; Jayaseelan et al., 2011). Several researchers have found that green-synthesized MNPs can inhibit the spread of tumor cells. Secondary metabolites found in root and tuber extracts are responsible for this action (Das et al., 2013). Dipankar and Murugan (2012) observed that both AgNPs and AuNPs were able to kill cancer cells. Below, we will go into detail about their few particularly relevant medicinal uses. Figure 3 displays several MNP usage examples.

4.1 Antibacterial

In pathogenic species, the AgNPs efficiently disrupt units of the cell membrane. Subsequently, MNPs disrupt the bacterial system by destroying the cell membrane and interfering with protein synthesis (Sondy & Sondy, 2004). Higher concentrations of silver nanoparticles are more effective at penetrating membranes and killing bacteria than lesser quantities. *Rhizophora apiculata-reduced* silver nanoparticles showed maximum conductivity but fewer bacterial colonies in the experiments compared with $AgNO_3$-treated cells. This distinction may be the result of the nanoparticles' tiny size and extent of surface area, both of which contribute to an increase in membrane permeability and cell death (Antony et al., 2011). AgNPs and AuNPs cleave to the active site of the cell membrane and suppress cell cycle processes in bacteria (Kim et al., 2007). *Citrus sinensis* root and tuber extracts facilitated the synthesis of AgNPs in a single step by acting as agents (reducing and

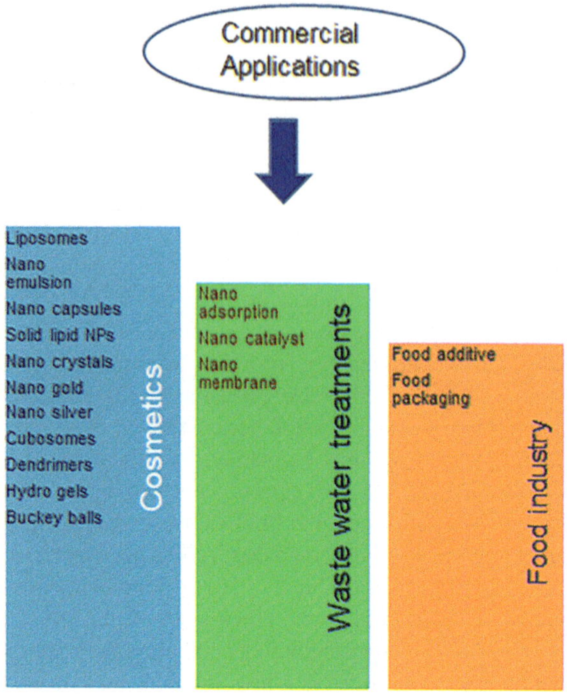

Fig. 3 Commercial application of root and tuber MNPs

capping). In vitro studies revealed that *C. sinensis* root and tuber extracts exhibit antimicrobial action against the gram-positive *S. aureus*, the gram-negative *P. aeruginosa*, and the gram-negative *E. coli* (Kaviya et al., 2011). Silver nanoparticles synthesized from the roots and tubers of *Acalypha indica* were found to successfully inhibit the growth of waterborne pathogenic bacteria at low concentrations (10 g/ml) of extracts (Krishnaraj et al., 2010). The root- and tuber-mediated antibacterial activities are mentioned in Table 4.

4.2 Antifungal

Biosynthesized metallic nanoparticles' fungicidal functions are more promising than commercial medicines such as fluconazole and amphotericin. *Candida* species bacterial membrane damage, fungal intercellular component damage, and cell function destruction have all been reported by AgNPs grown from the root and tuber of plants (Logeswari et al., 2012). Over-the-counter antifungal medications have minimal therapeutic utility, tend to have more side effects, and are less effective at curing the microbial condition. Commercial medications cause undesirable side effects, such as renal failure, fever, hepatic disorder, diarrheal problems, coughs, and colds.

Table 4 Root- and tuber-mediated antibacterial activities against various microorganisms

S. N.	Plant parts	Scientific name	Examined microorganism	References
1	Root	*Potentilla fulgens*	*E. coli, B. subtilis*	Mittal et al. (2015)
12		*Erythrina indica* L.	*S. aureus, M. luteus, E. coli, B. subtilis, S. typhi, S. paratyphi*	Sre et al. (2015)
3		*Diospyros paniculata*	*P. notatum, A. flavus, A. niger, C. albicans, S. cerevisiae*	Rao et al. (2016)
4		*Diospyros sylvatica*	*B. pumilus, P. aeruginosa, B. subtilis, S. aureus, K. pneumoniae, E. coli, S. pyogenes, P. vulgaris, A. niger, P. notatum, A. flavus, S. cerevisiae, C. albicans*	Pethakamsetty et al. (2017)
5		*Annona muricata*	*K. pneumoniae, S. aureus*	Esmaeilnejad et al. (2019). Witika et al. (2017)
6		*Cibotium barometz*	*S. aureus, S. enterica, P. aeruginosa*	Wang et al. (2016)
7		*Diospyros assimilis*	*B. pumilus, B. subtilis, S. aureus, S.pyogenes, K. pneumoniae, E. coli, P. aeruginosa, P. vulgaris, A. niger, A. flavus, C. albicans, P. notatum, S. cerevisiae*	Kantipudi et al. (2018)
8		*Pelargonium endlicherianum Fenzl.*	*P. aeruginosa, E. coli, S. epidermidis*	Karatoprak et al. (2017)
9		*Rheum palmatum*	*S. aureus, P. aeruginosa*	Arokiyaraj et al. (2016)
10		*Lepidium draba*	*S. aureus, B. cereus, S. typhimurium,E. coli*	Benakashani et al. (2017)
11		*Angelica pubescens Maxim*	*E. coli, S. aureus, P. aeruginosa*, and *S. enterica*	Markus et al. (2017)
12		*Phoenix dactylifera*	*E. coli, C. albicans*	Oves et al. (2018)
13	Root	*Arctium lappa*	*E. coli, A. tumefaciens, L. acidophilus, S. aureus*	Nguyen et al. (2018)
14		*Asparagus racemosus*	*E. coli, S. aureus, B. subtilis, K. pneumoniae, P. fluorescence, A. hydrophila, E. tarda, F.branchiophilum, Y. ruckeri*	Vijaykumar et al. (2019)
15		*Lysiloma acapulcensis*	*E. coli, P. aeruginosa, S. aureus, C. albicans*	Garibo et al. (2020)
16		*Raphanus sativus*	*S. aureus, E. coli, C. albicans, C. glabrata*	Khan et al. (2015)

(continued)

Table 4 (continued)

S. N.	Plant parts	Scientific name	Examined microorganism	References
		C. arborea	*S. aureus, M. luteus, P. aeruginosa, C. freundii, C. glabrata*	Ramadurga et al. (2019). Kashyap et al. (2022b)
17	Tuber	*Alpinia calcarata*	*P. mirabilis, E. coli, B. cereus, S. aureus*	Pugazhendhi et al. (2015)
18		*Allium cepa*	*E. coli, P. aeruginosa, B. subtilis, F. oxysporum*	Gomaa (2017)
19		*Helianthus tuberosus*	*R. solanacearum, X. axonopodis*	Aravinthan et al. (2015)
20		*Dioscorea alata*	*E. coli, S. auricularis*	Pugazhendhi et al. (2016)
21		*Crocus haussknechtii*	*S. aureus, P. aeruginosa*	Mosaviniya et al. (2021)
22		*Allium sativum*	*P. aeruginosa, B. licheniformis*	Saha and Bandyopadhyay (2017)
23		*Bergenia ciliata*	*M. luteus, S. aureus, E. aerogenes, B. bronchiseptica, A. niger, A. fumigatus, A. flavus, F. solani*	Phull et al. (2016)
24		*Dryopteris crassirhizoma*	*B. cereus* and *P. aeruginosa*	Lee et al. (2016)
25		*Coptis chinensis*	*E. coli, B. subtilis*	Lengke et al. (2006)
26	Tuber	*Coptidis rhizoma*	*E. coli, S. aureus*	Sharma et al. (2018)
27		*Curcuma longa*	*E. coli, L. monocytogenes*	Alsammarraie et al. (2018)
28		*Curcuma caesia*	*E. coli, S. aureus, S. pyogenes, A. niger, P. aeruginosa*	Kaur et al. (2018) and Bhardwaj et al. (2023b)
29		*Curcuma amada*	*E. coli, S. aureus, S. enterica, S. dysenteriae, B. subtilis, B. pumilus, V. cholerae, C. albicans, P. notatum, P. funiculosum, A. niger*	Kashyap et al. (2022a)
30		*Canna indica L*	*S. aureus, K. pneumoniae, E. coli*	Selvi et al. (2019)
31		*Ferula foetida*	*E. coli, K. pneumoniae, C. albicans*	Devanesan et al. (2020)
32		*Zingiber officinale*	*V. anguillarum, V. alginolyticus, A.punctate, V. parahaemolyticus, V.splendidus, V. harveyi*	Yang et al. (2017)

The synthesis of MNPs allowed for a new, very effective antimicrobial medication. Fatty acids and protein polymers make up most of the fungal cell wall. The fungus development is successfully destroyed by the multifunctional AgNPs, which also show promising efficacy against spore-producing fungi. After being exposed to MNPs, the membrane structure of fungal cells was found to have undergone significant alterations (Gardea-Torredey et al., 2002).

4.3 Anti-inflammatory

Anti-inflammatory treatment plays a crucial role in the healing process. Keratinocytes, T lymphocytes, B lymphocytes, and macrophages all release immune response compounds such as interleukins and cytokinins that can help fight inflammation (Jacob et al., 2012). The endocrine secretory system and inflammatory mediators, active enzymes, antibodies, and immunological organs also release cytokines, interleukins 1 and 2, and other active molecules with potential anti-inflammatory effects. According to Satyavani et al. (2011), anti-inflammatory mediators promote healing. Furthermore, inflammatory mediators regulate the spread of illnesses by participating in biochemical pathways. The inflammatory function of wounds was improved, and new tissues were grown to biosynthesize AuNPs. Inflammation can be treated with biosynthesized AuNPs and PtNPs, as shown in the investigations (Gurunathan et al., 2009).

4.4 Anticancer

Cancer is characterized by unregulated cell development as well as significant modifications in biochemical and enzymatic parameters, which are present in tumor cells. The use of biologically synthesized MNPs as innovative regulatory agents would stop overexpression of cellular growth and control it with methodical cell cycle processes in malignant cells (Akhtar et al., 2013). MNPs mediated by root and tuber are also highly effective against several human cancer cell types. The latest studies have shown that MNPs derived from plants' roots and tubers may be able to slow or stop the development of cancerous cells (Das et al., 2011). The heightened cytotoxic impact is due to the active molecules present in the synthesis medium and other nonmetal components (Raghunandan et al., 2011). AgNPs derived from roots and tubers are responsible for controlling both the cell cycle and the formation of the enzyme that circulates throughout the body. In addition to this, the MNPs synthesized from plants' roots and tubers may control the production of free radicals inside cells. It is not unusual for free radicals to promote cell development while also interfering with normal cellular function. A low concentration of AuNPs is sufficient to activate the apoptotic pathway in cancer cells, and the active molecules, phytochemicals, and other nonmetal components of the synthesis medium contribute to the heightened lethal impact (Alt et al., 2004; Dipankar & Murugan, 2012). In a similar manner, the concentration of biomolecules that was maintained in the cancer cell line MCF-7 had been treated with AgNPs, which led to metabolic control. The innovative medical uses of MNPs have potent results in the treatment of various cancers and other retroviral illnesses. The innovative nature of the MNPs lies in the fact that they may target dangerous tumors without causing damage to healthy cells. In comparison to synthetic drugs produced using chemical processes, AgNPs synthesized through environmentally friendly processes had a much greater cytotoxic impact on HeLa cell lines (Das et al., 2011; Suman et al., 2013).

4.5 Antiviral

Drugs mediated by plants offer an alternate approach to treating and containing viral infections. Viruses are extremely careless when they invade a host, and once inside, they speed up the translation process to increase their population. Nanoparticles of silver (AgNPs) synthesized in a lab can inhibit virus replication and spread and have wide antiviral activity. Suriyakalaa et al. (2013) reported that biologically synthesized AgNPs exhibit potent anti-HIV effects at the reverse transcription mechanism's initiation phase. MNPs are potent antiviral agents that prevent viruses from replicating inside the body. Multiple binding sites on the biosynthesized metallic nanoparticles allow them to connect with the viral membrane and regulate virus function. Cell-free and cell-associated viruses are both killed by the biologically synthesized MNPs (Sun et al., 2005). The HIV-1 life cycle is constantly slowed down by the AgNPs and AuNPs that have entered the cells. Thus, the metal nanoparticles will be a powerful medication against retroviruses.

4.6 Antidiabetics

When blood sugar levels rise unchecked, a condition known as diabetes mellitus (DM) sets in. While partial control of DM can be achieved with dietary restrictions, exercise, and synthetic insulin, eradicating the disease entirely remains a formidable issue. However, biosynthesized MNPs show promise as a potential "last resort" treatment for diabetes. Daisy and Saipriya (2012) have shown the beneficial therapeutic effects of AuNPs in diabetic animal models. Mice with diabetes that have been treated with AuNPs show a significant decrease in serum creatinine, uric acid, alanine transaminase, and uric acid. A diabetic model treated with AuNPs demonstrated normalization of HbA (glycosylated hemoglobin) levels. In a diabetes-induced animal model, Swarnalatha et al. (2012) reported that AgNPs biosynthesized from *Sphaeranthus amaranthoides* roots suppressed amylase and acarbose sugar because their primary components inhibited amylase. Pickup et al. (2008) also reported that MNPs were an effective therapeutic agent in the management of diabetes with fewer side effects. Sugar levels of 140 mg/dl were satisfactorily regulated in the clinical studies conducted on mice.

4.7 Antioxidant

Antioxidant agents, which may be either enzymatic or nonenzymatic substances, are responsible for controlling the generation of free radicals. Free radicals have been linked to a variety of different types of cell damage, including cancer, atherosclerosis, and brain impairment. Free radicals are the byproducts of ROS, which include SOD, HPx, and HFR. Researchers discovered that active organic molecules such as proteins, fats, FA, and phytochemicals strictly control the production of free radicals (Marambio-Jones & Hoek, 2010). Antioxidants, both enzymatic and nonenzymatic, are useful in the healing process of chronic diseases. When compared to other standard antioxidants such as AA (ascorbic acid), AgNPs and AuNPs have stronger antioxidant agents (Baratjmanikanth et al., 2010). Despite the increased phenolic and flavonoid concentrations in the root of *Careya arborea* Roxb., MNPs showed the greatest antioxidant action (Reichelt et al., 2012; Kashyap et al., 2022b).

5 Commercial Application of Root- and Tuber-Mediated MNPs

5.1 Wastewater Treatment

All forms of life depend critically on water. Water is essential to life, and its absence can lead to a cascade of catastrophes that could spell the end of humankind. Water supply protection and industrial water treatment are both issues that require several strategies. Wastewater treatment seems like a practical and cheap technique to deal with water shortages. There are a wide variety of options for recycling water, including physical, chemical, biological, and combined treatment systems. However, the limitations of these systems can be mitigated and improved upon by incorporating nanotechnology into the wastewater treatment process (Alizad Oghyanous, 2022). The use of nanomembranes in MBR technology is a very promising approach to delaying the onset of a water crisis by several years. There has been huge research on how nanotechnology can be used to clean wastewater. Based on the materials employed, nanotechnology can be broken into three broad categories: nano-adsorbents, nanocatalysts, and nanomembranes (Schodek et al., 2009).

I. Nano-adsorbents

Nanosized particles with a high affinity for adsorbing molecules can be formed from either organic or inorganic materials and are referred to as "adsorbent nanoparticles." They are effective at filtering several harmful substances. These nanoparticles may be advantageous in the removal of various pollutants as a result of their favorable qualities, such as catalytic potential, small size, high reactivity, increased energy, and larger surface. MNPs, oxide NPs, magnetic NPs, and metal oxide nanomaterials all suffer distinct adsorption processes (Schodek et al., 2009).

II. Nanocatalyst

In the field of nanocatalysis, when light energy interacts with MNPs, it leads to a wide range of powerful photocatalytic actions. The high and widespread photocatalytic activity of this therapy is driving its rising popularity. Hydroxyl (OH) radicals are responsible for the destruction of microorganisms and organic materials in a photocatalytic process. Inorganic materials such as semiconductors and metal oxides are frequently employed in nanocatalysts (Schodek et al., 2009). A nanophotocatalyst is considered to be a nanocatalyst if it is inert, has a low concentration in water and air, can form agglomerates and precipitates, and makes ordinary particles.

III. Nanomembrane

A nanomembrane acts as a filter to remove contaminants from the water. These filters are excellent at trapping pollutants such as dyes and metals. Nanomaterials such as nanotubes, nanoribbons, and nanofibers are employed as nanomembranes. Nanoparticles inserted into membranes are the most feasible and effective option when compared to nano-adsorbents, nanocatalysts, and nanomembranes. This is due to the technique that provides powerful physical therapy and also makes use of MNPs, thus improving the overall quality of the treatment. As a result of the fast development of nanotechnology, the fabrication of nanocomposite membranes has become a progressively noteworthy and systematic process. Polymeric nanocomposite membrane engineering properties have thus been studied in relation to a wide variety of nanoparticles (NPs); many of these NPs have shown considerable improvements in mechanical, thermal, and antifouling characteristics (Dasgupta et al., 2014). When nanoparticles (NPs) are distributed evenly throughout the matrix, it is known to enhance the mechanical and thermal characteristics of polymeric membranes because of the potent interfacial interactions that occur between MNPs and the polymeric matrix. Membranes have widespread application in membrane bioreactor (MBR) systems, which are used to treat a wide range of wastewater types. When activated sludge is combined with membrane filtration, as in the MBR process, effluent quality, footprint, organic loading rate, and sludge production are improved over conventional treatment methods. Membranes become contaminated with bacteria and their byproducts, as well as other organic debris, in this process. Operating circumstances have a significant impact on the likelihood of membrane fouling in MBRs, but membrane characteristics also play a crucial role in this process (Oghyanous et al., 2020).

5.2 Cosmetics

Many of the most successful firms in the cosmetic industry claim that their products include nanomaterials of varying sizes, including fullerenes, nanotubes, liposomes, and quantum dots, among others (Anonymous, 2007).

The types of nanomaterials used in cosmetics are the following:

I. Liposomes

Liposomes are bilayer vesicles made of GRAS (generally recognized as safe) natural or synthetic phospholipids that enclose a whole volume of water. Because their lipid bilayer may merge with other bilayers, including the cell membrane, liposomes are great for aesthetic delivery applications. They are ideal for cosmetic applications because of their ease of preparation, increased skin absorption of active substances, and sustained release of agents to cells over time (Morganti, 2010; Jain et al., 1998). Several vesicles are now in use with the purpose of enhancing the transdermal delivery of drugs. These include transferosomes (Cevc, 1996; Thong et al., 2007), niosomes (Uchegbua & Vyas, 1998; Balakrishnan et al., 2009), and ethosomes (Touitou et al., 2000).

II. Nanoemulsions

Nanoparticle-sized drops of one liquid are dispersed throughout another (Simonnet et al., 2001). They are metastable systems where the structure can be altered by changing the preparation procedure. All of the ingredients that went into making them are safe, approved by the FDA as GRAS (generally recognized as safe). Their increased stability and adaptability to transport active ingredients, as well as their longer shelf life, are the results of their reduced particle size (Fred et al., 2006; Sonnerville-Aubrun et al., 2004).

III. Nanocapsules

Nanocapsules are tiny particles that consist of a polymeric capsule around a watery or oily core. Hwang and Kim (2008) found that when nanocapsules were used to apply UV-filtered octyl methoxycinnamate to pig skin, the skin took in much less of the chemical than when traditional emulsions were used.

IV. Solid Lipid Nanoparticles

The lipids inside tiny particles are solid at body temperature and are bound together by surfactants. Solid lipid nanoparticles have the appearance of oily drops. In addition, more active chemicals are able to penetrate the stratum corneum (Muller et al., 2002). They are also able to prevent the encapsulated compounds from degrading and may be used to control the release of cosmetic ingredients over a prolonged period of time. According to in vivo trials, a formulation that contains SLN is superior to a placebo in terms of the amount of moisture it provides to the skin (Wissing & Muller, 2003). It was discovered that the addition of a molecular sunscreen, which was then analyzed, improved the UV-resistant characteristics of the material. By adding the UV absorber 3,4,5-trimethoxybenzoylchitin to SLNs, Song and Liu (2005) found that UV absorption was increased by a lot.

V. Nanocrystals

Nanocrystals are assemblages that may include anything from a few to millions of atoms and form a "cluster." These assemblies typically have diameters ranging from 10 to 400 nm, and their physicochemical characteristics are intermediate between bulk solids and molecules. Additionally, their sizes often fall in this range. They make it possible for a secure and efficient passage through the epidermis (Petersen, 2008).

VI. Nanosilver and Nanogold

Nanosilver and nanogold's increased antibacterial qualities are being put to use for a variety of cosmetic purposes by companies that produce the ingredients. Some companies are already developing perfumes on the basis that the nanosilver in the product will give up to 7 days of bacterial resistance. Nanosilver and nanogold have been included in toothpaste due to claims that they are particularly efficient at sanitizing the mouth against germs (Deshmukh et al., 2019).

VII. Dendrimers

Dendrimers are nanostructures that are single-molecular, monodisperse, and tiny particles in size. They have regularly branching, symmetrical structures with a high density of functional end groups at their perimeter. Dendrimers have a dimension of around 20 nanometers. They possess a significant number of external groups that are amenable to multifunctionalization (Michael, 2010; Tournilhac & Simon, 2001).

VIII. Cubosomes

Cubosomes are distinct nanostructured particles of the bi-continuous cubic liquid crystal phase. Their sizes range from submicron to several microns (Spicer et al., 2003). Self-assembled liquid crystalline particles of certain surfactants make up this material, together with water and a microstructure. The surface area, viscosity, and dilution range of cubosomes are favorable. They can transport both hydrophilic and hydrophobic compounds and are thermally stable. Because of their low production cost and the possibility of regulated release through functionalization, they are a promising candidate for use in both the cosmetics and pharmaceutical industries (Kesselman et al., 2007).

IX. Hydrogels

Hydrogels are three-dimensional hydrophilic polymer networks that expand into water molecules or other biological fluids without dissolving. This is possible because chemical or physical cross-links hold them together. They are able to anticipate modifications to the environment and adapt their qualities appropriately to protect it from harm (Morales et al., 2009).

X. Buckyballs

Buckminsterfullerene, often known as C60, is perhaps the most well-known nanomaterial. Its diameter is roughly 1 nanometer. It has made its way into several high-priced vanishing creams. Making use of its ability to act as a powerful FRS is the driving force behind this endeavor (Bakery et al., 2007).

5.3 Food Industry

The investigation of nanomaterials has the prospect of bringing about a change in the food industry. Because of its minuscule scale, nanotechnology offers a tremendous amount of potential for adjusting the color, flavor, and nutritional content of food, as well as prolonging its shelf life and checking its integrity via barcodes in the same manner as the cold chain, i.e., anytime there is even a minor change in the circumstances under which food is stored. It is a brand-new area of research that inspires novel food items and cooking techniques (Aigbogun et al., 2017). In most cases, there are two distinct sorts of uses for nanofood: first, food additives (inside), and second, food packaging (outside). Nanotechnology provides a robust foundation for a better understanding of interactions and assembly behavior at the microscopic level, which is likely to have an effect on the structure, rheology, and functional characteristics of food (Sanguansri & Augustin, 2006). Synthesized nanoparticles (Au, Ag, Fe, Cu, Zn, Pt, etc.) integrated into the packaging system may modify compound permeability. The goal of food processing is to take raw materials such as plants and animals and turn them into finished meals that people can eat. Food processing may refer to anything from the basic act of grinding grains into flour to the complex industrial procedures required to generate processed meals. Food processing has the potential to improve many aspects of the food system, including nutrients, food delivery, nanoencapsulation, protein bioseparation, organic and inorganic pollutant monitoring, solubilization, distribution, and coloration. Food processing has the potential to greatly enhance areas such as adaptive nutrients, nano-capsulation of pharmaceuticals, bio-removal of proteins, quick monitoring of pollutants, soluble nature, distribution, and coloration in the food sector (Ravichandran, 2010). To better meet the demands of the human body and assist in the effective distribution of nutrients, nanotechnology may help in the development of "functional foods." Modern techniques for "on-demand food" are also being developed in many laboratories; these foods may be stored in the body and their nutrients released into cells only when they are required. Thousands upon thousands of nanocapsules filled with flavor or color enhancers or additional nutritional elements would slumber in the meal, releasing their contents when activated by the customer (Amin et al., 2015). The creation of nanocapsules suitable for incorporation into food and the timely delivery of nutrients are major factors in this field. The application of nanoparticles in already processed meals to improve nutrient absorption is an example of the many recent developments in the food industry. One Western Australian bakery company has found that using nano-encapsulated tuna fish oil, which is a source of omega-3 fatty acids, and its "Tip-Top" bread have been enormously successful ventures (Bund, 2008). Microcapsules designed to dissolve only in the stomach may effectively cover up the fishy taste. Using nanosensors to detect even minute amounts of pesticides in a wide variety of food items might be a great solution to the issue of food protection. The ability to detect and identify even tiny amounts of organic compounds, diseases, and harmful chemicals makes these instruments invaluable. These devices are top-notch threat detectors and actuators (Otles & Yalcin, 2010; Yalcin & Otles, 2010). The primary application of nanosensors in agriculture, horticulture, and aquatic environments is the detection of pesticides. It is well-known that the higher permeability and solubility of pesticides pose health hazards. Agronomists rely heavily on them in their work. Trace analysis of highly reactive pollutants is of paramount importance (Mclaren et al., 2009). Nanosensor devices have advantages over GLC, LCMS, and mass spectrometry due to their larger surface ratios. These benefits include loading more antibodies and enzymes, a more sensitive interface, reduced recognition limits, exceptional selectivity despite their tiny particles, and a quicker response. It is well recognized that using nanosized materials may increase the sensitivity of the indications or signals produced by transducers. It has been discovered that several nanostructures constructed of carbon nanotubes, semiconductors, and metals exhibit amazing electrical or optical capabilities (Hwang et al., 2011). Coating a nano-biosensor to make it light differently depending on the sort of pathogen it comes into contact with is another aspect of innovative packaging. Tools for identifying and locating contaminants and illnesses in food are readily accessible. This sort of apparatus is excellent at picking up even trace amounts of microorganisms (Chen et al., 2008). Figure 3 depicts the commercial application of root and tuber MNPs.

6 Factors Influencing the Synthesis of MNPs

Several regulating elements are related to the nucleation and subsequent production of stable MNPs via the biological process. The pH value, reactant concentration, reaction time, and reaction temperature are significant variables of MNP synthesis (Fig. 4).

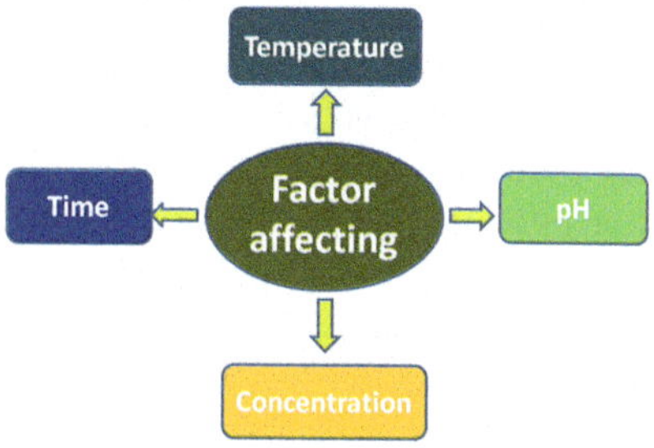

Fig. 4 Factor influencing the synthesis of root and tuber MNPs

- *pH:* Influences the shape and size of MNPs during biological process (Phull et al., 2016).
- *Concentration:* Influences the shape of MNPs during biological synthesis (Huang et al., 2007).
- *Time:* The reaction time upsurges the size of MNPs (Ahmad & Sharma, 2012).
- *Temperature:* It affects the size, shape, yield, and stability of MNPs (Sathishkumar et al., 2010; Song & Kim, 2009).

7 Conclusions

The adoption of nanotechnology in the food industry is expanding with advancements in this discipline. The relevance of widespread endorsements cannot be overstated. In addition to inorganic metals, metal oxides, and nanocomposites, nano-organic materials incorporating beneficial chemicals have also been adopted in the food sector. There is increasing public awareness regarding the potential impacts of human exposure to nanoparticles on both human health and the environment. The proliferation of nanostructured materials in both human populations and the natural environment has given rise to heightened apprehensions over health and safety, notwithstanding the numerous benefits associated with nanotechnology. Hence, it is imperative to establish a universally standardized framework of international regulations pertaining to nanotechnology. The physical characteristics of MNPs implemented in food applications can exhibit variability based on the synthesis techniques applied. Nevertheless, the prevailing sentiment among the general populace toward this groundbreaking technology remains ambiguous. Several regulatory bodies, including the USEPA, IOSOECD, NIOSH, HCPDEC, and FDA, have released guidelines regarding the potential hazards associated with nanoparticles. MNPs have the potential to improve food safety in a number of ways, including by increasing the efficiency of food packaging, prolonging food, and increasing the nutritional value of food through the use of additives. These advantages can be obtained without having an adverse effect on the flavor, aroma, or consistency of the final goods. Nanotechnology is steadily gaining pace, which makes it an important instrument for the food and bioprocessing industries as they work to meet the demands of a fast-rising global population. MNPs that are organic, inorganic, or carbon-based illustrate higher properties as compared to their heavier components. Due to their diminutive dimensions, nanoparticles exhibit superior characteristics such as increased reactivity, robustness, surface area, sensitivity, and stability. Many technologies used to synthesize MNPs for research and commercial applications have advanced significantly over the years. It almost changed the food sector and food processing. Also important is assessing metal nanoparticles' health and environmental concerns. The synthesis of inorganic nanosubstances and microfluids enables quick identification of infectious diseases, microbes, and chemical toxins. One more significant use for the nanosensor is in the food industry's efforts to reduce its contribution to environmental contamination. It is the responsibility of a functional food or nanosubstance to perform the function of a flavor or nutritional transporter, which in turn raises the food's nutritional value and ensures that it is safe to consume. The production of novel nanomaterials is progressing at an ever-increasing rate. In addition, nanomaterials with a wide variety of different compositions have been synthesized for use in a wide range of different contexts and applications. Simple procedures can be used to synthesize MNPs that have the desired size, shape, and characteristics; however, these processes could benefit from some more development and improvement. The administration of MNPs in conjunction with natural products and phytochemicals is particularly appealing as a means of combating biofilm formation, as it leads to the augmentation of the pharmacological properties inherent in the roots and tubers of plants. The implementation of MNPs in the food and beverage sector, with a focus on promoting health and sustainability, is currently undergoing continuous development. The concept of genetically altered food elicits a sense of unease among individuals. The responsible utilization of nanomaterials in food enterprises should be prioritized over unwarranted apprehension. The survival of nanotechnology-based products is contingent upon consumer acceptability, as their fate and potential toxicity remain uncertain. It is anticipated that nanofood technology will be a significant advancement in the coming decades, assuming that there will be no unforeseen repercussions for the health of humans. This chapter offers an introduction to MNPs as well as a description of their many different varieties, characteristics, synthesis processes, and applications in the environment.

References

Ahmad, N., & Sharma, S. (2012). Green synthesis of silver nanoparticles using extracts of ananas comosus. *Green and Sustainable Chemistry, 2*(4), 141–147. https://doi.org/10.4236/gsc.2012.24020

Ahmad, N., Sharma, S., & Alam, M. K. (2010). Rapid synthesis of silver nanoparticles using dried medicinal plant of basil. *Colloids and Surfaces B: Biointerfaces, 81*(1), 81–86. https://doi.org/10.1016/j.colsurfb.2010.06.029

Aigbogun, I. E., Mohammed, S. S. D., Orukotan, A. A., & Tanko, J. D. (2017). The role of nanotechnology in food industries—A review. *Journal of Advances in Microbiology, 7*(4), 1–9. https://doi.org/10.9734/jamb/2017/38175

Akhtar, M. S., Panwar, J., & Yun, Y. S. (2013). Biogenic synthesis of metallic nanoparticles by plant extracts. *ACS Sustainable Chemistry & Engineering, 1*(6), 591–602. https://doi.org/10.1021/sc300118u

Almaary, K. S., Sayed, S. R. M., Abd-Elkader, O. H., Dawoud, T. M., El-Orabi, N. F., & Elgorban, A. M. (2020). Complete green synthesis of silver-nanoparticles applying seed-borne Penicillium duclauxii. *Saudi Journal of Biological Sciences, 27*(5), 1333–1339. https://doi.org/10.1016/j.sjbs.2019.12.022

Alsammarraie, F. K., Wang, W., Zhou, P., Mustapha, A., & Lin, M. (2018). Green synthesis of silver nanoparticles using turmeric extracts and investigation of their antibacterial activities. *Colloids and Surfaces B: Biointerfaces, 171*(1), 398–405. https://doi.org/10.1016/j.colsurfb.2018.07.059

Alt, V., Bechert, T., & Steinrucke, P. (2004). An in vitro assessment of the antibacterial properties and cytotoxicity of nanoparticle silver bone cement. *Biomaterials, 25*(18), 4383–4391. https://doi.org/10.1016/j.biomaterials.2003.10.078

Amin, A. A., Salama, M. F., Seliem, E. I., & Mahmoud, K. F. (2015). Encapsulation of nano carotenoids; evaluation of stability and safety. *International Journal of Current Microbiology and Applied Sciences, 4*(2), 1017–1029.

Anonymous (2007). Safety committee on consumer products. *Opinion on Safety of Nanomaterials in Cosmetic Products*. SCCP/1147/07. https://ec.europa.eu/health/ph_risk/committees/04_sccp/docs/sccp_o_123.pdf

Antony, J. J., Sivalingam, P., Siva, D., Kamalakkannan, S., Anbarasu, K., & Sukirtha, R. (2011). Comparative evaluation of antibacterial activity of silver nanoparticles synthesized using Rhizophora apiculata and glucose. *Colloids and Surfaces B: Biointerfaces, 88*(1), 134–140. https://doi.org/10.1016/j.colsurfb.2011.06.022

Aravinthan, A., Govarthanan, M., Selvam, K., Praburaman, L., Selvankumar, T., Balamurugan, R., Kamala-Kannan, S., & Kim, J. H. (2015). Sunroot mediated synthesis and characterization of silver nanoparticles and evaluation of its antibacterial and rat splenocyte cytotoxic effects. *International Journal of Nanomedicine, 10*, 1977–1981. https://doi.org/10.2147/ijn.s79106

Aritonang, H. F., Koleangan, H. S. J., & Wuntu, A. D. (2019). Synthesis of silver nanoparticles using aqueous extract of medicinal plants' (*Impatiens balsamina* and *Lantana camara*) fresh leaves and analysis of antimicrobial activity. *International Journal of Microbiology, 2019*, 8642303. https://doi.org/10.1155/2019/8642303

Arokiyaraj, S., Vincent, S., Saravanan, M., Lee, Y., Oh, Y. K., & Kim, K. H. (2016). Green synthesis of silver nanoparticles using *Rheum palmatum* root extract and their antibacterial activity against *Staphylococcus aureus* and *Pseudomonas aeruginosa*. *Artificial Cells, Nanomedicine, and Biotechnology, 45*(2), 372–379. https://doi.org/10.3109/21691401.2016.1160403

Aromal, S. A., & Philip, D. (2012). Green synthesis of gold nanoparticles using *Trigonella foenum-graecum* and its size dependent catalytic activity. *Spectrochimica Acta A, 97*, 1–5. https://doi.org/10.1016/j.saa.2012.05.083

Bakery, R., Rainer, M. V., Muhammad, N., Matthias, R., Zoltan, S., & Christian, W. H. (2007). Medicinal applications of fullerenes. *International Journal of Nanomedicine, 2*(4), 639–649.

Balaji, D. S., Basavaraja, S., & Deshpande, R. (2009). Extracellular biosynthesis of functionalized silver nanoparticles by strains of Cladosporium cladosporioides fungus. *Colloids and Surfaces B: Biointerfaces, 68*(1), 88–92. https://doi.org/10.1016/j.colsurfb.2008.09.022

Balakrishnan, P., Shanmugam, S., Lee, W. S., & Lee, W. M. (2009). Formulation and in vitro assessment of minoxidil niosomes for enhanced skin delivery. *International Journal of Pharmaceutics, 377*(2), 1–8. https://doi.org/10.1016/j.ijpharm.2009.04.020

Bar, H., Bhui, D. K., Sahoo, G. P., Sarkar, P., De, S. P., & Misra, A. (2009). Green synthesis of silver nanoparticles using latex of *Jatropha curcas*. *Colloids and Surfaces A: Physicochemical and Engineering Aspects, 339*(1–3), 134–139. https://doi.org/10.1016/j.colsurfa.2009.02.008

Baratjmanikanth, S., Kalishwaralal, K., Sriram, M., Pandian, S. R. K., Hyung-seop, Y., Eom, S. H., & Gurunathan, S. (2010). Anti-oxidant effect of gold nanoparticles restrains hyperglycemic conditions in diabetic mice. *Journal of Nanbiotechnology, 8*, 16. https://doi.org/10.1186/1477-3155-8-16

Basavaraja, S., Balaji, S. D., & Lagashetty, A. (2008). Extracellular biosynthesis of silver nanoparticles using the fungus Fusarium semitectum. *Materials Research Bulletin, 43*(5), 1164–1170. https://doi.org/10.1016/j.materresbull.2007.06.020

Benakashani, F., Allafchian, A., & Jalali, S. A. H. (2017). Green synthesis, characterization and antibacterial activity of silver nanoparticles from root extract of *Lepidium draba* weed. *Green Chemistry Letters and Reviews, 10*(4), 324–330. https://doi.org/10.1080/17518253.2017.1363297

Bhainsa, K. C., & D'Souza, S. F. (2006). Extracellular biosynthesis of silver nanoparticles using the fungus aspergillus fumigatus. *Colloids and Surfaces B: Biointerfaces, 47*(2), 160–164. https://doi.org/10.1016/j.colsurfb.2005.11.026

Bhardwaj, A. K., Kashyap, N. K., Bera, S. K., Hait, M., & Dewangan, H. S. (2023a). Proximate composition and mineral content analysis of *Curcuma caesia* rhizome. *Biochemical Systematics and Ecology, 109*, 104661.

Bhardwaj, A. K., Kashyap, N. K., Hait, M., & Dewangan, H. S. (2023b). Physicochemical characterization of rhizome of *Curcuma caesia* Roxb. *ES Food Agroforestry, 11*, 813. https://doi.org/10.30919/esfaf813

Bhatt, A. I., Mechler, Á., Martin, L. L., & Bond, A. M. (2007). Synthesis of Ag and Au nanostructures in an ionic liquid: Thermodynamic and kinetic effects underlying nanoparticle, cluster and nanowire formation. *Journal of Materials Chemistry, 17*(21), 2241. https://doi.org/10.1039/b618036a

Bund. (2008). Ausdem Labor auf den Teller: Die Nutzung der Nanotechnologie imLebensmittelsektor. https://www.bund.net/fileadmin/user_upload_bund/publikationen/chemie/nanotechnologie_aus_dem_labor_auf_den_teller.pdf

Calderon-Jimenez, B., Johnson, M. E., Montoro Bustos, A. R., Murphy, K. E., Winchester, M. R., & Vega Baudrit, J. R. (2017). Silver nanoparticles: Technological advances, societal impacts, and metrological challenges. *Frontiers in Chemistry, 5*, 6. https://doi.org/10.3389/fchem.2017.00006

Cevc, G. (1996). Transfersomes, liposomes and other lipid suspensions on the skin: Permeation enhancement, vesicle penetration, and transdermal drug delivery. *Critical Reviews in Therapeutic Drug Carrier Systems, 13*(3–4), 257–388. https://doi.org/10.1615/critrevtherdrugcarriersyst.v13.i3-4.30

Chandran, S. P., Chaudhary, M., Pasricha, R., Ahmad, A., & Sastry, M. (2006). Synthesis of gold nanotriangles and silver nanoparticles using Aloevera plant extract. *Biotechnology Progress, 22*(2), 577–583. https://doi.org/10.1021/bp0501423

Chen, Y., Ji, C., & Creamer, K. S. (2008). Inhibition of anaerobic digestion process: A review. *Bioresource Technology, 99*(10), 4044–4064. https://doi.org/10.1016/j.biortech.2007.01.057

Chen, Y. L., Tuan, H. Y., & Tien, C. W. (2009). Augmented biosynthesis of cadmium sulfide nanoparticles by genetically engineered Escherichia coli. *Biotechnology Progress, 25*(5), 1260–1266. https://doi.org/10.1002/btpr.199

Cruz, D., Fale, P. L., Mourato, A., Vaz, P. D., Serralheiro, M. L., & Lino, A. R. L. (2010). Preparation and physicochemical characterization of Ag nanoparticles biosynthesized by *Lippia citriodora* (*Lemon Verbena*). *Colloids and Surfaces. B, Biointerfaces, 81*(1), 67–73. https://doi.org/10.1016/j.colsurfb.2010.06.025

Daisy, P., & Saipriya, K. (2012). Biochemical analysis of *Cassia fistula* aqueous extract and phytochemically synthesized gold nanoparticles as hypoglycemic treatment for diabetes mellitus. *International Journal of Nanomedicine, 7*, 1189–1202. https://doi.org/10.2147/ijn.s26650

Das, R. K., Gogoi, N., & Bora, U. (2011). Green synthesis of gold nanoparticles using *Nyctanthes arbortristis* flower extract. *Bioprocess and Biosystems Engineering, 34*(5), 615–619. https://doi.org/10.1007/s00449-010-0510-y

Das, S., Das, J., Samadder, A., Bhattacharyya, S. S., Das, D., & Khuda-Bukhsh, A. R. (2013). Biosynthesized silver nanoparticles by ethanolic extracts of *Phytolacca decandra*, *Gelsemium sempervirens*, *Hydrastis canadensis* and *Thuja occidentalis* induce differential cytotoxicity through G2/M arrest in A375 cells. *Colloids and Surfaces B: Biointerfaces, 101*(1), 325–336. https://doi.org/10.1016/j.colsurfb.2012.07.008

Dasgupta, J., Chakraborty, S., Sikder, J., Kumar, R., Pal, D., Curcio, S., & Drioli, E. (2014). The effects of thermally stable titanium silicon oxide nanoparticles on structure and performance of cellulose acetate ultrafiltration membranes. *Separation and Purification Technology, 133*, 55–68. https://doi.org/10.1016/j.seppur.2014.06.035

Deplanche, K., & Macaskie, L. E. (2008). Biorecovery of gold by *Escherichia coli* and *Desulfovibrio desulfuricans*. *Biotechnology and Bioengineering, 99*(5), 1055–1064. https://doi.org/10.1002/bit.21688

Deshmukh, S. P., Patil, S. M., Mullani, S. B., & Delekar, S. D. (2019). Silver nanoparticles as an effective disinfectant: A review. *Materials Science & Engineering. C, Materials for Biological Applications, 97*, 954–965. https://doi.org/10.1016/j.msec.2018.12.102

Devanesan, S., Ponmurugan, K., AlSalhi, M. S., & Al-Dhabi, N. A. (2020). Cytotoxic and antimicrobial efficacy of silver nanoparticles synthesized using a traditional phytoproduct, *Asafoetida* Gum. *International Journal of Nanomedicine, 15*, 4351–4362. https://doi.org/10.2147/ijn.s258319

Dipankar, C., & Murugan, S. (2012). The green synthesis, characterization and evaluation of the biological activities of silver nanoparticles synthesized from *Iresine herbstii* leaf aqueous extracts. *Colloids and Surfaces B: Biointerfaces, 98*, 112–119. https://doi.org/10.1016/j.colsurfb.2012.04.006

Dizaj, S. M., Lotfipour, F., Barzegar-Jalali, M., Zarrintan, M. H., & Adibkia, K. (2014). Antimicrobial activity of the metals and metal oxide nanoparticles. *Materials Science & Engineering. C, Materials for Biological Applications, 44*, 278–284. https://doi.org/10.1016/j.msec.2014.08.031

Du, L., Jiang, H., Liu, X., & Wang, E. (2007). Biosynthesis of gold nanoparticles assisted by *Escherichia coli* DH5α and its application on direct electrochemistry of haemoglobin. *Electrochemistry Communications, 9*(5), 1165–1170. https://doi.org/10.1016/j.elecom.2007.01.007

Du, L., Xu, Q., Huang, M., Xian, L., & Feng, J. X. (2015). Synthesis of small silver nanoparticles under light radiation by fungus *Penicillium oxalicum* and its application for the catalytic reduction of methylene blue. *Materials Chemistry and Physics, 160*, 40–47. https://doi.org/10.1016/j.matchemphys.2015.04.003

Dubey, S. P., Lahtinen, M., Sarkka, H., & Sillanpaa, M. (2010). Bioprospective of *Sorbus aucuparia* leaf extract in development of silver and gold nanocolloids. *Colloids and Surfaces. B, Biointerfaces, 80*(1), 26–33. https://doi.org/10.1016/j.colsurfb.2010.05.024

Esmaeilnejad, A., Mahmoudi, P., Zamanian, A., & Mozafari, M. (2019). Synthesis of titanium oxide nanotubes and their decoration by MnO nanoparticles for biomedical applications. *Ceramics International, 45*(15), 19275–19282. https://doi.org/10.1016/j.ceramint.2019.06.177

Esmail, R., Afshar, A., Morteza, M., Abolfazl, A., & Akhondi, E. (2022). Synthesis of silver nanoparticles with high efficiency and stability by culture supernatant of Bacillus ROM6 isolated from Zarshouran gold mine and evaluating its antibacterial effects. *BMC Microbiology, 22*(1), 97. https://doi.org/10.1186/s12866-022-02490-5

Flores, H. E., Hoy, M. W., & Pickard, J. J. (1987). Secondary metabolites from root cultures. *Trends in Biotechnology, 5*(3), 64–69. https://doi.org/10.1016/s0167-7799(87)80013-6

Fred, Z., Esther, B., Daniel, S., Christina, L., & Franz, S. (2006). *Preparation and properties of coenzyme Q10 Nanoemulsions*. Cosmetic Science Technology. https://cosmeticsciencetechnology.com/articles/samples/1111.pdf

Fu, M., Li, Q., & Sun, D. (2006). Rapid preparation process of silver nanoparticles by bioreduction and their characterizations. *Chinese Journal of Chemical Engineering, 14*(1), 114–117. https://doi.org/10.1016/s1004-9541(06)60046-3

Gade, A. K., Bonde, P., & Ingle, A. P. (2008). Exploitation of *Aspergillus Niger* for synthesis of silver nanoparticles. *Journal of Biobased Materials and Bioenergy, 2*(1), 243–247. https://doi.org/10.1166/jbmb.2008.401

Gan, P. P., Ng, S. H., Huang, Y., & Li, S. F. Y. (2012). Green synthesis of gold nanoparticles using palm oil mill effluent (POME): A low-cost and eco-friendly viable approach. *Bioresource Technology, 113*, 132–135. https://doi.org/10.1016/j.biortech.2012.01.015

Gardea-Torresdey, J. L., Parson, J. G., Gomez, E., Peralta-Videa, J., Troiani, H. E., Santiago, P., & Yacaman, M. J. (2002). Formation and growth of Au nanoparticles inside live alfalfa plants. *Nano Letters, 2*(4), 397–407. https://doi.org/10.1021/nl015673+

Garibo, D., Borbon-Nunez, H. A., De Leon, J. N. D., Garcia Mendoza, E., Estrada, I., Toledano-Magana, Y., Tiznado, H., Ovalle-Marroquin, M., Soto-Ramos, A. G., Blanco, A., Rodriguez, J. A., Romo, O. A., Chavez-Almazan, L. A., & Susarrey-Arce, A. (2020). Green synthesis of silver nanoparticles using Lysiloma acapulcensis exhibit high-antimicrobial activity. *Scientific Reports, 10*(1), 12805. https://doi.org/10.1038/s41598-020-69606-7

Ge, L., Chen, L., & Guo, R. (2007). Microstructure and lubrication properties of lamellar liquid crystal in BRIJ30/[BMIM]PF6/H2O system. *Tribology Letters, 28*(2), 123–130. https://doi.org/10.1007/s11249-007-9256-3

Ghiuta, I., Croitoru, C., Kost, J., Wenkert, R., & Munteanu, D. (2021). Bacteria-mediated synthesis of silver and silver chloride nanoparticles and their antimicrobial activity. *Applied Sciences, 11*(7), 3134. https://doi.org/10.3390/app11073134

Ghosh, S., Patil, S., Ahire, M., Kitture, R., Kale, S., Pardesi, K., Cameotra, S. S., Bellare, J., Dhavale, D. D., Jabgunde, A., & Chopade, B. A. (2012). Synthesis of silver nanoparticles using Dioscorea bulbifera tuber extract and evaluation of its synergistic potential in combination with antimicrobial agents. *International Journal of Nanomedicine, 7*(1), 483. https://doi.org/10.2147/ijn.s24793

Gnanadesigan, M., Anand, M., Ravikumar, S., Maruthupandy, M., & Vijayakumar, V. (2011). Biosynthesis of silver nanoparticles by using mangrove plant extract and their potential mosquito larvicidal property. *Asian Pacific Journal of Tropical Medicine, 4*(10), 799–803. https://doi.org/10.1016/s1995-7645(11)60197-1

Gomaa, E. Z. (2017). Antimicrobial, antioxidant and antitumor activities of silver nanoparticles synthesized by Allium cepa extract: A green approach. *Journal, Genetic Engineering & Biotechnology, 15*(1), 49–57. https://doi.org/10.1016/j.jgeb.2016.12.002

Gurunathan, S., Kyung-Jin, L., Kalishwaralal, K., Sheikpranbabu, R., Vaidyanathan, R., & Eom, S. H. (2009). Antiangiogenic properties of silver nanoparticles. *Biomaterials, 30*(31), 6341–6350. https://doi.org/10.1016/j.biomaterials.2009.08.008

Hait, M., Nemu, S. C., Kashyap, N. K., & Chaturvedi, A. (2018). Physicochemical and phytochemical exploration on flower of *Delonix regia*. *Journal of Medicinal Plants Studies, 6*(1), 15–18.

Hait, M., Kashyap, N. K., Deepak, J., & Patel, A. (2020). Seclusion of bioactive components from flower of *samanea saman*. *Plant Archives, 20*(1), 4581–4584.

He, S., et al. (2007). Biosynthesis of gold nanoparticles using the bacteria Rhodopseudomonas capsulate. *Materials Letters, 61*(18), 3984–3987. https://doi.org/10.1016/j.matlet.2007.01.018

Hemalatha K., Srinivasulu, S., & Satyanarayana, B. (2019). Green synthesis of silver nanoparticles using aqueous fruit and tuber extracts of Momordica cymbalaria. *Journal of Plant Biochemistry and Biotechnology, 30*(1), 196–204. https://doi.org/10.1007/s13562-019-00542-y

Holmes, J. D., Smith, P. R., & Richardson, D. J. (1995). Energy-dispersive X-ray analysis of the extracellular cadmium sulfide crystallites of *Klebsiella aerogenes*. *Archives of Microbiology, 163*(2), 143–147. https://doi.org/10.1007/BF00381789

Huang, J., Li, Q., Sun, D., Lu, Y., Su, Y., Yang, X., Wang, H., Wang, Y., Shao, W., He, N., Hong, J., & Chen, C. (2007). Biosynthesis of silver and gold nanoparticles by novel sundried *Cinnamomum camphora* leaf. *Nanotechnology, 18*(10), 105104. https://doi.org/10.1088/0957-4484/18/10/105104

Hulkoti, N. I., & Taranath, T. C. (2014). Biosynthesis of nanoparticles using microbes-A review. *Colloids and Surfaces. B, Biointerfaces, 121*, 474–483. https://doi.org/10.1016/j.colsurfb.2014.05.027

Hwang, S. L., & Kim, J. C. (2008). *In vivo* hair growth promotion effects of cosmetic preparations containing hinokitiol-loaded poly (ε-caprolactone) nanocapsules. *Journal of Microencapsulation, 25*(5), 351–356. https://doi.org/10.1080/02652040802000557

Hwang, S. H., Park, Y. B., Yoon, K. H., & Bang, D. S. (2011). *Smart materials and structures based on carbon nanotube composites*. InTech eBooks. https://doi.org/10.5772/17374

Ingle, A., Rai, M., & Gade, A. (2009). Fusarium solani: A novel biological agent for the extracellular synthesis of silver nanoparticles. *Journal of Nanoparticle Research, 11*(8), 2079–2085. https://doi.org/10.1007/s11051-008-9573-y

Iravani, S. (2011). Critical review—Green synthesis of metal nanoparticles using plants. *Green Chemistry, 13*, 2638–2650. https://doi.org/10.1039/C1GC15386B

Itoh, H., Naka, K., & Chujo, Y. (2004). Synthesis of gold nanoparticles modified with ionic liquid based on the imidazolium cation. *Journal of the American Chemical Society, 126*(10), 3026–3027. https://doi.org/10.1021/ja039895g

Jacob, S. J. P., Finub, J. S., & Narayanan, A. (2012). Synthesis of silver nanoparticles using *Piper longum* leaf extracts and its cytotoxic activity against Hep-2 cell line. *Colloids and Surfaces. B, Biointerfaces, 91*(1), 212–214. https://doi.org/10.1016/j.colsurfb.2011.11.001

Jain, S., Sapee, R., & Jain, N. K. (1998). *Proultraflexible lipid vesicles for effective transdermal delivery of norgesterol* (pp. 32–35). USA: Proceedings of 25th Conference of C.R.S.

Jayaseelan, C., Rahuman, A. A., Rajakumar, G., Kirthi, A. V., Santhoshkumar, T., & Marimuthu, S. (2011). Synthesis of pediculocidal and larvicidal silver nanoparticles by leaf extract from heart leaf moon seed plant *Tinospora cordifolia* Miers. *Parasitology Research, 109*(1), 185–194. https://doi.org/10.1007/s00436-010-2242-y

Joerger, R., Klaus, T., & Granqvist, C. G. (2000). Biologically produced silver–carbon composite materials for optically functional thin-film coatings. *Advanced Materials, 12*(1), 407–409. https://doi.org/10.1002/(SICI)1521-4095(200003)12:6<407::AID-ADMA407>3.0.CO;2-O

Kantipudi, S., Pethakamsetty, L., Kollana, S. M., Sunkara, J. R., Kollu, P., Parine, N. R., Rallabhandi, M., & Pammi, S. V. N. (2018). *Diospyros assimilis* root extract assisted biosynthesised silver nanoparticles and their evaluation of antimicrobial activity. *IET Nanobiotechnol., 12*(2), 133–137. https://doi.org/10.1049/iet-nbt.2017.0042

Karatoprak, G. S., Aydin, G., Altinsoy, B., Altinkaynak, C., Kos, M., & Ocsoy, I. (2017). The effect of *Pelargonium endlicherianum Fenzl* root extraction formation of nanoparticles and their antimicrobial activities. *Enzyme and Microbial Technology, 97*, 21–26. https://doi.org/10.1016/j.enzmictec.2016.10.019

Kashyap, N. K., Deepak, J., Bhardwaj, A. K., Hait, M., & Pal, D. K. (2022a). In-vitro antibacterial and antifungal activity of *Curcuma amada* Roxb. against human pathogens. *Trends in Sciences, 19*(22), 4839. https://doi.org/10.48048/tis.2022.4839

Kashyap, N. K., Hait, M., Roymahapatra, G. S., & Vaishnav, M. M. (2022b). Proximate and elemental analysis of *Careya arborea* Roxb. plant's root. *ES Food Agroforestry, 7*, 41–47. https://doi.org/10.30919/esfaf620

Kasthuri, J., Kathiravan, K., & Rajendiran, N. (2009). Phyllanthin assisted biosynthesis of silver and gold nanoparticles: A novel biological approach. *Journal of Nanoparticle Research, 11*(5), 1075–1085. https://doi.org/10.1007/s11051-008-9494-9

Kathiresan, K., Manivannan, S., Nabeel, M. A., & Dhivya, B. (2009). Studies on silver nanoparticles synthesized by a marine fungus, *Penicillium fellutanum* isolated from coastal mangrove sediment. *Colloids and Surfaces B: Biointerfaces, 71*(1), 133–137. https://doi.org/10.1016/j.colsurfb.2009.01.016

Kaur, R., Kaur, B., Suttee, A., & Kalsi, V. (2018). Comparative assessment of *invitro* antimicrobial activity of *Curcuma caesia* Roxb. and *Curcuma amada* Roxb. *Asian Journal of Pharmaceutical and Clinical Research, 11*(14), 94. https://doi.org/10.22159/ajpcr.2018.v11s2.28591

Kaviya, S., Santhanalakshmi, J., Viswanathan, B., Muthumary, J., Srinivasan, K. (2011) 'Biosynthesis of silver nanoparticles using citrus sinensis peel extract and its antibacterial activity.' *Spectrochimica Acta Part A:* Molecular and Biomolecular Spectroscopy, 79(3), pp. 594–598. doi:https://doi.org/10.1016/j.saa.2011.03.040.

Kesselman, E., Efrat, R., Garti, N., & Danino, D. (2007). *Formation of cubosomes as vehicles of biologically active substances*. https://www.yumpu.com/en/document/view/30962979/formation-of-cubosomes-as-vehicles-of-biologically-active-

Khaleghi, M., Khorrami, S., & Ravan, H. (2019). Identification of *Bacillus thuringiensis* bacterial strain isolated from the mine soil as a robust agent in the biosynthesis of silver nanoparticles with strong antibacterial and anti-biofilm activities. *Biocatalysis and Agricultural Biotechnology, 18*, 101047. https://doi.org/10.1016/j.bcab.2019.101047

Khan, M. N., Khan, T. A., Khan, Z., & Al-Thabaiti, S. A. (2015). Green synthesis of biogenic silver nanomaterials using *Raphanus sativus* extract, effects of stabilizers on the morphology, and their antimicrobial activities. *Bioprocess and Biosystems Engineering, 38*(12), 2397–2416. https://doi.org/10.1007/s00449-015-1477-5

Khan, S., Ansari, A. A., Malik, A., Chaudhary, A. A., Syed, J. B., & Khan, A. A. (2019). Preparation, characterizations and in vitro cytotoxic activity of nickel oxide nanoparticles on HT-29 and SW620 colon cancer cell lines. *Journal of Trace Elements in Medicine and Biology, 52*, 12–17. https://doi.org/10.1016/j.jtemb.2018.11.003

Khanal, L. N., Sharma, K. R., Pokharel, Y. R., & Kalauni, S. K. (2020). Assessment of phytochemical, antioxidant and antimicrobial activities of some medicinal plants from Kaski district of Nepal. *American Journal of Plant Sciences, 11*(9), 1383–1397. https://doi.org/10.4236/ajps.2020.119099

Khanal, L. N., Sharma, K. R., Paudyla, H., Parajuli, K., Dahal, B., Ganga, G. C., Pokharel, Y. R., & Kalauni, S. K. (2022). Green ynsthesis of silver nanoparticles from root extracts of *Rubus ellipticus* Sm. and comparison of antioxidant and antibacterial activity. *Journal of Nanomaterials, 2022*, 1–11. https://doi.org/10.1155/2022/1832587

Khare, V., Li, Z. H., & Mantion, A. (2010). Strong anion effects on gold nanoparticle formation in ionic liquids. *Journal of Materials Chemistry, 20*(7), 1332–1339. https://doi.org/10.1039/b917467b

Kim, J. S., Kuk, E., Yu, K. N., Jong-Ho, K., Park, S. J., Lee, H. J., & Kim, S. H. (2007). Antimicrobial effects of silver nanoparticles. *Nanomedicine: Nanotechnology, Biology and Medicine, 3*(1), 95–101. https://doi.org/10.1016/j.nano.2006.12.001

Kirsner, R., Orsted, H., & Wright, B. (2001). Matrix metalloproteinase in normal and impaired wound healing: A potential role of monocrystalline silver. *Wounds: a Compendium of Clinical Research and Practice, 13*(3), 4–12.

Konishi, Y., Tsukiyama, T., & Tachimi, T. (2007). Microbial deposition of gold nanoparticles by the metal-reducing bacterium Shewanella algae. *Electrochimica Acta, 53*(1), 186–192. https://doi.org/10.1016/j.electacta.2007.02.073

Korbekandi, H., Iravani, S., & Abbasi, S. (2012). Optimization of biological synthesis of silver nanoparticles using *lactobacillus casei* subsp. casei. *Journal of Chemical Technology and Biotechnology, 87*(7), 932–937. https://doi.org/10.1002/jctb.3702

Krishnaraj, C., Jagan, E. G., Rajasekar, S., Selvakumar, P., Kalaichelvan, P. T., & Mohan, N. (2010). Synthesis of silver nanoparticles using *Acalypha indica* leaf extracts and its antibacterial activity against water borne pathogens. *Colloids and Surfaces B: Biointerfaces, 76*(1), 50–56. https://doi.org/10.1016/j.colsurfb.2009.10.008

Lakshmipriya, T., & Gopinath, S. C. B. (2021). *Introduction to nanoparticles and analytical devices* (pp. 1–29). Elsevier eBooks. https://doi.org/10.1016/b978-0-12-821163-2.00001-7

Lazarus, L. L., Yang, A. S. J., & Chu, S. (2010). Flow-focused synthesis of monodisperse gold nanoparticles using ionic liquids on a microfluidic platform. *Lab on a Chip, 10*(24), 3377–3379. https://doi.org/10.1039/c0lc00297f

Lazarus, L. L., Riche, C. T., Malmstadt, N., & Brutchey, R. L. (2012). Effect of ionic liquid impurities on the synthesis of silver nanoparticles. *Langmuir, 28*(45), 15987–15993. https://doi.org/10.1021/la303617f

Lee, J. H., Lim, J. A., Velmurugan, P., Park, Y. J., Park, Y. J., Bang, K. S., & Oh, B. T. (2016). Photobiologic-mediated fabrication of silver nanoparticles with antibacterial activity. *Journal of Photochemistry and Photobiology B: Biology, 162*, 93–99. https://doi.org/10.1016/j.jphotobiol.2016.06.029

Lengke, M. F., Fleet, M. E., & Southam, G. (2006). Morphology of gold nanoparticles synthesized by filamentous cyanobacteria from gold(I)–thiosulfate and gold(III)–chloride complexes. *Langmuir, 22*(6), 2780–2787. https://doi.org/10.1021/la052652c

Logeswari, P., Silambarasan, S., & Abraham, J. (2012). Synthesis of silver nanoparticles using plant extracts and analysis of their antimicrobial property. *Journal of Saudi Chemical Society, 19*(3), 311–317. https://doi.org/10.1016/j.jscs.2012.04.007

Maduraiveeran, G., & Jin, W. (2020). *Functional nanomaterial-derived electrochemical sensor and biosensor platforms for biomedical applications* (pp. 297–327). Elsevier eBooks. https://doi.org/10.1016/b978-0-12-816699-4.00012-8

Majeed, S., Danish, M., & Norazmi, F. S. B. (2018). Fungal derived zinc oxide nanoparticles and their antibacterial and anticancer activities against human alveoli lung cancer A-549 cell line. *Advanced Science, Engineering and Medicine, 10*(6), 551–556. https://doi.org/10.1166/asem.2018.2187

Mandava, K. (2017). Biological and non-biological synthesis of metallic nanoparticles: Scope for current pharmaceutical research. *Indian Journal of Pharmaceutical Sciences, 79*(4). https://doi.org/10.4172/pharmaceutical-sciences.1000256

Mann, S., Frankel, R. B., & Blakemore, R. P. (1984). Structure, morphology and crystal growth of bacterial magnetite. *Nature, 310*(5976), 405–407. https://doi.org/10.1038/310405a0

Marambio-Jones, C., & Hoek, E. M. V. (2010). A review of the antibacterial effects of silver nanomaterials and potential implications for human health and the environment. *Journal of Nanoparticle Research, 12*(5), 1531–1551. https://doi.org/10.1007/s11051-010-9900-y

Markus, J., Wang, D., Kim, Y. J., Ahn, S., Mathiyalagan, R., Wang, C., & Yang, D. C. (2017). Biosynthesis, characterization, and bioactivities evaluation of silver and gold nanoparticles mediated by the roots of Chinese herbal angelica pubescens maxim. *Nanoscale Research Letters, 12*(1), 46. https://doi.org/10.1186/s11671-017-1833-2

Marshall, M. J., Beliaev, A. S., & Dohnalkova, A. C. (2006). C-type cytochrome-dependent formation of U(IV) nanoparticles by Shewanella oneidensis. *PLoS Biology, 4*(8), e268. https://doi.org/10.1371/journal.pbio.0040268

Mclaren, A., Valdes-Solis, T., Li, G., & Tsang, S. C. (2009). Shape and size effects of ZnO nanocrystals on photocatalytic activity. *Journal of the American Chemical Society, 131*(35), 12540–12541. https://doi.org/10.1021/ja9052703

Michael, F. (2010). *Cosmetic compositions for hair treatment containing dendrimers or dendrimer conjugates*. Patent 6068835.

Mittal, A. K., Chisti, Y., & Banerjee, U. C. (2013). Synthesis of metallic nanoparticles using plant extracts. *Biotechnology Advances, 31*(2), 346–356. https://doi.org/10.1016/j.biotechadv.2013.01.003

Mittal, A. K., Tripathy, D., Choudhary, A., Aili, P. K., Chatterjee, A., Singh, I. P., & Banerjee, U. C. (2015). Bio-synthesis of silver nanoparticles using Potentilla fulgens Wall. ex Hook. and its therapeutic evaluation as anticancer and antimicrobial agent. *Materials Science and Engineering: C, 53*, 120–127. https://doi.org/10.1016/j.msec.2015.04.038

Mody, V. V., Siwale, R., Singh, A., & Mody, H. R. (2010). Introduction to metallic nanoparticles. *Journal of Pharmacy & Bioallied Sciences, 2*(4), 282–289. https://doi.org/10.4103/0975-7406.72127

Monda, S., Roy, N., Laskar, R. A., Sk, I., Basu, S., Mandal, D., & Begum, N. A. (2011). Biogenic synthesis of Ag, Au and bimetallic Au/Ag alloy nanoparticles using aqueous extract of mahogany (*Swietenia mahagoni* JACQ) leaves. *Colloids and Surfaces. B, Biointerfaces, 82*(2), 497–504. https://doi.org/10.1016/j.colsurfb.2010.10.007

Morales, M. E., Gallardo, V., Clares, B., Garcia, M. B., & Ruiz, M. A. (2009). Study and description of hydrogels and organogels as vehicles for cosmetic active ingredients. *Journal of Cosmetic Science, 60*(6), 627–636.

Morganti, P. (2010). Use and potential of nanotechnology in cosmetic dermatology. *Clinical, Cosmetic and Investigational Dermatology, 3*, 5–13. https://doi.org/10.2147/ccid.s4506

Mosaviniya, S. Z., Mousavi, S. M., Khodanazary, A., & Hossenie, S. M. (2021). Effect of a chitosan-based nanocomposite containing ZnO and Zataria multiflora essential oil on quality properties of Asian sea bass (Lates calcarifer) fillet. *Journal of Food Science and Technology, 59*(3), 869–878. https://doi.org/10.1007/s13197-021-05082-9

Mukherjee, P., Ahmad, A., & Mandal, D. (2001). Fungus-mediated synthesis of silver nanoparticles and their immobilization in the mycelial matrix: A novel biological approach to nanoparticle synthesis. *Nano Letters, 1*(10), 515–519. https://doi.org/10.1021/nl0155274

Müller, R. H., Radtke, M., & Wissing, S. A. (2002). Solid lipid nanoparticles (SLN) and nanostructured lipid carriers (NLC) in cosmetic and dermatological preparations. *Advanced Drug Delivery Reviews, 54*, S131–S155. https://doi.org/10.1016/s0169-409x(02)00118-7

Nasr-Esfahani, P., & Ensafi, A. A. (2022). *Noble metals and nonnoble metal oxides based electrochemical sensors* (pp. 115–140). Elsevier eBooks. https://doi.org/10.1016/b978-0-12-823788-5.00023-5

Nassar, A., Atta, H. M., Abdel-Rahman, M. A., El Naghy, W. S., & Fouda, A. (2023). Myco-synthesized copper oxide nanoparticles using harnessing metabolites of endophytic fungal strain Aspergillus terreus: An insight into antibacterial, anti-Candida, biocompatibility, anti-cancer, and antioxidant activities. *BMC Complementary Medicine and Therapies, 23*(1), 261. https://doi.org/10.1186/s12906-023-04056-y

Nguyen, T. T. N., Vo, T. T., Nguyen, B. N. H., Nguyen, D. T., Dang, V. S., Dang, C. H., & Nguyen, T. D. (2018). Silver and gold nanoparticles biosynthesized by aqueous extract of burdock root, Arctium lappa as antimicrobial agent and catalyst for degradation of pollutants. *Environmental Science and Pollution Research, 25*(34), 34247–34261. https://doi.org/10.1007/s11356-018-3322-2

Noruzi, M., Zare, D., Khoshnevisan, K., & Davoodi, D. (2011). Rapid green synthesis of gold nanoparticles using Rosa hybrida petal extract at room temperature. *Spectrochimica Acta. Part A, Molecular and Biomolecular Spectroscopy, 79*(5), 1461–1465. https://doi.org/10.1016/j.saa.2011.05.001

Obliosca, J. M., Arellano, I. H. J., Huang, M. H., & Arco, S. D. (2010). Double layer micellar stabilization of gold nanocrystals by greener ionic liquid 1-butyl-3-methylimidazolium lauryl sulphate. *Materials Letters, 64*(9), 1109–1112. https://doi.org/10.1016/j.matlet.2010.02.029

Oghyanous, F. A. (2022). *Nanoparticles in wastewater treatment*. IntechOpen eBooks. https://doi.org/10.5772/intechopen.100239

Oghyanous, F. A., Etemadi, H., & Yegani, R. (2020). The effect of sludge retention time and organic loading rate on performance and membrane fouling in membrane bioreactor. *Journal of Chemical Technology and Biotechnology, 96*(3), 743–754. https://doi.org/10.1002/jctb.6588

Okaiyeto, K., Ojemaye, M. O., Hoppe, H., Mabinya, L. V., & Okoh, A. I. (2019). Phytofabrication of silver/silver chloride nanoparticles using aqueous leaf extract of *Oedera genistifolia*: Characterization and antibacterial potential. *Molecules, 24*(23), 4382. https://doi.org/10.3390/molecules24234382

Oliveira, M. M., Ugarte, D., Zanchet, D., & Zarbin, A. J. G. (2005). Influence of synthetic parameters on the size, structure, and stability of dodecanethiol-stabilized silver nanoparticles. *Journal of Colloid and Interface Science, 292*(2), 429–435. https://doi.org/10.1016/j.jcis.2005.05.068

Otles, S., & Yalcin, B. (2010). Nano-biosensors as new tool for detection of food quality and safety. *Electronics and Science Journal of Logistics, 6*(4), 67–70. http://www.logforum.net/vol6/issue4/no7

Oves, M., Aslam, M., Rauf, M. A., Qayyum, S., Qari, H. A., Khan, M. S., Alam, M. Z., Tabrez, S., Pugazhendhi, A., & Ismail, I. M. I. (2018). Antimicrobial and anticancer activities of silver nanoparticles synthesized from the root hair extract of *Phoenix dactylifera. Materials Science and Engineering, 89*, 429–443. https://doi.org/10.1016/j.msec.2018.03.035

Panigrahi, S., Kundu, S., Ghosh, S., Nath, S., & Pal, T. (2004). General method of synthesis for metal nanoparticles. *Journal of Nanoparticle Research, 6*, 411–414. https://doi.org/10.1007/s11051-004-6575-2

Pantic, S., Skodric, S. R., Loncar, Z., & Pantic, I. (2019). Zinc oxide nanoparticles: Potential novel applications in cellular physiology, pathology, neurosciences and cancer research. *Reviews on Advanced Materials Science, 58*(1), 17–21. https://doi.org/10.1515/rams-2019-0002

Patra, J. K., & Baek, K. H. (2016). Green synthesis of silver chloride nanoparticles using *Prunus persica* L. outer peel extract and investigation of antibacterial, anticandidal, antioxidant potential. *Green Chemistry Letters and Reviews, 9*(2), 132–142. https://doi.org/10.1080/17518253.2016.1192692

Petersen, R. (2008). *Nanocrystals for use in topical cosmetic formulations and method of production thereof*. Abbott GmbH and Co. US Patent 60/866233.

Pethakamsetty, L., Kothapenta, K., Nammi, H. R., Ruddaraju, L. K., Kollu, P., Yoon, S. G., & Pammi, S. V. N. (2017). Green synthesis, characterization and antimicrobial activity of silver nanoparticles using methanolic root extracts of Diospyros sylvatica. *Journal of Environmental Sciences (China), 55*, 157–163. https://doi.org/10.1016/j.jes.2016.04.027

Philipse, A. P., & Maas, D. (2002). Magnetic colloids from magnetotactic bacteria: Chain formation and colloidal stability. *Langmuir, 18*(25), 9977–9984. https://doi.org/10.1021/la0205811
Phull, A. R., Abbas, Q., Ali, A., Raza, H., Kim, S. J., Zia, M., & Haq, I. (2016). Antioxidant, cytotoxic and antimicrobial activities of green synthesized silver nanoparticles from crude extract of Bergenia ciliate. *Future Journal of Pharmaceutical Sciences, 2*(1), 31–36. https://doi.org/10.1016/j.fjps.2016.03.001
Pickup, J. C., Zhi, Z. L., Khan, F., Saxl, T., & Birch, D. J. (2008). Nanomedicine and its potential in diabetes research and practice. *Diabetes/Metabolism Research and Reviews, 24*(8), 604–610. https://doi.org/10.1002/dmrr.893
Pileni, M. P. (1997). Nanosized particles made in colloidal assemblies. *Langmuir, 13*(13), 3266–3276. https://doi.org/10.1021/la960319q
Pugazhendhi, S., Kirubha, E., Palanisamy, P. K., & Gopalakrishnan, R. (2015). Synthesis and characterization of silver nanoparticles from *Alpinia calcarata* by Green approach and its applications in bactericidal and nonlinear optics. *Applied Surface Science, 357*, 1801–1808. https://doi.org/10.1016/j.apsusc.2015.09.237
Pugazhendhi, S., Sathya, P., Palanisamy, P. K., & Gopalakrishnan, R. (2016). Synthesis of silver nanoparticles through green approach using *Dioscorea alata* and their characterization on antibacterial activities and optical limiting behaviour. *Journal of Photochemistry and Photobiology B: Biology, 159*, 155–160. https://doi.org/10.1016/j.jphotobiol.2016.03.043
Raghunandan, D., Ravishankar, B., Sharanbasava, G., Mahesh, D. B., Harsoor, V., Yalagatti, M. S., Bhagawanraju, M., & Venkataraman, A. (2011). Anti-cancer studies of noble metal nanoparticles synthesized using different plant extracts. *Cancer Nanotechnology, 2*, 57–65.
Rahman, A. U., Khan, A. U., Yuan, Q., Wei, Y., Ahmad, A., Ullah, S., Khan, Z. U. H., Shams, S., Tariq, M., & Ahmad, W. (2019). Tuber extract of Arisaema flavum eco-benignly and effectively synthesize silver nanoparticles: Photocatalytic and antibacterial response against multidrug resistant engineered *E. coli* QH4. *Journal of Photochemistry and Photobiology B: Biology, 193*, 31–38. https://doi.org/10.1016/j.jphotobiol.2019.01.018
Rai, M., & Ingle, A. P. (2012). Role of nanotechnology in agriculture with special reference to management of insect pests. *Applied Microbiology and Biotechnology, 94*(2), 287–293. https://doi.org/10.1007/s00253-012-3969-4
Rajakumar, G., Abdul, & Rahuman, A. (2011). Larvicidal activity of synthesized silver nanoparticles using *Eclipta prostrata* leaf extract against filariasis and malaria vectors. *Acta Tropica, 118*(3), 196–203. https://doi.org/10.1016/j.actatropica.2011.03.003
Ramadurga, B., Jat, R. K., & Badami, S. (2019). Pharmacognostic evaluation and antimicrobial activity of root of *Careya arborea*. *Pharmacognosy Journal, 11*(3), 608–612. https://doi.org/10.5530/pj.2019.11.97
Ramanathan, R., Field, M. R., O'Mullane, A. P., Smooker, P. M., Bhargava, S. K., & Bansal, V. (2013). Aqueous phase synthesis of copper nanoparticles: A link between heavy metal resistance and nanoparticle synthesis ability in bacterial systems. *Nanoscale, 5*(6), 2300–2306. https://doi.org/10.1039/c2nr32887a
Rao, N. H., Lakshmidevi, N., Pammi, S. V. N., Kollu, P., Ganapaty, S., & Lakshmi, P. (2016). Green synthesis of silver nanoparticles using methanolic root extracts of Diospyros paniculata and their antimicrobial activities. *Materials Science and Engineering: C, 62*, 553–557. https://doi.org/10.1016/j.msec.2016.01.072
Rathore, R. K., Kashyap, N. K., Annamareddy, S. H. K., Hait, M., Roymahapatra, G. S., & Vaisnav, M. M. (2023). Appraisal of physicochemical qualities of grazing land soil in Kartala Block, Chhattisgarh, India. *ES Food & Agroforestry, 11*, 856. https://doi.org/10.30919/esfaf856
Raut, D., Wankhede, K., & Vaidya, V. (2009). Copper nanoparticles in ionic liquids: Recyclable and efficient catalytic system for 1,3-dipolar cycloaddition reaction. *Catalysis Communications, 10*(8), 1240–1243. https://doi.org/10.1016/j.catcom.2009.01.027
Ravichandran, R. (2010). Nanotechnology applications in food and food processing: Innovative green approaches, opportunities and uncertainties for global market. *International*

Journal of Green Nanotechnology: Physics and Chemistry, 1(2), P72–P96. https://doi.org/10.1080/19430871003684440

Ravindra, B. K., & Rajasab, A. H. (2014). A comparative study on biosynthesis of silver nanoparticles using four different fungal species. *International Journal of Pharmacy and Pharmaceutical Sciences, 6*(1), 372–376.

Reichelt, K. V., Hoffmann-Lucke, P., Hartmann, B., Weber, B., Ley, J. P., Krammer, G. E., Swanepoel, K. M., & Engel, K. H. (2012). Phytochemical characterization of south African bush tea (*Athrixia phylicoides* DC). *South African Journal of Botany, 83*(58), 1–8. https://doi.org/10.1016/j.sajb.2012.07.006

Ren, G., Hu, D., Cheng, E. W., Vargas-Reus, M. A., Reip, P., & Allaker, R. P. (2009). Characterization of copper oxide nanoparticles for antimicrobial application. *International Journal of Antimicrobial Agents, 33*(6), 587–590. https://doi.org/10.1016/j.ijantimicag.2008.12.004

Saha, M., & Bandyopadhyay, P. K. (2017). Green biosynthesis of silver nanoparticle using garlic, allium sativum with reference to its antimicrobial activity against the pathogenic strain of bacillus sp. and pseudomonas sp. infecting goldfish, *Carassius auratus. Proceedings of the Zoological Society (Calcutta), 72*(2), 180–186. https://doi.org/10.1007/s12595-017-0258-3

Sangeetha, G., Rajeshwari, S., & Venckatesh, R. (2011). Green synthesis of zinc oxide nanoparticles by *Aloe barbadensis miller* leaf extract: Structure and optical properties. *Materials Research Bulletin, 46*(12), 2560–2566. https://doi.org/10.1016/j.materresbull.2011.07.046

Sangeetha, S., Ramamoorthy, U., Sreeram, K. J., & Nair, U. (2012). Enhancing collagen stability through nanostructures containing chromium (III) oxide. *Colloids and Surfaces. B, Biointerfaces, 100*, 36–41. https://doi.org/10.1016/j.colsurfb.2012.05.015

Sanghi, R., & Verma, P. (2009). Biomimetic synthesis and characterisation of protein capped silver nanoparticles. *Bioresource Technology, 100*(1), 501–504. https://doi.org/10.1016/j.biortech.2008.05.048

Sanguansri, P., & Augustin, M. A. (2006). Nanoscale materials development—A food industry perspective. *Trends in Food Science and Technology, 17*(10), 547–556. https://doi.org/10.1016/j.tifs.2006.04.010

Santhoshkumar, T., Rahuman, A. A., Rajakumar, G., Marimuthu, S., Bagavan, A., & Jayaseelan, C. (2011). Synthesis of silver nanoparticles using *Nelumbo nucifera* leaf extract and its larvicidal activity against malaria and filariasis vectors. *Parasitology Research, 108*(3), 693–702. https://doi.org/10.1007/s00436-010-2115-4

Sathishkumar, M., Sneha, K., Won, S. W., Cho, C. W., Kim, S., & Yun, Y. S. (2009). *Cinnamon zeylanicum* bark extract and powder mediated green synthesis of nano-crystalline silver particles and its bactericidal activity. *Colloids and Surfaces. B, Biointerfaces, 73*(2), 332–338. https://doi.org/10.1016/j.colsurfb.2009.06.005

Sathishkumar, M., Krishnamurthy, R., & Yun, Y. S. (2010). Immobilization of silver nanoparticles synthesized using *Curcuma longa* tuber powder and extract on cotton cloth for bactericidal activity. *Bioresource Technology, 101*(20), 7958–7965. https://doi.org/10.1016/j.biortech.2010.05.051

Satyavani, K., Gurudeeban, S., Ramanathan, T., & Balasubramanian, T. (2011). Biomedical potential of silver nanoparticles synthesized from calli cells of *Citrullus colocynthis* (L.). *Schrad Journal of Nanobiotechnology, 9*(1), 43–51. https://doi.org/10.1186/1477-3155-9-43

Schodek, D. L., Ferreira, P., & Ashby, M. F. (2009). *Nanomaterials, nanotechnologies and design: An introduction for engineers and architects* (1st ed.). Butterworth-Heinemann.

Schrand, A. M., Rahman, M. F., Hussain, S. M., Schlager, J. J., Smith, D. A., & Syed, A. F. (2010). Metal-based nanoparticles and their toxicity assessment. *Reviews-Nanomedicine and Nanobiotechnology, 2*(5), 544–568. https://doi.org/10.1002/wnan.103

Selvi, N. T., Navamathavan, R., Kim, H. Y., & Nirmala, R. (2019). Autoclave mediated synthesis of silver nanoparticles using aqueous extract of *Canna indica* L. Rhizome and evaluation of its antimicrobial activity. *Macromolecular Research, 27*(11), 1155–1160. https://doi.org/10.1007/s13233-019-7159-4

Setua, P., Pramanik, R., & Sarkar, S. (2011). Synthesis of silver nanoparticle in imidazolium and pyrolidium based ionic liquid reverse micelles: A step forward in nanostructure inorganic material in room temperature ionic liquid field. *Journal of Molecular Liquids, 162*(1), 33–37. https://doi.org/10.1016/j.molliq.2011.05.015

Shafey, A. M. E. (2020). Green synthesis of metal and metal oxide nanoparticles from plant leaf extracts and their applications: A review. *Green Processing and Synthesis, 9*(1), 304–339. https://doi.org/10.1515/gps-2020-0031

Shaligram, N. S., Bule, M., & Bhambure, R. (2009). Biosynthesis of silver nanoparticles using aqueous extract from the compactin producing fungal strain. *Process Biochemistry, 44*(8), 939–943. https://doi.org/10.1016/j.procbio.2009.04.009

Shameli, K., Ahmad, M., Al-Mulla, E. A. J., Ibrahim, N. A., Shabanzadeh, P., Rustaiyan, A., & Abdollahi, Y. (2012). Green biosynthesis of silver nanoparticles using *Callicarpa maingayi* stem bark extraction. *Molecules, 17*(7), 8506–8517. https://doi.org/10.3390/molecules17078506

Shanmuganathan, R., Karuppusamy, I., Saravanan, M., Muthukumar, H., Ponnuchamy, K., & Ramkumar, V. S. (2019). Synthesis of silver nanoparticles and their biomedical applications—A comprehensive review. *Current Pharmaceutical Design, 25*(24), 2650–2660. https://doi.org/10.2174/1381612825666190708185506

Shanthi, S., Jayaseelan, B. D., Velusamy, P., Vijayakumar, S., Chih, C. T., & Vaseeharan, B. (2016). Biosynthesis of silver nanoparticles using a probiotic Bacillus licheniformis Dahb1 and their antibiofilm activity and toxicity effects in *Ceriodaphnia cornuta*. *Microbial Pathogenesis, 93*(1), 70–77. https://doi.org/10.1016/j.micpath.2016.01.014

Sharma, G., Nam, J. S., Sharma, A. R., & Lee, S. S. (2018). Antimicrobial potential of silver nanoparticles synthesized using medicinal herb *Coptidis* rhizome. *Molecules, 23*(9), 2268. https://doi.org/10.3390/molecules23092268

Shi, H., Magaye, R., Castranova, V., & Zhao, J. (2013). Titanium dioxide nanoparticles: A review of current toxicological data. *Particle and Fibre Toxicology, 10*, 15. https://doi.org/10.1186/1743-8977-10-15

Shivaji, S., Sudan, M., & Singh, S. (2011). Extracellular synthesis of antibacterial silver nanoparticles using psychrophilic bacteria. *Process Biochemistry, 46*(9), 1800–1807. https://doi.org/10.1016/j.procbio.2011.06.008

Simonnet, J. T., Sonneville, O., & Legret, S. (2001). *Nanoemulsion based on phosphoric acid fatty acid esters and its uses in the cosmetics, dermatological, pharmaceutical, and/or ophthalmological fields*. L'Oreal, US Patent 6274150. https://patentimages.storage.googleapis.com/0a/96/b2/f3761a78961b8b/US6274150.pdf

Sondi, I., & Salopek-Sondi, B. (2004). Silver nanoparticles as antimicrobial agent: A case study on *E. coli* as a model for Gram-negative bacteria. *Journal of Colloid and Interface Science, 275*(1), 177–182. https://doi.org/10.1016/j.jcis.2004.02.012

Song, J. Y., & Kim, B. S. (2009). Biological synthesis of bimetallic Au/Ag nanoparticles using Persimmon (*Diospyros kaki*) leaf extract. *Korean Journal of Chemical Engineering, 25*(4), 808–811. https://doi.org/10.1007/s11814-008-0133-z

Song, C., & Liu, S. (2005). A new healthy sunscreen system for human: Solid lipid nanoparticles as carrier for 3,4,5-trimethoxybenzoylchitin and the improvement by adding vitamin E. *International Journal of Biological Macromolecules, 36*(1–2), 116–119. https://doi.org/10.1016/j.ijbiomac.2005.05.003

Sonneville-Aubrun, O., Simonnet, J. T., & L'Alloret, F. (2004). Nanoemulsions: A new vehicle for skincare products. *Advances in Colloid and Interface Science, 108–109*, 145–149. https://doi.org/10.1016/j.cis.2003.10.026

Southam, G., & Beveridge, T. J. (1994). The in vitro formation of placer gold by bacteria. *Geochimica et Cosmochimica Acta, 58*(1), 4527–4530. https://doi.org/10.1016/00167037(94)90355-7

Spicer, P. T., Lynch, M. L., Visscher, M., & Hoath, S. (2003). Bicontinuous cubic liquid crystalline phase and cubosome personal care delivery systems. In M. Rosen (Ed.), *Personal care delivery systems and formulations*. Noyes Publishing.

Sre, P. R. R., Reka, M., Poovazhagi, R., Arulkumar, M., & Murugesan, K. (2015). Antibacterial and cytotoxic effect of biologically synthesized silver nanoparticles using aqueous root extract of Erythrina indica L.' Spectrochimica Acta Part A. *Molecular and Biomolecular Spectroscopy, 135*, 1137–1144. https://doi.org/10.1016/j.saa.2014.08.019

Subramaniam, V. D., Ramachndran, M., Marotta, F., Banerjee, A., Sun, X. F., & Pathak, S. (2019). Comparative study on anti-proliferative potentials of zinc oxide and aluminium oxide nanoparticles in colon cancer cells. *Octa Biomedica, 90*(2), 241–247. https://doi.org/10.23750/abm.v90i2.6939

Sukumaran, P., & Poulose, E. K. (2012). Silver nanoparticles: Mechanism of antimicrobial action, synthesis, medical applications, and toxicity effects. *International Nano Letters, 2*, 32. https://doi.org/10.1186/2228-5326-2-32

Suman, T. Y., Rajasree, S. R., Kanchana, A., & Elizabeth, S. B. (2013). Biosynthesis, characterization and cytotoxic effect of plant mediated silver nanoparticles using *Morinda citrifolia* root extract. *Colloids and Surfaces. B, Biointerfaces, 106*, 74–78. https://doi.org/10.1016/j.colsurfb.2013.01.037

Sun, R. W., Rong, C., Chung, N. P. Y., Ho, C. M., Lin, C. L. S., & Che, C. M. (2005). Silver nanoparticles fabricated in Hepes buffer exhibit cytoprotective activities toward HIV-1 infected cells. *Chemical Communications, 40*, 5059–5061. https://doi.org/10.1039/b510984a

Suriyakalaa, U., Antony, J. J., Suganya, S., Siva, D., Sukirtha, R., Kamalakkannan, S., Pichiah, P. B. T., & Achiraman, S. (2013). Hepatocurative activity of biosynthesized silver nanoparticles fabricated using Andrographis paniculata. *Colloids and Surfaces. B, Biointerfaces, 102*, 189–194. https://doi.org/10.1016/j.colsurfb.2012.06.039

Swarnalatha, C., Rachela, S., Ranjan, P., & Baradwaj, P. (2012). Evaluation of *invitro* antidiabetic activity of *Sphaeranthus amaranthoides* silver nanoparticles. *International Journal of Nanomaterials and Biostructures, 2*(1), 25–29.

Szostak, K., Ostaszewski, P., Pulit-Prociak, J., & Banach, M. (2019). Bismuth oxide nanoparticles in drug delivery systems. *Pharmaceutical Chemistry Journal, 53*(1), 48–51. https://doi.org/10.1007/s11094-019-01954-9

Thakkar, K. N., Mhatre, S. S., & Rasesh, Y. P. (2010). Biological synthesis of metallic nanoparticles. *Nanomedicine, 6*(2), 257–262. https://doi.org/10.1016/j.nano.2009.07.002

Thong, H. Y., Zhai, H., & Maibach, H. I. (2007). Percutaneous penetration enhancers: An overview. *Skin Pharmacology and Physiology, 20*(6), 272–282. https://doi.org/10.1159/000107575

Tian, J., Wong, K. K. Y., Ho, C. M., Lok, C. N., Yu, W. Y., Che, C. M., Chiu, J. F., & Tam, P. K. H. (2007). Topical delivery of silver nanoparticles promotes wound healing. *ChemMedChem, 2*(1), 129–136. https://doi.org/10.1002/cmdc.200600171

Tong, S., Zhu, H., & Bao, G. (2019). Magnetic iron oxide nanoparticles for disease detection and therapy. *Materials Today, 31*, 86–99. https://doi.org/10.1016/j.mattod.2019.06.003

Touitou, E., Dayan, N., Bergelson, L., Godin, B., & Eliaz, M. (2000). Ethosomes-novel vesicular carriers for enhanced delivery: Characterization and skin penetration properties. *Journal of Controlled Release, 65*(3), 403–418. https://doi.org/10.1016/s0168-3659(99)00222-9

Tournilhac, F., & Simon, P. (2001). *Cosmetic or dermatological topical compositions comprising dendritic polyesters*. L'Oréal, US Patent 6287552. https://patentimages.storage.googleapis.com/22/88/84/3311a13fb6d408/US6287552

Tripathy, A., Raichur, A. M., Chandrasekaran, N., Prathna, T. C., & Mukherjee, A. (2010). Process variables in biomimetic synthesis of silver nanoparticles by aqueous extract of *Azadirachta indica* (neem) leaves. *Journal of Nanoparticle Research, 12*, 237–246. https://doi.org/10.1007/s11051-009-9602-5

Uchegbu, I. F., & Vyas, S. (1998). Non-ionic surfactant based vesicles (niosomes) in drug delivery. *International Journal of Pharmaceutics, 172*(1–2), 33–70. https://doi.org/10.1016/s0378-5173(98)00169-0

Vanaja, M., & Annadurai, G. (2012). Coleus aromaticus leaf extract mediated synthesis of silver nanoparticles and its bactericidal activity. *Applied Nanoscience, 3*(3), 217–223. https://doi.org/10.1007/s13204-012-0121-9

Vankar, P. S., & Bajpai, D. (2010). Preparation of gold nanoparticles from *Mirabilis jalapa* flowers. *Indian Journal of Biochemistry & Biophysics, 47*(3), 157–160.

Vanlalveni, C., Lallianrawna, S., Biswas, A., Selvaraj, M., Changmai, B., & Rokhum, S. L. (2021). Green synthesis of silver nanoparticles using plant extracts and their antimicrobial activities: A review of recent literature. *RSC Advances, 11*(5), 2804–2837. https://doi.org/10.1039/d0ra09941d

Vigneshwaran, N., Kathe, A. A., & Varadarajan, P. V. (2006). Biomimetics of silver nanoparticles by white rot fungus, Phanerochaete chrysosporium. *Colloids and Surfaces. B, Biointerfaces, 53*(1), 55–59. https://doi.org/10.1016/j.colsurfb.2006.07.014

Vigneshwaran, N., Ashtaputre, N. M., & Varadarajan, P. V. (2007). Biological synthesis of silver nanoparticles using the fungus Aspergillus flavus. *Materials Letters, 61*(6), 1413–1418. https://doi.org/10.1016/j.matlet.2006.07.042

Vijaykumar, P. P. N., Kalyanin, R. L., Veerla, S. C., Kollu, P., Shameem, U., & Rammi, S. V. N. (2019). Biogenic synthesis of stable silver nanoparticles via *Asparagus racemosus* root extract and their antibacterial efficacy towards human and fish bacterial pathogens. *Materials Research Express, 6*(10), 104008. https://doi.org/10.1088/2053-1591/ab3ce9

Vivek, M., Kumar, P. S., Steffi, S., & Sudha, S. (2011). Biogenic silver nanoparticles by *Gelidiella acerosa* extract and their antifungal effects. *Avicenna Journal Of Medical Biotechnology, 3*(3), 143–148.

Wang, D., Markus, J., Wang, C., Kim, Y. J., Mathiyalagan, R., Aceituno, V. C., Ahn, S., & Yang, D. C. (2016). Green synthesis of gold and silver nanoparticles using aqueous extract of *Cibotium barometz* root. *Artificial Cells, Nanomedicine, and Biotechnology, 45*(8), 1548–1555. https://doi.org/10.1080/21691401.2016.1260580

Wissing, S. A., & Müller, R. H. (2003). Cosmetic applications for solid lipid nanoparticles (SLN). *International Journal of Pharmaceutics, 254*(1), 65–68. https://doi.org/10.1016/s0378-5173(02)00684-1

Witika, B. A., Noundou, X. S., & Ukwueze, S. E. (2017). Green synthesis and characterization of monodispersed silver nanoparticles using root bark aqueous extract of Annona muricata Linn and their antimicrobial activity. *Applied Nanoscience, 7*(8), 905–911. https://doi.org/10.1007/s13204-017-0632-5

Wu, M., Hou, P., Dong, L., Cai, L., Chen, Z., Zhao, M., & Jingjing, L. I. (2019). Manganese dioxide nanosheets: From preparation to biomedical applications. *International Journal of Nanomedicine, 14*, 4781–4800. https://doi.org/10.2147/ijn.s207666

Yalcin, B., & Otles, S. (2010). Nanobiosensor and food pathogen interaction mechanisms. *Electronic Journal of Environmental, Agricultural and Food Chemistry, 9*(1), 1257–1273.

Yang, N., Li, F., Jian, T., Liu, C., Sun, H., Wang, L., & Xu, H. (2017). Biogenic synthesis of silver nanoparticles using ginger (*Zingiber officinale*) extract and their antibacterial properties against aquatic pathogens. *Acta Oceanologica Sinica, 36*(12), 95–100. https://doi.org/10.1007/s13131-017-1099-7

Zhang, Y., Yang, D., Kong, Y., Wang, X., Pandoli, O., & Gao, G. (2010). Synergetic antibacterial effects of silver nanoparticles Aloe Vera prepared via a green method. *Nano Biomedicine And Engineering, 2*(4), 252–257. https://doi.org/10.5101/nbe.v2i4.p252-257

Zhang, X. F., Liu, Z. G., Shen, W., & Gurunathan, S. (2016). Silver nanoparticles: Synthesis, characterization, properties, applications, and therapeutic approaches. *International Journal of Molecular Sciences, 17*(9), 1534. https://doi.org/10.3390/ijms17091534

Isolation of Various Carbon-Rich Materials from Bio-based Sources and Their Utilization

Kumar Gaurav, Arnab Kanti Giri, Jaya Verma, Tina Chakrabarty, and Milan Hait

Abbreviations

BCA	Biomass-derived carbon aerogel
CRT	Carbon compounds from rose tea
CVD	Chemical vapor deposition
FFSC	Flexible fibrous supercapacitor
GQDs	Graphene quantum dots
HTC	Hydrothermal carbonization
HTG	Hydrothermal gasification
HTL	Hydrothermal liquefaction
LIG	Laser-induced graphitization
NCN	Nitrogen-doped carbon nanosheets
SCWG	Supercritical water gasification
SSA	Specific surface area

K. Gaurav (✉)
Department of Chemistry, Indian Institute of Science Education and Research, Pune, Maharashtra, India

A. K. Giri
Metallurgical and Materials Engineering Division, IIT, Kharagpur, West Bengal, India

J. Verma
Center for Precision Engineering, Harbin Institute of Technology, Harbin, Heilongjiang, China

T. Chakrabarty
R&D Tata Steel Ltd., Jamshedpur (Jharkhand), India

M. Hait
Department of Chemistry, Dr. C.V. Raman University, Bilaspur, Chhattisgarh, India

A. K. Bhardwaj et al. (eds.), *Biogenic Wastes-Enabled Nanomaterial Synthesis*,
https://doi.org/10.1007/978-3-031-59083-2_3

1 Introduction

Carbonaceous materials transformed from bio-based sources are the most abundant natural carbon source, which can be produced from urban waste, forestry, and agriculture (Wang et al., 2017a, 2018; Antar et al., 2021). The classical treatment of waste biomaterial comprises either open burning or disposal of undesirable materials, which results in negative impacts on the ecosystem (Vassilev et al., 2013; Wang et al., 2017a, 2018). Hence, the transformation of bio-based material into high-value-added material products including biofuels, various carbon-rich materials, and several bioproducts (Ravi & Vadukumpully, 2016; Wang et al., 2017a, b, 2018; Chen et al., 2020; Antar et al., 2021) has gained special attention for biomass valorization (Cha & Choi, 2015; Song et al., 2021). Carbonaceous materials transformed from bio-based sources have a long history since the ancient times (Ravi & Vadukumpully, 2016). Biochar is the most commonly derived from bio-source, which can be synthesized by thermochemical conversion (Kambo & Dutta, 2015; Qin et al., 2020). In addition, biomass-derived carbonaceous material such as carbon fibers (Ragauskas et al., 2014; Ogale et al., 2016; Haile et al., 2021; Wen et al., 2021) and activated carbon (Mohamad Nor et al., 2013; Jain et al., 2016; Abuelnoor et al., 2021; Adesanya et al., 2021; Liu et al., 2021) has also been developed for several important applications. Carbonaceous materials derived from bio-based sources for the preparation of carbon nanostructures (or nanomaterials) such as fullerenes, carbon nanotubes, graphene, and graphene quantum dots have also gained considerable interest (Ravi & Vadukumpully, 2016; Wang et al., 2017b, 2018, 2021; Lyu et al., 2019; Chen et al., 2020; Chang et al., 2021; Hassan & Carr, 2021; Sun et al., 2021; Qiang et al., 2022).

According to the SciFinder search, for the number of publications published in this field over the last 25 years, fundamental research on the synthesis and uses of carbonaceous materials produced from bio-based sources has increased exponentially (Fig. 1), which clearly indicates that the exploration in research area of carbonaceous materials derived from bio-based sources has currently received considerable importance.

2 Synthesis of Carbonaceous Materials Derived from Biomass

2.1 *Typical Biomass Precursors*

In particular, the kind of biomass source used, the conversion techniques used, and the kinds of additives used throughout the preparation procedures, everything plays a significant role in the quality, characteristics, and applications of carbon-rich materials obtained from bio-based sources (Fig. 2).

Fig. 1 Data collected from SciFinder search of the annual number of published papers containing "Carbonaceous Materials from Bio-based Sources" and "Biomass" in the title, abstract, and keywords since 1998. It is showing the exponential growth in this field

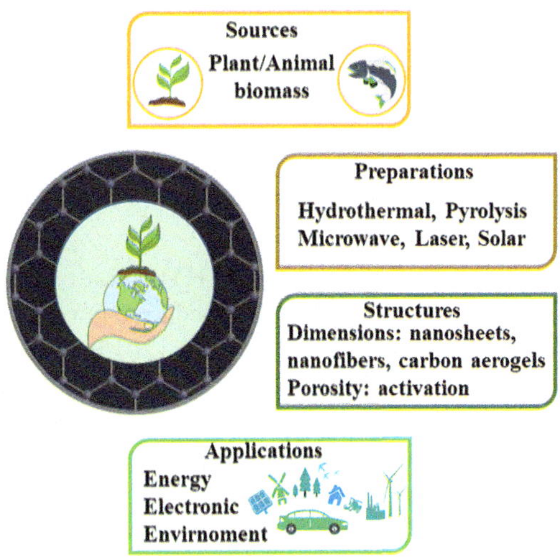

Fig. 2 Resources, preparation methods, structures, and diverse applications of carbonaceous materials from bio-based sources

Hydrothermal carbonization (HTC), pyrolysis, and laser-induced graphitization (LIG) are the three commonly used transformation methods to produce carbonaceous materials from bio-based sources (Titirici et al., 2012; Kambo & Dutta, 2015; Wang et al., 2017a; Qin et al., 2020). These techniques offer an extensive temperature range to encourage the graphitization and carbonization of biomass (Keiluweit et al., 2010; Yoo et al., 2018; Gomez-Martin et al., 2019; Ding et al., 2020). As shown in Fig. 3, laser (Ye et al., 2017; Han et al., 2018; Mahmood et al., 2019),

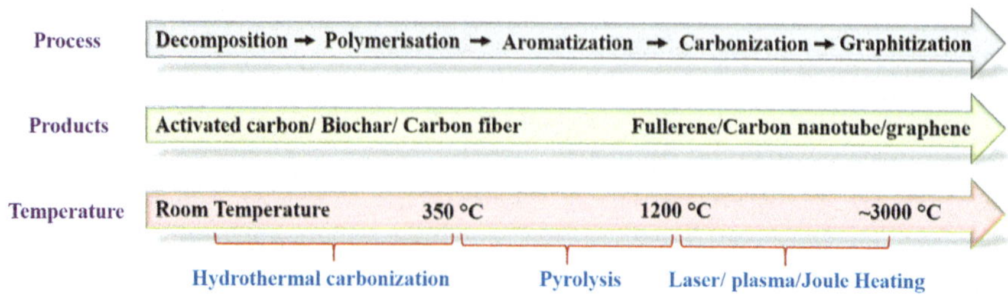

Fig. 3 An outline of the transformation of carbonaceous materials from precursors of biomass and common preparation techniques. Top: The conversion of carbon from biomass precursors. Bottom: The various preparation techniques, their respective operating temperatures, and the normal carbon byproducts

plasma (Seo et al., 2015; Wang et al., 2015; Kocjan et al., 2018; Shah et al., 2018), and joule heating (Luong et al., 2020; Chen et al., 2021) were also incorporated later to encourage the graphitization through the transient high-temperature medium. In addition to studying the structural transformation, ultrahigh-temperature pyrolysis (>1200 °C) was also performed, and several carbonaceous materials from bio-based sources were prepared (Emmerich, 1995; Shim et al., 2000; Baliga et al., 2003; Paris et al., 2005; Barin et al., 2013; Ding et al., 2013; Liu et al., 2018; Jiang et al., 2019). Particularly, carbon nanostructures (or nanomaterials) including fullerenes, carbon nanotubes, graphene, and graphene quantum dots (GQDs) have been obtained from carbonaceous materials transformed from bio-based source.

2.2 *Carbonization Driven by Hydrothermal Processes*

Hydrothermal carbonization (HTC) is a type of hydrothermal processing route in which transforms of biomass occurred into carbon-rich products such as hydrochars (Titirici et al., 2012; Hitzl et al., 2015; Kambo & Dutta, 2015; Benavente et al., 2017; Reißmann et al., 2018; Heidari et al., 2019; Nicolae et al., 2020; Qin et al., 2020; Sharma et al., 2020; Boutaieb et al., 2021) and graphene quantum dots (GQDs) (Wang et al., 2016; Ding et al., 2018). This process is typically performed at the temperature ranging from 180 to 260 °C under a certain pressure (2–6 MPa) for specific time ranging from 5 minutes to 4 hours. The process in which liquid products of hydrocarbons are generated when temperature is raised to above 260 °C is known as hydrothermal liquefaction (HTL). In order to modify the treatment process, supercritical water gasification (SCWG) or hydrothermal gasification (HTG) is used, with the formation of primarily gaseous products, if both the temperature and the pressure exceed to more than 374 °C and 22.1 MPa, respectively (Hitzl et al., 2015; Benavente et al., 2017; Reißmann et al., 2018; Heidari et al., 2019; Sharma et al., 2020). The hydrothermal carbonization technique is preferred to conventional pretreatment of dry-heat for converting bio-waste into carbonaceous materials due to its improved quality and direct usage of wet feedstock without the requirement for pre-drying. In order to produce high level of value-added carbonaceous materials, the HTC method also makes it easier for biomass to be carbonized as well as for carbon-containing components to be broken down, polymerized, and aromatized (Titirici et al., 2012; Kambo & Dutta, 2015; Nicolae et al., 2020). As a result, in recent years, HTC has been widely utilized as a form of treatment or pretreatment process for preparing carbonaceous materials obtained from biomass (Liu et al., 2014b; Parshetti et al., 2014; Long et al., 2015; Xing et al., 2016; Zeng et al., 2016; Ding et al., 2020; Nicolae et al., 2020; Boutaieb et al., 2021).

2.3 *Pyrolysis Methods*

Carbonization through pyrolysis of biomass is another feasible and effective way to synthesize carbonaceous materials from bio-based sources. It is usually performed under an inert (N_2/Ar) atmosphere at temperatures above 350 °C in a tubular furnace to produce various carbon-rich products, which can be tailored by controlling the temperature/heating parameter (Kambo & Dutta, 2015; Bhat et al., 2020; Qin et al., 2020). The rate of heating is one of the crucial operating factors that significantly affects the structure of carbonaceous material obtained from biomass.

Based on the temperature at which it operates, rate of heating, and residence time, pyrolysis can be classified into two major categories: (1) slow pyrolysis and (2) fast pyrolysis (Bhat et al., 2020; Qin et al., 2020). Slow pyrolysis refers to relatively low temperature (350–700 °C) for long durations with slow heating rate (0.01–2 °C s^{-1}), whereas fast pyrolysis refers to high temperature (300–1000 °C) for a short period of time ($<$10) with fast heating rate (–2 °C s^{-1}) (Qin et al., 2020). Several studies have found that the slow pyrolysis process favors high yield and leads to a good quality and highly organized carbon structure (Guerrero et al., 2005).

2.4 *Laser-Induced Graphitization*

Laser-induced graphitization (LIG) is another appealing method for turning biomass into carbon-based products. Unlike hydrothermal carbonization and pyrolysis, laser-induced graphitization procedures may carbonize a specific portion of the biomass and create conductive graphitic carbon. Photolysis and pyrolysis are just two examples of the optical and thermal processes that will occur in the laser-irradiated area. Therefore, laser-induced graphitization is a viable technique for making electronic devices from bio-based sources. The structural morphology and electrical characteristics of LIG produced carbonaceous materials obtained from bio-sources can be tuned by controlling manufacturing conditions (Wang et al., 2022). The laser radiation energy is one of the essential preconditions. In their investigation of the parameters necessary for LIG, Duy et al. discovered that a rise in laser energy generated by radiation causes the fluid dynamic production of graphitic carbon, when the morphology shifts from early carbon sheets to fibers and then to droplets (Duy et al., 2018). In accordance with the comprehensive study carried out by Abdulhafez and colleagues, laser fluence value is another aspect that will have an impact on the morphology of LIG products (Abdulhafez et al., 2021).

2.5 Other Thermal Treatment Methods

In addition to well-explored methods such as hydrothermal carbonization (HTC), pyrolysis, and laser-induced graphitization (LIG), several other thermal treatment techniques including microwave irradiation (Vivekanandhan et al., 2017; Omoriyekomwan et al., 2021; Zhang et al., 2021; Hoang et al., 2022), plasma-enhanced chemical vapor deposition (CVD) (Seo et al., 2015; Wang et al., 2015; Chinke et al., 2018; Shah et al., 2018), and Joule heating method (Jiang et al., 2019; Luong et al., 2020) have been developed in order to prepare the carbonaceous materials derived from bio-sources. The active bio-component responsible for producing carbon nanotubes from bio-waste sources by microwave radiation is cellulose (Omoriyekomwan et al., 2021). The biomass is pulverized and placed into a nickel case to be evacuated in a normal thermal treatment process using the CVD approach, which follows by deposition of plasma generated by microwaves. The graphitization process can proceed at temperatures that are lower (about 400 °C) because the plasma can produce a gas-rich environment that promotes the rapid formation of graphene (Shah et al., 2018). On the other hand, biomass can be transformed into graphene at extremely high temperatures using the Joule heating technique (Jiang et al., 2019). Jiang et al. show that annealing of lignin at 2500 K using Joule heating can remove impurities and further graphitize it into a confined structured and graphitic carbon (Jiang et al., 2019). Recently, researchers led by Luong et al. have found that using ash Joule heating, gram scale amounts of graphene could be made from a variety of cheap carbon sources, including coal, petroleum coke, biochar, carbon black, leftover food, rubber tires, and mixed plastic waste (Luong et al., 2020). Additionally, Ye et al. created laser-induced graphene from wood (rich in lignin) by using the CO_2 laser scribing technique (Ye et al., 2017). In general, increased laser power can result in better graphene structure quality (Ye et al., 2017). However, research indicates that when the intensity of the laser exceeds 90%, it may exhibit unfavorable effects and rupture the material's graphene structure (Ye et al., 2017; Mahmood et al., 2019).

2.6 Comparison Between Synthesis Methods

The aforementioned treatment techniques and strategies shed light on how to create carbonaceous compounds from biomass that have the desired qualities for real-world use. HTC is frequently utilized as a pyrolysis pretreatment technique and can be used to create graphene quantum dots (GQDs). HTC demonstrates significant advantages over other biomass treatment methods, when it comes to using wet and sludgy biogenic leftovers directly (Hitzl et al., 2015; Benavente et al., 2017; Reißmann et al., 2018; Heidari et al., 2019; Qin et al., 2020; Sharma et al., 2020). Furthermore, from a life cycle perspective, HTC techniques are being developed as a result of their great process energy efficiency and minimal potential for global warming (Benavente et al., 2017). However, HTC has yet to achieve large-scale and high-yielding production of hydrochars, graphene, and graphene quantum dots for use in practical applications. Additional study is being conducted to fill in knowledge gaps and address HTC technology's flaws (Hitzl et al., 2015; Benavente et al., 2017; Reißmann et al., 2018; Heidari et al., 2019; Nicolae et al., 2020; Sharma et al., 2020; Boutaieb et al., 2021). One of the most widely used techniques for creating various types of carbonaceous compounds from biomass is slow pyrolysis. The most significant aspects to consider when using pyrolysis-based processes to manufacture high-quality carbon-rich materials are the selection of the precursors of biomass and chemical additives and the optimization of pyrolysis parameters. Other thermal treatment techniques including laser scribing, Joule heating, and plasma-enhanced CVD have the potential to yield high-quality graphene derived from biomass. The brief ultrahigh temperature generated by Joule heating, plasma, and laser considerably improves the production of high-quality graphene derived from biomass. Additionally, the manufacture of carbon bers and carbon nanotubes from waste-bio-sources has been investigated through pyrolysis of cellulose-rich biomass using a microwave. The mass synthesis of carbon-rich materials obtained from bio-waste is however constrained by the capital and operating costs as well as the necessity to maintain fundamental conditions for plasma-enhanced CVD, Joule heating, and laser- and microwave-based methods. The approaches that have been condensed have shed light on the viability of converting bio-waste into high-value carbonaceous compounds, such graphene made from biomass. The thorough analysis of bio-waste-derived graphene in particular showed that its characteristics are on par with those of conventional synthetic graphene. Furthermore, a mass of graphene can be produced by employing a patch of graphene or graphene oxide as a template and using HTC, pyrolysis, and other thermal treatment procedures, breaking down the barrier between biomass-derived graphene and standard synthetic graphene.

3 Synthetic Mechanisms of Carbon-Rich Materials Derived from Bio-waste

A number of synthesis techniques have been developed, for the conversion of bio-waste into carbon-rich materials, such as biochars, hydrochars, carbon bers, activated carbon, GQDs, CNTs, and graphene/graphene-like products. The mechanisms of the synthesis process give information about the routes via which the bio-waste precursor turned into carbon-rich products. Further studies can be focused on developing specific biomass-derived nanomaterials, enriching the carbon nanostructure, forming the carbon nanostructure, and contrasting the properties of biomass-derived nanomaterials with those of conventional synthetic carbon nanomaterials now that the basic mechanistic understanding of how biomass can be converted into carbonaceous materials has been established. Different kinds of biomass precursors offer various ways to prepare carbonaceous materials derived from biomass. According to the synthetic techniques and the different types of biomass precursors (such as cellulose, lignin, and lignocellulosic/nonlignocellulosic biomass), studies on the mechanisms that facilitate the synthesis of carbonaceous substances from biomass are classified in the literature.

HTC was introduced as a primary approach to prepare graphene quantum dots from bio-sources in comparison to the conventional hydrochar. GQDs are a particular class of zero-dimensional carbon nanomaterial made up of graphene layers with a thickness ranging from 1 to 20 nm. In order to convert biomass to GQDs (Wang et al., 2016; Abbas et al., 2018), HTC has been used, in which the macromolecular biomass is first hydrolyzed at high temperature and pressure and then broken down into minute fragments. By aromatization, these pieces are then combined to create GQDs. In addition to GQDs made from biomass, Siddiqui et al. created magnetic nanomaterials composed of carbon with a graphene structure by solvothermal and hydrothermal carbonization-coprecipitation processes (Siddiqui et al., 2020). The identification of important intermediates of products revealed the following steps, even though the summarized formation mechanisms were not fully understood. During the hydrothermal carbonization treatment of iron ions and rice husk, the results of rice husk hydrolysis constructed a carbon matrix by the polymerization and aromatization process. Simultaneously, during the coprecipitation of iron ions, iron oxides were created through the nucleation of crystal growth. Furthermore, at the high temperature and pressure, the iron oxides encapsulated in the carbon matrix formed graphene structures. When compared to other HTC transformation pathways, microwave- and ultrasonication-based techniques could improve the conversion of bio-waste into graphene/graphene-like structures. In these techniques, the ultrasonication technique enhanced the decomposition of lignin, whereas microwave radiation was absorbed by the polar intermediate products in order to accelerate conversion into graphene structures (Xu et al., 2017; Gao et al., 2018).

The most prevalent aromatic biopolymer on Earth is lignin, which has significant potential as a prime candidate for pyrolysis-based hexagonal lattice formation in carbonaceous materials (Zhang et al., 2017). Given its chemical makeup, lignin has a high potential for producing carbon bers (Ogale et al., 2016; Haile et al., 2021) and graphene (Ding et al., 2020) through pyrolysis. The isolated lignin from black liquor was employed as a precursor to make carbon bers by incorporating proper spinning and carbonization throughout the pyrolysis process, which included melt spinning, thermal stabilization at 450–650 K, and carbonization about 1250–1300 K (Haile et al., 2021).

Compared to other sp^3-bonded structure biomass precursors, sp^2-bonded aromatic ring framework of lignin makes the transformation to the graphitic nanosheets much simpler. Waste lignin, which makes up the majority of black liquor, is typically dumped or burned in the pulp and paper sector (Jiang et al., 2019; Ding et al., 2020). Thus, there are several advantages and potential applications for converting lignin waste into high-value-added materials such as graphene/graphene-based products.

In addition to HTC and pyrolysis, additional techniques have been used to turn biomass into carbon-rich products, including microwave, plasma, laser, and ash Joule heating (Fig. 4). Omoriyekomwan et al. studied the pyrolysis of bio-wastes using a microwave to create CNTs and carbon nanoparticles (Omoriyekomwan et al., 2021). The active bio-components responsible for the synthesis of carbon

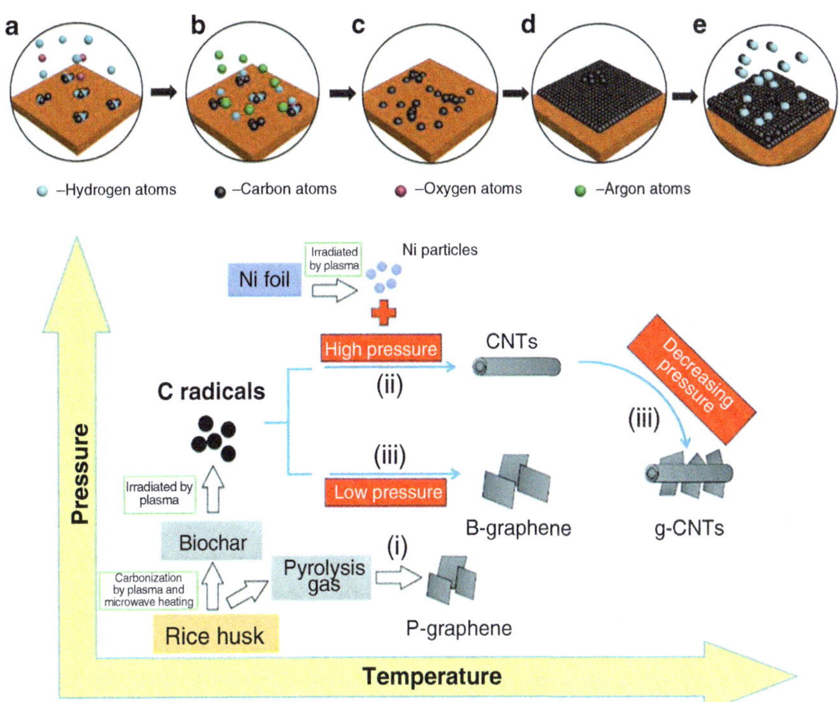

Fig. 4 Formation processes of graphene derived from biomass by plasma and flash Joule heating. (Reproduced with permission from Wang et al. (2015) and Shah et al. (2018). Copyright 2015, Elsevier; Copyright 2018, American Chemical Society)

nanoparticles and CNTs from biomass under the irradiation of microwave are mainly cellulose and polysaccharides (Omoriyekomwan et al., 2021).

4 Applications of Bio-waste-Derived Carbon Materials

The improved porosity, substantial surface area, and distinctive physicochemical features of carbonaceous materials derived from bio-sources have made them suitable for a wide range of applications in the fields of biology, materials science and engineering, and energy and the environment (Abbas et al., 2018; Qin et al., 2020). The literature has emphasized the important roles that hydrochars and biochars specifically play in natural soil and water settings (Titirici et al., 2012; Kambo & Dutta, 2015; Qin et al., 2020). In addition, the excellent adsorption and electrical properties of carbon bers and activated carbon derived from biomass support their productive uses in power generation and environmental applications (Mohamad Nor et al., 2013; Ragauskas et al., 2014; Jain et al., 2016; Ogale et al., 2016; Abuelnoor et al., 2021; Adesanya et al., 2021; Haile et al., 2021; Liu et al., 2021; Wen et al., 2021). Here, three important uses for carbonaceous materials obtained from bio-sources have heighted, specifically related to energy, electronics, and environment.

4.1 Applications in Energy and Related Field

Carbonaceous materials derived from bio-sources have shown great potential in various energy-related applications such as batteries (Liu et al., 2010, 2016; Hong et al., 2014; Yoon et al., 2018), supercapacitors (Sun et al., 2013; Gopalakrishnan et al., 2019; Zhang et al., 2019a, 2020), and the oxygen reduction reaction (ORR) for full cells (Liu et al., 2014a; Mondal et al., 2016; Zhou et al., 2019). This is because these materials have remarkable electrical conductivity, a high surface area, and a tuneable structure that can be altered by the addition of different dopants. The carbonaceous materials formed from biomass have more desirable qualities for energy-related applications as the degree of graphitization rises. Table 1 provides a summary of the features and specific performance of materials obtained from biomass in various energy applications.

Carbon-based substances derived from biomass with porous graphitic carbon nanostructures can accelerate ion diffusion and increase electrolyte transport. These materials have undergone substantial research in energy storage batteries for use in cars and portable electronic devices, especially lithium-ion (Li-ion) batteries (Liu et al., 2010, 2016; Yi et al., 2017), sodium-ion (Na-ion) batteries (Ding et al., 2013; Mun et al., 2013; Hong et al., 2014; Yoon et al., 2018), and zinc-air batteries (Liu et al., 2020).

In the perspective of reducing the rising energy consumption in a sustainable manner, recent investigations on the preparation of carbonaceous materials

Table 1 A comparison of selected energy-related applications of bio-waste-derived graphene with various parameters

Materials	Applications	Performance	Characteristics	References
$CuSi_2P_3$ at graphene	Li-ion batteries (LIBs)	Capacity: 2069 mA h/g with coulombic efficiency of ~91% after 500 cycles	Cation-disordered cubic ZnS structure	Li et al. (2021)
2D SnS_2/rGO as anode material	Na-ion batteries (NIBs)	Capacity: 956 mA h/g with coulombic efficiency of 60% after 200 cycles	Nanosheets with thickness of about 20 nm	Mao et al. (2021)
2D graphene oxide (GO)	LIBs	Capacity: 130 mA h/g with coulombic efficiency of ~98% after 100 cycles	Cubic spinel framework of $LiMn_2O_4$	Kumar et al. (2019)
Double shell Co_3Sn_2/SnO_2 at graphene	NIBs/LIBs	Capacity: 744 mA h/g with coulombic efficiency of 51% after 100 cycles	Co_3Sn_2/SnO_2 heterostructures	Hu et al. (2019)
Carbon with graphene planes	NIBs	Capacity: 297 mA h g^{-1} at 50 mA g^{-1} after 50 cycles	Graphene layers with a large interlayer spacing (0.403 nm) and a high micropore surface area (207.5 m^2 g^{-1})	Yoon et al. (2018)
N-doped graphene aerogel	LIBs	Capacity: 550 mA h g^{-1} at 5 A g^{-1}, 729 mA h g^{-1} at 0.1 A g^{-1} for 300 cycles	3D connected porous structure with an average shell thickness of ~6 nm	Liu et al. (2016)
Multivalent VO_x/graphene nanocomposite	Supercapacitors	Capacitance, 1110F/g; capacitance retention, 92% after 20,000 cycles	Graphene/vanadium oxide-based thin film	Huang et al. (2021)
$W_{18}O_{49}$ nanowires (NWs)-rGO	Supercapacitors	Capacitance, 365.5 F/g; capacitance retention, 96.7% after 12,000 cycles	rGO-nanocomposite based on nanowires of $W_{18}O_{49}$	Thalji et al. (2021)

(continued)

Table 1 (continued)

Materials	Applications	Performance	Characteristics	References
3D N-doped graphene	Supercapacitors	Capacitance, 1308.2 F g^{-1} at 1 A g^{-1}; capacitance retention rate of 91.6% after 10,000 cycles	3D porous graphene structure (10–30 nm), high surface area (832 m^2 g^{-1})	Zhang et al. (2020)
Graphene-like microporous carbon nanosheets	Supercapacitors	Capacitance, 390 F/g; capacitance retention, 93.3% after 3500 cycles	2D nanosheets with mesopores (2 nm), high surface area (320.7 m^2 g^{-1})	Gopalakrishnan et al. (2019)
Fe_3O_4-/Fe-doped graphene nanosheets	Oxygen reduction reaction	Current density: 4.4 mA cm^{-2} at 0.4 V	High electrical conductivity (2384.6 mS m^{-1}) high surface area (220 m^2 g^{-1})	Mondal et al. (2016)
N-doped graphene-like carbon sheets	Oxygen reduction reaction	Current density: 3.89 mA cm^{-2} at 0.3 V	Abundant mesoporosity (3.8 nm) and surface area (264 m^2 g^{-1})	Zhou et al. (2019)
Graphene-like N-doped carbon nanosheets	Oxygen reduction reaction	Current density: 37.04 mA cm^{-2} at −0.45 V	High surface area (1510 m^2 g^{-1})	Liu et al. (2014a)
MoS_2/N-GQDs/rGO	Solar cells	Photo-current efficiency: 2.01%	Ultrathin layered MoS_2	Silambarasan et al. (2021)
MoS_2/graphene	Solar cells	Photo-current efficiency: 8.92%	Interconnect MoS_2 nanoparticles in the 2D graphene nanosheets	Krishnamoorthy and Prakasam (2020)

generated from biomass for energy storage applications have attracted significant attention. The attractive properties of graphene/graphene-like materials derived from bio-waste, such as their high specific surface area (SSA), well-defined porous structure, magnificent electrical conductivity, and easily modifiable surface chemistry, make them an excellent candidate to be used as the foundation for supercapacitor electrodes (Benítez-Martínez et al., 2014).

Furthermore, carbon-rich materials derived from waste bio-waste can be utilized in fuel cells by promoting the oxygen reduction reaction via oxygen/hydrogen evolution processes to create sustainable energy (Liu et al., 2014a; Mondal et al., 2016; Zhou et al., 2019). Carbonaceous materials that have been heteroatom-doped with N, S, B, and P typically have competitive electrocatalytic activity and encourage the development of ORR active sites (Zhang et al., 2019b). When compared to conventional Pt-based cathodes, graphene-like nitrogen-doped carbon nanosheets (NCN) had superior electrocatalytic activity, better long-term stability, and the ability to tolerate methanol. These nanosheets were made by simply pyrolyzing urea and chitosan without using any catalyst or posttreatment procedure (Liu et al., 2014a). The

NCN was attractive because of its interconnected ultrathin nanosheet morphology and a high ratio of graphitic nitrogen species produced by the direct pyrolysis process, which made it suitable for additional potential uses in electronic capacitors or for the storage of hydrogen (Liu et al., 2014a). Zhang and colleagues were also able to generate heteroatom embedded graphene-like structures by pyrolyzing *Enteromorpha*, a common species of green algae that includes both N and S without adding any additional dopants at different temperatures (Zhang et al., 2019b).

4.2 Applications in Electronics

Soft electronics, whose components are small, light, flexible, and stretchable, have been the subject of burgeoning research in recent decades (Huang et al., 2019). Intelligent robotics, human-machine interfaces, and medical monitors can all benefit from the use of soft electronics (Wang et al., 2019). Carbon materials are thought to make an excellent matrix for soft electronics due to their high thermal and chemical stability, tremendous conductivity, and rapid transportation of electron. Biomass is frequently chosen as the carbon precursor in efforts to develop a sustainable method of creating carbon materials for soft electronics. Aside from being environmentally friendly, carbonaceous materials derived from bio-sources also have other benefits, such as integrating multiplex framework and morphological design, feasible functionalization, and customizable surface and electron structure (Jindal et al., 2022). Currently, carbonaceous materials transformed from waste biomass are utilized in a variety of soft electronics, such as sensors and e-textiles.

For instance, piezoresistive sensors, which can transform strain or pressure into resistance signals, are being employed in a variety of industrial applications, particularly in wearable technology (Trung & Lee, 2016). For instance, they can be affixed to clothing to track small movements on the human body. Based on these characteristics, materials utilized for piezoresistive sensors must possess a number of qualities, such as being lightweight, incredibly flexible, highly sensitive, and electrically conductive and having excellent biocompatibility. Biomass-derived carbon aerogels (BCAs), a subset of all carbonaceous materials derived from bio-sources, have drawn the most interest for use in piezoresistive sensors because of their extreme lightness and robustness. However, because of their random microstructure, BCAs have unsatisfactory strength and adaptability as well as a lack of linear sensitivity. As a result, the production of high-performance piezoresistive sensors benefits from the BCAs' rational structural design.

Furthermore, due to the growing importance of the chemical and engineering industries, gas detection, particularly dangerous gas detection, is critical for worker safety. The development of chemical detectors for the adsorption and identification of gases is presently a booming field of study. It has been shown that materials made of porous carbon that are generated from biomass are promising as chemical sensor materials (Zhao et al., 2020). Sun et al. developed porous carbon compounds derived from rose tea (CRT) as chemical detectors for NH3 gas screening, drawing inspiration from the pleated shape of the dog's maxillary turbinate (Sun et al., 2022).

Electronic-infused fabrics, often referred to as smart textiles, have received a lot of attention as wearable technology has developed quickly. This type of fabric has the ability to recognize environmental inputs, translate those stimuli into electric impulses, or even react to those stimuli (Grancarić et al., 2018). Conductive fibers, yarns, and fabrics are the main elements of smart textiles that enable various functions. Additionally, smart textiles should be washable because they are more likely to experience mechanical stresses or environmental corrosion than other wearable technologies. BCMs have a tremendous deal of potential in applications for smart textiles because of their high electrical conductivity, minimal toxicity, and long-term viability. For instance, Hsiao et al. developed flexible fibrous supercapacitors (FFSCs) in smart textiles by employing the honey bomb-like framework of *Tetrapanax papyrifer* (Tong Cao, TC). To wrap bundles of carbon fiber, TC is easily cut into paper. Commercially activated carbon is applied to and further carbonizes the composite yarn (Hsiao et al., 2020).

4.3 Environmental and Other Applications of Carbon Materials Derived from Biomass

Carbonaceous materials derived from bio-sources have not only been extensively studied for energy- and electronic-related applications but also for practical environmental ones such as physical removal of pollutants through adsorption, environmental sensing, and catalytic degradation of pollutants, particularly the use of biochars, hydrochars, and activated carbon to extract contaminants through adsorption (Titirici et al., 2012; Mohamad Nor et al., 2013; Kambo & Dutta, 2015; Jain et al., 2016; Qin et al., 2020; Abuelnoor et al., 2021; Adesanya et al., 2021; Liu et al., 2021). The environmental uses of carbon nanomaterials obtained from biomass, such as CNTs, GQDs, and graphene, are therefore highlighted. On account of the identification of metal ions and pollutants in the environment, bio-waste-derived GQDs with high water solubility, photostability, and adjustable fluorescence are particularly well suited for application as environmental sensors. Since aromatic organic pollutants interact strongly with the graphene layer, biomass-derived graphene that is both highly conductive and imperfection-free also has strong sensitivity to contaminants present in the atmosphere. The bio-waste-derived porous graphene synthesized by HTC and pyrolysis methods has favorable features for efficient adsorption of atmospheric pollutants, such as large specific surface area, effective porosity, and rich functional groups on the surface. It has considerable potential due to the substantial number of adsorption sites and favorable interactions with a wide range of contaminants, including gases, metal and nonmetal ions, and synthetic colors. Notably, the distinct p-bonding interactions occurring between the graphene framework and aromatic contaminants drive the creation of bio-waste-derived graphene with interlaced frameworks and a few defects as excellent adsorbents for the elimination of aromatic organic contaminants. Furthermore, graphene produced from bio-waste with a few defects has effective electron-accepting and micro-transport capabilities, making it appropriate for metal-free catalytic functionality. The inclusion of heteroatom additives or intrinsic components (such as N, P, and S) to a range of bio-waste precursors improves the synthesis of heteroatom-doped graphene, which can aid in the catalytic decomposition of pollutants.

5 Summery and Future Perspective

In summary, numerous biomass precursors and carbonaceous materials derived from natural bio-sources have been introduced in this chapter. Comprehensive discussion is provided with the representative techniques of synthetics and the characteristics they possess. In addition, the recent development of carbonaceous materials derived from biomass in a variety of applications is also discussed. These reviews and analyses show that tremendous progress has been made and that there is still a lot of interest in the study of carbonaceous compounds obtained from biomass. Sustainable development is aided by the conversion of biomass into useful carbon materials, which has significant economic value. Although there have been recent advances in the study of carbonaceous materials produced from biological sources, there are still a number of problems that need to be solved. The remaining difficulties as well as numerous potential future possibilities are enumerated in Scheme 1.

6 Conclusions

The biomass waste can be used to produce material composites for various uses. However, employing sustainable trash as a precursor does not reduce the negative environmental impact of waste; the resulting materials can act as a suitable sources of carbon and sources of inorganic materials. Regarding environmental impact and cost-effectiveness, materials for storing energy developed from commercial metal or metal oxide waste comprise the most significant research direction in waste-derived composites of graphene. In the case of bio-waste resource utilization for energy applications, the two crucial criteria for the functionality of composites are usually the amount of filler and its degree of dispersion of variance because they have a large influence on the systems through percolation barrier. Bio-waste-derived nanocomposites have outstanding mechanical characteristics and high conductivity to electricity, making them useful in a variety of applications including solar cells, supercapacitors, and LIBs/ NIBs.

This book chapter has provided an overview of recent developments in the research area of carbonaceous materials derived from bio-based sources,

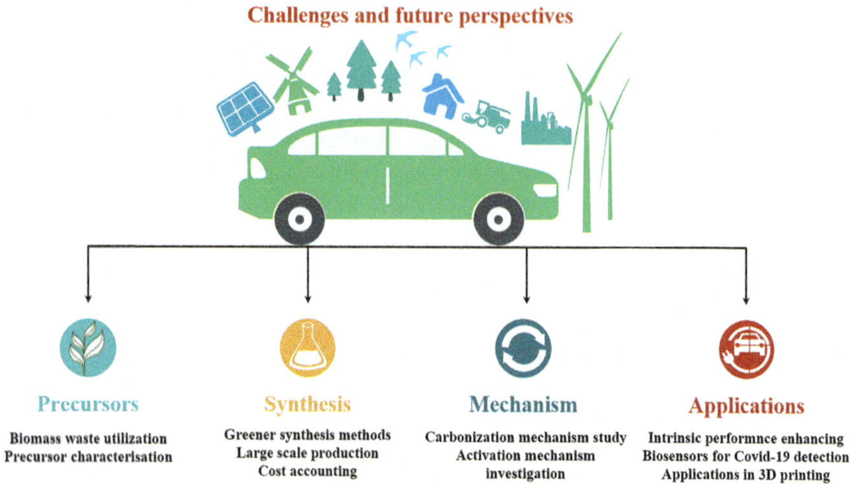

Scheme 1 An outline of challenges and prospects concerning the study of mechanisms, synthesis techniques, biomass precursors, and applications of carbonaceous materials obtained from bio-sources

preparation techniques, characteristics, and their applications, as well as existing difficulties and future directions for bio-based carbonaceous material development. We anticipate that this assessment will hasten the advancement of bio-based carbonaceous materials in various cutting-edge applications.

References

Abbas, A., Mariana, L. T., & Phan, A. N. (2018). Biomass-waste derived graphene quantum dots and their applications. *Carbon, 140*, 77–99. Available from: https://doi.org/10.1016/j.carbon.2018.08.016

Abdulhafez, M., Tomaraei, G. N., & Bedewy, M. (2021). Fluence-dependent morphological transitions in laser-induced graphene electrodes on polyimide substrates for flexible devices. *ACS Applied Nano Materials, 4*(3), 2973–2986. Available from: https://doi.org/10.1021/acsanm.1c00101

Abuelnoor, N., AlHajaj, A., Khaleel, M., Vega, L. F., & Abu-Zahra, M. R. M. (2021). Activated carbons from biomass-based sources for CO2 capture applications. *Chemosphere, 282*, 131111. Available from: https://doi.org/10.1016/j.chemosphere.2021.131111

Adesanya, E., Perumal, P., Luukkonen, T., Yliniemi, J., Ohenoja, K., Kinnunen, P., et al. (2021). Opportunities to improve sustainability of alkali-activated materials: A review of side-stream based activators. *Journal of Cleaner Production, 286*, 125558. Available from: https://doi.org/10.1016/j.jclepro.2020.125558

Antar, M., Lyu, D., Nazari, M., Shah, A., Zhou, X., & Smith, D. L. (2021). Biomass for a sustainable bioeconomy: An overview of world biomass production and utilization. *Renewable and Sustainable Energy Reviews, 139*, 110691. Available from: https://doi.org/10.1016/j.rser.2020.110691

Baliga, V., Sharma, R., Miser, D., McGrath, T., & Hajaligol, M. (2003). Physical characterization of pyrolyzed tobacco and tobacco components. *Journal of Analytical and Applied Pyrolysis, 66*(1–2), 191–215. Available from: https://doi.org/10.1016/S0165-2370(02)00114-6

Barin, G. B., Santos, Y. H., Rocha, J. A., Da Pereira Costa, L., Souza Filho, A. G., de Fátima Gimenez, I., et al. (2013). Graphene-like nanostructures obtained from biomass. *MRS Proceedings*, 1505. Available from: https://doi.org/10.1557/opl.2013.479

Benavente, V., Fullana, A., & Berge, N. D. (2017). Life cycle analysis of hydrothermal carbonization of olive mill waste: Comparison with current management approaches. *Journal of Cleaner Production, 142*, 2637–2648. Available from: https://doi.org/10.1016/j.jclepro.2016.11.013

Benítez-Martínez, S., López-Lorente, Á. I., & Valcárcel, M. (2014). Graphene quantum dots sensor for the determination of graphene oxide in environmental water samples. *Analytical Chemistry, 86*(24), 12279–12284. Available from: https://doi.org/10.1021/ac5035083

Bhat, V. S., Supriya, S., & Hegde, G. (2020). Review—Biomass derived carbon materials for electrochemical sensors. *Journal of the Electrochemical Society, 167*(3), 37526. Available from: https://doi.org/10.1149/2.0262003jes

Boutaieb, M., Román, S., Ledesma, B., Sabio, E., Guiza, M., & Ouederni, A. (2021). Towards a more efficient hydrothermal carbonization: Processing water recirculation under different conditions. *Waste Management (New York, N.Y.), 132*, 115–123. Available from: https://doi.org/10.1016/j.wasman.2021.07.023

Cha, H. G., & Choi, K.-S. (2015). Combined biomass valorization and hydrogen production in a photoelectrochemical cell. *Nature Chemistry, 7*(4), 328–333. Available from: https://doi.org/10.1038/nchem.2194

Chang, B. P., Rodriguez-Uribe, A., Mohanty, A. K., & Misra, M. (2021). A comprehensive review of renewable and sustainable biosourced carbon through pyrolysis in biocomposites uses: Current development and future opportunity. *Renewable and Sustainable Energy Reviews, 152*, 111666. Available from: https://doi.org/10.1016/j.rser.2021.111666

Chen, Q., Tan, X., Liu, Y., Liu, S., Li, M., Gu, Y., et al. (2020). Biomass-derived porous graphitic carbon materials for energy and environmental applications. *Journal of Materials Chemistry A, 8*(12), 5773–5811. Available from: https://doi.org/10.1039/C9TA11618D&newsite=1

Chen, W., Li, J. T., Wang, Z., Algozeeb, W. A., Luong, D. X., Kittrell, C., et al. (2021). Ultrafast and controllable phase evolution by flash joule heating. *ACS Nano, 15*(7), 11158–11167. Available from: https://doi.org/10.1021/acsnano.1c03536

Chinke, S. L., Sandhu, I. S., Saroha, D. R., & Alegaonkar, P. S. (2018). Graphene-like nanoflakes for shock absorption applications. *ACS Applied Nano Materials, 1*(11), 6027–6037. Available from: https://doi.org/10.1021/acsanm.8b01061

Ding, J., Wang, H., Li, Z., Kohandehghan, A., Cui, K., Xu, Z., et al. (2013). Carbon nanosheet frameworks derived from peat moss as high performance sodium ion battery anodes. *ACS Nano, 7*(12), 11004–11015. Available from: https://doi.org/10.1021/nn404640c

Ding, Z., Li, F., Wen, J., Wang, X., & Sun, R. (2018). Gram-scale synthesis of single-crystalline graphene quantum dots derived from lignin biomass. *Green Chemistry, 20*(6), 1383–1390. Available from: https://doi.org/10.1039/C7GC03218H&newsite=1

Ding, Z., Yuan, T., Wen, J., Cao, X., Sun, S., Xiao, L.-P., et al. (2020). Green synthesis of chemical converted graphene sheets derived from pulping black liquor. *Carbon, 158*, 690–697. Available from: https://doi.org/10.1016/j.carbon.2019.11.041

Duy, L. X., Peng, Z., Li, Y., Zhang, J., Ji, Y., & Tour, J. M. (2018). Laser-induced graphene fibers. *Carbon, 126*, 472–479. Available from: https://doi.org/10.1016/j.carbon.2017.10.036

Emmerich, F. G. (1995). Evolution with heat treatment of crystallinity in carbons. *Carbon, 33*(12), 1709–1715. Available from: https://doi.org/10.1016/0008-6223(95)00127-8

Gao, Y., Liu, Y., Zhu, G., Xu, J., Xu, H., Yuan, Q., et al. (2018). Microwave-assisted hydrothermal carbonization of dairy manure: Chemical and structural properties of the products. *Energy, 165*, 662–672. Available from: https://doi.org/10.1016/j.energy.2018.09.185

Gomez-Martin, A., Martinez-Fernandez, J., Ruttert, M., Winter, M., Placke, T., & Ramirez-Rico, J. (2019). Correlation of structure and performance of hard carbons as anodes for sodium ion batteries. *Chemistry of Materials, 31*(18), 7288–7299. Available from: https://doi.org/10.1021/acs.chemmater.9b01768

Gopalakrishnan, A., Kong, C. Y., & Badhulika, S. (2019). Scalable, large-area synthesis of heteroatom-doped few-layer graphene-like microporous carbon nanosheets from biomass for high-capacitance supercapacitors. *New Journal of Chemistry, 43*(3), 1186–1194. Available from: https://doi.org/10.1039/C8NJ05128C&newsite=1

Grancarić, A. M., Jerković, I., Koncar, V., Cochrane, C., Kelly, F. M., Soulat, D., et al. (2018). Conductive polymers for smart textile applications. *Journal of Industrial Textiles, 48*(3), 612–642. Available from: https://doi.org/10.1177/1528083717699368

Guerrero, M., Ruiz, M. P., Alzueta, M. U., Bilbao, R., & Millera, A. (2005). Pyrolysis of eucalyptus at different heating rates: Studies of char characterization and oxidative reactivity. *Journal of Analytical and Applied Pyrolysis, 74*(1–2), 307–314. Available from: https://doi.org/10.1016/j.jaap.2004.12.008

Haile, A., Gelebo, G. G., Tesfaye, T., Mengie, W., Mebrate, M. A., Abuhay, A., et al. (2021). Pulp and paper mill wastes: Utilizations and prospects for high value-added biomaterials. *Bioresources and Bioprocessing, 8*(1). Available from: https://doi.org/10.1186/s40643-021-00385-3

Han, X., Ye, R., Chyan, Y., Wang, T., Zhang, C., Shi, L., et al. (2018). Laser-induced graphene from wood impregnated with metal salts and use in electrocatalysis. *ACS Applied Nano Materials, 1*(9), 5053–5061. Available from: https://doi.org/10.1021/acsanm.8b01163

Hassan, M. M., & Carr, C. M. (2021). Biomass-derived porous carbonaceous materials and their composites as adsorbents for cationic and anionic dyes: A review. *Chemosphere, 265*, 129087. Available from: https://doi.org/10.1016/j.chemosphere.2020.129087

Heidari, M., Dutta, A., Acharya, B., & Mahmud, S. (2019). A review of the current knowledge and challenges of hydrothermal carbonization for biomass conversion. *Journal of the Energy Institute, 92*(6), 1779–1799. Available from: https://doi.org/10.1016/j.joei.2018.12.003

Hitzl, M., Corma, A., Pomares, F., & Renz, M. (2015). The hydrothermal carbonization (HTC) plant as a decentral biorefinery for wet biomass. *Catalysis Today, 257*, 154–159. Available from: https://doi.org/10.1016/j.cattod.2014.09.024

Hoang, A. T., Nižetić, S., Cheng, C. K., Luque, R., Thomas, S., Banh, T. L., et al. (2022). Heavy metal removal by biomass-derived carbon nanotubes as a greener environmental remediation: A comprehensive review. *Chemosphere, 287*(Pt 1), 131959. Available from: https://doi.org/10.1016/j.chemosphere.2021.131959

Hong, K., Qie, L., Zeng, R., Yi, Z., Zhang, W., Wang, D., et al. (2014). Biomass derived hard carbon used as a high performance anode material for sodium ion batteries. *Journal of Materials Chemistry A, 2*(32), 12733. Available from: https://doi.org/10.1039/C4TA02068E&newsite=1

Hsiao, C., Lee, C., & Tai, N. (2020). Biomass-derived three-dimensional carbon framework for a flexible fibrous supercapacitor and its application as a wearable smart textile. *Advances, 10*(12), 6960–6972. Available from: https://doi.org/10.1039/c9ra07441d

Hu, X., Wang, G., Wang, B., Liu, X., & Wang, H. (2019). Co3Sn2/SnO2 heterostructures building double shell micro-cubes wrapped in three-dimensional graphene matrix as promising anode materials for lithium-ion and sodium-ion batteries. *Chemical Engineering Journal, 355*, 986–998. Available from: https://doi.org/10.1016/j.cej.2018.07.173

Huang, S., Liu, Y., Zhao, Y., Ren, Z., & Guo, C. F. (2019). Flexible electronics: Stretchable electrodes and their future. *Advanced Functional Materials, 29*(6), 1805924. Available from: https://doi.org/10.1002/adfm.201805924

Huang, A., El-Kady, M. F., Chang, X., Anderson, M., Lin, C.-W., Turner, C. L., et al. (2021). Facile fabrication of multivalent VO x /graphene nanocomposite electrodes for high-energy-density symmetric supercapacitors. *Advanced Energy Materials, 11*(26), 2100768. Available from: https://doi.org/10.1002/aenm.202100768

Jain, A., Balasubramanian, R., & Srinivasan, M. P. (2016). Hydrothermal conversion of biomass waste to activated carbon with high porosity: A review. *Chemical Engineering Journal, 283*, 789–805. Available from: https://doi.org/10.1016/j.cej.2015.08.014

Jiang, F., Yao, Y., Natarajan, B., Yang, C., Gao, T., Xie, H., et al. (2019). Ultrahigh-temperature conversion of biomass to highly conductive graphitic carbon. *Carbon, 144*, 241–248. Available from: https://doi.org/10.1016/j.carbon.2018.12.030

Jindal, S., Anand, R., Sharma, N., Yadav, N., Mudgal, D., Mishra, R., et al. (2022). Sustainable approach for developing graphene-based materials from natural resources and biowastes for electronic applications. *ACS Applied Electronic Materials, 4*(5), 2146–2174. Available from: https://doi.org/10.1021/acsaelm.2c00097

Kambo, H. S., & Dutta, A. (2015). A comparative review of biochar and hydrochar in terms of production, physico-chemical properties and applications. *Renewable and Sustainable Energy Reviews, 45*, 359–378. Available from: https://doi.org/10.1016/j.rser.2015.01.050

Keiluweit, M., Nico, P. S., Johnson, M. G., & Kleber, M. (2010). Dynamic molecular structure of plant biomass-derived black carbon (biochar). *Environmental Science & Technology, 44*(4), 1247–1253. Available from: https://doi.org/10.1021/es9031419

Kocjan, A., Schmidt, R., Lazar, A., Prado-Gonjal, J., Kovač, J., Logar, M., et al. (2018). In situ generation of 3D graphene-like networks from cellulose nanofibres in sintered ceramics. *Nanoscale, 10*(22), 10488–10497. Available from: https://doi.org/10.1039/C8NR00717A&newsite=1

Krishnamoorthy, D., & Prakasam, A. (2020). Preparation of MoS2/graphene nanocomposite-based photoanode for dyesensitized solar cells (DSSCs). *Inorganic Chemistry Communications,118*,108016. Available from: https://doi.org/10.1016/j.inoche.2020.108016

Kumar, N., Rodriguez, J. R., Pol, V. G., & Sen, A. (2019). Facile synthesis of 2D graphene oxide sheet enveloping ultrafine 1D LiMn2O4 as interconnected framework to enhance cathodic property for Li-ion battery. *Applied Surface Science, 463*, 132–140. Available from: https://doi.org/10.1016/j.apsusc.2018.08.210

Li, W., Ma, Q., Shen, P., Zhou, Y., Soule, L., Li, Y., et al. (2021). Yolk-shell structured CuSi2P3@ Graphene nanocomposite anode for long-life and high-rate lithium-ion batteries. *Nano Energy, 80*, 105506. Available from: https://doi.org/10.1016/j.nanoen.2020.105506

Liu, T., Luo, R., Qiao, W., Yoon, S.-H., & Mochida, I. (2010). Microstructure of carbon derived from mangrove charcoal and its application in Li-ion batteries. *Electrochimica Acta, 55*(5), 1696–1700. Available from: https://doi.org/10.1016/j.electacta.2009.10.051

Liu, Q., Duan, Y., Zhao, Q., Pan, F., Zhang, B., & Zhang, J. (2014a). Direct synthesis of nitrogen-doped carbon nanosheets with high surface area and excellent oxygen reduction performance. *Langmuir: The ACS Journal of Surfaces and Colloids, 30*(27), 8238–8245. Available from: https://doi.org/10.1021/la404995y

Liu, F., Peng, H., You, C., Fu, Z., Huang, P., Song, H., et al. (2014b). High-performance doped carbon catalyst derived from nori biomass with melamine promoter. *Electrochimica Acta, 138*, 353–359. Available from: https://doi.org/10.1016/j.electacta.2014.06.098

Liu, L., Yang, X., Lv, C., Zhu, A., Zhu, X., Guo, S., et al. (2016). Seaweed-derived route to Fe2O3 hollow nanoparticles/N-doped graphene aerogels with high lithium ion storage performance. *ACS Applied Materials & Interfaces, 8*(11), 7047–7053. Available from: https://doi.org/10.1021/acsami.5b12427

Liu, B., Yang, M., Chen, H., Liu, Y., Yang, D., & Li, H. (2018). Graphene-like porous carbon nanosheets derived from salvia splendens for high-rate performance supercapacitors. *Journal of Power Sources, 397*, 1–10. Available from: https://doi.org/10.1016/j.jpowsour.2018.06.100.

Liu, Y., Sun, K., Cui, X., Li, B., & Jiang, J. (2020). Defect-rich, graphenelike carbon sheets derived from biomass as efficient electrocatalysts for rechargeable zinc–air batteries. *ACS Sustainable Chemistry & Engineering, 8*(7), 2981–2989. Available from: https://doi.org/10.1021/acssuschemeng.9b07621

Liu, H., Cheng, C., & Wu, H. (2021). Sustainable utilization of wetland biomass for activated carbon production: A review on recent advances in modification and activation methods. *The Science of the Total Environment, 790*, 148214. Available from: https://doi.org/10.1016/j.scitotenv.2021.148214

Long, C., Chen, X., Jiang, L., Zhi, L., & Fan, Z. (2015). Porous layer-stacking carbon derived from in-built template in biomass for high volumetric performance supercapacitors. *Nano Energy, 12*, 141–151. Available from: https://doi.org/10.1016/j.nanoen.2014.12.014

Luong, D. X., Bets, K. V., Algozeeb, W. A., Stanford, M. G., Kittrell, C., Chen, W., et al. (2020). Gram-scale bottom-up flash graphene synthesis. *Nature, 577*(7792), 647–651. Available from: https://doi.org/10.1038/s41586-020-1938-0

Lyu, L., Seong, K., Ko, D., Choi, J., Lee, C., Hwang, T., et al. (2019). Recent development of biomass-derived carbons and composites as electrode materials for supercapacitors. *Materials Chemistry Frontiers, 3*(12), 2543–2570. Available from: https://doi.org/10.1039/C9QM00348G&newsite=1

Mahmood, F., Zhang, C., Xie, Y., Stalla, D., Lin, J., & Wan, C. (2019). Transforming lignin into porous graphene via direct laser writing for solid-state supercapacitors. *RSC Advances, 9*(39), 22713–22720. Available from: https://doi.org/10.1039/C9RA04073K&newsite=1

Mao, W., Ding, Y., Li, M., Ma, C., Gong, H., Pan, J., et al. (2021). Synthesis and electrochemical characterization of 2D SnS2/RGO as anode material in sodium-ion batteries. *Journal of Alloys and Compounds, 855*, 157209. Available from: https://doi.org/10.1016/j.jallcom.2020.157209

Mohamad Nor, N., Lau, L. C., Lee, K. T., & Mohamed, A. R. (2013). Synthesis of activated carbon from lignocellulosic biomass and its applications in air pollution control—A review. *Journal of Environmental Chemical Engineering, 1*(4), 658–666. Available from: https://doi.org/10.1016/j.jece.2013.09.017

Mondal, D., Sharma, M., Wang, C.-H., Lin, Y.-C., Huang, H.-C., Saha, A., et al. (2016). Deep eutectic solvent promoted one step sustainable conversion of fresh seaweed biomass to functionalized graphene as a potential electrocatalyst. *Green Chemistry, 18*(9), 2819–2826. Available from: https://doi.org/10.1039/C5GC03106K&newsite=1

Mun, S. P., Cai, Z., & Zhang, J. (2013). Fe-catalyzed thermal conversion of sodium lignosulfonate to graphene. *Materials Letters, 100*, 180–183. Available from: https://doi.org/10.1016/j.matlet.2013.02.101

Nicolae, S. A., Au, H., Modugno, P., Luo, H., Szego, A. E., Qiao, M., et al. (2020). Recent advances in hydrothermal carbonisation: From tailored carbon materials and biochemicals to applications and bioenergy. *Green Chemistry, 22*(15), 4747–4800. Available from: https://doi.org/10.1039/D0GC00998A&newsite=1

Ogale, A. A., Zhang, M., & Jin, J. (2016). Recent advances in carbon fibers derived from bio-based precursors. *Journal of Applied Polymer Science, 133*(45). Available from: https://doi.org/10.1002/app.43794

Omoriyekomwan, J. E., Tahmasebi, A., Dou, J., Wang, R., & Yu, J. (2021). A review on the recent advances in the production of carbon nanotubes and carbon nanofibers via microwave-assisted pyrolysis of biomass. *Fuel Processing Technology, 214*, 106686. Available from: https://doi.org/10.1016/j.fuproc.2020.106686

Paris, O., Zollfrank, C., & Zickler, G. A. (2005). Decomposition and carbonisation of wood biopolymers—A microstructural study of softwood pyrolysis. *Carbon, 43*(1), 53–66. Available from: https://doi.org/10.1016/j.carbon.2004.08.034

Parshetti, G. K., Chowdhury, S., & Balasubramanian, R. (2014). Plant derived porous graphene nanosheets for efficient CO 2 capture. *RSC Advances, 4*(84), 44634–44643. Available from: https://doi.org/10.1039/C4RA05522E&newsite=1

Qiang, R., Feng, S., Chen, Y., Ma, Q., & Chen, B. (2022). Recent progress in biomass-derived carbonaceous composites for enhanced microwave absorption. *Journal of Colloid and Interface Science, 606*(Pt 1), 406–423. Available from: https://doi.org/10.1016/j.jcis.2021.07.144

Qin, C., Wang, H., Yuan, X., Xiong, T., Zhang, J., & Zhang, J. (2020). Understanding structure-performance correlation of biochar materials in environmental remediation and electrochemical devices. *Chemical Engineering Journal, 382*, 122977. Available from: https://doi.org/10.1016/j.cej.2019.122977

Ragauskas, A. J., Beckham, G. T., Biddy, M. J., Chandra, R., Chen, F., Davis, M. F., et al. (2014). Lignin valorization: Improving lignin processing in the biorefinery. *Science (New York, N.Y.), 344*(6185), 1246843. Available from: https://doi.org/10.1126/science.1246843

Ravi, S., & Vadukumpully, S. (2016). Sustainable carbon nanomaterials: Recent advances and its applications in energy and environmental remediation. *Journal of Environmental Chemical Engineering, 4*(1), 835–856. Available from: https://doi.org/10.1016/j.jece.2015.11.026.

Reißmann, D., Thrän, D., & Bezama, A. (2018). Hydrothermal processes as treatment paths for biogenic residues in Germany: A review of the technology, sustainability and legal aspects. *Journal of Cleaner Production, 172*, 239–252. Available from: https://doi.org/10.1016/j.jclepro.2017.10.151

Seo, D. H., Pineda, S., Yick, S., Bell, J., Han, Z. J., & Ostrikov, K. (2015). Plasma-enabled sustainable elemental lifecycles: Honeycomb-derived graphenes for next-generation biosensors and supercapacitors. *Green Chemistry, 17*(4), 2164–2171. Available from: https://doi.org/10.1039/C4GC02135E&newsite=1

Shah, J., Lopez-Mercado, J., Carreon, M. G., Lopez-Miranda, A., & Carreon, M. L. (2018). Plasma synthesis of graphene from mango peel. *ACS Omega, 3*(1), 455–463. Available from: https://doi.org/10.1021/acsomega.7b01825

Sharma, H. B., Sarmah, A. K., & Dubey, B. (2020). Hydrothermal carbonization of renewable waste biomass for solid biofuel production: A discussion on process mechanism, the influence of process parameters, environmental performance and fuel properties of hydrochar. *Renewable and Sustainable Energy Reviews, 123*, 109761. Available from: https://doi.org/10.1016/j.rser.2020.109761

Shim, H.-S., Hurt, R. H., & Yang, N. Y. (2000). A methodology for analysis of 002 lattice fringe images and its application to combustion-derived carbons. *Carbon, 38*(1), 29–45. Available from: https://doi.org/10.1016/S0008-6223(99)00096-2

Siddiqui, M., Ahmed Baloch, H., Nizamuddin, S., Haris, M., Mubarak, N., Czajka, M., et al. (2020). Synthesis of novel magnetic carbon nano-composite from waste biomass: A comparative study of industrially adoptable hydro/solvothermal co-precipitation route. *Journal of Environmental Chemical Engineering, 8*(2), 103519. Available from: https://doi.org/10.1016/j.jece.2019.103519

Silambarasan, K., Harish, S., Hara, K., Archana, J., & Navaneethan, M. (2021). Ultrathin layered MoS2 and N-doped graphene quantum dots (N-GQDs) anchored reduced graphene oxide (rGO) nanocomposite-based counter electrode for dye-sensitized solar cells. *Carbon, 181*, 107–117. Available from: https://doi.org/10.1016/j.carbon.2021.01.162

Song, B., Lin, R., Lam, C. H., Wu, H., Tsui, T.-H., & Yu, Y. (2021). Recent advances and challenges of inter-disciplinary biomass valorization by integrating hydrothermal and biological techniques. *Renewable and Sustainable Energy Reviews, 135*, 110370. Available from: https://doi.org/10.1016/j.rser.2020.110370

Sun, L., Tian, C., Li, M., Meng, X., Wang, L., Wang, R., et al. (2013). From coconut shell to porous graphene-like nanosheets for high-power supercapacitors. *Journal of Materials Chemistry A, 1*(21), 6462. Available from: https://doi.org/10.1039/C3TA10897J&newsite=1

Sun, J., Wu, Z., Ma, C., Xu, M., Luo, S., Li, W., et al. (2021). Biomass-derived tubular carbon materials: Progress in synthesis and applications. *Journal of Materials Chemistry A, 9*(24), 13822–13850. Available from: https://doi.org/10.1039/D1TA02412D&newsite=1

Sun, Q., Wu, Z., Qin, Z., Chen, X., Zhang, C., Cao, B., et al. (2022). A dog nose-inspired high-performance NH 3 gas sensor of biomass carbon materials with a pleated structure derived from rose tea. *Journal of Materials Chemistry A, 10*(27), 14326–14335. Available from: https://doi.org/10.1039/d2ta02670h

Thalji, M. R., Ali, G. A., Liu, P., Zhong, Y. L., & Chong, K. F. (2021). W18O49 nanowires-graphene nanocomposite for asymmetric supercapacitors employing AlCl3 aqueous electrolyte. *Chemical Engineering Journal, 409*, 128216. Available from: https://doi.org/10.1016/j.cej.2020.128216

Titirici, M.-M., White, R. J., Falco, C., & Sevilla, M. (2012). Black perspectives for a green future: Hydrothermal carbons for environment protection and energy storage. *Energy & Environmental Science, 5*(5), 6796. Available from: https://doi.org/10.1039/C2EE21166A&newsite=1

Trung, T. Q., & Lee, N.-E. (2016). Flexible and stretchable physical sensor integrated platforms for wearable human-activity monitoringand personal healthcare. *Advanced Materials (Deerfield Beach, Fla.), 28*(22), 4338–4372. Available from: https://doi.org/10.1002/adma.201504244

Vassilev, S. V., Baxter, D., & Vassileva, C. G. (2013). An overview of the behaviour of biomass during combustion: Part I. Phase-mineral transformations of organic and inorganic matter. *Fuel, 112*, 391–449. Available from: https://doi.org/10.1016/j.fuel.2013.05.043

Vivekanandhan, S., Schreiber, M., Muthuramkumar, S., Misra, M., & Mohanty, A. K. (2017). Carbon nanotubes from renewable feedstocks: A move toward sustainable nanofabrication. *Journal of Applied Polymer Science, 134*(4). Available from: https://doi.org/10.1002/app.44255

Wang, Z., Ogata, H., Morimoto, S., Ortiz-Medina, J., Fujishige, M., Takeuchi, K., et al. (2015). Nanocarbons from rice husk by microwave plasma irradiation: From graphene and carbon nanotubes to graphenated carbon nanotube hybrids. *Carbon, 94*, 479–484. Available from: https://doi.org/10.1016/j.carbon.2015.07.037

Wang, Z., Yu, J., Zhang, X., Li, N., Liu, B., Li, Y., et al. (2016). Large-scale and controllable synthesis of graphene quantum dots from rice husk biomass: A comprehensive utilization strategy. *ACS Applied Materials & Interfaces, 8*(2), 1434–1439. Available from: https://doi.org/10.1021/acsami.5b10660

Wang, S., Dai, G., Yang, H., & Luo, Z. (2017a). Lignocellulosic biomass pyrolysis mechanism: A state-of-the-art review. *Progress in Energy and Combustion Science, 62*, 33–86. Available from: https://doi.org/10.1016/j.pecs.2017.05.004

Wang, J., Nie, P., Ding, B., Dong, S., Hao, X., Dou, H., et al. (2017b). Biomass derived carbon for energy storage devices. *Journal of Materials Chemistry A, 5*(6), 2411–2428. Available from: https://doi.org/10.1039/C6TA08742F&newsite=1

Wang, Z., Shen, D., Wu, C., & Gu, S. (2018). State-of-the-art on the production and application of carbon nanomaterials from biomass. *Green Chemistry, 20*(22), 5031–5057. Available from: https://doi.org/10.1039/C8GC01748D&newsite=1

Wang, C., Xia, K., Zhang, Y., & Kaplan, D. L. (2019). Silk-based advanced materials for soft electronics. *Accounts of Chemical Research, 52*(10), 2916–2927. Available from: https://doi.org/10.1021/acs.accounts.9b00333

Wang, Y., Zhang, M., Shen, X., Wang, H., Wang, H., Xia, K., et al. (2021). Biomass-derived carbon materials: Controllable preparation and versatile applications. *Small (Weinheim an Der Bergstrasse, Germany), 17*(40), e2008079. Available from: https://doi.org/10.1002/smll.202008079

Wang, H., Zhao, Z., Liu, P., & Guo, X. (2022). Laser-induced graphene based flexible electronic devices. *Biosensors, 12*(2). Available from: https://doi.org/10.3390/bios12020055

Wen, Y., Kok, M. D., Tafoya, J. P. V., Sobrido, A. B. J., Bell, E., Gostick, J. T., et al. (2021). Electrospinning as a route to advanced carbon fibre materials for selected low-temperature electrochemical devices: A review. *Journal of Energy Chemistry, 59*, 492–529. Available from: https://doi.org/10.1016/j.jechem.2020.11.014

Xing, H., Zhang, F., Lu, Y., Zhai, B., Zhai, S., An, Q., et al. (2016). Facile synthesis of carbon nanoparticles/graphene composites derived from biomass resources and their application in lithium ion batteries. *RSC Advances, 6*(83), 79366–79371. Available from: https://doi.org/10.1039/C6RA15690H&newsite=1

Xu, M., Xing, L., Zhang, Q., & Pu, J. (2017). Ultrasonic-assisted method of graphite preparation from wheat straw. *BioResources, 12*(3). Available from: https://doi.org/10.15376/biores.12.3.6405-6417

Ye, R., Chyan, Y., Zhang, J., Li, Y., Han, X., Kittrell, C., et al. (2017). Laser-induced graphene formation on wood. *Advanced Materials (Deerfield Beach, Fla.)*, 29(37). Available from: https://doi.org/10.1002/adma.201702211

Yi, X., He, W., Zhang, X., Yue, Y., Yang, G., Wang, Z., et al. (2017). Graphene-like carbon sheet/Fe3O4 nanocomposites derived from soda papermaking black liquor for high performance lithium ion batteries. *Electrochimica Acta, 232*, 550–560. Available from: https://doi.org/10.1016/j.electacta.2017.02.130

Yoo, S., Kelley, S. S., Tilotta, D. C., & Park, S. (2018). Structural characterization of loblolly pine derived biochar by X-ray diffraction and electron energy loss spectroscopy. *ACS Sustainable Chemistry & Engineering, 6*(2), 2621–2629. Available from: https://doi.org/10.1021/acssuschemeng.7b04119

Yoon, D., Hwang, J., Chang, W., & Kim, J. (2018). Carbon with expanded and well-developed graphene planes derived directly from condensed lignin as a high-performance anode for sodium-ion batteries. *ACS Applied Materials & Interfaces, 10*(1), 569–581. Available from: https://doi.org/10.1021/acsami.7b14776

Zeng, X., Leng, L., Liu, F., Wang, G., Dong, Y., Du, L., et al. (2016). Enhanced Li-O2 battery performance, using graphene-like nori-derived carbon as the cathode and adding LiI in the electrolyte as a promoter. *Electrochimica Acta, 200*, 231–238. Available from: https://doi.org/10.1016/j.electacta.2016.03.161

Zhang, X., Yan, Q., Hassan, E. B., Li, J., Cai, Z., & Zhang, J. (2017). Temperature effects on formation of carbon-based nanomaterials from Kraft lignin. *Materials Letters, 203*, 42–45. Available from: https://doi.org/10.1016/j.matlet.2017.05.125

Zhang, Y., He, J., Gao, Z., & Li, X. (2019a). Converting eggs to flexible, all-solid supercapacitors. *Nano Energy, 65*, 104045. Available from: https://doi.org/10.1016/j.nanoen.2019.104045

Zhang, F., Miao, J., Liu, W., Xu, D., & Li, X. (2019b). Heteroatom embedded graphene-like structure anchored on porous biochar as efficient metal-free catalyst for ORR. *International Journal of Hydrogen Energy, 44*(59), 30986–30998. Available from: https://doi.org/10.1016/j.ijhydene.2019.09.239

Zhang, M., Song, Z., Liu, H., & Ma, T. (2020). Biomass-derived highly porous nitrogen-doped graphene orderly supported NiMn2O4 nanocrystals as efficient electrode materials for asymmetric supercapacitors. *Applied Surface Science, 507*, 145065. Available from: https://doi.org/10.1016/j.apsusc.2019.145065

Zhang, Y. S., Zhu, H. L., Yao, D., Williams, P. T., Wu, C., Xu, D., et al. (2021). Thermochemical conversion of carbonaceous wastes for CNT and hydrogen production: A review. *Sustainable Energy & Fuels, 5*(17), 4173–4208. Available from: https://doi.org/10.1039/D1SE00619C&newsite=1

Zhao, L., Xiao, P.-W., Chen, Q., Fu, C., & Han, B.-H. (2020). Polycarbazole and biomass-derived flexible nitrogen-doped porous carbon materials for gas adsorption and sensing. *Journal of Materials Chemistry A, 8*(14), 6804–6811. Available from: https://doi.org/10.1039/c9ta13910a

Zhou, T., Ma, R., Zhang, T., Li, Z., Yang, M., Liu, Q., et al. (2019). Increased activity of nitrogen-doped graphene-like carbon sheets modified by iron doping for oxygen reduction. *Journal of Colloid and Interface Science, 536*, 42–52. Available from: https://doi.org/10.1016/j.jcis.2018.10.021

Green Fabrication of Magnetic Nanomaterial and Their Application

Hadia Shoaib, Ahmed Raza Sidhu, Umaima Ismail, Saba Naz, and Hamide Filiz Ayyildiz

1 Introduction

Nanotechnology is an evolving and multidisciplinary field focused on researching, designing, manipulating, and creating ultrafine nanomaterials (NMs) with dimensions typically ranging from 1 to 100 nanometers. This field has a broad range of applications aimed at enhancing human well-being. The term "nano" originates from the Greek term meaning "dwarf" or "small" (Singh et al., 2021). Nanoparticles (NPs) have garnered significant interest over the last few decades due to their exceptional chemical and physical characteristics, which encompass electrical, catalytic, optical, and biomedical properties. These unique attributes make nanoparticles highly versatile and valuable in a wide range of applications, sparking intensive research and exploration across various scientific disciplines (Bhardwaj et al., 2018). In the present era, nanoparticles (NPs) are regarded as promising solutions for addressing a multitude of technological and environmental challenges across various domains. They hold significant potential in fields such as biosensing, medicine, clinical diagnostics, harnessing solar energy, and water treatment, all aimed at benefiting humanity (Bhardwaj & Naraian, 2021). NPs offer innovative approaches

H. Shoaib (✉)
National Centre of Excellence in Analytical Chemistry, University of Sindh, Jamshoro, Pakistan

A. R. Sidhu · S. Naz
Dr. M. A. Kazi Institute of Chemistry, University of Sindh, Jamshoro, Pakistan

U. Ismail
Faculty of Pharmacy, Department of Pharmacognosy, University of Sindh, Jamshoro, Pakistan

H. F. Ayyildiz
Faculty of Pharmacy, Department of Basic Pharmaceutical Sciences, Selcuk University, Konya, Türkiye

A. K. Bhardwaj et al. (eds.), *Biogenic Wastes-Enabled Nanomaterial Synthesis*,
https://doi.org/10.1007/978-3-031-59083-2_4

and materials that can revolutionize these sectors, making them more efficient, sustainable, and environmentally friendly, thereby contributing to advancements with wide-reaching positive impacts on society. The wide range of potential applications for nanoparticles (NPs) is largely achievable, thanks to their unique attributes, particularly their high surface area-to-volume ratio and nanoscale dimensions. NPs can assume various nanostructures, encompassing different dimensions, including zero-dimensional (0D), one-dimensional (1D), two-dimensional (2D), and three-dimensional (3D) arrays. Importantly, these diverse nanostructures can be precisely engineered and fabricated in a controlled manner (Singh et al., 2021).

"Top-down" and "bottom-up" are the two methods for the synthesis of NPs as represented in Fig. 1.

The top-down approach involves the fragmentation of bulk materials into particles with nano-scale dimensions by using a physical method such as laser ablation, high-energy ball milling, or wire explosion method. (Gul et al., 2019). Ball milling is considered a convenient process to produce MNPs made from bulk material. It entails the crushing of coarse-textured particles mechanically by using steel balls to

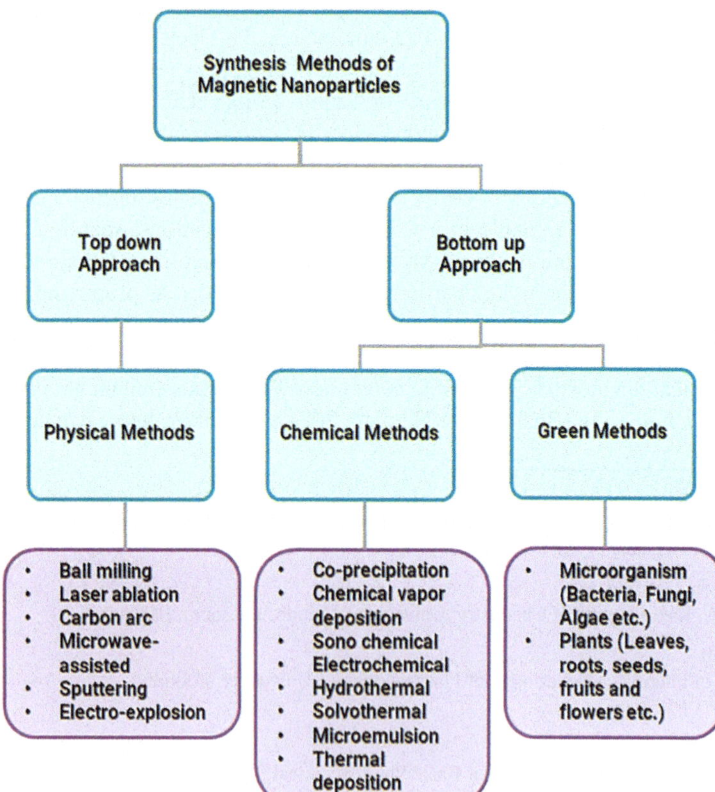

Fig. 1 Synthesis methods of magnetic nanoparticles

form fine-textured particles. The primary disadvantage of this method is that there is a probability of contamination of the product and wide particle size distribution in comparison to synthesis by chemical methods (Mohamed & Mohamed, 2019). Laser ablation also known as laser evaporation is a simple technique that includes the expenditure of high-energy laser production of MNPs (Biehl et al., 2018). This technique does not produce any dangerous waste nor require any expensive chemicals, that is why it is considered an effective one but it consumes a high amount of energy. However, in the bottom-up approach, chemical and biological processes use the self-processing of atoms to produce new nuclei, which develop particles of a nanoscale (Yousaf et al., 2020).

The synthesis of nanoparticles (NPs) within a short timeframe can be achieved using chemical methods such as sol-gel, hydrothermal, thermal decomposition, and microemulsion, as highlighted (Ali et al., 2021). These chemical methods offer convenience in producing NPs with a wide range of compositions. However, this approach is favored primarily for its simplicity. It is important to note that while it allows for efficient NP production, there can be occasional challenges in precisely controlling the shape of the NPs (Flores-Rojas et al., 2022).

Above all benefits, the main disadvantage of this method is the use of hazardous chemicals that ultimately produce unsustainable byproducts that are destructive to the environment and living creatures, which can never be part of the eco-friendly policy. Therefore, the research has been focused on a broad consideration of biogenic or biological techniques using microorganisms and various plant components for the green synthesis of NPs in light of the aforementioned challenges (Bhardwaj et al., 2018; Katas et al., 2019; Sanjivkumar et al., 2019; Tomer et al., 2019). These synthesis techniques are straightforward, biocompatible, cost-effective, and environmentally friendly (Naraian & Abhishek, 2020).

Earlier reports showed that different types of NPs such as organic and inorganic and semiconductor NPs can be synthesized by green methods due to their diverse characteristics, flexibility, and various applications. Emerging attention has been observed for noble metal NPs such as Ag and Au, as they are equipped with superior qualities with practical flexibility (Vadlapudi & Kaladhar, 2014). However, MNPs possess unique and distinguished magnetic properties, i.e., superparamagnetic, ferrimagnetic, and ferromagnetic properties. For this reason, MNPs have been employed in various fields such as biomedicine, agriculture, catalysis, engineering, and environmental applications. Ferromagnetic particles possess important particle size (<15 nm) and contain a single magnetic domain, i.e., a particle that possesses a homogenous magnetic condition in any field was first anticipated by Frenkel and Dorfman (Akbarzadeh et al., 2012).

The magnetization behavior of these particles is comparable to that of atomic paramagnets (superparamagnetic), except for their high susceptibilities (Thorat et al., 2017; Zhang et al., 2019). Various metal elements and their oxides that possess magnetic properties are used to create the MNPs (Kefeni et al., 2017). Nowadays, iron oxide magnetic nanoparticles (IOMNPs) are gaining significant interest for their application in various aspects (Bansal et al., 2017). By modifying the physicochemical characteristics of superparamagnetic IOMNPs, it can be applied for numerous different purposes, which include cell separation, drug delivery, magnetic resonance imaging, immunoassays, and hyperthermia (Jiang et al., 2023; Palzer et al., 2021; Rezaei et al., 2019). MNPs have a large surface area, metal-rich compounds, and adjustable architectures, which makes this field so intriguing due to its many potential uses. Additionally, form and size have become distinct research areas in recent decades due to their tuning properties (Purbia & Paria, 2015). MNPs now become a subject of importance and a developing area of nanoscience and nanobiotechnology as a result of numerous publications in recent years (Ali et al., 2021).

Therefore, this chapter is entirely devoted to the information about the synthesis of MNPs through green methods and discusses their diverse applications in numerous fields, i.e., biosensing, biomedicine, environment, agriculture, and catalysis.

2 Green Synthesis

In recent days, the green synthesis of NPs has gained a lot of interest in comparison to conventional methods. Green synthesis is gradually replacing physical and chemical processes because of the disadvantages such as the use of energy at a large scale, the manufacture of hazardous or poisonous substances, the implementation of sophisticated machinery, and conditions of synthesis (Ahmed & Ikram, 2015). Researchers have assessed the precise green methods, or the natural sources and their products, to address these issues. In comparison to chemical and physical approaches, green synthesis has numerous benefits, including the fact that it is not harmful, clean, without pollution, ecologically safe, inexpensive, feasible, and biologically sustainable (Alsammarraie et al., 2018; Devi et al., 2019; Kataria & Garg, 2018).

MNPs have emerged as a fascinating category among all metallic NPs because of their exceptional characteristics such as high coercivity and superparamagnetic nature. The synthesis of metal nanoparticles (MNPs) using conventional methods has several limitations, particularly in biomedical applications. One major issue is the use of nonpolar organic solvents, which are not suitable for biomedical use. Additionally, nanoscale zero-valent iron nanoparticles produced by physical or chemical methods are highly reactive and tend to aggregate. This aggregation reduces their reactivity, making them less effective. To address these problems, researchers are developing alternative synthesis techniques and stabilizing agents to prevent aggregation and maintain the reactivity of MNPs in aqueous or biologically compatible environments (Lu et al., 2017; Mohammed et al., 2017). Additionally, IOMNPs obtained by such processes do not possess magnetism and dispersibility upon exposure to air (Jiang et al., 2023).

In recent times, biological materials such as plants (roots, seeds, leaves, fruits, and flowers) and microbes (bacteria, fungi, and algae) are used for the green synthesis of MNPs. Figure 2 shows a schematic representation of the mechanism of

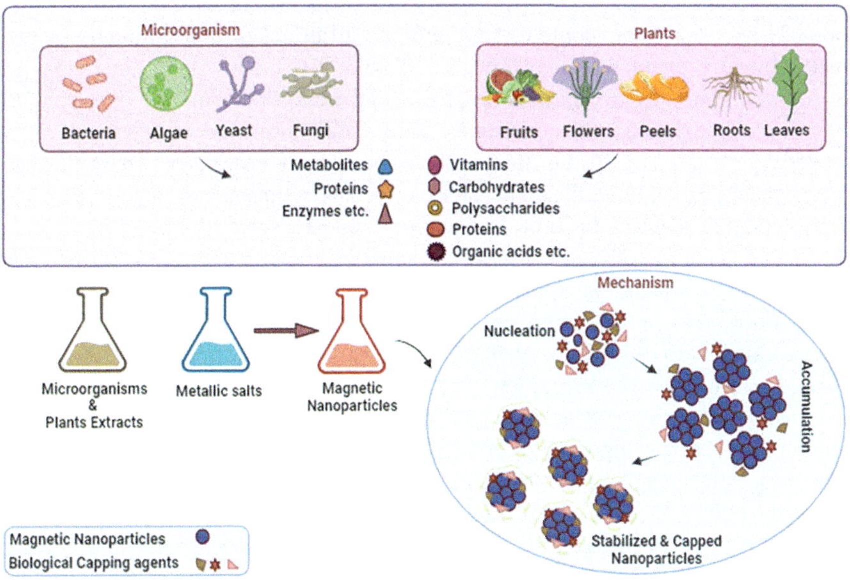

Fig. 2 Mechanism for the synthesis of magnetic nanoparticles using green methods. (Created with BioRender.com)

biologically synthesized MNPs. Further sections contain a description of the fabrication of MNPs using a green method.

3 Synthesis of Magnetic Nanoparticles by Microorganism

In recent years, microorganism-based NP production has gained attention over traditional chemical synthesis as it offers many advantages that include their ability to survive in harsh environment, less chemical consumption, and ambient temperature synthesis, which is an energy-efficient method (Park et al., 2016). NPs can be created by microorganisms such as fungi, bacteria, algae, and yeast through mechanisms outside and inside the cells. In these methods, metal ions are reduced enzymatically, producing NPs that are uniformly distributed with a narrow size range. Natural capping agents employed in these processes are proteins, peptides, and genes that ensure stability and prevent NP aggregation (Singh et al., 2016). Contrary to the extracellular processes i-e, electrostatic reduction of metal ions enzymatically linked to the microorganism's cell wall, the intracellular mechanism involves metal ion diffusion, where interaction with enzymes takes place to generate NPs. An essential step in the process of synthesis is the reduction of metal ions which is influenced by several variables, including the type of strain, environmental circumstances such as pH, the temperature of synthesis, media for growth, the concentration of salt, and cell wall functional groups (necessary in biomineralization). NP's size, shape, content, and yield are all affected, either directly or indirectly, by such characteristics (Iravani & Varma, 2020; Saratale et al., 2018). The most important ones are temperature, pH, and the length of time the microbial culture is incubated. Since this environmentally friendly method of synthesis is controlled by an enzyme system that reduces the ions of metals (NADH reductase), biosynthesis depends on the right environmental circumstances. The enzymes are fully rendered inactive by high temperature and pH, which halts their work and inhibits the creation of NPs (Jubran et al., 2020). In the process of synthesis, along with enzymes, organic molecules such as proteins, peptides, cofactors, and organic compounds serve as stabilizing, reducing, and capping agents. A series of studies on the production of NPs show that extracellular polysaccharides and cytoplasmic extracts are used as reduction activator molecules (Singh et al., 2016). In general, two distinct pathways: Intracellular and Extracellular have been explored in bacteria for the production of nanoparticles NPs (Gahlawat & Choudhury, 2019). In microbial cells, ion transportation serves as the foundation for intracellular production. Since the cell wall enzymes are engaged in the process of reduction of metals, the cell wall is considered an important element in this approach, whereas extracellular processes rely on the interaction of metal ions electrostatically and negatively charged enzymes on the cell wall and produce soluble enzymes by microorganisms to assemble the ions of metal on the microbial cell's surface (Mughal et al., 2021; Punjabi et al., 2018; Singh et al., 2018). Since the former synthesis does not require downstream processing, sonication, centrifugation, or washing processes to extract the product, it is more efficient and successful than intracellular approaches and makes this process economical (Singh et al., 2016). Due to its great efficiency and simple extraction technique, the extracellular approach has many industrial uses.

Fungi can bioaccumulate metals and have more tolerance, high binding capacity, and intracellular absorption properties similar to bacteria, which are employed in the synthesis of metallic NPs. Moreover, it offers improved polydispersity, stable architectures, and variation in NP dimensions. In addition to providing a lot of surface area for interaction, fungi's cell walls also aid in metal ion reduction and absorption to create metal NPs (Khandel & Shahi, 2018).

Organisms that synthesize their food by the process of photosynthesis and do not possess roots and leaves are referred to as algae. Both unicellular or multicellular organisms (microalgae or macroalgae) are both widely employed in the field of nanotechnology to produce various kinds of metallic NPs, i.e., silver, gold, iron, copper, and palladium (Agarwal et al., 2017). Algal bodies contain a significant proportion of biomolecules such as carbohydrates, proteins, peptides, terpenes, alkaloids, fats, glycoproteins, cell wall polysaccharides, and enzymes that are essential in the process of fabrication, capping, reduction, and stabilization of NPs (Siddiqi & Husen, 2016).

Abdeen et al. (2016) have reported that *Aspergillus niger* was extracted and cultivated for 7 days to produce MNPs (Abdeen et al., 2016). The literature states that precursor salts ($FeSO_4$ and $FeCl_3$) were combined with 250 mg/mL homogenized mycelia, cultured for 6 days, subjected to centrifugation, and dried to produce the pellets of FeS and Fe_2O_3 NPs.

Furthermore, *Alternaria alternata*, a phytopathogenic fungus, has been employed in the extracellular synthesis of IOMNPs (Sarkar et al., 2017). In this method, IOMNPs were created by using the cultivated fungus filtrate as a reducing agent.

Mukherjee (2017) reported that cultivated strains of the *Microbacterium marinilacus* were used to produce IOMNPs. In this method, the isolated bacterium was mixed with 1 mM ferrie chloride solution, and after incubation, the culture's color changed to dark brown indicating the creation of NPs (Mukherjee, 2017).

Fe_3O_4 NPs are produced using the brown seaweed *Sargassum muticum* extract, an algal species. Sulfated polysaccharides were the main component that helped the formation of NPs (Mahdavi et al., 2013).

Salvadori et al. (2016) used the yeast *Rhodotorula mucilaginosa* to create spherical nickel nanoparticles (NiNPs) of particle size of 5.5 nm. It is a synergistic strategy that uses green technology to cut costs and minimize environmental effects (Salvadori et al., 2016). Table 1 shows the lists of microorganisms that are utilized to create MNPs.

4 Green Synthesis of Magnetic Nanoparticles by Plants

In comparison to alternative physical or chemical processes, using natural plants in the production of MNPs is a cost-effective, environmentally friendly, manageable, and convenient procedure (Abinaya et al., 2021). The biosynthesis process involves utilizing various parts of plants, such as petals, flowers, peels, seeds, leaves, fruits, roots, whole plants, and husks, which are rich in biomolecular substances such as protein, flavonoids, amino acids, carbohydrates, terpenoids, and saponins. During the creation of NPs, these biomolecules serve as redox mediators, reducers, stabilizers, and capping agents (Mirza et al., 2018; Mondal et al., 2020; Vasantharaj et al., 2019). In recent times, numerous studies are currently underway to create IOMNPs with plants; this is primarily because bacteria and fungi require extended incubation periods, whereas the phytochemical in plants can reduce metal ions in a shorter amount of time (Singh et al., 2018).

To prepare plant extracts from various plant parts, including leaves, fruit pulp, bark, flowers, roots, and fruit peels, distilled water or ethanol is typically employed as a solvent. The process of preparing the extracts begins with washing the selected plant parts (leaves, roots, flowers, bark, fruit pulp, and peels), followed by drying and cutting or chopping them into powder form. The next stage is to boil the plant components that have been reduced to powder form at a certain temperature with a predetermined amount of the desired solvents (ethanol, methanol, or deionized water). Following a predetermined amount of boiling time, filtration is necessary to separate the phytochemicals such as flavonoids, alkaloids, phenolic acids, sugars, and their derivative compounds with hydroxyl, amino, steroids, allyl, carboxylic, and alkoxy groups available in particular plant parts in the desired solvents.

Table 1 Microorganisms used for the synthesis of magnetic nanoparticles

Microorganism	Types of MNPs	Precursor salt	Size (nm)	Shape	Application	References
Bacterial species						
Bacillus cereus	Fe	$FeCl_3.6H_2O$	18.8–28.3	Spherical	Anticancer activity	Fatemi et al. (2018)
Proteus mirabilis	Fe	$Fe(NO_3)_3.9H_2$	1.44–1.92	Spherical	Wastewater treatment	Zaki et al. (2019)
Pseudomonas stutzeri	Fe	Ferric quinate	10–20	Spherical	Cellulase immobilization and photocatalytic activities	Desai and Pawar (2020)
Proteus vulgaris	Fe	$FeCl_3$	19.23–30.51	Spherical	Antibacterial, antioxidant, and cytocompatibility activities	Majeed et al. (2021)
Fungal species						
Aspergillus niger	Fe	$FeCl_3$. $6H_2O$ and $FeSO_4.7H_2O$	20–40	Orthorhombic	Removal of chromium	Chatterjee et al. (2020)
Aspergillus brasiliensis	Co	$CoSO_4.7H_2O$	20–27	Quasi-spherical	Antimicrobial activity	Omran et al. (2020)
Penicillium oxalicum	Fe	$FeSO_4$	140	Spherical	Decolorization of methylene blue	Mathur et al. (2021)
Algal species						
a. *Colpomenia sinuosa* (brown seaweed) b. *Pterocladia capillacea* (red seaweed)	Fe	$FeCl_3{\cdot}6H_2O$	a. 11.2–33.7 b. 16.8–22.4	Nanospheres	Antimicrobial activity	Salem et al. (2019)

a. *Sargassum vulgare* b. *Ulva fasciata* c. *Jania rubens*	Fe	$FeCl_3 \cdot 6H_2O$	a.17.05–34.09 b. 22.73–39.77 c. 22.22–33.33	Nanospheres	Antibiofilm activity	Salem et al. (2020)
Rhodophyta (marine red algae)	Ni	$NiCl_2 \cdot 6H_2O$	32.64	Spherical	Catalytic activity	Moavi et al. (2021)
Rhodophyta (red algae)	Co	$Co(NO_3)_2 \cdot 6H_2O$	29.8 ± 8.6	Spherical	Antioxidant, antimicrobial, anticancer, and anticoagulant activities	Ajarem et al. (2021)
Ulva lactuca (green algae)	Fe	Ferric sulphate	20–40	Spherical	Bactericidal activity	Bensy et al. (2022)
Grateloupia sparsa (red algae)	Co	$Co(NO_3)_2 \cdot 6H_2O$	23.2–11.8	Spherical	Antioxidant, antibacterial, anticancer, and hemolytic activities	Hajri et al. (2022)

The resulting plant extract was then mixed with a specific volume and concentration of metal salt precursors. The phytochemicals found in the extract can serve as agents that cap, reduce, and stabilize, aiding in the creation of MNPs from metal ions in the precursor solution (Singh, 2022).

To obtain a tiny and small MNP with the desired qualities, researchers optimized specific parameters based on these highlighted features (Ebrahiminezhad et al., 2018; Harshiny et al., 2015). The generated NPs were then characterized using a variety of techniques. UV-vis spectrophotometry is used for preliminary characterization. Further analysis of NPs is carried out using Fourier transform infrared spectroscopy, transmission electron microscopy, field emission scanning electron microscopy, X-ray diffraction, X-ray photoelectron spectroscopy, light scattering, energy dispersive spectroscopy, and dynamic and thermogravimetric analysis (Kaur & Sidhu, 2021; Singh, 2022). Leaves are the most widely used plant parts in the production of iron nanoparticles (INPs) via plant-mediated synthesis. Tea extract, which is easily available and enriched with many polyphenols that can serve as reducing as well as capping agents, is frequently employed in the biosynthesis of MNPs.

Eucalyptus leaves are a popular choice for INP synthesis due to their large number of active ingredients (Mo et al., 2015). Notably, the polyphenols in eucalyptus leaves may function as antioxidants that can control NP aggregation.

The use of different leaf extracts for the biosynthesis of INPs is becoming more widespread. Recent research has made it clear that plant extracts from *Ziziphora tenuior*, *Abutilon indicum*, *Solanum trilobatum*, *Persea americana*, *Camellia sinensis*, and *Azadirachta indica* were used to create IONPs of various sizes (Arularasu et al., 2018). According to recent research, nanoparticles created using plant extracts are significantly more stabilized than those created using standard methods (Bibi et al., 2019) and typically have unique shapes including stems, needles, prisms, and dendrites in addition to being cubical, spherical, and cylindrical (Mo et al., 2015).

Furthermore, NiNPs were also produced using the extract *Ocimum sanctum* and *Annona squamosa* leaves (Jeyaraj Pandian et al., 2016; Mamuru et al., 2015).

Fruit extract contains a variety of reducing agents, including flavonoids, phenolic compounds, vitamins, and saccharides, that are also used for the synthesis of MNPs (Timoszyk, 2018). Cobalt oxide nanoparticles (CoONPs) were created by using the extract of citrus lemon fruit as a bio-reductive and stabilizing agent (Anuradha & Raji, 2021).

Seeds also contain edible oil and beneficial antibacterial, biological, and chemical components, which are also employed for the fabrication of INPs. Synthesis of MNPs based on seeds is the simplest way since NPs created by just stirring at a moderate temperature. For instance, IOMNPs were synthesized using *Syzygium cumini* and pomegranate seeds seed extract (Bibi et al., 2019; Venkateswarlu et al., 2014).

Apart from the leaves, fruit, and seeds, other parts of plants such as roots, peels, pods, bark/husk, and flowers have also been extensively utilized for synthesizing MNPs (Clarina et al., 2018; Nasiri et al., 2019; Sebastian et al., 2018). A summary of the biosynthesis of MNPs using different parts of plants is presented in tabular form in Table 2.

Table 2 Various Plants parts used for the synthesized of magnetic nanoparticles

Plant	Types of MNPs	Precursor salt	Size (nm)	Shape	Magnetization	References
Leaves						
Dodonaea viscosa	Fe	$FeCl_3$	50–60	Spherical	Antibacterial activity	Kiruba Daniel et al. (2013)
Aegle marmelos	Ni	$NiCl_2$	80–100	Triangular	In vitro anti-inflammatory and mosquito larvicidal activities	Angajala et al. (2014)
Green tea	Fe	$FeSO_4$	70–80	Spherical amorphous	Degradation of malachite green	Huang et al. (2015)
Calotropis gigantea	Co	$Co(NO_3)_2{\cdot}6H_2O$	60	Spherical	Thermal decomposition of ammonium perchlorate and dye-sensitized solar cells	Sharma et al. (2015)
Raphanus sativus var. *longipinnatus*	Co	Cobalt acetate	80	Spherical	Antibacterial and cytotoxic activities	Koyyati et al. (2016)
Ocimum sanctum	Ni	Nickel nitrate	12–36	Spherical	Antimicrobial activity	Jeyaraj Pandian et al. (2016)
Lagenaria siceraria	Fe	$FeCl_3.6H_2O$	30–100	Cube	Antimicrobial activity	Kanagasubbulakshmi and Kadirvelu (2017)
Sageretia thea	Fe	$FeSO_4.7H_2O$	29	Tetragonal crystalline	Antimicrobial and cytotoxic activities	Khalil et al. (2017)
Kappaphycus alvarezii whole plant	Fe	$FeCl_2.4H_2O$ $FeCl_2.6H_2O$	10–30	Irregular or hexagonal	Photocatalytic activity	Arularasu et al. (2018)
Daphne mezereum	Fe	$FeCl_3.6H_2O$	6.5–14.9	Spherical	Dye removal	Beheshtkhoo et al. (2018)
Taraxacum officinale	Co	Cobalt sulfate	50–100	Spherical	Catalytic reductions of dyes	Rasheed et al. (2019)
Mangifera indica	Co	$CoCl_2.6H_2O$	25–40	Irregular, cubic, and pentagonal	Detection of manganese (II) ions in water	Okwunodulu et al. (2019)
Gardenia resinifera	Fe	$FeCl_3.6H_2O$	3–8	Spherical	In vitro hyperthermia activity	Karade et al. (2019)

(continued)

Table 2 (continued)

Plant	Types of MNPs	Precursor salt	Size (nm)	Shape	Magnetization	References
Plantago major	Fe	$FeCl_3.6H_2O$	4.6–30.6	Spherical	Catalytic activity for decolorization of azo dye	Lohrasbi et al. (2019)
Platanus orientalis	Fe	$Fe(NO_3)_3.9H_2O$	38	Spherical	Antifungal activity	Devi et al. (2019)
Sageretia thea	Co	Cobalt acetate	20.03	Cubic	Antibacterial, antioxidant, antileishmanial cytotoxicity, and hemolytic assay	Khalil et al. (2020)
Ziziphora clinopodioides Lam	Co	$Co(NO_3)_2{\cdot}6H_2O$	29.08	Spherical	Antioxidant, cytotoxicity, antimicrobial, antidiabetic, anticholinergics, and wound healing activities	Hou et al. (2020)
Citrus leaves	Fe	Iron chloride	15–80	Spherical	Antibacterial activity	Hooda and Sharma (2020)
Psidium guajava	Fe	$FeCl_3.6H_2O$	1–5	Spherical	Antibacterial activity	Madubuonu et al. (2020)
Syzygium cumini	Fe	$FeCl_3$	40–52	Spherical	Adsorption of aflatoxin B_1 and antimicrobial activity	Asghar et al. (2020)
Cymbopogon citratus	Fe	$FeCl_3.6H_2O$	9	Irregular cubic	Superparamagnetic	Patiño-Ruiz et al. (2020)
Green tea leaves	Fe	$FeSO_4{\cdot}7H_2O$	50–80	Spherical	Environmental remediation	Hao et al. (2021)
Cleistocalyx operculatus	Fe	$FeCl_3$	100	Spherical	Treatment of dye	Le et al. (2022)
Ficus carica	Fe	$FeCl_3.6H_2O$	43–57	Irregular	Antioxidant activity	Üstün et al. (2022)
Murraya koenigii	Fe	$FeCl_3$	4–9	Spherical	Antimicrobial, antioxidant, anti-inflammatory, and antidiabetic assays	Sivakami et al. (2022)
Solanum tuberosum	Co	$Co(NO_3)_2.6H_2O$	21.05–27.98	Spherical	Electrochemical sensor and antibacterial applications	Bekele et al. (2022)
Aspalathus linearis	Fe	Anhydrous ferric chloride	18	Spherical	Drought abiotic stress of the plant	Ndou et al. (2023)
Rosmarinus officinalis	Fe	$Fe_2(SO_4)_3$	23	Spherical	Antibacterial effect	Jazar (2023)

Calotropis gigantea	Ni	$NiCl_2$	17.06	Hexagonal	Antibacterial and antioxidant assay	Venkatalakshmi et al. (2023)
Fruits						
Passiflora tripartita	Fe	$FeCl_3.6H_2O$	18.23–24.65	Spherical Crystalline	Catalytic activity	Kumar et al. (2014)
Citrus Limon	Co	$CoCl_2.6H_2O$	46–56	Pyramid-like	Antimicrobial activity	Anuradha and Raji (2021)
Terminalia chebula	Ni	Nickel nitrite	15.6	Spherical	Anticancer, antibacterial, antioxidant, and photocatalytic activities	Kareem et al. (2023)
Flowers						
Hibiscus rosa-sinensis	Fe	$FeCl_2.4H_2O$	51	Spherical	Antibacterial activity	Buarki et al. (2022)
Cassia auriculata	Fe	$FeCl_3.6H_2O$	15–35	Spherical	Photocatalytic degradation and larvicidal effect	Sudhakar et al. (2022)
Melia azedarach	Fe	$FeSO_4.7H_2O$	231.43	Spherical	Antimicrobial and antioxidant activities	Muzafar et al. (2022)
Chromolaena odorata	Fe	$FeCl_3.6H_2O$	3.14	Cubic	Removal of hexavalent chromium from water bodies	Garvasis et al. (2023)
Seeds						
Punica granatum	Fe	Iron chloride	25–55	Variable, semispherical	Photocatalytic activity for the degradation of textile dye	Bibi et al. (2019)
Phoenix dactylifera L.	Fe	$FeCl_2$ and $FeCl_3$	30	Spherical crystalline	Antibacterial applications	Majid et al. (2021)
Salvia hispanica L.	Ni	$Ni(NO_3)_2.6H_2O$	30	Spherical	Photocatalytic activity and cytotoxicity effects	Sabouri et al. (2021)
Nigella sativa	Fe	$FeCl_2$ and $FeCl_3$	31.45	Spherical	Cytotoxicity and antimicrobial activities	Al-Karagoly et al. (2022)
Nigella sativa	Fe	$FeCl_2·4H_2O$	5–10	Spherical	Antibacterial and cytotoxic activities	Aliya et al. (2023)

(continued)

Table 2 (continued)

Plant	Types of MNPs	Precursor salt	Size (nm)	Shape	Magnetization	References
Phoenix dactylifera	Co	$Co(NO_3)_2.6H_2O$	80	Spherical	Antimicrobial and photocatalytic activities	Rajeswari et al. (2023)
Moringa	Co	$Co(NO_3)_2.6H_2O$	24.2	Spherical	Photocatalytic activity	Savitha et al. (2023)
Peels						
Musa paradisiaca	Fe	$FeCl_3$	60	Cylindrical	Arsenic remediation in water and sludge	Majumder et al. (2019)
Malus pumila	Fe	$FeCl_2.4H_2O$ $FeCl_2.6H_2O$	50–100	Elliptical and spherical	Decolorization of malachite green dye	Ting and Chin (2020)
Pyrus sinkiangensis	Fe	$FeSO_4.7H_2O$	10–90	Irregularly shaped	Removal of aqueous chromium	Rong et al. (2020)
Artocarpus heterophyllus	Fe	$FeCl_3.6H_2O$	33	Spherical	Catalytic activity for the degradation of fuchsin basic dye	Jain et al. (2021)
Husk						
Juglans regia green	Fe	$FeCl_3.6H_2O$ $FeCl_3.4H_2O$	12.6	Cubic	Cytotoxicity assay	Izadiyan et al. (2020)
Stem						
Berberis baluchistanica	Ni	$NiNO_3$	31.44	Rhombohedral	Antioxidant, antibacterial, antifungal, cytotoxic, and phytotoxicity assay	Uddin et al. (2021)
Parkia biglobosa	Co	$Co(NO_3)_2.6H_2O$	–	Spherical	Antibacterial activity	Yoro et al. (2022)

5 Application of Magnetic Nanoparticles

Over the past decade, the MNPs have attracted significant attention for their favorable results in different sectors due to their exceptional magnetic character, unique size, shape, volume ratio, high surface area, and compatibility with living organisms, which further enhance their potential applications. These characteristics have drawn many scholars from various domains to it. We have outlined the uses of MNPs in some well-known disciplines in this review, including biosensing, biomedicine, agriculture, catalysis, and the environment. The potential uses of MNPs in these sectors are briefly described below (Fig. 3).

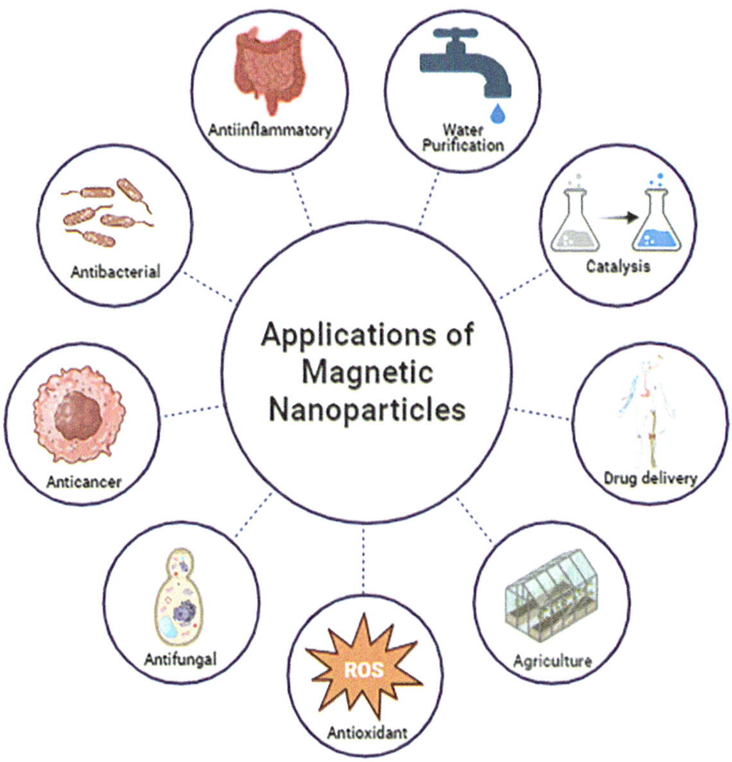

Fig. 3 Applications of green-synthesized magnetic nanoparticles. (Developed using BioRender.com)

5.1 Biological Activities

Due to their numerous physical and chemical characteristics, simplicity in manufacture, stability, and biocompatibility, MNPs have become increasingly prevalent in numerous biological applications. MNPs can communicate with magnetic fields outside. The externally supplied magnetic field at the dipoles generates various types of force and torque, leading to energy loss through rotation, translation, and dissipation. These phenomena have a variety of applications, as well as drug release triggering, magnetic drug delivery, cell separation/biomarker, biomedical imaging, bacterial theranostics, magneto-mechanical activation of cell surface receptors, and hyperthermia. However, one must take into account their possible toxicity or biocompatibility in biomedical research as the most crucial issue (Kong et al., 2011).

5.2 Antibacterial Activity

The antibacterial effectiveness of a material or substance depends on its ability to inhibit bacterial growth or destroy bacteria without causing harm to nearby tissues. Antimicrobial agents are classified as either bacteriostatic, which slows the growth rate of bacteria, or bactericidal, which kills bacteria, based on their properties (Singh et al., 2019). Numerous studies have demonstrated that INPs synthesized via plants can serve as an alternative to antibiotics for antibacterial purposes. INPs synthesized using extracts from *Gardenia jasminoides* and *Lawsonia inermis* leaves exhibited significant antimicrobial activity against a range of pathogens. The findings revealed that *Gardenia jasminoides*-synthesized INPs processed a broader zone of inhibition and greater potential antimicrobial efficiency against pathogens compared to INPs synthesized by *Lawsonia inermis* (Naseem & Farrukh, 2015).

INPs made from *Sageretia thea* leaf extract also possessed effective antibacterial properties against various human pathogenic microorganisms (Khalil et al., 2017). Similarly to this, Helan et al. (2016) proposed that NiNPs have been used to decrease microbial infections (Helan et al., 2016). Khalil et al. (2020) reported that CoONPs produced by using an extract of *Sageretia thea* possess antibacterial efficiency against numerous bacteria (Khalil et al., 2020). Other studies have also used extracts from various plants, i.e., *Sesbania sesban*, *Celosia argentea*, and *Phoenix dactylifera*, to produce CoONPs and test their effectiveness against bacterial strains (Ghadi et al., 2018; Rajeswari et al., 2023; Shahzadi et al., 2019). The results of these experiments indicated that all of the synthesized cobalt nanoparticles exhibited excellent antibacterial properties.

5.3 Antifungal Activity

Antibiotic and medication resistance among bacteria and fungi is increasing alarmingly quickly. Strong antifungal treatments are therefore required to eradicate fungi that are immune to the existing medications (Ahmad et al., 2019). The antifungal properties of cobalt nanoparticles (CoNPs) were investigated by Hou et al. (2020), and the findings revealed that the CoNPs have potent antifungal activity (Hou et al., 2020). Likewise, CoONP synthesis through the flower extract of *Hibiscus rosa sinensis* and IOMNPs synthesized by the leaf extract of *Platanus orientalis* show significant antifungal activity against various pathogens (Devi et al., 2019; Khan et al., 2021).

5.4 Larvicidal Activities

Vector and vector-borne diseases are becoming a major health problem for the public (Klempner et al., 2007). The microbes and the vectors are the key components for human vector-borne infectious diseases such as dengue, malaria, fever, and yellow fever that are spread by mosquitoes. Marimuthu et al. (2013) conducted an in vitro test on CoNPs that were synthesized using *Bacillus thuringiensis* bacteria. The finding indicated that the CoNPs synthesized in a green way demonstrated remarkable activity (Marimuthu et al., 2013). As a result, there is a potential for using the CoNPs in drug formulation for these parasites.

5.5 Antioxidant Activities

Oxidation is one of the main causes of chemical spoilage, which has an impact on food flavor, texture, nutritional content, and safety. Shahzadi et al. (2019) studied the antioxidant and radical-scavenging abilities of bioinspired CoNPs and found that an increase in activity was correlated with an increase in CoNP concentration (Shahzadi et al., 2019). Similarly, CoNPs were synthesized using *Ziziphora clinopodioides Lam* leaf extract, and INPs using the waste of *Camellia sinensis* L. show impressive antioxidant activity (Hou et al., 2020; Periakaruppan et al., 2021).

6 Environmental Remediation

Due to the rising anthropogenic emission of dangerous and poisonous chemicals and substances, contamination of the water, soil, and atmosphere has become a major environmental issue. These contaminants' bodies create serious health issues for humans (Govan, 2020; Jin et al., 2014). Nanotechnology is currently one of the better and more dependable alternatives to traditional remedies.

Green synthetic IOMNPs from pomegranate leaves have been used in adsorption investigations on the color Congo red (Prasad et al., 2017). Cadmium, which is often present in water, was utilized for adsorption using IOMNPs made from tangerine peel extract (Ehrampoush et al., 2015).

7 Catalysis

The transformation of reactants into products has been achieved through the development of numerous catalytic systems and processes. Catalysts assisted by MNPs have recently been used to alleviate the heterogeneous catalyst's shortcomings. These catalysts have high surface areas that support active sites for the conversion of reactants into products, and they can be easily separated (Ali et al., 2021).

The *Amaranthus spinosus* leaf extract was utilized to synthesize IONPs and demonstrated excellent catalytic ability in decolorizing methylene blue and methyl orange. These NPs exhibited superior performance compared to chemically synthesized nanoparticles (Muthukumar & Matheswaran, 2015). Another study found that NiNPs possess excellent catalyst properties in various applications such as separating cleaning oil spills, emulsion, removing impurities, and purifying water from samples. The strong magnetic response and interfacial properties of NiNPs play a vital role in facilitating efficient adsorption and rapid separation (Simonsen et al., 2018). Additionally, Nie et al. (2011) developed a nonenzymatic amperometric sensor using NiNPs. This sensor exhibited a significant surface area and high electrocatalytic action for glucose oxidation (Nie et al., 2011).

8 Agriculture

Numerous studies have been done demonstrating the effective use of MNPs in soil quality improvement, plant protection, and seed germination (El-Temsah et al., 2016). IOMNPs, for instance, can be employed as soil nutrition to boost output with the fewest adverse effects (Mishra et al., 2017). Many crops, including peanuts, show iron shortage. Several research studies have been conducted to address the shortage and increase iron consumption using IOMNPs in this context (Zia-ur-Rehman et al., 2018).

Most of the time, NPs are used as nanofertilizers in hydroponic environments rather than in the field and are thought to be the most plentiful nutrients in the soil (Ju et al., 2019).

Wang et al. (2017) built an immune-electrochemistry sensor based on a thin layer electrode to establish a sensitive, quick, and easy detection approach for organophosphorus pesticides such as chlorpyrifos. The CoONP-based sensor allowed for the voltammetric monitoring of the chlorpyrifos concentration in agricultural products (Wang et al., 2017). It was made of polyaniline-coated nanoparticles. This method allowed for the efficient detection of chlorpyrifos in agricultural products.

9 Other Biological Applications

MNPs have several biological and medicinal uses, including the fact that cancer is the second leading cause of human suffering after cardiovascular disorders (Wu & Huang, 2017). CoNPs have shown promising anticancer activities indicating their potential in cancer treatment.

Devi et al. (2019) have demonstrated the potential of CoONPs to combat MCF-7 breast cancer cells. Furthermore, another researcher used *Euphorbia tirucalli* extract to synthesize CoONPs and assessed their antiproliferative activity on MCF-7 breast cancer cells using the MTT assay (Oktri Mulya Dewi et al., 2019). Additionally, researchers revealed numerous other uses of MNPs, including their usage in NMR imaging to diagnose medication function by enhancing the contrast between healthy and diseased tissues and indicating organ function or blood flow (Akbarzadeh et al., 2012). Biogenic MNPs have been shown to exhibit photothermal therapy (Shen et al., 2013), enzyme inhibition (Wang et al., 2013), antidiabetic activity (Badragheh et al., 2018), and anticholinergic activity (Waris et al., 2021).

10 Present Status, Challenges, and Future Remarks

The green synthesis of nanomaterial is considered a promising area in the current scenario for the development of new science and technology. Researchers are trying to open a new edge of safe, economical, green synthesis methods of various functional nanomaterials. These synthesis methods consume less energy, are inexpensive, are eco-friendly, and produce high yields. Several green materials are used as a source for the production of MNPs such as parts of different plants, green waste, fungi, bacteria, and yeast. Green syntheses typically involve the use of natural and biocompatible materials as reducing agents and stabilizers for MNPs and are well-suited for biomedical applications due to their biocompatibility. They have been explored for drug delivery, magnetic hyperthermia, and magnetic resonance imaging (MRI). MNP-based drug administration can minimize the doses of drugs and subsequently reduce the possible side effects. Moreover, the MNPs could be a proper clinical obstacle when combined with antimicrobial agents, which have synergistic effects in therapies that lead to improving the efficacy of antimicrobial drugs. Similarly, the purification of wastewater is being practiced by using MNPs including removal of organic/inorganic pollutants, dye degradation, and killing or separation of microbes. These methods contribute to sustainable and eco-friendly approaches to addressing pollution. The MNPs can be used as soil fertilizers to increase production as well as for plant disease management.

The MNPs show tremendous potential in heterogeneous catalysis as well as in other applied fields. The coating of the catalysts with MNPs has been used in different kinds of catalysis and clean energy technology. The MNPs can provide numerous active sites for the reactants to be converted into products.

However, several challenges need to be addressed for the optimal utilization of MNPs. Firstly, ensuring the long-term safety and biocompatibility of MNPs, especially in medical applications, remains a critical concern. Researchers must work actively to address this challenge by developing biocompatible coatings and surface modifications that reduce potential toxicity and enhance the stability of MNPs in physiological environments. Secondly, achieving precise control over the size, shape, and magnetic properties of MNPs during synthesis is essential. Variations in these parameters can significantly affect their performance and applicability. To overcome this issue, precise control over MNP synthesis could be achieved through cutting-edge techniques, such as microfluidic synthesis or the use of templates, enabling custom-designed MNPs with tailored properties.

Thirdly, MNPs are prone to agglomeration, which hampers their dispersibility and effectiveness. To address the challenge of agglomeration in MNPs, researchers must actively pursue several strategies that include the incorporation of organic or inorganic stabilizers, which create a protective layer around MNPs, preventing them from clustering and enhancing their dispersibility and overall effectiveness. Another approach that implicates altering the chemical properties of MNPs via surface modification is used to minimize attractive forces between particles and enhance their stability. Furthermore, the introduction of dispersants, such as surfactants or polymers, is widely employed to increase the repulsive forces between MNPs, promoting their even dispersion in diverse media. This multifaceted approach collectively aims to mitigate agglomeration challenges and unlock the full potential of MNPs across a range of applications.

To successfully overcome the substantial challenges associated with MNPs and facilitate their responsible integration into various applications, the scientific community needs to prioritize streamlining and expediting the clinical trial process. This can be achieved by fostering collaborative efforts among researchers, healthcare professionals, and regulatory agencies to develop standardized protocols, robust safety assessments, and efficient regulatory pathways. By simplifying and accelerating the clinical trial phase, we can expedite the translation of MNP-based innovations into real-world solutions, ultimately contributing to a brighter and more promising future in healthcare and beyond.

11 Conclusion

This chapter comprehensively addressed the fabrication, characterization, and applications of MNPs through environmentally friendly greener routes with the help of microorganisms and different parts of plants such as peels, fruits, roots, and seeds. The green method of synthesis is an effective practice for the fabrication of different molecular structures with their countless bioactivity. With the use of biological resources in green methods, the synthesis of NPs becomes economical, less time-consuming, and a one-step synthesis, increases product yield, and decreases formation of waste while improving the chemical and physical properties of NPs. In addition, the green method is an environment-friendly synthesis by eradicating the use and production of hazardous chemicals and raw and waste materials. This comprehensive section not only provides the information about synthesis and characterization of MNPS but also emphasizes their miscellaneous applications. The MNPs possess antibacterial, antifungal, antimalarial, antioxidant, anticancer, photocatalytic, and metal toxicity properties. This study will provide the path of forthcoming research in green synthesis of NPs and their myriad environmental and biological applications.

References

Abdeen, M., Sabry, S., Ghozlan, H., El-Gendy, A. A., & Carpenter, E. E. (2016). Microbial-physical synthesis of Fe and Fe3O4 magnetic nanoparticles using aspergillus Niger YESM1 and supercritical condition of ethanol. *Journal of Nanomaterials, 2016*, 1.

Abinaya, S., Kavitha, H. P., Prakash, M., & Muthukrishnaraj, A. (2021). Green synthesis of magnesium oxide nanoparticles and its applications: A review. *Sustainable Chemistry and Pharmacy, 19*, 100368.

Agarwal, H., Kumar, S. V., & Rajeshkumar, S. (2017). A review on green synthesis of zinc oxide nanoparticles–An eco-friendly approach. *Resource-Efficient Technologies, 3*(4), 406–413.

Ahmad, S., Munir, S., Zeb, N., Ullah, A., Khan, B., Ali, J., Bilal, M., Omer, M., Alamzeb, M., & Salman, S. M. (2019). Green nanotechnology: A review on green synthesis of silver nanoparticles—An ecofriendly approach. *International Journal of Nanomedicine, 14*, 5087–5107.

Ahmed, S., & Ikram, S. (2015). Silver nanoparticles: One pot green synthesis using Terminalia arjuna extract for biological application. *Journal of Nanomedicine & Nanotechnology, 6*(4), 309.

Ajarem, J. S., Maodaa, S. N., Allam, A. A., Taher, M. M., & Khalaf, M. (2021). Benign synthesis of cobalt oxide nanoparticles containing red algae extract: Antioxidant, antimicrobial, anticancer, and anticoagulant activity. *Journal of Cluster Science, 33*, 1–12.

Akbarzadeh, A., Samiei, M., & Davaran, S. (2012). Magnetic nanoparticles: Preparation, physical properties, and applications in biomedicine. *Nanoscale Research Letters, 7*, 1–13.

Ali, A., Shah, T., Ullah, R., Zhou, P., Guo, M., Ovais, M., Tan, Z., & Rui, Y. (2021). Review on recent progress in magnetic nanoparticles: Synthesis, characterization, and diverse applications. *Frontiers in Chemistry, 9*, 629054.

Aliya, S., Rethinasabapathy, M., Yoo, J., Kim, E., Chung, J.-Y., Cha, J.-H., & Huh, Y. S. (2023). Phytogenic fabrication of iron oxide nanoparticles and evaluation of their in vitro antibacterial and cytotoxic activity. *Arabian Journal of Chemistry, 16*(6), 104703.

Al-Karagoly, H., Rhyaf, A., Naji, H., Albukhaty, S., AlMalki, F. A., Alyamani, A. A., Albaqami, J., & Aloufi, S. (2022). Green synthesis, characterization, cytotoxicity, and antimicrobial activity of iron oxide nanoparticles using Nigella sativa seed extract. *Green Processing and Synthesis, 11*(1), 254–265.

Alsammarraie, F. K., Wang, W., Zhou, P., Mustapha, A., & Lin, M. (2018). Green synthesis of silver nanoparticles using turmeric extracts and investigation of their antibacterial activities. *Colloids and Surfaces B: Biointerfaces, 171*, 398–405.

Angajala, G., Ramya, R., & Subashini, R. (2014). In-vitro anti-inflammatory and mosquito larvicidal efficacy of nickel nanoparticles phytofabricated from aqueous leaf extracts of Aegle marmelos Correa. *Acta Tropica, 135*, 19–26.

Anuradha, C., & Raji, P. (2021). Citrus limon fruit juice-assisted biomimetic synthesis, characterization and antimicrobial activity of cobalt oxide (Co_3O_4) nanoparticles. *Applied Physics A, 127*, 1–9.

Arularasu, M., Devakumar, J., & Rajendran, T. (2018). An innovative approach for green synthesis of iron oxide nanoparticles: Characterization and its photocatalytic activity. *Polyhedron, 156*, 279–290.

Asghar, M. A., Zahir, E., Asghar, M. A., Iqbal, J., & Rehman, A. A. (2020). Facile, one-pot biosynthesis and characterization of iron, copper and silver nanoparticles using Syzygium cumini leaf extract: As an effective antimicrobial and aflatoxin B1 adsorption agents. *PLoS One, 15*(7), e0234964.

Badragheh, S., Zeeb, M., & Olyai, M. R. T. B. (2018). Silica-coated magnetic iron oxide functionalized with hydrophobic polymeric ionic liquid: A promising nanoscale sorbent for simultaneous extraction of antidiabetic drugs from human plasma prior to their quantitation by HPLC. *RSC Advances, 8*(53), 30550–30561.

Bansal, R., Nagórniewicz, B., Storm, G., & Prakash, J. (2017). Relaxin-coated superparamagnetic iron-oxide nanoparticles as a novel theranostic approach for the diagnosis and treatment of liver fibrosis. *Journal of Hepatology, 1*(66), S43.

Beheshtkhoo, N., Kouhbanani, M. A. J., Savardashtaki, A., Amani, A. M., & Taghizadeh, S. (2018). Green synthesis of iron oxide nanoparticles by aqueous leaf extract of Daphne mezereum as a novel dye removing material. *Applied Physics A, 124*, 1–7.

Bekele, E. T., Murthy, H., Muniswamy, D., Lemenh, Y. A., Shume, M. S., Tadesse Ayanie, G., Kumar, A. P., Ravikumar, C., Balachandran, R., & Roy, A. (2022). Solanum tuberosum leaf extract templated synthesis of Co3O4 nanoparticles for electrochemical sensor and antibacterial applications. *Bioinorganic Chemistry and Applications, 2022*, 8440756.

Bensy, A. D., Christobel, G. J., Muthusamy, K., Alfarhan, A., & Anantharaman, P. (2022). Green synthesis of iron nanoparticles from Ulva lactuca and bactericidal activity against enteropathogens. *Journal of King Saud University-Science, 34*(3), 101888.

Bhardwaj, A. K., & Naraian, R. (2021). Cyanobacteria as biochemical energy source for the synthesis of inorganic nanoparticles, mechanism and potential applications: A review. *3 Biotech, 11*(10), 445.

Bhardwaj, A. K., Shukla, A., Maurya, S., Singh, S. C., Uttam, K. N., Sundaram, S., Singh, M. P., & Gopal, R. (2018). Direct sunlight enabled photo-biochemical synthesis of silver nanoparticles and their Bactericidal Efficacy: Photon energy as key for size and distribution control. *Journal of Photochemistry and Photobiology B: Biology, 188*, 42–49.

Bibi, I., Nazar, N., Ata, S., Sultan, M., Ali, A., Abbas, A., Jilani, K., Kamal, S., Sarim, F. M., & Khan, M. I. (2019). Green synthesis of iron oxide nanoparticles using pomegranate seeds extract and photocatalytic activity evaluation for the degradation of textile dye. *Journal of Materials Research and Technology, 8*(6), 6115–6124.

Biehl, P., Von der Lühe, M., Dutz, S., & Schacher, F. H. (2018). Synthesis, characterization, and applications of magnetic nanoparticles featuring polyzwitterionic coatings. *Polymers, 10*(1), 91.

Buarki, F., AbuHassan, H., Al Hannan, F., & Henari, F. (2022). Green synthesis of iron oxide nanoparticles using Hibiscus rosa sinensis flowers and their antibacterial activity. *Journal of Nanotechnology, 2022*, 1–6.

Chatterjee, S., Mahanty, S., Das, P., Chaudhuri, P., & Das, S. (2020). Biofabrication of iron oxide nanoparticles using manglicolous fungus Aspergillus niger BSC-1 and removal of Cr (VI) from aqueous solution. *Chemical Engineering Journal, 385*, 123790.

Clarina, T., Flomina, P. J., Thangeswari, P., & Rama, V. (2018). Polpala flower extract mediated one step green synthesis and characterization of magnetite (Fe3O4) nanoparticles. *Asian Journal of Research in Chemistry, 11*(2), 459–462.

Desai, M. P., & Pawar, K. D. (2020). Immobilization of cellulase on iron tolerant Pseudomonas stutzeri biosynthesized photocatalytically active magnetic nanoparticles for increased thermal stability. *Materials Science and Engineering: C, 106*, 110169.

Devi, H. S., Boda, M. A., Shah, M. A., Parveen, S., & Wani, A. H. (2019). Green synthesis of iron oxide nanoparticles using Platanus orientalis leaf extract for antifungal activity. *Green Processing and Synthesis, 8*(1), 38–45.

Ebrahiminezhad, A., Zare-Hoseinabadi, A., Sarmah, A. K., Taghizadeh, S., Ghasemi, Y., & Berenjian, A. (2018). Plant-mediated synthesis and applications of iron nanoparticles. *Molecular Biotechnology, 60*, 154–168.

Ehrampoush, M. H., Miria, M., Salmani, M. H., & Mahvi, A. H. (2015). Cadmium removal from aqueous solution by green synthesis iron oxide nanoparticles with tangerine peel extract. *Journal of Environmental Health Science and Engineering, 13*, 1–7.

El-Temsah, Y. S., Sevcu, A., Bobcikova, K., Cernik, M., & Joner, E. J. (2016). DDT degradation efficiency and ecotoxicological effects of two types of nano-sized zero-valent iron (nZVI) in water and soil. *Chemosphere, 144*, 2221–2228.

Fatemi, M., Mollania, N., Momeni-Moghaddam, M., & Sadeghifar, F. (2018). Extracellular biosynthesis of magnetic iron oxide nanoparticles by Bacillus cereus strain HMH1: Characterization and in vitro cytotoxicity analysis on MCF-7 and 3T3 cell lines. *Journal of Biotechnology, 270*, 1–11.

Flores-Rojas, G. G., López-Saucedo, F., Vera-Graziano, R., Mendizabal, E., & Bucio, E. (2022). Magnetic nanoparticles for medical applications: Updated review. *Macromolecules, 2*(3), 374–390.

Gahlawat, G., & Choudhury, A. R. (2019). A review on the biosynthesis of metal and metal salt nanoparticles by microbes. *RSC Advances, 9*(23), 12944–12967.

Garvasis, J., Prasad, A. R., Shamsheera, K., Roy, T. N., & Joseph, A. (2023). A facile one-pot synthesis of phyto-conjugate superparamagnetic magnetite nanoparticles for the rapid removal of hexavalent chromium from water bodies. *Materials Research Bulletin, 160*, 112130.

Ghadi, F. E., Ghara, A. R., & Naeimi, A. (2018). Phytochemical fabrication, characterization, and antioxidant application of copper and cobalt oxides nanoparticles using Sesbania sesban plant. *Chemical Papers, 72*(11), 2859–2869.

Govan, J. (2020). Recent advances in magnetic nanoparticles and nanocomposites for the remediation of water resources. *Magnetochemistry, 6*(4), 49.

Gul, S., Khan, S. B., Rehman, I. U., Khan, M. A., & Khan, M. (2019). A comprehensive review of magnetic nanomaterials modern day theranostics. *Frontiers in Materials, 6*, 179.

Hajri, A. K., Albalawi, M. A., Alsharif, I., & Jamoussi, B. (2022). Marine algae extract (Grateloupia Sparsa) for the green synthesis of Co_3O_4 NPs: Antioxidant, antibacterial, anticancer, and hemolytic activities. *Bioinorganic Chemistry and Applications, 2022*, 1.

Hao, R., Li, D., Zhang, J., & Jiao, T. (2021). Green synthesis of iron nanoparticles using green tea and its removal of hexavalent chromium. *Nanomaterials, 11*(3), 650.

Harshiny, M., Iswarya, C. N., & Matheswaran, M. (2015). Biogenic synthesis of iron nanoparticles using Amaranthus dubius leaf extract as a reducing agent. *Powder Technology, 286*, 744–749.

Helan, V., Prince, J. J., Al-Dhabi, N. A., Arasu, M. V., Ayeshamariam, A., Madhumitha, G., Roopan, S. M., & Jayachandran, M. (2016). Neem leaves mediated preparation of NiO nanoparticles and its magnetization, coercivity and antibacterial analysis. *Results in Physics, 6*, 712–718.

Hooda, R., & Sharma, M. (2020). Green synthesis, characterization and antibacterial activity of iron oxide nanoparticles. *Plant Archives, 20*(1), 1196–1200.

Hou, H., Mahdavi, B., Paydarfard, S., Zangeneh, M. M., Zangeneh, A., Sadeghian, N., Taslimi, P., Erduran, V., & Sen, F. (2020). Retracted article: Novel green synthesis and antioxidant, cytotoxicity, antimicrobial, antidiabetic, anticholinergics, and wound healing properties of cobalt nanoparticles containing Ziziphora clinopodioides Lam leaves extract. *Scientific Reports, 10*(1), 12195.

Huang, L., Luo, F., Chen, Z., Megharaj, M., & Naidu, R. (2015). Green synthesized conditions impacting on the reactivity of Fe NPs for the degradation of malachite green. *Spectrochimica Acta Part A: Molecular and Biomolecular Spectroscopy, 137*, 154–159.

Iravani, S., & Varma, R. S. (2020). Bacteria in heavy metal remediation and nanoparticle biosynthesis. *ACS Sustainable Chemistry & Engineering, 8*(14), 5395–5409.

Izadiyan, Z., Shameli, K., Miyake, M., Hara, H., Mohamad, S. E. B., Kalantari, K., Taib, S. H. M., & Rasouli, E. (2020). Cytotoxicity assay of plant-mediated synthesized iron oxide nanoparticles using Juglans regia green husk extract. *Arabian Journal of Chemistry, 13*(1), 2011–2023.

Jain, R., Mendiratta, S., Kumar, L., & Srivastava, A. (2021). Green synthesis of iron nanoparticles using Artocarpus heterophyllus peel extract and their application as a heterogeneous Fenton-like catalyst for the degradation of Fuchsin Basic dye. *Current Research in Green and Sustainable Chemistry, 4*, 100086.

Jazar, Z. H. (2023). Antibacterial effect of iron nanoparticles synthesized from rosemary leaves extract (Rosmarinus officinalis L). *South Asian Research Journal of Agriculture and Fisheries, 5*(1), 10–15.

Jeyaraj Pandian, C., Palanivel, R., & Dhanasekaran, S. (2016). Screening antimicrobial activity of nickel nanoparticles synthesized using Ocimum sanctum leaf extract. *Journal of Nanoparticles, 2016*, 1–13.

Jiang, L., Zheng, R., Zeng, N., Wu, C., & Su, H. (2023). In situ self-assembly of amphiphilic dextran micelles and superparamagnetic iron oxide nanoparticle-loading as magnetic resonance imaging contrast agents. *Regenerative Biomaterials, 10*, rbac096.

Jin, Y., Liu, F., Shan, C., Tong, M., & Hou, Y. (2014). Efficient bacterial capture with amino acid modified magnetic nanoparticles. *Water Research, 50*, 124–134.

Ju, M., Navarreto-Lugo, M., Wickramasinghe, S., Milbrandt, N. B., McWhorter, A., & Samia, A. C. S. (2019). Exploring the chelation-based plant strategy for iron oxide nanoparticle uptake in garden cress (Lepidium sativum) using magnetic particle spectrometry. *Nanoscale, 11*(40), 18582–18594.

Jubran, A. S., Al-Zamely, O. M., & Al-Ammar, M. H. (2020). A study of iron oxide nanoparticles synthesis by using bacteria. *International Journal of Pharmaceutical Quality Assurance, 11*(01), 01–08.

Kanagasubbulakshmi, S., & Kadirvelu, K. (2017). Green synthesis of iron oxide nanoparticles using Lagenaria siceraria and evaluation of its antimicrobial activity. *Defence Life Science Journal, 2*(4), 422–427.

Karade, V., Parit, S., Dawkar, V., Devan, R., Choudhary, R., Kedge, V., Pawar, N., Kim, J., & Chougale, A. (2019). A green approach for the synthesis of α-Fe2O3 nanoparticles from Gardenia resinifera plant and it's In vitro hyperthermia application. *Heliyon, 5*(7), e02044.

Kareem, M. M., Babu, M. H., & Lakshmi, G. V. (2023). Anticancer, antibacterial, antioxidant, and photo-catalytic activities of eco-friendly synthesized Ni nanoparticles. *Inorganic Chemistry Communications, 148*, 110274.

Kataria, N., & Garg, V. (2018). Green synthesis of Fe3O4 nanoparticles loaded sawdust carbon for cadmium (II) removal from water: Regeneration and mechanism. *Chemosphere, 208*, 818–828.

Katas, H., Lim, C. S., Azlan, A. Y. H. N., Buang, F., & Busra, M. F. M. (2019). Antibacterial activity of biosynthesized gold nanoparticles using biomolecules from Lignosus rhinocerotis and chitosan. *Saudi Pharmaceutical Journal, 27*(2), 283–292.

Kaur, K., & Sidhu, A. K. (2021). Green synthesis: An eco-friendly route for the synthesis of iron oxide nanoparticles. *Frontiers in Nanotechnology, 3*, 655062.

Kefeni, K. K., Msagati, T. A., & Mamba, B. B. (2017). Ferrite nanoparticles: Synthesis, characterisation and applications in electronic device. *Materials Science and Engineering: B, 215*, 37–55.

Khalil, A. T., Ovais, M., Ullah, I., Ali, M., Shinwari, Z. K., & Maaza, M. (2017). Biosynthesis of iron oxide (Fe2O3) nanoparticles via aqueous extracts of Sageretia thea (Osbeck.) and their pharmacognostic properties. *Green Chemistry Letters and Reviews, 10*(4), 186–201.

Khalil, A. T., Ovais, M., Ullah, I., Ali, M., Shinwari, Z. K., & Maaza, M. (2020). Physical properties, biological applications and biocompatibility studies on biosynthesized single phase cobalt oxide (Co3O4) nanoparticles via Sageretia thea (Osbeck.). *Arabian Journal of Chemistry, 13*(1), 606–619.

Khan, M. A., Ali, F., Faisal, S., Rizwan, M., Hussain, Z., Zaman, N., Afsheen, Z., Uddin, M. N., & Bibi, N. (2021). Exploring the therapeutic potential of Hibiscus rosa sinensis synthesized cobalt oxide (Co3O4-NPs) and magnesium oxide nanoparticles (MgO-NPs). *Saudi Journal of Biological Sciences, 28*(9), 5157–5167.

Khandel, P., & Shahi, S. K. (2018). Mycogenic nanoparticles and their bio-prospective applications: Current status and future challenges. *Journal of Nanostructure in Chemistry, 8*, 369–391.

Kiruba Daniel, S., Vinothini, G., Subramanian, N., Nehru, K., & Sivakumar, M. (2013). Biosynthesis of Cu, ZVI, and Ag nanoparticles using Dodonaea viscosa extract for antibacterial activity against human pathogens. *Journal of Nanoparticle Research, 15*, 1–10.

Klempner, M. S., Unnasch, T. R., & Hu, L. T. (2007). Taking a bite out of vector-transmitted infectious diseases. *New England Journal of Medicine, 356*(25), 2567–2569.

Kong, B., Seog, J. H., Graham, L. M., & Lee, S. B. (2011). Experimental considerations on the cytotoxicity of nanoparticles. *Nanomedicine, 6*, 929–941.

Koyyati, R., Kudle, K. R., & Padigya, P. R. M. (2016). Evaluation of antibacterial and cytotoxic activity of green synthesized cobalt nanoparticles using Raphanus sativus var. longipinnatus leaf extract. *International Journal of Pharmtech Research, 9*(3), 466–472.

Kumar, B., Smita, K., Cumbal, L., & Debut, A. (2014). Biogenic synthesis of iron oxide nanoparticles for 2-arylbenzimidazole fabrication. *Journal of Saudi Chemical Society, 18*(4), 364–369.

Le, N. T., Dang, T.-D., Binh, K. H., Nguyen, T. M., Xuan, T. N., La, D. D., Nadda, A. K., Chang, S. W., & Nguyen, D. D. (2022). Green synthesis of highly stable zero-valent iron nanoparticles for organic dye treatment using Cleistocalyx operculatus leaf extract. *Sustainable Chemistry and Pharmacy, 25*, 100598.

Lohrasbi, S., Kouhbanani, M. A. J., Beheshtkhoo, N., Ghasemi, Y., Amani, A. M., & Taghizadeh, S. (2019). Green synthesis of iron nanoparticles using plantago major leaf extract and their application as a catalyst for the decolorization of azo dye. *BioNanoScience, 9*, 317–322.

Lu, Y., Xu, Y.-J., Zhang, G.-B., Ling, D., Wang, M.-Q., Zhou, Y., Wu, Y.-D., Wu, T., Hackett, M. J., & Hyo Kim, B. (2017). Iron oxide nanoclusters for T 1 magnetic resonance imaging of non-human primates. *Nature Biomedical Engineering, 1*(8), 637–643.

Madubuonu, N., Aisida, S. O., Ahmad, I., Botha, S., Zhao, T.-K., Maaza, M., & Ezema, F. I. (2020). Bio-inspired iron oxide nanoparticles using Psidium guajava aqueous extract for antibacterial activity. *Applied Physics A, 126*, 1–8.

Mahdavi, M., Namvar, F., Ahmad, M. B., & Mohamad, R. (2013). Green biosynthesis and characterization of magnetic iron oxide (Fe3O4) nanoparticles using seaweed (Sargassum muticum) aqueous extract. *Molecules, 18*(5), 5954–5964.

Majeed, S., Danish, M., Mohamad Ibrahim, M. N., Sekeri, S. H., Ansari, M. T., Nanda, A., & Ahmad, G. (2021). Bacteria mediated synthesis of iron oxide nanoparticles and their antibacterial, antioxidant, cytocompatibility properties. *Journal of Cluster Science, 32*, 1083–1094.

Majid, A., Naz, F., Jamro, H. A., Ansari, B., Abbasi, S., Lal, S., & Ujjan, S. A. (2021). Facile green synthesis of iron oxide nanoparticles using Phoenix dactylifera L. seed extract and their antibacterial applications. *Journal of Pharmaceutical Research International, 33*, 21–29.

Majumder, A., Ramrakhiani, L., Mukherjee, D., Mishra, U., Halder, A., Mandal, A. K., & Ghosh, S. (2019). Green synthesis of iron oxide nanoparticles for arsenic remediation in water and sludge utilization. *Clean Technologies and Environmental Policy, 21*, 795–813.

Mamuru, S. A., Bello, A. S., & Hamman, S. B. (2015). Annona squamosa leaf extract as an efficient bioreducing agent in the synthesis of chromium and nickel nanoparticles. *International Journal of Applied Sciences and Biotechnology, 3*(2), 167–169.

Marimuthu, S., Rahuman, A. A., Kirthi, A. V., Santhoshkumar, T., Jayaseelan, C., & Rajakumar, G. (2013). Eco-friendly microbial route to synthesize cobalt nanoparticles using Bacillus thuringiensis against malaria and dengue vectors. *Parasitology Research, 112*, 4105–4112.

Mathur, P., Saini, S., Paul, E., Sharma, C., & Mehtani, P. (2021). Endophytic fungi mediated synthesis of iron nanoparticles: Characterization and application in methylene blue decolorization. *Current Research in Green and Sustainable Chemistry, 4*, 100053.

Mirza, A. U., Kareem, A., Nami, S. A., Khan, M. S., Rehman, S., Bhat, S. A., Mohammad, A., & Nishat, N. (2018). Biogenic synthesis of iron oxide nanoparticles using Agrewia optiva and Prunus persica phyto species: Characterization, antibacterial and antioxidant activity. *Journal of Photochemistry and Photobiology B: Biology, 185*, 262–274.

Mishra, S., Keswani, C., Abhilash, P., Fraceto, L. F., & Singh, H. B. (2017). Integrated approach of agri-nanotechnology: Challenges and future trends. *Frontiers in Plant Science, 8*, 471.

Mo, Y.-Y., Tang, Y.-K., Wang, S.-Y., Lin, J.-M., Zhang, H.-B., & Luo, D.-Y. (2015). Green synthesis of silver nanoparticles using eucalyptus leaf extract. *Materials Letters, 144*, 165–167.

Moavi, J., Buazar, F., & Sayahi, M. H. (2021). Algal magnetic nickel oxide nanocatalyst in accelerated synthesis of pyridopyrimidine derivatives. *Scientific Reports, 11*(1), 1–14.

Mohamed, A. E.-M. A., & Mohamed, M. A. (2019). Nanoparticles: Magnetism and applications. In *Magnetic nanostructures: Environmental and agricultural applications* (pp. 1–12). Springer.

Mohammed, L., Gomaa, H. G., Ragab, D., & Zhu, J. (2017). Magnetic nanoparticles for environmental and biomedical applications: A review. *Particuology, 30*, 1–14.

Mondal, P., Anweshan, A., & Purkait, M. K. (2020). Green synthesis and environmental application of iron-based nanomaterials and nanocomposite: A review. *Chemosphere, 259*, 127509.

Mughal, B., Zaidi, S. Z. J., Zhang, X., & Hassan, S. U. (2021). Biogenic nanoparticles: Synthesis, characterisation and applications. *Applied Sciences, 11*(6), 2598.

Mukherjee, P. (2017). Stenotrophomonas and microbacterium: Mediated biogenesis of copper, silver and iron nanoparticles—Proteomic insights and antibacterial properties versus biofilm formation. *Journal of Cluster Science, 28*, 331–358.

Muthukumar, H., & Matheswaran, M. (2015). Amaranthus spinosus leaf extract mediated FeO nanoparticles: Physicochemical traits, photocatalytic and antioxidant activity. *ACS Sustainable Chemistry & Engineering, 3*(12), 3149–3156.

Muzafar, W., Kanwal, T., Rehman, K., Perveen, S., Jabri, T., Qamar, F., Faizi, S., & Shah, M. R. (2022). Green synthesis of iron oxide nanoparticles using Melia azedarach flowers extract and evaluation of their antimicrobial and antioxidant activities. *Journal of Molecular Structure, 1269*, 133824.

Naraian, R., & Abhishek, A. K. B. (2020). Green synthesis and characterization of silver NPs using oyster mushroom extract for antibacterial efficacy. *Journal of Chemistry, Environmental Sciences and its Applications, 7*(1), 13–18-13–18.

Naseem, T., & Farrukh, M. A. (2015). Antibacterial activity of green synthesis of iron nanoparticles using Lawsonia inermis and Gardenia jasminoides leaves extract. *Journal of Chemistry, 2015*, 1.

Nasiri, J., Motamedi, E., Naghavi, M. R., & Ghafoori, M. (2019). Removal of crystal violet from water using β-cyclodextrin functionalized biogenic zero-valent iron nanoadsorbents synthesized via aqueous root extracts of Ferula persica. *Journal of Hazardous Materials, 367*, 325–338.

Ndou, N., Rakgotho, T., Nkuna, M., Doumbia, I. Z., Mulaudzi, T., & Ajayi, R. F. (2023). Green synthesis of iron oxide (Hematite) nanoparticles and their influence on Sorghum bicolor growth under drought stress. *Plants, 12*(7), 1425.

Nie, H., Yao, Z., Zhou, X., Yang, Z., & Huang, S. (2011). Nonenzymatic electrochemical detection of glucose using well-distributed nickel nanoparticles on straight multi-walled carbon nanotubes. *Biosensors and Bioelectronics, 30*(1), 28–34.

Oktri Mulya Dewi, N., Yulizar, Y., & Oky Bagus Apriandanu, D. (2019). Green synthesis of Co3O4 nanoparticles using Euphorbia heterophylla L. leaves extract: Characterization and photocatalytic activity. *Materials Science and Engineering Conference Series, 509*, 012105.

Okwunodulu, F. U., Chukwuemeka-Okorie, H. O., & Okorie, F. C. (2019). Biological synthesis of cobalt nanoparticles from Mangifera indica leaf extract and application by detection of manganese (II) ions present in industrial wastewater. *Chemical Science International Journal, 27*(1), 1–8.

Omran, B., Nassar, H., Younis, S., El-Salamony, R., Fatthallah, N., Hamdy, A., El-Shatoury, E., & El-Gendy, N. S. (2020). Novel mycosynthesis of cobalt oxide nanoparticles using Aspergillus brasiliensis ATCC 16404—optimization, characterization and antimicrobial activity. *Journal of applied microbiology, 128*(2), 438–457.

Palzer, J., Eckstein, L., Slabu, I., Reisen, O., Neumann, U. P., & Roeth, A. A. (2021). Iron oxide nanoparticle-based hyperthermia as a treatment option in various gastrointestinal malignancies. *Nanomaterials, 11*(11), 3013.

Park, T. J., Lee, K. G., & Lee, S. Y. (2016). Advances in microbial biosynthesis of metal nanoparticles. *Applied Microbiology and Biotechnology, 100*, 521–534.

Patiño-Ruiz, D., Sánchez-Botero, L., Tejeda-Benitez, L., Hinestroza, J., & Herrera, A. (2020). Green synthesis of iron oxide nanoparticles using Cymbopogon citratus extract and sodium carbonate salt: Nanotoxicological considerations for potential environmental applications. *Environmental Nanotechnology, Monitoring & Management, 14*, 100377.

Periakaruppan, R., Chen, X., Thangaraj, K., Jeyaraj, A., Nguyen, H. H., Yu, Y., Hu, S., Lu, L., & Li, X. (2021). Utilization of tea resources with the production of superparamagnetic biogenic iron oxide nanoparticles and an assessment of their antioxidant activities. *Journal of Cleaner Production, 278*, 123962.

Prasad, C., Karlapudi, S., Venkateswarlu, P., Bahadur, I., & Kumar, S. (2017). Green arbitrated synthesis of Fe3O4 magnetic nanoparticles with nanorod structure from pomegranate leaves and Congo red dye degradation studies for water treatment. *Journal of Molecular Liquids, 240*, 322–328.

Punjabi, K., Mehta, S., Yedurkar, S., Jain, R., Mukherjee, S., Kale, A., & Deshpande, S. (2018). Extracellular synthesis of silver nanoparticle by Pseudomonas hibiscicola–Mechanistic approach. *Advances in Nano Research, 6*(1), 81.

Purbia, R., & Paria, S. (2015). Yolk/shell nanoparticles: Classifications, synthesis, properties, and applications. *Nanoscale, 7*(47), 19789–19873.

Rajeswari, V. D., Khalifa, A. S., Elfasakhany, A., Badruddin, I. A., Kamangar, S., & Brindhadevi, K. (2023). Green and ecofriendly synthesis of cobalt oxide nanoparticles using Phoenix dactylifera L: Antimicrobial and photocatalytic activity. *Applied Nanoscience, 13*(2), 1367–1375.

Rasheed, T., Nabeel, F., Bilal, M., & Iqbal, H. M. (2019). Biogenic synthesis and characterization of cobalt oxide nanoparticles for catalytic reduction of direct yellow-142 and methyl orange dyes. *Biocatalysis and Agricultural Biotechnology, 19*, 101154.

Rezaei, M., Hosseini, S. N., Khavari-Nejad, R. A., Najafi, F., & Mahdavi, M. (2019). HBs antigen and mannose loading on the surface of iron oxide nanoparticles in order to immunotargeting: Fabrication, characterization, cellular and humoral immunoassay. *Artificial Cells, Nanomedicine, and Biotechnology, 47*(1), 1543–1558.

Rong, K., Wang, J., Zhang, Z., & Zhang, J. (2020). Green synthesis of iron nanoparticles using Korla fragrant pear peel extracts for the removal of aqueous Cr (VI). *Ecological engineering, 149*, 105793.

Sabouri, Z., Rangrazi, A., Amiri, M. S., Khatami, M., & Darroudi, M. (2021). Green synthesis of nickel oxide nanoparticles using Salvia hispanica L.(chia) seeds extract and studies of their photocatalytic activity and cytotoxicity effects. *Bioprocess and Biosystems Engineering, 44*, 2407–2415.

Salem, D. M., Ismail, M. M., & Aly-Eldeen, M. A. (2019). Biogenic synthesis and antimicrobial potency of iron oxide (Fe3O4) nanoparticles using algae harvested from the Mediterranean Sea, Egypt. *The Egyptian Journal of Aquatic Research, 45*(3), 197–204.

Salem, D. M., Ismail, M. M., & Tadros, H. R. (2020). Evaluation of the antibiofilm activity of three seaweed species and their biosynthesized iron oxide nanoparticles (Fe3O4-NPs). *The Egyptian Journal of Aquatic Research, 46*(4), 333–339.

Salvadori, M. R., Ando, R. A., Muraca, D., Knobel, M., Nascimento, C. A. O., & Corrêa, B. (2016). Magnetic nanoparticles of Ni/NiO nanostructured in film form synthesized by dead organic matrix of yeast. *RSC Advances, 6*(65), 60683–60692.

Sanjivkumar, M., Vaishnavi, R., Neelakannan, M., Kannan, D., Silambarasan, T., & Immanuel, G. (2019). Investigation on characterization and biomedical properties of silver nanoparticles synthesized by an actinobacterium Streptomyces olivaceus (MSU3). *Biocatalysis and Agricultural Biotechnology, 17*, 151–159.

Saratale, R. G., Karuppusamy, I., Saratale, G. D., Pugazhendhi, A., Kumar, G., Park, Y., Ghodake, G. S., Bharagava, R. N., Banu, J. R., & Shin, H. S. (2018). A comprehensive review on green nanomaterials using biological systems: Recent perception and their future applications. *Colloids and Surfaces B: Biointerfaces, 170*, 20–35.

Sarkar, J., Mollick, M. M. R., Chattopadhyay, D., & Acharya, K. (2017). An eco-friendly route of γ-Fe2O3 nanoparticles formation and investigation of the mechanical properties of the HPMC-γ-Fe2O3 nanocomposites. *Bioprocess and Biosystems Engineering, 40*, 351–359.

Savitha, S., Surendhiran, S., Jagan, K., Karthik, A., Kalpana, B., & Senthilmurugan, R. (2023). Evaluation of the physicochemical characteristics and photocatalytic activity of cobalt oxide nanoparticles derived from Moringa seed extract. *Journal of Materials Science: Materials in Electronics, 34*(2), 1–19.

Sebastian, A., Nangia, A., & Prasad, M. (2018). A green synthetic route to phenolics fabricated magnetite nanoparticles from coconut husk extract: Implications to treat metal contaminated water and heavy metal stress in Oryza sativa L. *Journal of Cleaner Production, 174*, 355–366.

Shahzadi, T., Zaib, M., Riaz, T., Shehzadi, S., Abbasi, M. A., & Shahid, M. (2019). Synthesis of eco-friendly cobalt nanoparticles using Celosia argentea plant extract and their efficacy studies as antioxidant, antibacterial, hemolytic and catalytical agent. *Arabian Journal for Science and Engineering, 44*, 6435–6444.

Sharma, J., Srivastava, P., Singh, G., Akhtar, M. S., & Ameen, S. (2015). Green synthesis of Co3O4 nanoparticles and their applications in thermal decomposition of ammonium perchlorate and dye-sensitized solar cells. *Materials Science and Engineering: B, 193*, 181–188.

Shen, S., Kong, F., Guo, X., Wu, L., Shen, H., Xie, M., Wang, X., Jin, Y., & Ge, Y. (2013). CMCTS stabilized Fe 3 O 4 particles with extremely low toxicity as highly efficient near-infrared photothermal agents for in vivo tumor ablation. *Nanoscale, 5*(17), 8056–8066.

Siddiqi, K. S., & Husen, A. (2016). Fabrication of metal and metal oxide nanoparticles by algae and their toxic effects. *Nanoscale Research Letters, 11*(1), 1–11.

Simonsen, G., Strand, M., & Øye, G. (2018). Potential applications of magnetic nanoparticles within separation in the petroleum industry. *Journal of Petroleum Science and Engineering, 165*, 488–495.

Singh, A. K. (2022). A review on plant extract-based route for synthesis of cobalt nanoparticles: Photocatalytic, electrochemical sensing and antibacterial applications. *Current Research in Green and Sustainable Chemistry, 5*, 100270.

Singh, P., Kim, Y.-J., Zhang, D., & Yang, D.-C. (2016). Biological synthesis of nanoparticles from plants and microorganisms. *Trends in Biotechnology, 34*(7), 588–599.

Singh, J., Dutta, T., Kim, K.-H., Rawat, M., Samddar, P., & Kumar, P. (2018). 'Green' synthesis of metals and their oxide nanoparticles: applications for environmental remediation. *Journal of nanobiotechnology, 16*(1), 1–24.

Singh, K., Mishra, A., Sharma, D., & Singh, K. (2019). Antiviral and antimicrobial potentiality of nano drugs. In *Applications of targeted nano drugs and delivery systems* (pp. 343–356). Elsevier.

Singh, M. P., Bhardwaj, A. K., Bharati, K., Singh, R. P., Chaurasia, S. K., Kumar, S., Singh, R. P., Shukla, A., Naraian, R., & Vikram, K. (2021). Biogenic and non-biogenic waste utilization in the synthesis of 2D materials (graphene, h-BN, g-C2N) and their applications. *Frontiers in Nanotechnology, 3*, 685427.

Sivakami, M., Renuka Devi, K., & Renuka, R. (2022). Phytomediated synthesis of magnetic nanoparticles by Murraya koenigii leaves extract and its biomedical applications. *Applied Physics A, 128*(4), 272.
Sudhakar, C., Poonkothai, M., Selvankumar, T., & Selvam, K. (2022). Facile synthesis of iron oxide nanoparticles using Cassia auriculata flower extract and accessing their photocatalytic degradation and larvicidal effect. *Journal of Materials Science: Materials in Electronics, 33*(14), 11434–11445.
Thorat, N. D., Bohara, R. A., Noor, M. R., Dhamecha, D., Soulimane, T., & Tofail, S. A. (2017). Effective cancer theranostics with polymer encapsulated superparamagnetic nanoparticles: Combined effects of magnetic hyperthermia and controlled drug release. *ACS Biomaterials Science & Engineering, 3*(7), 1332–1340.
Timoszyk, A. (2018). A review of the biological synthesis of gold nanoparticles using fruit extracts: Scientific potential and application. *Bulletin of Materials Science, 41*, 1–11.
Ting, A. S. Y., & Chin, J. E. (2020). Biogenic synthesis of iron nanoparticles from apple peel extracts for decolorization of malachite green dye. *Water, Air, & Soil Pollution, 231*(6), 278.
Tomer, A. K., Rahi, T., Neelam, D. K., & Dadheech, P. K. (2019). Cyanobacterial extract-mediated synthesis of silver nanoparticles and their application in ammonia sensing. *International Microbiology, 22*, 49–58.
Uddin, S., Safdar, L. B., Anwar, S., Iqbal, J., Laila, S., Abbasi, B. A., Saif, M. S., Ali, M., Rehman, A., & Basit, A. (2021). Green synthesis of nickel oxide nanoparticles from Berberis balochistanica stem for investigating bioactivities. *Molecules, 26*(6), 1548.
Üstün, E., Önbaş, S. C., Çelik, S. K., Ayvaz, M. Ç., & Şahin, N. (2022). Green synthesis of iron oxide nanoparticles by using Ficus carica leaf extract and its antioxidant activity. *Biointerface Research in Applied Chemistry, 12*, 2108–2116.
Vadlapudi, V., & Kaladhar, D. (2014). Green synthesis of silver and gold nanoparticles. *Middle-East Journal of Scientific Research, 19*(6), 834–842.
Vasantharaj, S., Sathiyavimal, S., Senthilkumar, P., LewisOscar, F., & Pugazhendhi, A. (2019). Biosynthesis of iron oxide nanoparticles using leaf extract of Ruellia tuberosa: antimicrobial properties and their applications in photocatalytic degradation. *Journal of Photochemistry and Photobiology B: Biology, 192*, 74–82.
Venkatalakshmi, N., Kini, H. J., & Naik, H. B. (2023). Green-synthesized Nickel oxide nanoparticles: Magnetic and biomedical applications. *Inorganic Chemistry Communications, 151*, 110490.
Venkateswarlu, S., Kumar, B. N., Prasad, C., Venkateswarlu, P., & Jyothi, N. (2014). Bio-inspired green synthesis of Fe3O4 spherical magnetic nanoparticles using Syzygium cumini seed extract. *Physica B: Condensed Matter, 449*, 67–71.
Wang, S., Su, P., Huang, J., Wu, J., & Yang, Y. (2013). Magnetic nanoparticles coated with immobilized alkaline phosphatase for enzymolysis and enzyme inhibition assays. *Journal of Materials Chemistry B, 1*(12), 1749–1754.
Wang, W., Han, Z., Liang, P., Guo, D., Xiang, Y., Tian, M., Song, Z., & Zhao, H. (2017). Co 3O4/PAn magnetic nanoparticle-modified electrochemical immunosensor for chlorpyrifos. *Digest Journal of Nanomaterials & Biostructures (DJNB), 12*(1), 1–9.
Waris, A., Din, M., Ali, A., Afridi, S., Baset, A., Khan, A. U., & Ali, M. (2021). Green fabrication of Co and Co3O4 nanoparticles and their biomedical applications: A review. *Open Life Sciences, 16*(1), 14–30.
Wu, M., & Huang, S. (2017). Magnetic nanoparticles in cancer diagnosis, drug delivery and treatment. *Molecular and Clinical Oncology, 7*(5), 738–746.
Yoro, M., Joshua, J., Isiyaku, A., Jonah, J. W. K., & Bello, P. D. (2022). Green synthesis, characterization and antibacterial activity of cobalt nanoparticles from *Parkia biglobosa* aqueous stem extract. *Sch Int J Chem Mater Sci, 5*(5), 79–85.
Yousaf, H., Mehmood, A., Ahmad, K. S., & Raffi, M. (2020). Green synthesis of silver nanoparticles and their applications as an alternative antibacterial and antioxidant agents. *Materials Science and Engineering: C, 112*, 110901.

Zaki, S. A., Eltarahony, M. M., & Abd-El-Haleem, D. A. (2019). Disinfection of water and wastewater by biosynthesized magnetite and zerovalent iron nanoparticles via NAP-NAR enzymes of Proteus mirabilis 10B. *Environmental Science and Pollution Research, 26*, 23661–23678.
Zhang, Q., Yang, X., & Guan, J. (2019). Applications of magnetic nanomaterials in heterogeneous catalysis. *ACS Applied Nano Materials, 2*(8), 4681–4697.
Zia-ur-Rehman, M., Naeem, A., Khalid, H., Rizwan, M., Ali, S., & Azhar, M. (2018). Responses of plants to iron oxide nanoparticles. In *Nanomaterials in plants, algae, and microorganisms* (pp. 221–238). Elsevier.

A Synthesis of Biogenic Nanoparticles (NPs) for the Treatment of Wastewater and Its Application: A Review

Prangya Rath, Laxmi Kant Bhardwaj, Poornima Yadav, and Abhishek Kumar Bhardwaj

1 Introduction

In recent years, the escalating global concern over water pollution has driven intensive research into novel and sustainable solutions for wastewater treatment. Among these solutions, the synthesis of biogenic nanoparticles (NPs) has emerged as a promising avenue of exploration. Nanomaterials (NMs) are substances with at least one dimension as ≤100 nm. They could be of various shapes which are particles, tubes, rods, and fibers. Nanoparticles (NPs) are categorized into four types. These are (1) inorganic, (2) organic, (3) carbon-based, and (4) composite-based (Bhardwaj et al., 2022a). Various R&D and industries have focused on the synthesis of NPs for many purposes. The synthesis of NPs broadly follows two approaches: a top-down approach and a bottom-up approach (Baig et al., 2021). During the process of the top-down approach, the bulk materials are divided to produce nanostructured substances, while during the process of the bottom-up approach, NPs are synthesized by means of chemical reactions involving atoms/ions/molecules (Singh et al., 2020; Baig et al., 2021). There have been few studies that have reported the side effects and challenges due to excessive NP usage (Buzea et al., 2007; Verma & Kumar,

P. Rath
Amity Institute of Environmental Sciences (AIES), Amity University, Noida, Uttar Pradesh, India

L. K. Bhardwaj (✉)
Amity Institute of Environmental, Toxicology, Safety, & Management (AIETSM), Amity University, Noida, Uttar Pradesh, India

P. Yadav
Fair Quality Institute, New Delhi, India

A. K. Bhardwaj
Department of Environmental Science, Amity School of Life Sciences (ASLS), Amity University Madhya Pradesh (AUMP), Gwalior, Madhya Pradesh, India

A. K. Bhardwaj et al. (eds.), *Biogenic Wastes-Enabled Nanomaterial Synthesis*,
https://doi.org/10.1007/978-3-031-59083-2_5

2019). The conventional routes of NPs synthesis have been observed to be associated with hazardous and volatile chemicals; thus, NPs further create secondary pollution (Gautam et al., 2019). For example, chemicals such as sodium borohydride ($NaBH_4$), hydrazine hydrate (N_2H_4), amines (RNH_2), and thiols (R–SH) during the synthesis of NPs have been utilized as agents for reduction, capping, and stabilization. These chemicals have also been reported to be causing hazardous impacts on the environment and the health of living beings (Hannah & Thompson, 2008). Toxic hydrazine compounds have been reported to cause disruption to the aquatic ecosystem while affecting organs and body systems such as the liver, central nervous system (CNS), kidneys, and cardiovascular system of animals (Oh & Shin, 2015).

Studies have, therefore, now focused on the biosynthesis of NPs using plants, microbes, etc. Biogenic NPs are synthesized using biological organisms. Such organisms could be cultured easily and have high metal uptake intracellularly. The metabolic activities of such microorganisms enable extracellular/intracellular synthesis of NPs utilizing a different route of synthesis (Khan & Jamil Khan, 2017). The biogenic methods provide an ecological and economically feasible way and replace the hazardous chemicals used with the natural products (Vaseghi et al., 2018). These natural compounds such as amino acids, classes of vitamins, polyphenols, and carbohydrate moieties are derived from plants and microbes such as bacteria, fungi, and yeast. The synthesis, processing, and application of biogenic NPs have been considered to be safe, simple, cost-effective, sustainable, and eco-friendly techniques for the fabrication of NPs that would comply with "green chemistry" objectives (Iravani & Varma, 2020). These biogenic NPs have been observed to exhibit remarkable efficiency in removing a wide range of contaminants, including heavy metals, inorganic pollutants, pharmaceutical residues, and organic compounds. According to the study of Klaus-Joerger and team, the usage of biological organisms as potential bio-nanofactories has been a novel approach for the production of various nano-sized particles (Klaus-Joerger et al., 2001). For example, cadmium sulfide nanoparticles (CdS-NPs) of the dimensional range of 20–200 nm were produced inside the cells of *Klebsiella aerogenes* bacterium. Similarly, gold (Au) and silver (Ag) nano-sized particles were amalgamated by using *Verticillium* sp. and the fungus *Fusarium oxysporum*, and Ag nanocrystals of specific size and morphology were amalgamated by using *Pseudomonas stutzeri* bacterium (Klaus-Joerger et al., 2001).

Previous research studies have been carried out to tackle water pollution problems using NPs as they possess beneficial properties such as high surface area, high chemical reactivity, low cost, and lesser energy, effectively regenerating for reusage (Gautam et al., 2019). The applications of biogenic NPs for the treatment of wastewater have been a rapid escalating area of research. A study reported that tin oxide-NPs (SnO_2-NPs) were synthesized from bacterial biomass. *Escherichia herbicola* effectively sequestrate dyes such as methylene blue, Eriochrome Black T, and methyl orange, from samples of wastewater through the process of photocatalytic degradation (Srivastava & Mukhopadhyay, 2014). Similarly, a study comprised removing 98% of the toxic organic dye bromophenol blue within approx. 12 min of treatment of Ag-NPs of size 8–10 nm which were synthesized using the aqueous extract of *Cercidiphyllum japonicum* (Khan et al., 2016). Similar works have been carried out with a marine alga (*Lemanea fluviatilis*); it can biosynthesized Au-NPs of size 5.9 nm and these NPs have been utilized for water purification (Rauwel et al., 2015; AlNadhari et al., 2021). Plant extracts of *Mentha spicata* L. have been used to synthesize iron nanoparticles (Fe-NPs) of size 20–45 nm that effectively sequestered arsenic (As III and V) from wastewater (Prasad et al., 2014). Similarly, the leaves of mango (*Mangifera indica*) have been used to biosynthesize polycrystalline Fe nanorods that effectively treated 38–49% of heavy viscosity of oil (Al-Ruqeishi et al., 2019).

Such studies have highlighted the successful applications of biogenic NPs for the treatment of wastewater. This review provides an insight about biosynthesis of various NPs using plants, bacteria, fungi, yeast, and its potential applications for wastewater treatment. This synthesis aims to provide an in-depth exploration of biogenic NP synthesis techniques, their applications in wastewater treatment, and the challenges and opportunities inherent in their commercialization. By delving into the cutting-edge research and potential of biogenic NPs, we embark on a journey toward cleaner and more sustainable water resources for our planet. The authors have suggested some recommendations for successful application of biogenic NPs for wastewater remediation purposes.

2 Routes of Synthesis of Biogenic Nanoparticles (NPs)

Many physical, biological, and chemical techniques have been studied and employed to synthesize NPs. Biological techniques utilize plants, bacteria, fungi, yeasts, and viruses to synthesize metal as well as metal-oxide NPs (Akhtar et al., 2013). The brief procedure of biosynthesis of NPs has been depicted in Fig. 1.

2.1 Plants

As an eco-friendly and reliable method of synthesis of NPs, plants have successfully been explored for quick and extracellular biosynthesis of metal NPs such as Au- and Ag-NPs (Akhtar et al., 2013). NPs of platinum (Pt) and palladium (Pd) has been synthesized using the extracts of various parts of different plant species. Sugar-derived moieties undergo cross-coupling reactions in the presence of isopropanol under controlled thermal and microwave heating conditions to yield metal NPs. Addition of 5 mol % (or calculated as mole fraction of a compound multiplied to 100 times) of glucose into the above reaction mixture leads to an increase in yield of the products (Camp et al., 2014). A recent study has highlighted the synthesis of Ag- and Au-NPs from extracts of lemongrass leaves, for example, synthesis of Au-NPs by using leaf/stem extracts of *Hyptis capitata* with aqueous tetrachloroaurate ($AuCl_4^-$) ions (Revathy et al., 2022). Similarly, Ag-NPs were prepared using

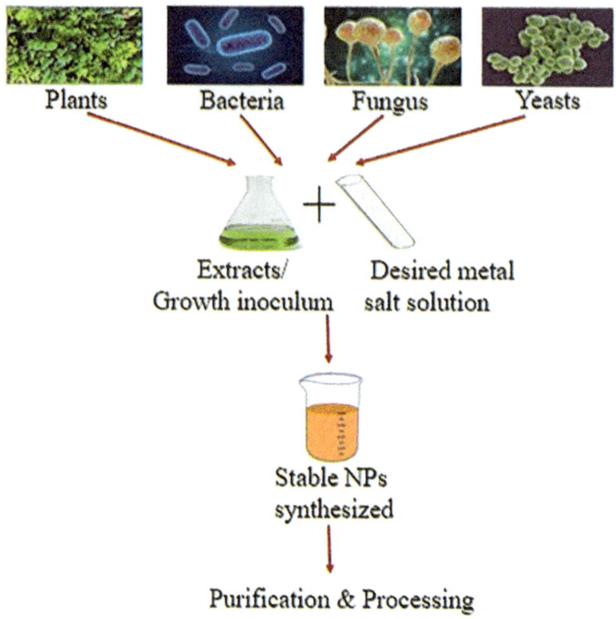

Fig. 1 Procedure for synthesis of biogenic nanoparticles (NPs)

10 mL of the leaf extract with 90 mL of aqueous silver nitrate ($AgNO_3$) solution and subjecting to microwave irradiation (Revathy et al., 2022). Stable and crystalline Au-NPs (16–40 nm) were synthesized by exposing the aqueous extracts of the leaves of geranium and fruit extracts of amla (*Emblica officinalis*) with $AgNO_3$ solution (Tu et al., 2022). Similarly, leaf extracts of tamarind, neem (*Azadirachta indica*), geranium, and aloe vera have been observed to reduce Au ions into Au-NPs and Ag ions into Ag-NPs (Jannathul Firdhouse & Lalitha, 2022).

2.2 *Microorganisms*

Researchers are focusing on synthesis of NPs using microbial sources that have emerged as a promising field in nanobiotechnology. Microorganisms, such as bacteria, fungi, yeasts and viruses, have been studied to possess intra- and extracellular potentiality to produce metal NPs and are therefore considered as bio-factories (Zambonino et al., 2023). Biosynthesis of NPs by using microbes is a rapid, multistep, and comparatively costlier process as it includes microbial isolation, culturing process, maintenance, etc.

2.2.1 Bacteria

Bacterial cells have been studied to possess properties related to biomineralization, bioaccumulation, bioleaching, etc. The biotransformation of metals by bacteria has been observed by researchers and has generated interest in fabrication of NPs using bacteria (Ramanathan et al., 2013; Bhardwaj et al., 2019). A study conducted on *Lactobacillus* sp. and *Saccharomyces cerevisiae* successfully synthesized titanium nanoparticles (Ti-NPs) of size 8–35 nm under controlled pH, partial pressure of hydrogen gas, and redox potential possessed by culture solutions (Jha et al., 2009). Biosynthesis of NPs of Pd, Ag, Fe, cobalt (Co), rhodium (Rh), nickel (Ni), Pt, lithium (Li), etc. has been achieved using the bacterial strain *Pseudomonas aeruginosa*. To obtain the respective NPs, the species were grown under static conditions in uniform solutions of 0.001 M of sodium tetrachloropalladate (Na_2PdCl_4), $AgNO_3$, iron(III) nitrate ($Fe(NO_3)_3$), cobalt(II) chloride hexahydrate ($CoCl_2.6H_2O$), rhodium sodium chloride ($Na_3RhCl_6.2H_2O$), nickel(II) chloride hydrate ($NiCl_2.6H_2O$), ammonium hexachloroplatinate ($(NH_4)_2PtCl_6$), and lithium chloride (LiCl) (Srivastava & Constanti, 2012).

Similarly, bacterial species such as *E. coli*, *Staphylococcus aureus*, Enterobacteria, *P. stutzeri*, and *Geobacter metallireducens* GS-15 have been reported to synthesize metallic NPs of various sizes (Mandal et al., 2006; Narayanan & Sakthivel, 2010; Jiang et al., 2020). For example, *E. coli* K-12 strain was utilized for synthesizing Au-NPs that were used for catalytically degrading 4-nitrophenol pollutants from water bodies (Srivastava et al., 2013). A study established a simple method for synthesizing Au-NPs from *Aspergillum* sp. WL-Au strain was incubated with different concentrations of chloroauric acid ($HAuCl_4$) in phosphate-buffered saline (PBS) systems at 30 °C, 150 rpm for 7 days (Zhang et al., 2021).

2.2.2 Fungi

Myco-nanotechnology has been used for synthesizing NPs from kingdom fungi. Fungi possess diversity, tolerance to high metals, ability to accumulate metal ions, etc. (Moghaddam et al., 2015). Mechanisms of biosynthesis of NPs are intracellular/extracellular. During intracellular mechanisms, metal precursors are mixed with fungal culture. It then gets internalized by biomass and is synthesized. The NPs are subsequently extracted using chemical treatments, centrifugation, filtration, etc. (Molnár et al., 2018). During extracellular synthesis, desired metal precursors are mixed with aqueous filtrate that consists of fungal biomolecules, which results in the formation of free-formed NPs in dispersion/solution (Costa Silva et al., 2017; Gudikandula et al., 2017).

Filamentous fungus, *Verticillium* sp., *Fusarium oxysporum*, etc. have been utilized for the biosynthesis of metallic NPs, such as Ag-NPs of dimensions 25 ± 12 nm. This was achieved by growing the biomass in respective metallic ion solutions (Priyabrata Mukherjee et al., 2001). Similarly, *Aspergillus fumigatus* has been used for synthesizing zinc oxide nanoparticles (ZnO-NPs), and *Rhizopus*, *Fusarium*,

Schizophyllum radiatum, *Candida albicans*, etc. have been used for synthesizing Au- and Ag-NPs (Moghaddam et al., 2015; Naraian & Abhishek, 2020). Nicotinamide adenine dinucleotide (NADH) as well as its dependent enzymes such as nitrate reductase has been studied to be one of the key steps in the biosynthesis of metallic NPs (Baymiller et al., 2017). The proposed method includes pathways of NADH-dependent reductases along with shuttle quinone extracellular procedures. It was reported that depending upon the quantity of NADH formed, the synthesis of alloy of Au-Ag-NPs with a variety of compounds was possible (Mubarakali et al., 2012; Bhardwaj & Naraian, 2021). The size of NPs formed has been observed to be dependent on synthesis conditions such as fungal species used, temperature, pH, and dispersion medium (Lee & Jun, 2019).

2.2.3 Yeast

Marine yeast species *Yarrowia lipolytica* has been observed to biosynthesize Au-NPs upon exposure to chloroauric acid ($HAuCl_4$) (Agnihotri et al., 2009). In another study, the marine yeast strain *Rhodosporidium diobovatum* has been studied to synthesize lead sulfide nanoparticles (PbS-NPs) of size 2–5 nm with a ratio of 1:2 Pb/S (Seshadri et al., 2011). Various strains of *Saccharomyces* genus have been used to synthesize NPs of zinc sulfide (ZnS), Ag, and titanium dioxide (TiO_2) (Sandana Mala & Rose, 2014). Research conducted by Venkat and team successfully synthesized intracellular phytochelatin-coated cadmium sulfide nanoparticles (CdS-NPs) of size 50–60 nm by using *C. albicans* and CdS solution. The synthesized CdS-NPs showed antibacterial activities against bacterial species *Salmonella typhi* and *S. aureus* (Kumar et al., 2019). Biogenesis of selenium nanoparticles (Se-NPs) was obtained using yeast-derived extracts of *Magnusiomyces ingens* LH-F1 by growing it in 2 mM SeO_2, 500 mg/L protein at pH 7 (Lian et al., 2019). *Saccharomyces cerevisiae* has been studied to synthesize NPs of ZnS, TiO_2, and CdS, using solutions of zinc sulfate ($ZnSO_4$), titanic acid ($TiO(OH)_2$), and cadmium chloride ($CdCl_2$) with hydrogen sulfide (H_2S), respectively (Patel et al., 2021). Another study was conducted on baker's yeast *S. cerevisiae*-biosynthesized Ag-NPs of size 3–60 nm using cell extracts. The synthesized Ag-NPs were observed to possess antimicrobial activities against pathogens such as *E. coli*, *S. aureus*, *P. aeruginosa*, *Bacillus subtilis*, and *C. albicans* (Bhardwaj et al., 2022b; Salem, 2022).

3 Applications of Biogenic Nanoparticles (NPs) for the Treatment of Wastewater

Biogenic NPs have broad applications in context to chemical engineering, manufacturing of textiles, tissue engineering, clinical diagnostics (nanobots), nanomedicines, and electronics, as well as organ implantations (Santhoshkumar et al., 2011), biosensors (Probin Phanjom et al., 2012), biological imaging (Nagaraj et al., 2021), biomarkers, cell labeling, etc. (Le et al., 2011; Singh & Nalwa, 2011). Gautam and team studied the removal of inorganic/organic pollutants, pharmaceutical pollutants, and toxic heavy metals found in wastewater by using biogenic NPs (Gautam et al., 2019; Singh et al., 2021). The diagrammatic representation of the use of NPs for the treatment of wastewater is shown in Fig. 2.

3.1 Removal of Organic Pollutants

Due to the expansion of urbanization, and the activities of industries and agriculture, water is polluted (Gupta et al., 2023). Organic pollutants are made up of mostly carbon and hydrogen. These pollutants are pesticides, insecticides, synthetic dyes, polychlorinated biphenyls (PCBs), polycyclic aromatic hydrocarbons (PAHs), and phenols (Lapworth et al., 2012). Persistent organic pollutants (POPs) are harmful in nature and have been reported as hazardous to the environment (Bhardwaj & Jindal, 2019; Bhardwaj et al., 2020, 2021). These organic pollutants enter aquatic bodies through agricultural runoff, industrial effluents, and sewages of domestic sectors. After that, these pollutants enter the food chain and accumulate in the tissue of aquatic plants and animals and then human beings.

Nanotechnology (NT) is an emerging technique for the removal of various pollutants from different commodities (Bhardwaj et al., 2022a). Nano-adsorbents, nano-films, and nano-catalysts are NT-based products that are used to remove organic pollutants. Wastewater contaminated with monochlorobenzene, for example, was treated using degradation techniques based upon Fenton-like oxidation that

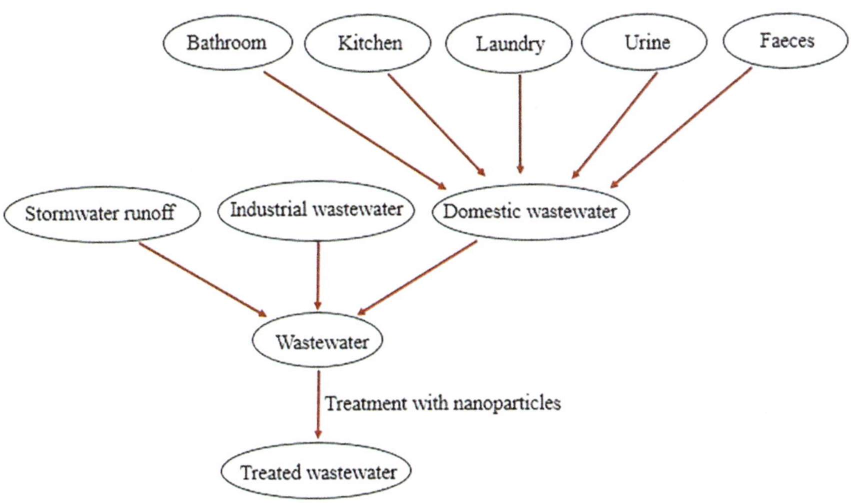

Fig. 2 Use of nanoparticles (NPs) for the treatment of wastewater

was mediated by Fe-NPs that were simultaneously reduced or capped by polyphenols from tea (Kuang et al., 2013). Research showed synthesis of Fe and Fe/Pb bimetallic NPs using tea extracts. The formed NPs are sized between 20 and 30 nm (Smuleac et al., 2011). The NPs were used for eliminating highly toxic pollutants, e.g., trichloroethane (TCE), from wastewater through the reductive degradative mechanisms. Biogenic Ag nanocatalysts have substantiated their potency in removing/degrading many organic pollutants. Besides Ag, other biogenic NPs have also been studied for their effectiveness in disassembling some toxic organic contaminants found in wastewater (Smuleac et al., 2011).

Photocatalytically removing naphthalene found in aqueous phase was carried out using biologically fabricated NPs doped with Fe and Zn (Muthukumar et al., 2017). The photocatalytic efficiencies of Fe-/Zn-NPs were observed to be superior as compared to Zn-NPs. Aqueous extracts of fungal strains *Trichoderma viride* and *Hypocrea lixii* were utilized for synthesizing extracellular Au-NPs for treating 4-nitrophenol present in wastewater (Mishra et al., 2014). Au-NPs were used to convert 4-nitrophenol into 4-aminophenol. A research study developed biocompatible Au-NPs from *Aspergillum* sp. for effectively removing the presence of aromatic water pollutants such as *o*-nitroaniline, 2-nitrophenol, *m*-nitroaniline 3-nitrophenol, and 4-nitrophenol (Qu et al., 2017).

Arora and team studied phenolic pollutants and their toxicity and stated that they are carcinogens and can damage the liver as well as red blood cells (RBCs) (Arora et al., 2014). After interaction with the microorganisms, they may generate other compounds that could be similarly toxic to the original compound. Wang and team described the mechanism of the production of manganese oxide (BioMnOx) nanobiocomposite from the green algal species *Desmodesmus* sp. WR1 and reported that it is helpful for removing bisphenol from samples of wastewater (Wang et al., 2017). A study reported synthesizing manganese nanocatalysts using the manganese (Mn)-oxidizing bacterial species *Pseudomonas* sp. G7, and it is helpful in the oxidative degradation of 2,4-dichlorophenol, 2,4,6-trichlorophenol, and 2-chlorophenol (Tu et al., 2015). A list of different sources of biogenic NPs which were used in various applications have been highlighted in Table 1.

3.2 Removal of Pharmaceutical Pollutants

Several pharmaceutical compounds such as hormones, antibiotics, endocrine-disrupting compounds (EDCs), and steroids might create problems for the health of the wildlife and humans if these compounds enter in the water supply system (Kim et al., 2007). Malik and team suggested the nano-based approach for the removal of these compounds (Malik et al., 2017). A team led by Forrez developed biogenic bio-palladium (bio-Pd) and manganese oxides (bioMnOx) that were nano-sized particles and are being used for removing different pharmaceutical compounds from wastewater (Forrez et al., 2011). There are various pharmaceutical pollutants that were reported in sewage runoff and were successfully removed

Table 1 Different sources (bacteria/fungi/algae) of biogenic nanoparticles (NPs) with their applications

S. no.	Biogenic sources (bacteria/fungi/algae)	Nanoparticles (NPs)	Applications	References
Bacteria				
1	*Lactobacillus kimchicus*	Gold	Drug delivery, cancer diagnostic	Mathivanan et al. (2019)
2	*Staphylococcus epidermidis*		Catalysts	Pantidos and Horsfall (2014)
3	*Paracoccus haeundaensis*		Antioxidants	Srinath and Rai (2015)
4	*Cupriavidus* sp.	Silver	Antibacterial	Markus et al. (2016)
5	*Bacillus subtilis*			Patil et al. (2019)
6	*Staphylococcus aureus*	Zinc oxide		Rauf et al. (2017)
7	*Bacillus subtilis*		Synthetic dyes	Dhandapani et al. (2020)
8	*Bacillus* sp.	Nanoselenium	Biomedical	Bharathi et al. (2020)
9	*Desulfovibrio vulgaris*	Platinum and palladium	Catalysts	Martins et al. (2017)
Fungi				
10	*Aspergillus niger*	Iron oxide	Wastewater treatment	Mughal et al. (2021)
11	*Aspergillus terreus*	Silver	Antibacterial, anticancer	Hulikere and Joshi (2019)
12	*Cladosporium cladosporioides*		Antioxidant, antimicrobial	Zhang et al. (2020)
13	*Aspergillus niger*	Copper	Antidiabetic and antibacterial	Zhang et al. (2019)
14	*Mariannaea* sp. HJ	Selenium	Medicinal and electronics	Vijayanandan and Balakrishnan (2018)
15	*Aspergillus nidulans*	Cobalt oxide	Energy storage	Joshi et al. (2017)
16	*Cladosporium cladosporioides*	Gold	Antioxidant, antimicrobial	Bhargava et al. (2016)
17	*Cladosporium oxysporum*		Catalysis	Srivastava et al. (2019)
18	*Fusarium oxysporum*	Platinum	Nanomedicine	Chatterjee et al. (2020)
Algae				
19	*Botryococcus braunii*	Silver	Catalysis	Manikandakrishnan et al. (2019)
20	*Portieria hornemannii*		Antibacterial	Arya et al. (2019)
21	*Colpomenia sinuosa* and *Pterocladia capillacea*	Iron oxide		Colin et al. (2018)
22	*Caulerpa racemosa*	Gold	Biomedical	Salem et al. (2019)
23	*Egregia* sp.			Gonzalez-Ballesteros et al. (2017)
24	*Cystoseira baccata*		Cancer	Mourdikoudis et al. (2018)

by bioMnOx-MBR such as diclofenac (86%), triclosan (78%), iopromide (68%), codeine (93%), clarithromycin (75%), N-acetyl-sulfamethoxazole (92%), sulfamethoxazole (52%), diuron (94%), mecoprop (81%), iomeprol (63%), naproxen (95%), ibuprofen (95%), iohexol (72%), and chlorophene (89%) (Forrez et al., 2011). Martins and team synthesized bio-Pd and bio-Pt using *Desulfovibrio vulgaris* for the removal of 17-β-estradiol, ciprofloxacin, and sulfamethoxazole (Martins et al., 2017).

De Corte and team synthesized bio-Pd-NPs from the metal-reducing bacteria which were used in the treatment of diclofenac-bearing water (De Corte et al., 2012). De Gusseme and team synthesized Pd-NPs using *Shewanella oneidensis* for the removal of diatrizoate from the effluent (De Gusseme et al., 2011). Au/Pd nanocatalyst was prepared using *S. oneidensis* and used to purify trichloroethylene- and diclofenac-contaminated water (De Corte et al., 2012). De Corte and team reported that bio-Au and bio-Pd nanocatalysts failed to remove the pollutants individually, while the combined effectiveness of Au and Pd nanocatalysts was notable and was able to remove >78% of pollutants (De Corte et al., 2012). Furgal and team synthesized bio-manganese oxide nanoparticles (MnO_2-NPs) using *P. putida* for the elimination of various micropollutants from aquatic samples (Furgal et al., 2015). Xu and team studied the combined effectiveness of biosynthesized Pd-NPs, that were chemically developed MnO_2 and iron oxide (Fe_3O_4) NPs for degrading various pharmaceutical pollutants (Xu et al., 2018).

3.3 Removal of Inorganic and Radioactive Pollutants

Due to several nuclear activities, the aquatic bodies have been contaminated by radioactive waste. These wastes may pollute the surface water as well as ground water through leaching from mines and runoff from nuclear reactors and nuclear power plants. These wastes can cause chronic and acute toxic effects on living beings (Gawande & Jenkins-Smith, 2001). Researchers studied the contamination level in the PET bottled drinking water and lake water (Bhardwaj & Sharma, 2021; Bhardwaj, 2022). Handley-Sidhu and team developed biogenic hydroxyapatite materials (bio-HAPs) using *Serratia* sp. for the remediation of such waste (Handley-Sidhu et al., 2011a). The remediation efficiency of bio-HAP for strontium (Sr^{2+}) was higher than commercial HAP. Bio-HAP-NPs successfully scavenged some biohazard materials such as Sr, Co, uranium (U), and europium (Eu) from samples of groundwater (Handley-Sidhu et al., 2014). Handley-Sidhu and team used bio-HAP-NPs (20–90 nm) for the elimination of Sr^{2+}, Co^{2+}, etc. from the water (Handley-Sidhu et al., 2011b). The surface area of bio-HAP-NPs was found to be more than the commercial HAP-NPs. Bio-HAP-NPs with a size <40 nm and a surface area >70 m^2g^{-1} showed greater removal of radioactive waste. It was stated that calcium ions (Ca^{2+}), magnesium ions (Mg^{2+}), and sodium ions (Na^+) showed a minimal impact on removing the ability of bio-HAP-NPs. Nanoscale bio-HAP-NPs were also described for the elimination of radionuclides from the water (Gangappa et al., 2017). Choi and team synthesized biogenic Au-NPs (bio-Au-NPs) using a bacterial species, *Deinococcus radiodurans*, for eliminating radioactive iodine ($^{125}I^-$) (Choi et al., 2017).

Nitrate (NO_3^-) contamination in samples of water is a serious problem because of its great solubility and persistent nature. Higher concentrations of NO_3^- can cause methemoglobinemia, congenital malformations, and cancers (Fewtrell, 2004). NO_3^- can enter water bodies through agricultural activities and cause eutrophication which is harmful to aquatic ecosystems (Camargo et al., 2005). A research study stated that various nanotechnological approaches are present to remove the contamination of NO_3^- from samples of wastewater (Tyagi et al., 2018). Wang and team studied the batch adsorption method and synthesized iron nano-adsorbents from extracts of eucalyptus leaves and green tea (Wang et al., 2014). Copper nanoparticles (Cu-NPs) synthesized from the flower extract of *Hibiscus sabdariffa* were impregnated with activated carbon which was prepared using the biomass of babassu coconut. These carbon complexes were utilized for removing NO_3^- at varying temperatures (Paixão et al., 2017). The capacities of adsorption of the nano-carbon complexes were expected by Langmuir and Freundlich isotherm and were superior than simple activated carbon. Katata-Seru and team synthesized eco-friendly Fe-NPs from the leaf and seed of *Moringa oleifera* for the elimination of NO_3^- from the ground water and surface water samples (Katata-Seru et al., 2017).

Phosphorus (P) is also known to cause eutrophication by promoting algal growth. Yong and team synthesized hydroxyapatite crystalline nanomaterial using *Serratia* sp. for the remedy of phosphate (PO_4^{3-}) present in wastewater (Yong et al., 2004). Cao and team synthesized Fe-NPs treated with a cationic surfactant (CTAB) from eucalyptus leaf extract for the elimination of PO_4^{3-}. The PO_4^{3-} elimination efficiencies of CTAB-coated NPs were higher as compared to uncoated NPs (Cao et al., 2016). Gan and team reported the efficient elimination of PO_4^{3-} via Fe_3O_4-NPs that were synthesized from the extract of eucalyptus leaf (Gan et al., 2018).

3.4 Removal of Heavy Metals

Heavy metals are well acknowledged for their toxic effect on human health and the environment. Lead (Pb), mercury (Hg), nickel (Ni), arsenic (As), cadmium (Cd), and chromium (Cr) are extremely toxic and poisonous due to their affinity for bioaccumulation (Chipasa, 2003; Sadre Alam et al., 2023; Bhardwaj et al., 2023a, b). Domestic and industrial effluents, agricultural runoff, and acid mine drainage are the main contributors to heavy metal pollution of the aquatic environment. Mukherjee and team synthesized Fe_2O_3-NPs by using *Aloe vera* leaf extract for the removal of As from the aqueous phase (Mukherjee et al., 2016). A study organically synthesized magnetic Fe_3O_4-NPs from tea waste for the elimination of As(III) and As(V) (Lunge et al., 2014). Martinez-Cabans and team manufactured Fe_3O_4-NPs by using eucalyptus extract for the elimination of As from the wastewater (Martínez-Cabanas et al., 2016). Andjelkovic and team extracted Fe_3O_4 nanowires from the chemoautotrophic bacteria *Mariprofundus ferrooxydans* for the removal of As(V) and As(III) (Andjelkovic et al., 2017).

Cr is well acknowledged for its lethal effects on living beings. Rao and team synthesized Fe-NPs using *Punica granatum* extract for eliminating Cr(VI) from the wastewater. The NPs' surface was altered by two strains (NCIM 3590 and NCIM 3589) of yeast and *Yarrowia lipolytica* used for the removal of Cr(VI) present in aqueous solution (Rao et al., 2013). Fe-NPs sized between 50 and 80 nm were synthesized using the leaf extract of eucalyptus for eliminating Cr(VI) from samples of wastewater (Madhavi et al., 2013). Fe-NPs were synthesized using the extracts of fruit and plant (e.g., *Mentha spicata*, *P. granatum*, *Syzygium aromaticum*, and *Camellia sinensis*) for the elimination of Cr(VI) from the aqueous phase (Mystrioti et al., 2016). Xiao and team reported that the sorption of Cr was greater in acidic medium because of the charge on the surface of Fe-NPs (Xiao et al., 2016). Lingamdinne and team synthesized Fe-NPs from the seed extract of *Cnidium monnieri* (L.) Cuss for eliminating Pb(II) and Cr(III) from the aqueous medium (Lingamdinne et al., 2017).

Cd toxicity in water may be the source of itai-itai and other medical or health concerns by damaging the normal action of the cardiovascular system, liver, and kidney (Nordberg, 2004). Biogenic NPs have been extensively used to clean the water from toxicity of Cd. Raj and team synthesized CdS-NPs from the marine bacterium *P. aeruginosa* JP-11 for the elimination of Cd from samples of wastewater (Raj et al., 2016). Another study reported the elimination of Cd from the aqueous phase through the Fe-NPs that were synthesized from the extract of tangerine peel (Ehrampoush et al., 2015).

Some heavy metals such as Zn and Cu are considered micronutrients, but they may become poisonous for aquatic life and humans at high concentration. Jain and team reported biogenic Se-NPs by mixing and incubating Se with cultured anaerobic sludge for minimizing Zn(II) in samples of wastewater (Jain et al., 2015). A team led by Kandasamy synthesized biogenic Fe-NPs from *Streptomyces thermolineatus* for the elimination of Cu ions from the effluents of the pigment industry (Kandasamy, 2017). A group of researchers engineered Se-NPs from bacterial extracellular polymeric substances (EPS) for removing elemental Hg from groundwater (Wang et al., 2018). Srivastava and team synthesized MgO-NPs by utilizing biocompatible acacia gum for eliminating Cd(II), Mn(II), Pb(II), Ni(II), Zn(II), Cu(II), and Co(II) from synthetic wastewater (Srivastava et al., 2015). Zhou and team informed the synthesis of BMO-NPs by using *P. putida* MnB1 for removing Zn(II), Cd(II), and Pb(II) (Zhou et al., 2018).

4 Conclusions and Future Recommendations

The escalating and persistent increase in water body pollution, stemming from various organic and inorganic pollutants, is a growing source of concern. Novel technologies have been extensively researched to combat water pollution, with a particular focus on wastewater treatment. Traditional methods of NP synthesis often involve the use of volatile and hazardous chemicals, leading to the generation of secondary pollution. Consequently, there is a strong emphasis on exploring the biogenic synthesis of NPs, known for their environmental safety and cost-effectiveness. Biogenic NPs are synthesized using natural compounds such as carbohydrates, polymers, proteins, flavonoids, alkaloids, polyphenols, and more, sourced from fungi, plants, yeast, and bacteria. These compounds have demonstrated their efficacy as stabilizing and capping agents during NP synthesis. This study highlights the significance of and recent advancements in biogenic NP synthesis. It explores various biosynthesis routes utilizing plants, bacteria, fungi, and yeast, showcasing the potential applications of these NPs in wastewater treatment. Biogenic NPs show promise in the removal of heavy metals, inorganic and radioactive pollutants, pharmaceutical residues, and organic contaminants from wastewater. While this field of research is rapidly expanding, the commercial application of these techniques remains underexplored. Challenges, primarily related to cost-effectiveness, hinder their large-scale implementation for wastewater purification. Additionally, there is a pressing need to develop modified NPs that are more efficient, easily manageable, sustainable, and environmentally friendly. In summary, the rise in water pollution necessitates innovative solutions. Biogenic NPs, synthesized from natural compounds, hold great potential for addressing this issue. Although their applications in wastewater treatment are still emerging, their environmental benefits are substantial. However, scaling up and commercializing these techniques present challenges that must be addressed to fully harness their potential for sustainable water purification.

Acknowledgments The authors are grateful to Amity University for providing the platform to do this study.

Statement Declaration Funding: This study was not supported by any funding agency.

Conflicts of Interest/Competing Interest: The authors declare that they have no competing financial interest or personal relationship that could have appeared to influence the work reported in this chapter.

Availability of Data and Materials: Not applicable.

Code Availability: Not applicable.

Author's Contributions: All authors have equal contributions.

References

Agnihotri, M., et al. (2009). Biosynthesis of gold nanoparticles by the tropical marine yeast Yarrowia lipolytica NCIM 3589. *Materials Letters, 63*(15), 1231–1234. Available at:. https://doi.org/10.1016/J.MATLET.2009.02.042

Akhtar, M. S., Panwar, J., & Yun, Y. S. (2013). Biogenic synthesis of metallic nanoparticles by plant extracts. *ACS Sustainable Chemistry and Engineering, 1*(6), 591–602. Available at: https://doi.org/10.1021/SC300118U/ASSET/IMAGES/MEDIUM/SC-2012-00118U_0002.GIF

Alam, S., et al. (2023). Estimation of heavy metals and fluoride ion in vegetables grown nearby the stretch of river Yamuna, Delhi (NCR), India. *Indian Journal of Environmental Protection, 43*(1), 64–73. Available at: https://www.e-ijep.co.in/43-1-64-73/. Accessed 21 Mar 2023.

AlNadhari, S., et al. (2021). A review on biogenic synthesis of metal nanoparticles using marine algae and its applications. *Environmental Research, 194*, 110672. Available at: https://doi.org/10.1016/J.ENVRES.2020.110672

Al-Ruqeishi, M. S., Mohiuddin, T., & Al-Saadi, L. K. (2019). Green synthesis of iron oxide nanorods from deciduous Omani mango tree leaves for heavy oil viscosity treatment. *Arabian Journal of Chemistry, 12*(8), 4084–4090. Available at: https://doi.org/10.1016/J.ARABJC.2016.04.003

Andjelkovic, I., et al. (2017). Bacterial iron-oxide nanowires from biofilm waste as a new adsorbent for the removal of arsenic from water. *RSC Advances, 7*(7), 3941–3948. Available at: https://doi.org/10.1039/C6RA26379H

Arora, P. K., Srivastava, A., & Singh, V. P. (2014). Bacterial degradation of nitrophenols and their derivatives. *Journal of Hazardous Materials, 266*, 42–59. Available at: https://doi.org/10.1016/j.jhazmat.2013.12.011

Arya, A., Mishra, V., & Chundawat, T. S. (2019). Green synthesis of silver nanoparticles from green algae (Botryococcus braunii) and its catalytic behavior for the synthesis of benzimidazoles. *Chemical Data Collections, 20*. Available at:. https://doi.org/10.1016/J.CDC.2019.100190

Baig, N., et al. (2021). Nanomaterials: A review of synthesis methods, properties, recent progress, and challenges. *Materials Advances, 2*(6), 1821–1871. Available at: https://doi.org/10.1039/D0MA00807A

Baymiller, M., Huang, F., & Rogelj, S. (2017). Rapid one-step synthesis of gold nanoparticles using the ubiquitous conzyme NADH. *Matters* [Preprint]. Available at: https://doi.org/10.19185/MATTERS.201705000007

Bharathi, S., et al. (2020). Extracellular synthesis of nanoselenium from fresh water bacteria Bacillus sp., and its validation of antibacterial and cytotoxic potential. *Biocatalysis and Agricultural Biotechnology, 27*, 101655. Available at: https://doi.org/10.1016/J.BCAB.2020.101655

Bhardwaj, L. K. (2022). Evaluation of bis (2-ethylhexyl) phthalate (DEHP) in the PET bottled mineral water of different brands and impact of heat by GC-MS/MS. *Chemistry Africa, 5*(4), 929–942. Available at: https://doi.org/10.21203/RS.3.RS-298354/V1

Bhardwaj, L. K., & Jindal, T. (2019). Contamination of Lakes in Broknes Peninsula, East Antarctica through the Pesticides and PAHs. *Asian-Journal of Chemistry, 31*(7), 1574–1580. Available at: https://doi.org/10.14233/ajchem.2019.22022

Bhardwaj, A. K., & Naraian, R. (2021). Cyanobacteria as biochemical energy source for the synthesis of inorganic nanoparticles, mechanism and potential applications: A review. *3 Biotech, 11*, 10. Available at: https://doi.org/10.1007/S13205-021-02992-5

Bhardwaj, L. K., & Sharma, A. (2021). Estimation of physico-chemical, trace metals, microbiological and phthalate in PET bottled water. *Chemistry Africa, 4*(4), 981–991. Available at: https://doi.org/10.1007/S42250-021-00267-3/METRICS

Bhardwaj, A. K., et al. (2019). Bacterial killing efficacy of synthesized rod shaped cuprous oxide nanoparticles using laser ablation technique. *SN Applied Sciences, 1*(11), 1–8. Available at: https://doi.org/10.1007/S42452-019-1283-9/FIGURES/7

Bhardwaj, L. K., et al. (2020). Persistent organic pollutants in lakes of Broknes peninsula at Larsemann Hills area, East Antarctica. *Ecotoxicology, 28*(5), 589–596. Available at: https://doi.org/10.1007/s10646-019-02045-x

Bhardwaj, L. K., Sharma, S., & Jindal, T. (2021). Occurrence of polycyclic aromatic hydrocarbons (PAHs) in the lake water at Grovnes Peninsula over East Antarctica. *Chemistry Africa, 4*(4), 965–980. Available at: https://doi.org/10.1007/S42250-021-00278-0

Bhardwaj, L. K., Rath, P., & Choudhury, M. (2022a). A comprehensive review on the classification, uses, sources of nanoparticles (NPs) and their toxicity on health. *Aerosol Science and Engineering*, 1–18. Available at: https://doi.org/10.1007/S41810-022-00163-4

Bhardwaj, A. K., et al. (2022b). Biogenic and non-biogenic waste for the synthesis of nanoparticles and their applications. In *Bioremediation* (pp. 207–218). CRC Press. Available at: https://doi.org/10.1201/9781003181224-13

Bhardwaj, L. K., Kumar, D., & Kumar, A. (2023a). Phytoremediation potential of Ocimum sanctum: A sustainable approach for remediation of heavy metals. Available at: https://doi.org/10.20944/PREPRINTS202308.0593.V1

Bhardwaj, L. K., Sharma, S., & Jindal, T. (2023b). Estimation of physico-chemical and heavy metals in the lakes of Grovnes & Broknes Peninsula, Larsemann Hill, East Antarctica. *Chemistry Africa*, 1–18. Available at: https://doi.org/10.1007/S42250-023-00668-6/METRICS

Bhargava, A., et al. (2016). Utilizing metal tolerance potential of soil fungus for efficient synthesis of gold nanoparticles with superior catalytic activity for degradation of rhodamine B. *Journal of Environmental Management, 183*, 22–32. Available at: https://doi.org/10.1016/J.JENVMAN.2016.08.021

Buzea, C., Pacheco, I. I., & Robbie, K. (2007). Nanomaterials and nanoparticles: Sources and toxicity. *Biointerphases, 2*(4), MR17. Available at: https://doi.org/10.1116/1.2815690

Camargo, J. A., Alonso, A., & Salamanca, A. (2005). Nitrate toxicity to aquatic animals: A review with new data for freshwater invertebrates. *Chemosphere, 58*(9), 1255–1267. Available at: https://doi.org/10.1016/J.CHEMOSPHERE.2004.10.044

Camp, J. E., et al. (2014). Glucose-derived palladium(0) nanoparticles as in situ-formed catalysts for suzuki-miyaura cross-coupling reactions in isopropanol. *ACS Sustainable Chemistry and Engineering, 2*(3), 500–505. Available at: 10.1021/SC400410V/SUPPL_FILE/SC400410V_SI_001.PDF.

Cao, D., et al. (2016). Removal of phosphate using iron oxide nanoparticles synthesized by eucalyptus leaf extract in the presence of CTAB surfactant. *Chemosphere, 159*, 23–31. Available at: https://doi.org/10.1016/J.CHEMOSPHERE.2016.05.080

Chatterjee, S., et al. (2020). Biofabrication of iron oxide nanoparticles using manglicolous fungus Aspergillus niger BSC-1 and removal of Cr(VI) from aqueous solution. *Chemical Engineering Journal, 385*, 123790. Available at: https://doi.org/10.1016/J.CEJ.2019.123790

Chipasa, K. B. (2003). Accumulation and fate of selected heavy metals in a biological wastewater treatment system. *Waste Management (New York, N.Y.), 23*(2), 135–143. Available at: https://doi.org/10.1016/S0956-053X(02)00065-X

Choi, M. H., et al. (2017). Efficient bioremediation of radioactive iodine using biogenic gold nanomaterial-containing radiation-resistant bacterium, Deinococcus radiodurans R1. *Chemical Communications, 53*(28), 3937–3940. Available at: https://doi.org/10.1039/C7CC00720E

Colin, J. A., et al. (2018). Gold nanoparticles synthesis assisted by marine algae extract: Biomolecules shells from a green chemistry approach. *Chemical Physics Letters, 708*, 210–215. Available at: https://doi.org/10.1016/J.CPLETT.2018.08.022

Costa Silva, L. P., et al. (2017). Extracellular biosynthesis of silver nanoparticles using the cell-free filtrate of nematophagous fungus Duddingtonia flagrans. *International Journal of Nanomedicine, 12*, 6373. Available at: https://doi.org/10.2147/IJN.S137703

De Corte, S., et al. (2012). Biosupported bimetallic Pd-Au nanocatalysts for dechlorination of environmental contaminants. *Environmental Science & Technology, 45*(19), 8506–8513. Available at: https://doi.org/10.1021/ES2019324

De Gusseme, B., et al. (2011). Biogenic palladium enhances diatrizoate removal from hospital wastewater in a microbial electrolysis cell. *Environmental Science and Technology, 45*(13), 5737–5745. Available at: https://doi.org/10.1021/ES200702M/SUPPL_FILE/ES200702M_SI_001.PDF

Dhandapani, P., et al. (2020). Ureolytic bacteria mediated synthesis of hairy ZnO nanostructure as photocatalyst for decolorization of dyes. *Materials Chemistry and Physics, 243*, 122619. Available at: https://doi.org/10.1016/J.MATCHEMPHYS.2020.122619

Ehrampoush, M. H., et al. (2015). Cadmium removal from aqueous solution by green synthesis iron oxide nanoparticles with tangerine peel extract. *Journal of Environmental Health Science and Engineering, 13*(1), 1–7. Available at: https://doi.org/10.1186/S40201-015-0237-4/TABLES/2

Fewtrell, L. (2004). Drinking-water nitrate, methemoglobinemia, and global burden of disease: A discussion. *Environmental Health Perspectives, 112*(14), 1371. Available at: https://doi.org/10.1289/EHP.7216

Forrez, I., et al. (2011). Biogenic metals for the oxidative and reductive removal of pharmaceuticals, biocides and iodinated contrast media in a polishing membrane bioreactor. *Water Research, 45*(4), 1763–1773. Available at: https://doi.org/10.1016/J.WATRES.2010.11.031

Furgal, K. M., Meyer, R. L., & Bester, K. (2015). Removing selected steroid hormones, biocides and pharmaceuticals from water by means of biogenic manganese oxide nanoparticles in situ at ppb levels. *Chemosphere, 136*, 321–326. Available at: https://doi.org/10.1016/J.CHEMOSPHERE.2014.11.059

Gan, L., et al. (2018). Effects of cetyltrimethylammonium bromide on the morphology of green synthesized Fe3O4 nanoparticles used to remove phosphate. *Materials Science & Engineering. C, Materials for Biological Applications, 82*, 41–45. Available at: https://doi.org/10.1016/J.MSEC.2017.08.073

Gangappa, R., Farrier, A., & Macaskie, L. E. (2017). Eu3+ sequestration by biogenic nano-hydroxyapatite synthesized at neutral and alkaline pH. *Geomicrobiology Journal, 34*(9), 753–759. Available at: https://doi.org/10.1080/01490451.2016.1261966/SUPPL_FILE/UGMB_A_1261966_SM9701.DOCX

Gautam, P. K., et al. (2019). Synthesis and applications of biogenic nanomaterials in drinking and wastewater treatment. *Journal of Environmental Management, 231*, 734–748. Available at: https://doi.org/10.1016/J.JENVMAN.2018.10.104

Gawande, K., & Jenkins-Smith, H. (2001). Nuclear waste transport and residential property values: Estimating the effects of perceived risks. *Journal of Environmental Economics and Management, 42*(2), 207–233. Available at: https://doi.org/10.1006/JEEM.2000.1155

González-Ballesteros, N., et al. (2017). Green synthesis of gold nanoparticles using brown algae Cystoseira baccata: Its activity in colon cancer cells. *Colloids and Surfaces. B, Biointerfaces, 153*, 190–198. Available at: https://doi.org/10.1016/J.COLSURFB.2017.02.020

Gudikandula, K., Vadapally, P., & Charya, M. A. S. (2017). Biogenic synthesis of silver nanoparticles from white rot fungi: Their characterization and antibacterial studies. *OpenNano, 2*, 64–78. Available at: https://doi.org/10.1016/j.onano.2017.07.002

Gupta, P., Sharma, A., & Bhardwaj, L. K. (2023). Solid waste management (SWM) and its effect on environment & human health. Available at: https://doi.org/10.20944/PREPRINTS202309.0384.V1

Handley-Sidhu, S., et al. (2011a). Nano-crystalline hydroxyapatite bio-mineral for the treatment of strontium from aqueous solutions. *Biotechnology Letters, 33*(1), 79–87. Available at: https://doi.org/10.1007/S10529-010-0391-9

Handley-Sidhu, S., et al. (2011b). Uptake of Sr 2+ and Co 2+ into biogenic hydroxyapatite: Implications for biomineral ion exchange synthesis. *Environmental Science and Technology, 45*(16), 6985–6990. Available at: https://doi.org/10.1021/ES2015132/SUPPL_FILE/ES2015132_SI_001.PDF

Handley-Sidhu, S., et al. (2014). Bacterially produced calcium phosphate nanobiominerals: Sorption capacity, site preferences, and stability of captured radionuclides. *Environmental Science and Technology, 48*(12), 6891–6898. Available at: https://doi.org/10.1021/ES500734N/SUPPL_FILE/ES500734N_SI_001.PDF

Hannah, W., & Thompson, P. B. (2008). Nanotechnology, risk and the environment: A review. *Journal of Environmental Monitoring: JEM, 10*(3), 291–300. Available at: https://doi.org/10.1039/B718127M

Hulikere, M. M., & Joshi, C. G. (2019). Characterization, antioxidant and antimicrobial activity of silver nanoparticles synthesized using marine endophytic fungus-Cladosporium cladosporioides. *Process Biochemistry, 82*, 199–204.

Iravani, S., & Varma, R. S. (2020). Greener synthesis of lignin nanoparticles and their applications. *Green Chemistry, 22*(3), 612–636. Available at: https://doi.org/10.1039/C9GC02835H

Jain, R., et al. (2015). Adsorption of zinc by biogenic elemental selenium nanoparticles. *Chemical Engineering Journal, 260*, 855–863. Available at: https://doi.org/10.1016/J.CEJ.2014.09.057

Jannathul Firdhouse, M., & Lalitha, P. (2022). Biogenic green synthesis of gold nanoparticles and their applications – A review of promising properties. *Inorganic Chemistry Communications, 143*, 109800. Available at: https://doi.org/10.1016/J.INOCHE.2022.109800

Jha, A. K., Prasad, K., & Kulkarni, A. R. (2009). Synthesis of TiO2 nanoparticles using microorganisms. *Colloids and surfaces. B, Biointerfaces, 71*(2), 226–229. Available at: https://doi.org/10.1016/J.COLSURFB.2009.02.007

Jiang, Z., Shi, M., & Shi, L. (2020). Degradation of organic contaminants and steel corrosion by the dissimilatory metal-reducing microorganisms Shewanella and Geobacter spp. *International Biodeterioration & Biodegradation, 147*, 104842. Available at: https://doi.org/10.1016/J.IBIOD.2019.104842

Joshi, C. G., et al. (2017). Biogenic synthesis of gold nanoparticles by marine endophytic fungus-Cladosporium cladosporioides isolated from seaweed and evaluation of their antioxidant and antimicrobial properties. *Process Biochemistry, 63*, 137–144.

Kandasamy, R. (2017). A novel single step synthesis and surface functionalization of iron oxide magnetic nanoparticles and thereof for the copper removal from pigment industry effluent. *Separation and Purification Technology, 188*, 458–467. Available at: https://doi.org/10.1016/J.SEPPUR.2017.07.059

Katata-Seru, L., et al. (2017). Green synthesis of iron nanoparticles using Moringa oleifera extracts and their applications: Removal of nitrate from water and antibacterial activity against Escherichia coli. *Journal of Molecular Liquids, 256*, 296–304. Available at: https://doi.org/10.1016/J.MOLLIQ.2017.11.093

Khan, N. T., & Jamil Khan, M. (2017). Biogenic nanoparticles: An introduction to what they are and how they are produced. *International Journal of Biotechnology and Bioengineering, 3*(3), 66–70. Available at: https://doi.org/10.25141/2475-3432-2017-3.0066

Khan, Z. U. H., et al. (2016). Enhanced photocatalytic and electrocatalytic applications of green synthesized silver nanoparticles. *Journal of Molecular Liquids, 220*, 248–257. Available at: https://doi.org/10.1016/J.MOLLIQ.2016.04.082

Kim, S. D., et al. (2007). Occurrence and removal of pharmaceuticals and endocrine disruptors in South Korean surface, drinking, and waste waters. *Water Research, 41*(5), 1013–1021. Available at: https://doi.org/10.1016/J.WATRES.2006.06.034

Klaus-Joerger, T., et al. (2001). Bacteria as workers in the living factory: Metal-accumulating bacteria and their potential for materials science. *Trends in Biotechnology, 19*(1), 15–20. Available at: https://doi.org/10.1016/S0167-7799(00)01514-6

Kuang, Y., et al. (2013). Heterogeneous Fenton-like oxidation of monochlorobenzene using green synthesis of iron nanoparticles. *Journal of Colloid and Interface Science, 410*, 67–73. Available at: https://doi.org/10.1016/J.JCIS.2013.08.020

Kumar, S. V., et al. (2019). Preparation of yeast mediated semiconductor nanoparticles by candida albicans and its bactericidal potential against Salmonella typhi and Staphylococcus aureus. *International Journal of Research in Pharmaceutical Sciences, 10*(2), 861–864. Available at: https://doi.org/10.26452/ijrps.v10i2.262

Lapworth, D. J., et al. (2012). Emerging organic contaminants in groundwater: A review of sources, fate and occurrence. *Environmental Pollution (Barking, Essex: 1987), 163*, 287–303. Available at: https://doi.org/10.1016/J.ENVPOL.2011.12.034

Le, A. T., et al. (2011). Novel silver nanoparticles: Synthesis, properties and applications. *International Journal of Nanotechnology, 8*(3–5), 278–290. Available at: https://doi.org/10.1504/IJNT.2011.038205

Lee, S. H., & Jun, B. H. (2019). Silver Nanoparticles: Synthesis and Application for Nanomedicine. *International Journal of Molecular Sciences, 20*(4), 865. Available at: https://doi.org/10.3390/IJMS20040865

Lian, S., et al. (2019). Characterization of biogenic selenium nanoparticles derived from cell-free extracts of a novel yeast Magnusiomyces ingens. *3 Biotech, 9*(6), 1–8. Available at: https://doi.org/10.1007/S13205-019-1748-Y/METRICS

Lingamdinne, L. P., et al. (2017). Biogenic reductive preparation of magnetic inverse spinel iron oxide nanoparticles for the adsorption removal of heavy metals. *Chemical Engineering Journal, 307*, 74–84. Available at: https://doi.org/10.1016/J.CEJ.2016.08.067

Lunge, S., Singh, S., & Sinha, A. (2014). Magnetic iron oxide (Fe3O4) nanoparticles from tea waste for arsenic removal. *Journal of Magnetism and Magnetic Materials, 356*, 21–31. Available at: https://doi.org/10.1016/J.JMMM.2013.12.008

Madhavi, V., et al. (2013). Application of phytogenic zerovalent iron nanoparticles in the adsorption of hexavalent chromium. *Spectrochimica Acta. Part A, Molecular and Biomolecular Spectroscopy, 116*, 17–25. Available at: https://doi.org/10.1016/J.SAA.2013.06.045

Malik, B., et al. (2017). Biosynthesis of nanoparticles and their application in pharmaceutical industry. *Nanotechnology: Food and Environmental Paradigm*, 235–252. Available at: https://doi.org/10.1007/978-981-10-4678-0_13

Mandal, D., et al. (2006). The use of microorganisms for the formation of metal nanoparticles and their application. *Applied Microbiology and Biotechnology, 69*(5), 485–492. Available at: https://doi.org/10.1007/S00253-005-0179-3/METRICS

Manikandakrishnan, M., et al. (2019). Facile green route synthesis of gold nanoparticles using Caulerpa racemosa for biomedical applications. *Journal of Drug Delivery Science and Technology, 54*, 101345. Available at: https://doi.org/10.1016/J.JDDST.2019.101345

Markus, J., et al. (2016). Intracellular synthesis of gold nanoparticles with antioxidant activity by probiotic Lactobacillus kimchicus DCY51T isolated from Korean kimchi. *Enzyme and Microbial Technology, 95*, 85–93. Available at: https://doi.org/10.1016/J.ENZMICTEC.2016.08.018

Martínez-Cabanas, M., et al. (2016). Green synthesis of iron oxide nanoparticles. Development of magnetic hybrid materials for efficient As(V) removal. *Chemical Engineering Journal, 301*, 83–91. Available at: https://doi.org/10.1016/J.CEJ.2016.04.149

Martins, M., et al. (2017). Biogenic platinum and palladium nanoparticles as new catalysts for the removal of pharmaceutical compounds. *Water research, 108*, 160–168. Available at: https://doi.org/10.1016/J.WATRES.2016.10.071

Mathivanan, K., et al. (2019). Biologically synthesized silver nanoparticles against pathogenic bacteria: Synthesis, calcination and characterization. *Biocatalysis and Agricultural Biotechnology, 22*, 101373. Available at: https://doi.org/10.1016/J.BCAB.2019.101373

Mishra, A., et al. (2014). Biocatalytic and antimicrobial activities of gold nanoparticles synthesized by Trichoderma sp. *Bioresource Technology, 166*, 235–242. Available at: https://doi.org/10.1016/J.BIORTECH.2014.04.085

Moghaddam, A. B., et al. (2015). Nanoparticles biosynthesized by fungi and yeast: A review of their preparation, properties, and medical applications. *Molecules, 20*(9), 16540. Available at: https://doi.org/10.3390/MOLECULES200916540

Molnár, Z., et al. (2018). Green synthesis of gold nanoparticles by thermophilic filamentous fungi. *Scientific Reports, 8*(1), 1–12. Available at: https://doi.org/10.1038/s41598-018-22112-3

Mourdikoudis, S., Pallares, R. M., & Thanh, N. T. K. (2018). Characterization techniques for nanoparticles: comparison and complementarity upon studying nanoparticle properties. *Nanoscale, 10*(27), 12871–12934. Available at: https://doi.org/10.1039/C8NR02278J

Mubarakali, D., et al. (2012). Synthesis and characterization of CdS nanoparticles using C-phycoerythrin from the marine cyanobacteria. *Materials Letters, 74*, 8–11. Available at: https://doi.org/10.1016/J.MATLET.2012.01.026

Mughal, B., et al. (2021). Biogenic Nanoparticles: Synthesis, Characterisation and Applications. *Applied Sciences, 11*(6), 2598. Available at: https://doi.org/10.3390/APP11062598

Mukherjee, P., et al. (2001). Fungus-mediated synthesis of silver nanoparticles and their immobilization in the mycelial matrix: A novel biological approach to nanoparticle synthesis. *Nano Letters, 1*(10), 515–519. Available at: https://doi.org/10.1021/NL0155274

Mukherjee, D., et al. (2016). Green synthesis of α-Fe2O3 nanoparticles for arsenic(V) remediation with a novel aspect for sludge management. *Journal of Environmental Chemical Engineering, 4*(1), 639–650. Available at: https://doi.org/10.1016/J.JECE.2015.12.010

Muthukumar, H., et al. (2017). Biogenic synthesis of nano-biomaterial for toxic naphthalene photocatalytic degradation optimization and kinetics studies. *International Biodeterioration & Biodegradation, 119*, 587–594. Available at: https://doi.org/10.1016/J.IBIOD.2016.10.036

Mystrioti, C., et al. (2016). Comparative evaluation of five plant extracts and juices for nanoiron synthesis and application for hexavalent chromium reduction. *The Science of the Total Environment, 539*, 105–113. Available at: https://doi.org/10.1016/J.SCITOTENV.2015.08.091

Nagaraj, S., Cheirmadurai, K., & Thanikaivelan, P. (2021). Visible-light active collagen-TiO2 nanobio-sponge for water remediation: A sustainable approach. *Cleaner Materials, 1*, 100011. Available at: https://doi.org/10.1016/J.CLEMA.2021.100011

Naraian, R., & Abhishek, A. K. B. (2020). Green synthesis and characterization of silver NPs using oyster mushroom extract for antibacterial efficacy. *Journal of Chemistry, Environmental Sciences and its Applications, 7*(1), 13–18. Available at: https://doi.org/10.15415/JCE.2020.71003

Narayanan, K. B., & Sakthivel, N. (2010). Biological synthesis of metal nanoparticles by microbes. *Advances in Colloid and Interface Science, 156*(1–2), 1–13. Available at: https://doi.org/10.1016/J.CIS.2010.02.001

Nordberg, G. F. (2004). Cadmium and health in the 21st century--historical remarks and trends for the future. *Biometals: An International Journal on the Role of Metal Ions in Biology, Biochemistry, and Medicine, 17*(5), 485–489. Available at: https://doi.org/10.1023/B:BIOM.0000045726.75367.85

Oh, J. A., & Shin, H. S. (2015). Simple and sensitive determination of hydrazine in drinking water by ultra-high-performance liquid chromatography–tandem mass spectrometry after derivatization with naphthalene-2,3-dialdehyde. *Journal of Chromatography A, 1395*, 73–78. Available at: https://doi.org/10.1016/J.CHROMA.2015.03.051

Paixão, R. M., et al. (2017). Activated carbon of Babassu coconut impregnated with copper nanoparticles by green synthesis for the removal of nitrate in aqueous solution. *Environmental Technology, 39*(15), 1994–2003. Available at: https://doi.org/10.1080/09593330.2017.1345990

Pantidos, N., & Horsfall, L. E. (2014). Biological Synthesis of Metallic Nanoparticles by Bacteria, Fungi and Plants. *Journal of Nanomedicine & Nanotechnology, 5*(5), 1000233. Available at: https://doi.org/0.4172/2157-7439.1000233

Patel, A., et al. (2021). Integrating biometallurgical recovery of metals with biogenic synthesis of nanoparticles. *Chemosphere, 263*, 128306. Available at: https://doi.org/10.1016/J.CHEMOSPHERE.2020.128306

Patil, M. P., et al. (2019). Extracellular synthesis of gold nanoparticles using the marine bacterium Paracoccus haeundaensis BC74171T and evaluation of their antioxidant activity and antiproliferative effect on normal and cancer cell lines. *Colloids and Surfaces. B, Biointerfaces, 183*. Available at: https://doi.org/10.1016/J.COLSURFB.2019.110455

Phanjom, P., et al. (2012). Green synthesis of silver nanoparticles using leaf extract of Myrica esculenta. *International Journal of NanoScience and Nanotechnology, 3*(2), 73–79. Available at: https://p.urbanpro.com/tv-prod/documents%2Fnull-Paper+publication.pdf. Accessed 21 Mar 2023.

Prasad, K. S., Gandhi, P., & Selvaraj, K. (2014). Synthesis of green nano iron particles (GnIP) and their application in adsorptive removal of As(III) and As(V) from aqueous solution. *Applied Surface Science, 317*, 1052–1059. Available at: https://doi.org/10.1016/J.APSUSC.2014.09.042

Qu, Y., et al. (2017). Biosynthesis of gold nanoparticles by Aspergillum sp. WL-Au for degradation of aromatic pollutants. *Physica E: Low-Dimensional Systems and Nanostructures, 88*, 133–141. Available at: https://doi.org/10.1016/J.PHYSE.2017.01.010

Raj, R., et al. (2016). Extracellular polymeric substances of a marine bacterium mediated synthesis of CdS nanoparticles for removal of cadmium from aqueous solution. *Journal of Colloid and Interface Science, 462*, 166–175. Available at: https://doi.org/10.1016/J.JCIS.2015.10.004

Ramanathan, R., et al. (2013). Aqueous phase synthesis of copper nanoparticles: A link between heavy metal resistance and nanoparticle synthesis ability in bacterial systems. *Nanoscale* [Preprint]. Available at: https://pubmed.ncbi.nlm.nih.gov/23223802/. Accessed 21 Feb 2023.

Rao, A., et al. (2013). Removal of hexavalent chromium ions by Yarrowia lipolytica cells modified with phyto-inspired Fe0/Fe3O4 nanoparticles. *Journal of Contaminant Hydrology, 146*, 63–73. Available at: https://doi.org/10.1016/J.JCONHYD.2012.12.008

Rauf, M. A., et al. (2017). Biomimetically synthesized ZnO nanoparticles attain potent antibacterial activity against less susceptible S. aureus skin infection in experimental animals. *RSC Advances, 7*(58), 36361–36373. Available at: https://doi.org/10.1039/C7RA05040B

Rauwel, P., et al. (2015). Silver nanoparticles: Synthesis, properties, and applications. *Advances in Materials Science and Engineering, 2015*. Available at: https://doi.org/10.1155/2015/624394

Revathy, R., et al. (2022). Synthesis and catalytic applications of silver nanoparticles: a sustainable chemical approach using indigenous reducing and capping agents from Hyptis capitata. *Environmental Science: Advances, 1*(4), 491–505. Available at: https://doi.org/10.1039/D2VA00044J

Salem, S. S. (2022). Baker's yeast-mediated silver nanoparticles: Characterisation and antimicrobial biogenic tool for suppressing pathogenic microbes. *BioNanoScience, 12*(4), 1220–1229. Available at: https://doi.org/10.1007/S12668-022-01026-5/METRICS

Salem, D. M. S. A., Ismail, M. M., & Aly-Eldeen, M. A. (2019). Biogenic synthesis and antimicrobial potency of iron oxide (Fe3O4) nanoparticles using algae harvested from the Mediterranean Sea, Egypt. *The Egyptian Journal of Aquatic Research, 45*(3), 197–204. Available at: https://doi.org/10.1016/J.EJAR.2019.07.002

Sandana Mala, J. G., & Rose, C. (2014). Facile production of ZnS quantum dot nanoparticles by Saccharomyces cerevisiae MTCC 2918. *Journal of Biotechnology, 170*(1), 73–78. Available at: https://doi.org/10.1016/J.JBIOTEC.2013.11.017

Santhoshkumar, T., et al. (2011). Synthesis of silver nanoparticles using Nelumbo nucifera leaf extract and its larvicidal activity against malaria and filariasis vectors. *Parasitology Research, 108*(3), 693–702. Available at: https://doi.org/10.1007/S00436-010-2115-4

Seshadri, S., Saranya, K., & Kowshik, M. (2011). Green synthesis of lead sulfide nanoparticles by the lead resistant marine yeast, Rhodosporidium diobovatum. *Biotechnology Progress, 27*(5), 1464–1469. Available at: https://doi.org/10.1002/BTPR.651

Singh, R., & Nalwa, H. S. (2011). Medical applications of nanoparticles in biological imaging, cell labeling, antimicrobial agents, and anticancer nanodrugs. *Journal of Biomedical Nanotechnology, 7*(4), 489–503. Available at: https://doi.org/10.1166/JBN.2011.1324

Singh, J. P., et al. (2020). Bottom-up and top-down approaches for MgO. *Sonochemical Reactions* [Preprint]. Available at: https://doi.org/10.5772/INTECHOPEN.91182

Singh, M. P., et al. (2021). Biogenic and non-biogenic waste utilization in the synthesis of 2D materials (Graphene, h-BN, g-C2N) and their applications. *Frontiers in Nanotechnology, 3*, 685427. Available at: https://doi.org/10.3389/FNANO.2021.685427/BIBTEX

Smuleac, V., et al. (2011). Green synthesis of Fe and Fe/Pd bimetallic nanoparticles in membranes for reductive degradation of chlorinated organics. *Journal of Membrane Science, 379*(1–2), 131–137. Available at: https://doi.org/10.1016/J.MEMSCI.2011.05.054

Srinath, B. S., & Rai, V. R. (2015). Rapid biosynthesis of gold nanoparticles by Staphylococcus epidermidis: its characterisation and catalytic activity. *Materials Letters, 146*, 23–25.

Srivastava, S. K., & Constanti, M. (2012). Room temperature biogenic synthesis of multiple nanoparticles (Ag, Pd, Fe, Rh, Ni, Ru, Pt, Co, and Li) by Pseudomonas aeruginosa SM1. *Journal of Nanoparticle Research, 14*(4), 1–10. Available at: https://doi.org/10.1007/S11051-012-0831-7/METRICS

Srivastava, N., & Mukhopadhyay, M. (2014). Biosynthesis of SnO2 nanoparticles using bacterium Erwinia herbicola and their photocatalytic activity for degradation of dyes. *Industrial and*

Engineering Chemistry Research, 53(36), 13971–13979. Available at: https://doi.org/10.1021/IE5020052

Srivastava, S. K., et al. (2013). Biogenic synthesis and characterization of gold nanoparticles by Escherichia coli K12 and its heterogeneous catalysis in degradation of 4-nitrophenol. *Nanoscale Research Letters, 8*(1), 70. Available at: https://doi.org/10.1186/1556-276X-8-70

Srivastava, V., Sharma, Y. C., & Sillanpää, M. (2015). Green synthesis of magnesium oxide nanoflower and its application for the removal of divalent metallic species from synthetic wastewater. *Ceramics International, 41*(5 Part B), 6702–6709. Available at: https://doi.org/10.1016/J.CERAMINT.2015.01.112

Srivastava, S., et al. (2019). Production, characterization and antibacterial activity of silver nanoparticles produced by Fusarium oxysporum and monitoring of protein-ligand interaction through in-silico approaches. *Microbial Pathogenesis, 129*, 136–145. Available at: https://doi.org/10.1016/J.MICPATH.2019.02.013

Tu, J., Yang, Z., & Hu, C. (2015). Efficient catalytic aerobic oxidation of chlorinated phenols with mixed-valent manganese oxide nanoparticles. *Journal of Chemical Technology & Biotechnology, 90*(1), 80–86. Available at: https://doi.org/10.1002/jctb.4289

Tu, W., et al. (2022). The CO_3O_4 nanosheet hybridized with silver nanoparticles affords long-acting synergetic antimicrobial and catalytic degradation activity. *Journal of Alloys and Compounds, 914*, 165284. Available at: https://doi.org/10.1016/J.JALLCOM.2022.165284

Tyagi, S., et al. (2018). Strategies for Nitrate removal from aqueous environment using Nanotechnology: A Review. *Journal of Water Process Engineering, 21*, 84–95. Available at: https://doi.org/10.1016/j.jwpe.2017.12.005

Vaseghi, Z., Nematollahzadeh, A., & Tavakoli, O. (2018). Green methods for the synthesis of metal nanoparticles using biogenic reducing agents: A review. *Reviews in Chemical Engineering, 34*(4), 529–559. Available at: https://doi.org/10.1515/REVCE-2017-0005/ASSET/GRAPHIC/J_REVCE-2017-0005_CV_003.JPG

Verma, N., & Kumar, N. (2019). Synthesis and biomedical applications of copper oxide nanoparticles: An expanding horizon. *ACS Biomaterials Science and Engineering, 5*(3), 1170–1188. Available at: https://doi.org/10.1021/ACSBIOMATERIALS.8B01092/ASSET/IMAGES/MEDIUM/AB-2018-010925_0012.GIF

Vijayanandan, A. S., & Balakrishnan, R. M. (2018). Biosynthesis of cobalt oxide nanoparticles using endophytic fungus Aspergillus nidulans. *Journal of Environmental Management, 218*, 442–450. Available at: https://doi.org/10.1016/J.JENVMAN.2018.04.032

Wang, T., et al. (2014). Green synthesized iron nanoparticles by green tea and eucalyptus leaves extracts used for removal of nitrate in aqueous solution. *Journal of Cleaner Production, 83*, 413–419. Available at: https://doi.org/10.1016/J.JCLEPRO.2014.07.006

Wang, R., et al. (2017). Biogenic manganese oxides generated by green algae Desmodesmus sp. WR1 to improve bisphenol A removal. *Journal of Hazardous Materials, 339*, 310–319. Available at: https://doi.org/10.1016/J.JHAZMAT.2017.06.026

Wang, X., et al. (2018). Interactions between biogenic selenium nanoparticles and goethite colloids and consequence for remediation of elemental mercury contaminated groundwater. *The Science of the Total Environment, 613–614*, 672–678. Available at: https://doi.org/10.1016/J.SCITOTENV.2017.09.113

Xiao, Z., et al. (2016). Plant-mediated synthesis of highly active iron nanoparticles for Cr (VI) removal: Investigation of the leading biomolecules. *Chemosphere, 150*, 357–364. Available at: https://doi.org/10.1016/J.CHEMOSPHERE.2016.02.056

Xu, H., et al. (2018). Effect of anodes decoration with metal and metal oxides nanoparticles on pharmaceutically active compounds removal and power generation in microbial fuel cells. *Chemical Engineering Journal, 335*, 539–547. Available at: https://doi.org/10.1016/J.CEJ.2017.10.159

Yong, P., et al. (2004). Synthesis of nanophase hydroxyapatite by a Serratia sp. from waste-water containing inorganic phosphate. *Biotechnology Letters, 26*(22), 1723–1730. Available at: https://doi.org/10.1007/S10529-004-3744-4

Zambonino, M. C., et al. (2023). Biogenic selenium nanoparticles in biomedical sciences: Properties, Current trends, novel opportunities and emerging challenges in theranostic nanomedicine. *Nanomaterials, 13*(3), 424. Available at: https://doi.org/10.3390/NANO13030424

Zhang, H., et al. (2019). Biosynthesis of selenium nanoparticles mediated by fungus Mariannaea sp. HJ and their characterization. *Colloids and Surfaces A: Physicochemical and Engineering Aspects, 571*, 9–16. Available at: https://doi.org/10.1016/J.COLSURFA.2019.02.070

Zhang, C. L., et al. (2020). The controlled synthesis of Fe3C/Co/N-doped hierarchically structured carbon nanotubes for enhanced electrocatalysis. *Applied Catalysis B: Environmental, 261*, 118224.

Zhang, M., et al. (2021). High stability Au NPs: From design to application in nanomedicine. *International Journal of Nanomedicine, 16*, 6067. Available at: https://doi.org/10.2147/IJN.S322900

Zhou, N. Q., et al. (2018). Continuous degradation of ciprofloxacin in a manganese redox cycling system driven by Pseudomonas putida MnB-1. *Chemosphere, 211*, 345–351. Available at: https://doi.org/10.1016/J.CHEMOSPHERE.2018.07.117

Fungi's Involvement in Metal NPS Synthesis and Environmentally Sustainable Practices

Maneesh Kumar, Bibekananda Bhoi, Harish Kumar, and Himanshu Bhardwaj

1 Introduction

Nanotechnology is rapidly developing, making it crucial in many facets of human life. It involves controlling materials at the nanoscale (1–100 nanometers). It is a multidisciplinary field involving physics, chemistry, biology, and materials science. Because of its ability to enhance and even generate brand-new items, nanotechnology has emerged as a pivotal industry. It improves energy sources, materials, and computers. It can also improve biotechnology, health care, and medical diagnostics. In electronics, manufacturing, and aerospace, it can be utilized to develop new products and technologies (Subashini & Bhuvaneswari, 2018). It also improves sustainability and the environment. In medicine, agriculture, cosmetics, energy research, geology, astronomy, geography, and photochemistry, it is becoming increasingly essential. It is used to make nanosensors, tailored drug delivery systems, and nanoscale implants to detect and cure diseases. Nanotechnology improves fertilizers, insecticides, and other agricultural products. It makes smaller, more effective skincare products. It improves the performance of energy storage devices such as solar cells and batteries (Bardov et al., 2022). It also improves geological map resource measurements, telescopes, instruments, dyes, pigments, and coatings in photochemistry and astronomy. Fungal nanoparticles, or "myconanoparticles," are those that have been synthesized in a lab. Fungi have long been utilized in the food and beverage industries, both as a source of nutrition and as a means of preservation

M. Kumar (✉) · H. Kumar
Department of Biotechnology, Magadh University, Bodh Gaya, Bihar, India

B. Bhoi
Department of Microbiology, Dr. Harisingh Gour Vishwavidyalaya (A Central University), Sagar, Madhya Pradesh, India

H. Bhardwaj
Department of Biochemistry, Patna University, Patna, Bihar, India

A. K. Bhardwaj et al. (eds.), *Biogenic Wastes-Enabled Nanomaterial Synthesis*,
https://doi.org/10.1007/978-3-031-59083-2_6

through fermentation and storage (Guilger-Casagrande & Lima, 2019; Mitchell et al., 2021; Wang et al., 2021). More recently, the ability of fungi to produce a variety of compounds and bioactive substances has made them a source of interest for the industrial production of nanoparticles. Myconanoparticles have the potential to be used in a variety of medical applications, such as targeted drug delivery, tissue engineering, and toxin removal (Khandel & Shahi, 2018). Fungal nanoparticles also have potential applications in the food and beverage industry, including improving the shelf life and nutritional value of food and drink products, as well as aiding in the production of healthier food and beverage items (Dikshit et al., 2021; Rai et al., 2021a).

Metal nanoparticles (NPs) can be synthesized using fungi. Fungal biosynthesis of metal nanoparticles is one of the emerging fields of research that has demonstrated the potential of fungi with much therapeutic potential. Fungi can produce a wide range of enzymes and metabolites, which can be used to reduce metal ions and form nanoparticles. They can produce these particles in the form of nanospheres, nanorods, nanoflowers, nanowires, and nanocomposites (Khalil et al., 2022; Salem & Fouda, 2021). Fungal-based synthesis of metal NPs is a green and cost-effective process, as it does not require any additional chemical reagents and can be performed under mild conditions. Fungal-based synthesis of metal NPs typically involves the use of fungi that have been genetically modified to produce enzymes that can reduce metals from their salts (Khandel & Shahi, 2018; Ratan et al., 2020). The metal salts are then mixed with the fungus, and the enzymes catalyze the reduction of the metal ions to produce the desired metal NPs. This process is advantageous because it is a clean and green process that does not produce any hazardous by-products and can be used to produce NPs of various sizes and shapes. Additionally, the process is cost-effective as the reagents used are generally cheap and readily available. Fungal-based synthesis of metal NPs involves the use of fungal species such as *Aspergillus niger*, *Penicillium purpurogenum*, and *Trichoderma viride* to reduce metal salts into gold NPs (Guilger-Casagrande & Lima, 2019; Khandel & Shahi, 2018; Raj et al., 2021). This is done by the production of extracellular enzymes and proteins, which activate the reduction process and stabilize the metal NPs. The metal NPs can then be harvested and purified using various methods such as centrifugation, filtration, and dialysis. These fungi also secrete reducing agents such as polyphenol oxidase, laccase, and peroxidase enzymes that can convert metal salts into colloidal metal NPs (Rai et al., 2021b). The metal NPs can then be stabilized with biopolymers such as chitosan, alginate, and proteins.

Fungal-based synthesis of metal NPs has several advantages over traditional chemical methods, such as low cost, environmentally friendly production, and the ability to produce particles of different sizes and shapes. Additionally, the use of biopolymers can improve the biocompatibility of the NPs, making them suitable for medical applications. Additionally, the fungus *Trametes versicolor* can be used to produce silver and zinc oxide nanoparticles (Jamkhande et al., 2019; Pour et al., 2022). Fungi can also be used to produce copper oxide, iron oxide, and other metal oxide nanoparticles. Fungal-made nanoparticles have several advantages compared to conventional methods such as being nontoxic, eco-friendly, cost-effective, and less time-consuming. The synthesis process involves the fungal mycelia releasing reducing agents and enzymes, which reduce the metal ions and form the NPs (Jeevanandam et al., 2022). The NPs can then be collected from the fungal broth and separated for further use. The size and shape of the NPs can be controlled by manipulating the growth conditions and optimizing the synthesis process. Fungal synthesis of metal NPs is an environmentally friendly and cost-effective method compared to other chemical and physical methods (Beyene et al., 2017). There has been a significant uptick in the use of nanoparticles (NPs) across many fields in recent decades, ranging from drug delivery to catalysis. Fungal synthesis of metal NPs is well compared to other chemical and physical methods. Fungal synthesis has the advantage of being able to produce metal NPs with a range of different shapes, sizes, and compositions. It is also able to produce metal NPs with specific properties, such as surface charge, catalytic activity, and optical properties. It is also a renewable and sustainable method, as the fungal cells can be reused (Singh et al., 2018). Finally, the use of fungi eliminates the need for toxic chemical pesticides as they do not need to be reapplied frequently, making it a safer and more environmentally friendly process (Dikshit et al., 2021; Saratale et al., 2018).

In the past few decades, nanoparticles (NPs) have been used more and more in a wide range of fields, such as precision agriculture and food production, biological imaging, catalysis, cleaning up the environment, optoelectronics, magnetooptics, personal care products, nano-microbiology, and pharmacology. The unique properties of NPs, such as their small size, large surface-to-volume ratio, and chemical reactivity, make them useful and versatile in many fields. NPs can be synthesized and engineered for specific purposes, such as to increase the efficacy of a drug or to better target a certain cell or tissue (Calzoni et al., 2019; Mitchell et al., 2021). In addition, NPs have been found to have minimal side effects, making them an attractive alternative to traditional treatments. The potential health implications of NPs have been studied extensively in recent years. Research has shown that NPs may be able to pass through the cell membrane and enter the body, potentially causing harm to the organs, tissues, and cells within the body. In particular, the long-term effects of ingesting, inhaling, or injecting NPs into the body are still unknown. Moreover, interactions between NPs and biological systems have not been fully understood. It is therefore important to further explore the potential health risks of using NPs in various applications (Kik et al., 2020; Wang et al., 2021). To ensure safety, it is important to develop and implement regulations regarding the use of NPs, as well as conduct more research on the potential health risks. Due to their small size, NPs can penetrate various tissue types and cells and can be used to deliver drugs and other treatments directly to the target area. Furthermore, NPs can be engineered to provide specific functions such as the controlled release of drugs, enhanced solubility, and prolonged stability. This makes them attractive for use in various medical and industrial applications (Siddique & Chow, 2020). For example, they are used to create biocompatible materials, improve the performance of drug delivery systems, and create novel nanomaterials for drug delivery. Additionally, NPs can be used to create novel nanostructured materials with enhanced optical, thermal, and electrical properties. These materials can be used in various applications, including sensors, solar cells, and displays (Kumar, 2021; Said et al., 2022). As technology advances, NPs have the potential to disrupt numerous industries by providing new approaches to treating illnesses and improving performance. The potential of fungi NPSs as potent antibacterial, antioxidant, anti-inflammatory, anticancer, and antifungal agents was extensively detailed in this chapter. The study also revealed the inflammatory impact of several microbial infections.

2 Mycogenic Nanoparticles and Nanominerals

Mycogenic nanoparticles and nanominerals are nanomaterials that are produced from fungi or their by-products. They are composed of nanoparticles and nanoscale minerals and can range from 10 to 100 nanometers in size (Bhardwaj et al., 2018). These particles are used in a range of applications, from drug delivery to carbon sequestration. Mycogenic nanoparticles and nanominerals have unique physical and chemical properties that make them suitable for a variety of applications. For instance, they have high surface area-to-volume ratios, which make them ideal for drug delivery and mass transfer processes. They are nontoxic and biocompatible, making them safe for use in medical applications. They have a high degree of biodegradability, making them an environmentally friendly option for many applications (Siddique & Chow, 2020; Yu et al., 2020).

About 30 elements of the periodic table are currently known to be synthesized by fungi in the form of nanoparticles and nanominerals. These elements span the periodic table from alkaline earth metals (Sr, Mg, Ba) to post-transition metals (Al, Zn, Cd, Bi) to metalloids (Si, Te) to nonmetals (H, C, N, O, P, S, Se) to lanthanides (La, Ce) to an actinide (U). Numerous fungi are capable of producing nanoscale elemental particles of gold (via reduction of Au(III) to Au(0)), silver (via reduction of Ag(I) to Ag(0)), palladium (via reduction of Pd(IV, II) to Pd(0)), and metalloids such as selenium and tellurium (via reduction of Se/Te oxyanions (Se(IV, VI)/Te(IV, VI) to Se(0)/Te(0)). Oxides, carbonates, phosphates, selenides, tellurides, and sulfides can all be nanominerals, as can metalloids, including copper, zinc, manganese, nickel, boron, barium, tin, and tin dioxide. Due to the abundance of nanoparticle-forming fungi and potentially vulnerable elements, chemicals, and minerals, this list is expected to grow (Naraian & Abhishek, 2020). Nanoparticles of minerals and elements can form both inside living cells and outside of them, on living biomass, and on decomposing matter. It should be pointed out that some of the procedures that lead to nanomineral formation may be the same as those involved in the production of secondary minerals on a larger scale, with scale differences resulting from chemical and biological circumstances, protein and metabolism excretion, proximity to substrate, and so on. Perhaps reflecting peculiarities in their manner of formation, some nanoparticles appear to be synthesized solely at the nanoscale, especially elemental forms such as Ag(0), Au(0), Se(0), and Te(0) (Li et al., 2022; Litwin et al., 2020).

3 Fungal-Mediated Synthesis of Nanoparticles

Fungi can be used to synthesize nanoparticles by first growing them in a nutrient-rich medium. Then, the nanoparticles can be extracted from the fungal biomass. Fungal systems can be used to generate nanoparticles from various metals, such as gold, silver, copper, cobalt, palladium, and iron, as well as from organic compounds, such as polysaccharides and proteins (Madkour, 2018). It can be used to functionalize the nanoparticles, allowing for further customization of their properties. Fungal cells can be used as bioreactors to produce metallic nanoparticles such as gold, silver, and copper, or semiconductor nanoparticles such as quantum dots. Fungal-mediated synthesis of nanoparticles involves the use of fungal cells or enzymes to produce nanoparticles with desirable properties (Kumar & Akhtar, 2019; Patel et al., 2021; Saravanan et al., 2021). This procedure has the potential to be more cost-effective and environmentally friendly than traditional methods of nanoparticle production. Fungi are being explored as a source of biocatalysts for the synthesis of nanoparticles due to their ability to produce a variety of secondary metabolites with the potential for nanoparticle formation (Alghuthaymi et al., 2022; Litwin et al., 2020). It involves the use of fungal cells or enzymes to produce nanoparticles with desirable properties (Kumar et al., 2017). The fungal enzymes can be used to generate nanoparticles from metal salts or organic compounds. In some cases, the size, shape, surface area, and composition of the nanoparticles can be manipulated by varying the reaction conditions, such as temperature, pH, and the type and concentration of reactants and substrates (Siddique & Chow, 2020).

Fungal-mediated synthesis of nanoparticles is attractive due to its potential to produce uniform and monodisperse nanoparticles with controllable size and composition. In addition, this approach offers the potential for low-cost and environmentally friendly production of nanoparticles. This utilizes the ability of certain fungi to induce the formation of nanoparticles from the surrounding environment (Mitchell et al., 2021; Litwin et al., 2020). Fungal enzymes, such as laccase, are used to catalyze the oxidation of organic compounds, resulting in the formation of nanoparticles. In addition, fungal cells can be used as scaffolds on which to grow and assemble nanoparticles. This process can be used to synthesize a variety of nanoparticles, including metals, oxides, and polymers. The resulting nanoparticles can be used for a variety of applications, including drug delivery, biosensors, and catalysis. The benefits of fungal-mediated synthesis of nanoparticles include cost savings, increased control over size and shape, and decreased environmental impacts. This has potential applications in a variety of fields, including pharmaceuticals, cosmetics, and catalysis. For example, fungi such as *Aspergillus fumigatus* and *Fusarium oxysporum* have been used to form silver nanoparticles (Roy et al., 2021). In contrast, viruses are typically used to form gold nanoparticles. Finally, enzymes have been used to form iron oxide nanoparticles. Enzymes such as laccase, tyrosinase, and peroxidase have been used to reduce iron ions into iron oxide nanoparticles. Overall, different biological agents will behave differently with different metal solutions to form nanoparticles. Depending on the type of metal, a variety of bacteria, fungi, viruses, and enzymes can be used to form metal nanoparticles (Marouzi et al., 2021; Rana et al., 2020). Fungi produce a wide variety of natural compounds in their metabolic processes, which are potentially useful as NPs. They are capable of metabolizing a range of organic compounds, including lignocellulosic biomass, industrial wastes, and crude oil. Fungal NPs can be obtained from the fungal cell wall or secretome or produced by fungal cell cultures. These NPs can be used as potential drugs, cosmetics, and food additives. Additionally, many fungal species can produce NPs with therapeutic properties, such as antifungal, anti-inflammatory, antioxidant, and antibacterial activities. The ease of producing and manipulating fungal cells makes them an attractive option for NP synthesis. Fungal nanoparticles have also been used in the production of photovoltaic devices, fuel cells, sensors, and other advanced materials (Jampílek & Kráľová, 2023; Kumari et al., 2019). These applications point to the potential of fungi to revolutionize nanotechnology, and their unique properties have made them attractive to researchers in the field. Fungal nanomaterials are being explored as potential replacements for expensive, hard-to-manufacture synthetic nanomaterials. They can be used as nanoscale building blocks for the fabrication of novel materials, such as nanofibers, nanosheets, nanobeads, and nanocapsules (Siddique & Chow, 2020; Litwin et al., 2020). Fungal nanomaterials also have potential applications in drug delivery, sensing, filtration, and energy storage. They have been used to create high-efficiency fuel cells, solar cells, and batteries and have even been explored for use in biotechnology and data storage. In addition, fungi are being studied for their ability to degrade pollutants, with the potential to be used in water and soil remediation (Marouzi et al., 2021).

4 Formation of Fungal Nanoparticles

The chemical synthesis of nanoparticles often involves reducing agents that are either toxic to the environment or physiologically hazardous; thus, there has been a search for safer, more sustainable production methods. Recent studies have uncovered biological routes for the synthesis of nanoparticles by microbes, plant extracts, and fungi (Kumar et al., 2018). Fungi are used to produce a wide variety of nanoparticles; silver and gold are the most prevalent, but zinc oxide, platinum, magnetite, zirconia, silica, titanium, and cadmium sulfide and cadmium selenide quantum dots have all been synthesized by fungi (Patel et al., 2021; Rehman et al., 2021). Fungi synthesize nanoparticles through a variety of mechanisms, including the reduction of metal ions and the biosynthesis of organic molecules (Fig. 1). Fungal nanoparticles are particles that are produced by fungal cells. These particles are typically smaller than 100 nanometers in size and are composed of proteins, polysaccharides, and other bioactive compounds. Fungal nanoparticles can be formed by various techniques, including electrostatic precipitation, precipitation with organic solvents, precipitation with supercritical fluids, and sonication. In each case, the process

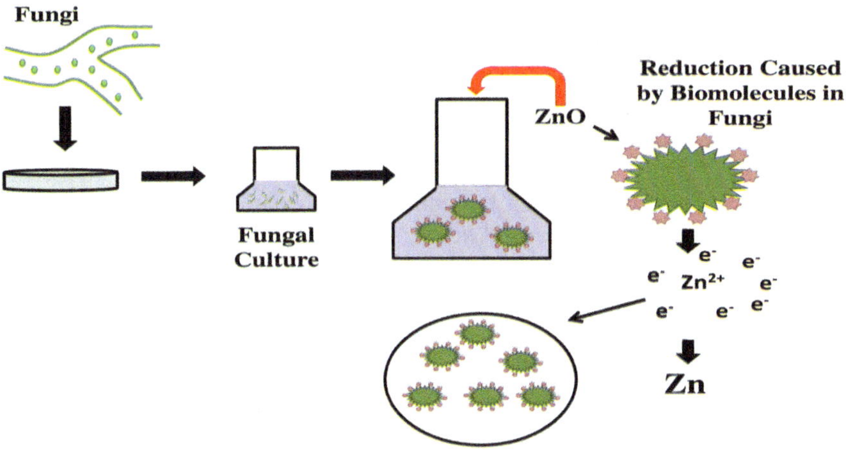

Fig. 1 Mechanisms for the naturally occurring production of zinc nanoparticles

involves breaking down the fungal cell walls and releasing the nanoparticles into the solution.

These particles can then be collected and used for a variety of applications, such as drug delivery, gene therapy, and cell imaging. Here are the primary steps for the formation of nanoparticles:

1. *Isolation of fungi:* The first step in making fungal nanoparticles is to isolate the fungal species from a natural source, such as soil, water, or air, and culture it. Next, the fungal biomass must be harvested and purified. After this process, the fungal biomass must be processed using physical or chemical methods to break it down into smaller particles. This can include high-pressure homogenization, sonication, or chemical extraction. Finally, the fungal nanoparticles must be characterized and purified (Kumar et al., 2019; Mishra et al., 2019). This can involve size separation, purification, and characterization techniques such as electron microscopy, dynamic light scattering, and atomic force microscopy. Once the culture has grown to the desired size, the mycelium can be harvested and processed either by mechanical or chemical methods to break it down into nanoparticles. The nanoparticles can then be further modified to improve their physical characteristics or used as-is in various applications.
2. *Extraction of fungal biomass:* Once the fungal species has been isolated, the fungal biomass must be extracted from the culture. This can be done using mechanical, enzymatic, or chemical methods. Mechanical methods involve physically crushing the cells, while enzymatic methods involve the use of enzymes to break down the cell walls. Mechanical methods include grinding, homogenization, sonication, and high-pressure homogenization. Enzymatic methods include the use of enzymes such as lysozyme and cellulase to break down the cell walls of bacterial and plant cells. Chemical methods involve the use of organic solvents or surfactants to dissolve the fungal biomass (Surendran

& Sherje, 2022). These methods are often used in combination with physical methods, such as grinding or sonication, to increase the efficiency of the extraction process. Commonly used solvents include ethanol, acetone, and chloroform. Surfactants, including sodium dodecyl sulfate, may also be used to disrupt the fungal cell wall and facilitate the extraction of intracellular components. Once the biomass has been extracted, it can then be used for various applications, such as the production of enzymes, biopesticides, or biofuel.

3. *Processing of fungal biomass:* After the extraction of fungal biomass, it must be processed to extract the nanoparticles. This typically involves homogenization and purification processes. Homogenization is a process that breaks down the biomass into smaller particles, allowing for easier extraction of the nanoparticles. This process can be done in a variety of ways, such as homogenization with high pressure, ultrasonication, and cell lysis. It can also be done chemically, by using enzymes to break down the biomass into individual molecules. This process is used to produce nanoparticles from a variety of materials, including proteins, lipids, carbohydrates, and even minerals. It can also be used to increase the surface area of a material, allowing for better absorption or release of a substance. Purification involves the removal of contaminants and other impurities from the extracted biomass. This step is important to ensure that the final product contains a high concentration of nanoparticles and is free from impurities. It can be achieved using various techniques, including centrifugation, filtration, chromatography, and precipitation. Once the biomass is adequately processed, it can be used to produce the nanoparticles.
4. *Characterization of nanoparticles:* Once the nanoparticles have been extracted, they must be characterized to determine their size, shape, and other properties. This can be done through a variety of techniques such as X-ray diffraction, electron microscopy, and atomic force microscopy. X-ray diffraction can be used to measure the size and shape of the particles, while electron microscopy can be used to observe the surface features of the particles. Atomic force microscopy can be used to measure the mechanical properties of the particles, such as stiffness and elasticity. Also, spectroscopic techniques such as UV-visible and infrared spectroscopy can be used to measure the optical and vibrational properties of the particles. These techniques can provide detailed information on the structure and properties of the nanoparticles, which can then be used to optimize their applications.
5. *Application of the nanoparticles:* The fungal nanoparticles can then be used in various applications, such as drug delivery, gene therapy, and tissue engineering. Fungal nanoparticles can be designed to carry therapeutic agents, such as drugs or proteins, to specific areas of the body. They can also be used to deliver genetic material to targeted cells, allowing for gene therapy. Moreover, they can be used to deliver scaffolds and other materials to aid in tissue engineering. Fungal nanoparticles have also been proposed for use in the development of biosensors, as well as for the manufacture of biofuel cells.

5 Different Methods of Nanoparticle Synthesis

Nanoparticles can be synthesized by a variety of methods including chemical, physical, and biological processes. Chemical synthesis involves the use of precursors such as inorganic salts, organometallic compounds, and polymers. These precursors can be reacted in solution or in the vapor phase to form nanoparticles. In solution-phase synthesis, nanoparticles are typically formed by salt reduction, precipitation, and chemical reduction. Salt reduction is a process in which a metal salt is reduced to its elemental form by a reducing agent, such as sodium borohydride. In precipitation, the metal salt is dissolved in a solvent and then a precipitating agent is added to form the nanoparticle. Chemical reduction involves the use of a reducing agent such as hydrazine to reduce the metal salt to its metallic form. In the vapor phase, nanoparticles are formed through a process called chemical vapor deposition (CVD). In this process, a metal-containing gas is introduced into a chamber, and the metal is then deposited onto a substrate. The metal is then transformed into nanoparticles by heating and cooling the substrate in a controlled manner. In addition to these methods, nanoparticles can also be synthesized using physical methods such as laser ablation and sonication. In laser ablation, a laser is used to vaporize a target material, which is then condensed onto a substrate to form nanoparticles. Sonication involves the use of ultrasonic waves to break down a target material into nanoparticles (Hachem et al., 2022). Physical methods of synthesis use techniques such as laser ablation, arc discharge, and sputtering to produce nanoparticles. Biological synthesis involves the use of living organisms such as fungi to produce nanoparticles through metabolic processes. This process is often used to create various kinds of nanoparticles, such as metal nanoparticles, quantum dots, and carbon nanotubes. These nanoparticles can be used for various applications, such as medical diagnostics, renewable energy, and nanoelectronics. The process of biological synthesis is relatively simple and requires minimal equipment and expenses (Khan et al., 2022). It is also a green process that does not generate hazardous by-products, making it an attractive option for many industries.

1. *Laser Ablation:* Laser ablation is a process by which materials can be synthesized in the form of nanoparticles. It involves the use of high-energy laser beams to break down bulk materials into nanoparticles. This process can be used to produce particles of various sizes, shapes, and compositions. It is a relatively simple and cost-effective method for synthesizing nanoparticles and has proven to be particularly useful for synthesizing metal nanoparticles. Laser ablation is often used in the production of semiconductor materials, such as for the manufacture of thin-film solar cells. In addition, it can be used to create nanoparticles for medical applications, such as drug delivery and cancer therapy. Laser ablation is used to prepare nanoparticles using fungi extracts for biomedical applications. It is a process where a focused laser beam is used to evaporate and break down material into small particles, usually in the nanometer or micrometer range (Mandal et al., 2022). This technique can be used to prepare nanoparticles from a variety of materials, including fungal extracts. Fungal extracts contain a wide

range of bioactive compounds, such as polysaccharides, proteins, lipids, and polyphenols, which are beneficial for various biomedical applications. Laser ablation is used to break down these components into small particles, allowing them to be used in a wide range of biomedical applications, such as targeted drug delivery, tissue engineering, and regenerative medicine.

2. *Hydrothermal Synthesis:* Hydrothermal synthesis is a technique used to produce nanoparticles by dissolving the bulk material in an aqueous solution. The solution is then subjected to high temperatures and pressures in an autoclave, which facilitates the formation of nanoparticles. The nanoparticles can be further modified by adding different catalysts to the solution. This method is used to synthesize a wide range of materials, including metal oxides, metal nitrides, and other nanoparticles. The high temperature and pressure environment of hydrothermal synthesis allows for the formation of well-defined nanoparticles with controlled size and shape. This method is particularly useful for synthesizing materials with high melting points and materials that are difficult to synthesize using other methods. Fungal extracts can be used as a source of organic material in the hydrothermal synthesis of nanoparticles. The extracts contain a variety of biomolecules, proteins, and enzymes that can be used as a catalyst and stabilizers for the formation of nanoparticles. The extracts can also provide a variety of nutrients, such as nitrogen and phosphorus, which are essential for the growth of nanomaterials. It can also provide a variety of minerals, such as iron, copper, and zinc, which can be used as a source of metal ions for the synthesis of nanoparticles (Saravanan et al., 2021). Additionally, the extract can also provide a variety of other essential elements, such as sulfur, potassium, magnesium, and calcium, which can be used to tailor the physical and chemical properties of the nanoparticles.
3. *Sol-Gel and Pechini Process:* Sol-gel and Pechini processes are two methods used to synthesize metal oxide nanoparticles. The sol-gel process involves the hydrolysis of metal alkoxides, while the Pechini process involves the precipitation of metal salts. The sol-gel process involves the hydrolysis of metal alkoxides in an organic solvent such as alcohol or acetone, followed by an acid-catalyzed condensation reaction. This forms a colloidal solution of metal oxide nanoparticles. The Pechini process involves the precipitation of metal salts in an aqueous solution, followed by a controlled thermal treatment. This process forms metal oxide nanoparticles that are then calcined. The major difference between the two processes is the starting materials used (Tamboli et al., 2023). The sol-gel process uses metal alkoxides, while the Pechini process uses metal salts. The products of the two processes are also different, with the sol-gel process producing colloidal solutions and the Pechini process producing metal oxide nanoparticles. Both processes involve mixing the fungal extract with a liquid or solid precursor material such as a metal salt or organometallic compound to form a gel. For a chemical reaction, this mixture is heated. After the reaction, solid nanoparticles are collected and filtered. To create a homogenous solution, the sol-gel technique mixes fungal extract with an alkoxide precursor such as ethoxylated alkoxide. Hydrolysis and polycondensation processes occur when this solution is heated.

Collect and purify the nanoparticles. The sol-gel technique uses alkoxide precursors, but the Pechini process uses polyalkoxides (Wonoputri et al., 2023). Hydrolysis and polycondensation processes create nanoparticles when polyalkoxides are heated. Collect and purify nanoparticles. Fungal extracts can make nanoparticles using sol-gel and Pechini methods. Both methods involve heating a fungal extract and precursor material to create nanoparticles. Purified nanoparticles are used for different applications.

4. *Chemical Reduction:* Chemical reduction is a process in which metal ions are reduced to metal atoms using a reducing agent. This process can be used to synthesize metal nanoparticles with a wide range of sizes and shapes. The reducing agent can be a chemical such as hydrazine, and sodium borohydride or a metal such as zinc or iron. The reducing agent reacts with the metal ions in a solution and reduces them to metal atoms. The resulting metal atoms can then form metal nanoparticles through a process called nucleation and growth. The size, shape, and composition of the nanoparticles can be controlled by adjusting the concentration of reactants, temperature, and pH of the solution. The chemical reduction process involves the use of fungal extracts to produce nanoparticles. This process involves the reduction of metal ions by the fungal extract (Das et al., 2017). The metal ions are first dissolved in an aqueous solution and then the fungal extract is added. The fungal extract contains enzymes that reduce the metal ions to the nanoparticle form. The nanoparticles are then collected, washed, and dried. The size of the nanoparticles depends on the type of fungal extract used. The advantages of using fungal extracts in nanoparticle production include the ability to produce nanoparticles without the use of hazardous chemicals, as well as the ability to produce nanoparticles with a narrow size range.
5. *Green Synthesis:* Green synthesis is a process by which nanoparticles are synthesized using renewable sources, such as plant extracts. It is a process of synthesizing nanoparticles using eco-friendly, inexpensive, and safe materials such as plant extracts, microorganisms, and natural minerals. This process involves various processes such as wet chemical methods, sol-gel methods, physical methods, and chemical methods. These processes are used to synthesize nanoparticles such as silver, gold, copper, iron oxide, and silica. The green synthesis of nanoparticles offers many advantages such as low cost, high efficiency, and more environmentally friendly compared to traditional methods. The green synthesis of nanoparticles has been used in various applications such as in nanomedicine, nanoelectronics, and nanomaterials. Furthermore, it is much faster than traditional methods, as it can be completed in a matter of hours rather than days or weeks (Bandeira et al., 2020). The process is an emerging field in nanotechnology that uses biological sources such as plants, fungi, algae, and bacteria to synthesize nanoparticles. The fungal extract is a potential source for the green synthesis of nanoparticles. The fungal extract contains a variety of metabolites, enzymes, and other biomolecules that can act as reducing and stabilizing agents in the synthesis of nanoparticles. The fungal extract contains biomolecules that can reduce the metal ions and stabilize the formed nanoparticles. The use of fungal extract also eliminates the need for any toxic chemical reagents. In the

green synthesis of nanoparticles, the fungal extract is first dispersed in an aqueous solution, and then a metal salt solution is added (Singh et al., 2020). The metal salt is reduced by the biomolecules present in the fungal extract and forms the nanoparticles. The nanoparticles are then stabilized by the biomolecules present in the fungal extract, resulting in the formation of stable nanoparticles (Chugh et al., 2021; Singh et al., 2020). Finally, it produces nanoparticles with fewer toxic by-products, making them more suitable for industrial applications.

6 Present Status of Fungal NPS

The synthesis of metal nanoparticles (NPs) using fungi is of great importance in today's environmental landscape. This innovative approach provides a sustainable and environmentally friendly alternative to conventional chemical synthesis methods and fits perfectly into the global quest for green technologies (Bandeira et al., 2020). Metal NPs synthesized by fungi offer promising solutions to mitigate environmental problems. They can help remove heavy metals from contaminated soils and waters, contribute to sustainable agricultural practices, purify drinking water, and advance nanotechnology applications while reducing energy consumption and promoting biodiversity conservation (Mandal et al., 2022). This field, while still at an early stage of development, shows great potential to address pressing environmental issues and promote a more harmonious relationship between technology and the environment (Fig. 2).

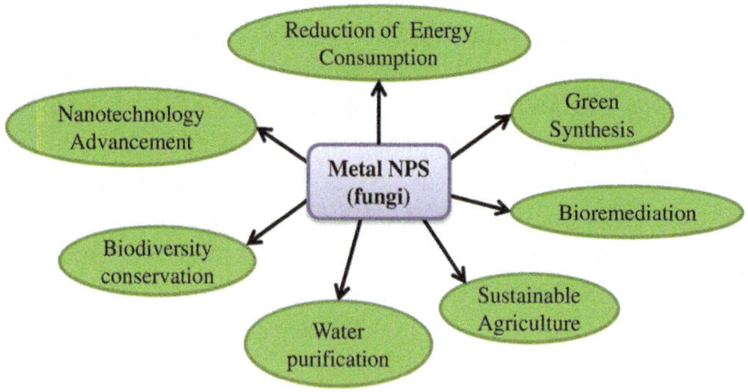

Fig. 2 Current role of metal nanoparticles (NPs) using fungi

7 Exclusive Fungi Involve in Nanoparticles' Role

Nanoparticle production, stabilization, and surface functionalization are all processes that can benefit from the involvement of fungi. Nanoparticles can be synthesized using enzymes derived from fungi, such as lipases, and fungal polysaccharides and proteins can serve as stabilizers and defenders for the nanoparticles (Bhardwaj & Naraian, 2021). Nanoparticles' surfaces are modified with fungal extracellular polysaccharides such as chitosan to facilitate specific molecular interactions. Magnetic nanoparticles made from fungi have numerous potential medical and industrial applications (Table 1).

Researchers have discovered that the fungi *Aspergillus niger* (Rajeshkumar & Rinitha, 2018), *Penicillium purpurogenum* (Muthuchamy et al., 2020), *Trametes versicolor* (Pour et al., 2022), and *Saccharomyces cerevisiae* (Sanusi et al., 2019) are all involved in the production of nanoparticles. Reductase enzymes in *A. niger* are responsible for the reduction of gold ions, which results in the production of gold nanoparticles. It is an ubiquitous fungus that can be found in soil and decomposing organic material. This species has a great potential for biotechnological applications including the production of enzymes, metabolites, and nanoparticles. *A. niger* has been widely studied for its ability to produce various types of nanoparticles including gold, silver, iron oxide, and zinc oxide (Belava et al., 2017). The main advantages of using *A. niger* for the synthesis of nanoparticles include its low cost, high tolerance to various environmental conditions, and ability to produce nanoparticles with a wide range of sizes and morphologies. The synthesis of nanoparticles using *A. niger* involves the reduction of metal ions in the presence of fungal enzymes. This process is often referred to as "fungal-mediated synthesis," and it is a promising alternative to chemical routes for the production of nanoparticles (Rajeshkumar & Rinitha, 2018). Silver nanoparticles have been made using *P. purpurogenum*, and zinc oxide nanoparticles have been made using *Trametes versicolor* and *Saccharomyces cerevisiae*. *P. purpurogenum* is a species of fungus in the family *Trichocomaceae* and is known for its ability to synthesize nanoparticles from various substrates. This species has been used in the development of some other metal-based nanoparticles, such as gold, and copper, with other materials, such as carbon nanotubes (Muthuchamy et al., 2020). They produce nanoparticles based on fungal-mediated synthesis. This method involves the use of a fungal strain to catalyze the formation of nanoparticles from a substrate, such as an organic compound, inorganic salt, or metal. *P. purpurogenum* can also be used to produce magnetic nanoparticles, which have been studied for use in drug delivery, tissue engineering, and other biomedical applications. This species is very promising for the production of nanoparticles as it is easy to grow and can produce large quantities (Muthuchamy et al., 2020; Sagar & Naraian, 2018). *T. versicolor* is another capable fungus that produces a variety of other nanoparticles, including zinc oxide, titanium dioxide, and iron oxide nanoparticles. *T. versicolor* is capable of producing nanoparticles with different sizes, shapes, and chemical compositions, making it a versatile platform for the synthesis of nanoparticles. The potential role of *T. versicolor*

Table 1 Different fungi involved in the fungal-mediated synthesis of nanoparticles

Fungus	Metal (nanoparticles)	Role	References
Alternaria alternata	Gold, iron oxide, and zinc oxide	Antibacterial effects against Gram-positive and Gram-negative bacteria, antifungal	Sardar et al. (2022) and Amuthavalli et al. (2021)
Aspergillus flavus	Silver, gold, zinc, and copper	Diagnosis and treatment of cancer and bacterial infections	Rai et al. (2022), Das et al. (2020), and Muthuchamy et al. (2020)
Aspergillus fumigatus	Silver, gold, iron oxide, titanium dioxide, zinc oxide, and copper oxide	Medical and industrial applications, antimicrobial and anticancer properties	Rai et al. (2022), Lalitha et al. (2020), Jogee and Rai (2020), and Belava et al. (2017)
Aspergillus niger	Gold, silver, iron oxide, and zinc oxide	Biomineralization	Rai et al. (2022), Rajeshkumar and Rinitha (2018), and Belava et al. (2017)
Candida albicans	Gold, silver, and copper	Drug delivery, sensors, and environmental remediation	Carmo et al. (2023), Shinde et al. (2023), and Ashrafi et al. (2020)
Cladosporium cladosporioides	Silver, gold, and copper	Nanomedicine, drug delivery, and environmental protection	Sharma and Tripathi (2022)
Fusarium acuminatum	Silver, gold, iron, titanium dioxide, and other metals	Catalysts, drug delivery, biosensors, and photovoltaics	Rai et al. (2021a), and Lalitha et al. (2020)
Fusarium oxysporum	Silver and gold	Drug delivery and bioimaging	Vazhacharickal et al. (2022), and Rai et al. (2021b)
Helminthosporium solani	Silver, gold, copper, and zinc	Antibacterial	Rizwana et al. (2021)
Mucor hiemalis	Gold, silver, iron oxide, and zinc oxide	Bioreduction	Yadav et al. (2020)
Novosphingobium oryzae	Carbon nanotubes, silver, gold, and magnetic nanoparticles	Enhanced optical, electrical, and magnetic properties	Shinde et al. (2023)
Paecilomyces variotii	Gold, silver, iron oxide, and titanium dioxide	Drug delivery, water purification, and imaging	Rai et al. (2022)
Phanerochaete chrysosporium	Gold ions	Biodegradation of lignocellulose	Rajeshkumar and Sivapriya (2020)
Penicillium purpurogenum	Gold, silver, and copper, as well as other materials, such as carbon nanotubes	Biomedicine and bioremediation	Muthuchamy et al. (2020) and Sagar and Naraian (2018)

(continued)

Table 1 (continued)

Fungus	Metal (nanoparticles)	Role	References
Rhizopus arrhizus	Gold, silver, iron oxide, and titanium dioxide	Drug delivery and bioimaging	Shah et al. (2022) and Jogee and Rai (2020)
Saccharomyces cerevisiae	Gold, silver, and iron oxide	Drug delivery and sensing	Sanusi et al. (2019)
Sclerotium rolfsii	Silver, gold, and zinc oxide	Industrial application	Narayanan and Sakthivel (2011)
Trametes versicolor	Zinc oxide, titanium dioxide, and iron oxide	Catalysis, drug delivery, biosensors, and materials science	Pour et al. (2022)
Trichoderma viride	Gold, silver, and copper	Drug delivery, diagnostics, and cancer therapy	El-Wakil (2020)

nanoparticles in various fields and applications such as catalysis, drug delivery, biosensors, and materials science are being explored (Pour et al., 2022). The ability of these fungi to create enzymes that can convert metal salts into metal nanoparticles makes them all useful in the creation of nanoparticles. In addition, these fungi are versatile in that they may be cultivated on agar, liquid cultures, or solid media, all of which make them excellent candidates for the production of nanoparticles. Sanusi et al. (2019) prepared nanoparticles from *Saccharomyces cerevisiae* fungi. *S. cerevisiae* is a commonly used yeast species in the synthesis of nanoparticles. It is used as a bioreactor to produce nanoparticles with a wide variety of shapes, sizes, and surface properties. In particular, *S. cerevisiae* has been used to synthesize gold nanoparticles, silver nanoparticles, and iron oxide nanoparticles. Additionally, *S. cerevisiae* has been used to synthesize composite nanoparticles, including polysaccharide-coated gold nanoparticles and polysaccharide-coated silver nanoparticles. These nanoparticles can be used for many applications, including drug delivery and sensing.

Aspergillus flavus (Muthuchamy et al., 2020), *Alternaria alternate* (Sardar et al., 2022), *Aspergillus fumigates* (Lalitha et al., 2020), *Candida albicans* (Ashrafi et al., 2020), *Cladosporium cladosporioides* (Sharma & Tripathi, 2022), *Fusarium acuminatum* (Lalitha et al., 2020), *Fusarium oxysporum* (Vazhacharickal et al., 2022), *Helminthosporium solani* (Rizwana et al., 2021), *Mucor hiemalis* (Yadav et al., 2020), *Novosphingobium oryzae* (Shinde et al., 2023), *Paecilomyces variotii* (Rai et al., 2022), *Phanerochaete chrysosporium* (Rajeshkumar & Sivapriya, 2020), *Rhizopus arrhizus* (Jogee & Rai, 2020), *Sclerotium rolfsii* (Narayanan & Sakthivel, 2011), and *Trichoderma viride* (El-Wakil, 2020) are some of the significant fungi that have been involved in the synthesis of nanoparticles. These fungi have the capability of synthesizing a wide variety of nanoparticles, including gold, silver, iron oxide, titanium dioxide, and zinc oxide, among others. Fungi utilize enzymes, proteins, and other molecules in the reduction and oxidation of metal ions, resulting in the production of nanostructures during the synthesis of nanoparticles. Unique features of fungal nanoparticles, such as their high surface area, biocompatibility, and adsorption capacity, make them useful in a wide range of biomedical and environmental remediation settings. Interestingly, *Aspergillus flavus* produces nanoparticles by secreting extracellular enzymes and other metabolites, which can be used to reduce metals to their nanosized form. *Aspergillus flavus* is one of the commonly used fungi for this purpose (Das et al., 2020). In addition, it has been used to produce nanoparticles of ceramics, such as silicon dioxide and carbon. It is also known for its ability to produce nanoparticles with low toxicity and high biocompatibility (Rai et al., 2022). Amuthavalli et al. (2021) utilized the *Alternaria alternata* cell wall and extracellular enzymes for the production of nanomaterials. One study published in 2018 showed that *A. alternata* fungi were able to generate silver nanoparticles. The study showed that the nanoparticles produced were mainly spherical in shape and had an average size of 10–20 nanometers. The study also noted that the synthesized nanoparticles had antibacterial effects against Gram-positive and Gram-negative bacteria. *Cladosporium cladosporioides* fungal-mediated synthesis of nanoparticles involves the production of metal ions, which can be used for this type of synthesis because they contain enzymes that can reduce metal ions to form nanoparticles and have potential applications in various fields, such as nanomedicine, drug delivery, and environmental protection (Sharma & Tripathi, 2022). In the same way, the nanoparticles from *Fusarium acuminatum* can be used in various applications, such as catalysts, drug delivery, biosensors, and photovoltaics (Rai, Bonde, et al., 2021a). *Novosphingobium oryzae* are actively capable of producing various nanoparticles such as carbon nanotubes, gold nanoparticles, magnetic nanoparticles, and silver nanoparticles. The nanoparticles synthesized via this process have been found to have improved functionalities such as enhanced optical, electrical, and magnetic properties (Shinde et al., 2023). *Helminthosporium solani* fungi are involved in making an attractive option for the production of nanoparticles that possess promising role in curing many illnesses (Rizwana et al., 2021). Several potential fungal nanoparticles are extensively used in industry.

8 Future Prospects of Fungal NPS and Sustainability

Microorganisms such as fungi, characterized by remarkable adaptability, are now considered viable options for the production of metal nanoparticles (NPs), a topic of crucial importance for nanotechnology and materials research. Their future contribution in this field gives hope for environmentally friendly methods. First, the use of fungi in the synthesis of metal NPs has an environmental advantage (Krishnani et al., 2022). The environmental impact of fungal-mediated synthesis is much lower because it takes place under safe conditions and biocompatible molecules are used instead of the aggressive molecules required by conventional chemical methods. Fungal-based synthesis NP is also a viable option for industrial production due to its scalability and low cost. As technology evolves, biosynthesis of NP using the inherent disposition of fungi could pave the way to environmentally friendly production methods, where waste products from one sector are reused as inputs in another. In addition, the surface chemistry of NPs formed from fungi can be modified for specific purposes such as catalysis or drug delivery, making them an important component of environmentally friendly and sustainable technologies. The unique ability of fungi to heal polluted environments has promising implications for environmental sustainability as a whole (Tyagi et al., 2023). Some fungi can remove metals from polluted soils and waters, which they then enrich. Fungi play an important role in promoting sustainable practices consistent with circular economy concepts, as they are involved in both synthesizing NP and repairing the environment (Teijido et al., 2023).

9 Conclusion

Fungal-mediated nanoparticle synthesis is gaining popularity as an alternative to conventional chemical synthesis due to its many potential uses in fields such as nanomedicine and catalysis medicine and its low environmental and economic impact. This also involves the use of certain species of fungi to reduce metal ions into metallic nanoparticles, which can be used in a variety of applications. This process typically involves using an aqueous solution containing a metal ion, an oxidizing agent, and a reducing agent. The fungi then act as a catalyst to reduce the metal ion into a metallic nanoparticle, which can then be used in various applications. Some potential applications of the nanoparticles produced through fungal-mediated synthesis include drug delivery, biosensors, catalysis, and water purification. In drug delivery, the nanoparticles can provide a carrier for drugs, allowing them to target specific areas of the body for a more effective treatment. Biosensors can detect different molecules in a sample, such as proteins, hormones, and other biomolecules. In catalysis, the nanoparticles can increase the efficiency of chemical reactions, and water purification systems can use the particles to remove pollutants from water. It is a viable alternative to chemical synthesis methods, offering a low environmental and economic impact. This method is also more environmentally friendly than conventional chemical synthesis methods, as it avoids the use of harsh chemicals.

References

Alghuthaymi, M. A., Abd-Elsalam, K. A., Abo Dalam, H. M., Ahmed, F. K., Ravichandran, M., Kalia, A., & Rai, M. (2022). Trichoderma: An eco-friendly source of nanomaterials for sustainable agroecosystems. *Journal of Fungi, 8*(4), 367.

Amuthavalli, P., Hwang, J. S., Dahms, H. U., Wang, L., Anitha, J., Vasanthakumaran, M., Gandhi, A. D., Murugan, K., Subramaniam, J., Paulpandi, M., & Chandramohan, B. (2021). Zinc oxide nanoparticles using plant Lawsonia inermis and their mosquitocidal, antimicrobial, anticancer applications showing moderate side effects. *Scientific Reports, 11*(1), 1–13.

Ashrafi, M., Bayat, M., Mortazavi, P., Hashemi, S. J., & Meimandipour, A. (2020). Antimicrobial effect of chitosan–silver–copper nanocomposite on Candida albicans. *Journal of Nanostructure in Chemistry, 10*, 87–95.

Bandeira, M., Giovanela, M., Roesch-Ely, M., Devine, D. M., & da Silva Crespo, J. (2020). Green synthesis of zinc oxide nanoparticles: A review of the synthesis methodology and mechanism of formation. *Sustainable Chemistry and Pharmacy, 15*, 100223.

Bardov, V. G., Omelchuk, S. T., Merezhkina, N. V., Aleksiichuk, V. D., Anisimov, Y. M., Antonenko, A. M., Bardov, G. P., Biletska, E. M., Blagaia, A. V., Borysenko, A. A., & Vavrinevych, O. P. (2022). Hygiene and ecology.

Bhardwaj, A. K., & Naraian, R. (2021). Cyanobacteria as biochemical energy source for the synthesis of inorganic nanoparticles, mechanism and potential applications: A review. 3. *Biotech, 11*(10), 445.

Bhardwaj, A. K., Shukla, A., Maurya, S., Singh, S. C., Uttam, K. N., Sundaram, S., Singh, M. P., & Gopal, R. (2018). Direct sunlight enabled photo-biochemical synthesis of silver nanoparticles and their bactericidal efficacy: Photon energy as key for size and distribution control. *Journal of Photochemistry and Photobiology B: Biology, 188*, 42–49.

Belava, V. N., Panyuta, O. O., Yakovleva, G. M., Pysmenna, Y. M., & Volkogon, M. V. (2017). The effect of silver and copper nanoparticles on the wheat—Pseudocercosporella herpotrichoides Pathosystem. *Nanoscale Research Letters, 12*, 1–10.

Beyene, H. D., Werkneh, A. A., Bezabh, H. K., & Ambaye, T. G. (2017). Synthesis paradigm and applications of silver nanoparticles (AgNPs), a review. *Sustainable Materials and Technologies, 13*, 18–23.

Calzoni, E., Cesaretti, A., Polchi, A., Di Michele, A., Tancini, B., & Emiliani, C. (2019). Biocompatible polymer nanoparticles for drug delivery applications in cancer and neurodegenerative disorder therapies. *Journal of Functional Biomaterials, 10*(1), 4.

Carmo, P. H. F. D., Garcia, M. T., Figueiredo-Godoi, L. M. A., Lage, A. C. P., Silva, N. S. D., & Junqueira, J. C. (2023). Metal nanoparticles to combat *Candida albicans* infections: An update. *Microorganisms, 11*(1), 138.

Chugh, D., Viswamalya, V. S., & Das, B. (2021). Green synthesis of silver nanoparticles with algae and the importance of capping agents in the process. *Journal of Genetic Engineering and Biotechnology, 19*(1), 1–21.

Das, P. E., Abu-Yousef, I. A., Majdalawieh, A. F., Narasimhan, S., & Poltronieri, P. (2020). Green synthesis of encapsulated copper nanoparticles using a hydroalcoholic extract of Moringa oleifera leaves and assessment of their antioxidant and antimicrobial activities. *Molecules, 25*(3), 555.

Das, R. K., Pachapur, V. L., Lonappan, L., Naghdi, M., Pulicharla, R., Maiti, S., Cledon, M., Dalila, L. M. A., Sarma, S. J., & Brar, S. K. (2017). Biological synthesis of metallic nanoparticles: Plants, animals and microbial aspects. *Nanotechnology For Environmental Engineering, 2*, 1–21.

Dikshit, P. K., Kumar, J., Das, A. K., Sadhu, S., Sharma, S., Singh, S., Gupta, P. K., & Kim, B. S. (2021). Green synthesis of metallic nanoparticles: Applications and limitations. *Catalysts, 11*(8), 902.

El-Wakil, D. A. (2020). Antifungal activity of silver nanoparticles by Trichoderma species: Synthesis, characterization and biological evaluation. *Egyptian Journal of Phytopathology, 48*(1), 71–80.

Guilger-Casagrande, M., & Lima, R. D. (2019). Synthesis of silver nanoparticles mediated by fungi: A review. *Frontiers in Bioengineering and Biotechnology, 7*, 287.

Hachem, K., Ansari, M. J., Saleh, R. O., Kzar, H. H., Al-Gazally, M. E., Altimari, U. S., Hussein, S. A., Mohammed, H. T., Hammid, A. T., & Kianfar, E. (2022). Methods of chemical synthesis in the synthesis of nanomaterial and nanoparticles by the chemical deposition method: A review. *BioNanoScience, 12*(3), 1032–1057.

Jamkhande, P. G., Ghule, N. W., Bamer, A. H., & Kalaskar, M. G. (2019). Metal nanoparticles synthesis: An overview on methods of preparation, advantages and disadvantages, and applications. *Journal of Drug Delivery Science and Technology, 53*, 101174.

Jampílek, J., & Kráľová, K. (2023). Mycosynthesis of metal-based nanoparticles and their perspectives in Agri-food and veterinary/medical applications. In *Fungal cell factories for sustainable nanomaterials productions and agricultural applications* (pp. 423–482). Elsevier.

Jeevanandam, J., Kiew, S. F., Boakye-Ansah, S., Lau, S. Y., Barhoum, A., Danquah, M. K., & Rodrigues, J. (2022). Green approaches for the synthesis of metal and metal oxide nanoparticles using microbial and plant extracts. *Nanoscale, 14*(7), 2534–2571.

Jogee, P., & Rai, M. (2020). Application of nanoparticles in inhibition of mycotoxin-producing fungi. In *Nanomycotoxicology* (pp. 239–250). Academic Press.

Khalil, A. T., Ovais, M., Iqbal, J., Ali, A., Ayaz, M., Abbas, M., Ahmad, I., & Devkota, H. P. (2022). Microbes-mediated synthesis strategies of metal nanoparticles and their potential role in cancer therapeutics. In *Seminars in cancer biology* (Vol. 86, pp. 693–705). Academic Press.

Khan, S., Mansoor, S., Rafi, Z., Kumari, B., Shoaib, A., Saeed, M., Alshehri, S., Ghoneim, M. M., Rahamathulla, M., Hani, U., & Shakeel, F. (2022). A review on nanotechnology: Properties, applications, and mechanistic insights of cellular uptake mechanisms. *Journal of Molecular Liquids, 348*, 118008.

Khandel, P., & Shahi, S. K. (2018). Mycogenic nanoparticles and their bio-prospective applications: Current status and future challenges. *Journal of Nanostructure in Chemistry, 8*, 369–391.

Kik, K., Bukowska, B., & Sicińska, P. (2020). Polystyrene nanoparticles: Sources, occurrence in the environment, distribution in tissues, accumulation and toxicity to various organisms. *Environmental Pollution, 262*, 114297.

Krishnani, K. K., Boddu, V. M., Chadha, N. K., Chakraborty, P., Kumar, J., Krishna, G., & Pathak, H. (2022). Metallic and non-metallic nanoparticles from plant, animal, and fisheries wastes: Potential and valorization for application in agriculture. *Environmental Science and Pollution Research, 29*(54), 81130–81165.

Kumar, M. (2021). In silico efficacy of [S]-8-gingerol (a derivative) with 6-gingerol against PT-domain of polyketide synthase a (PksA). *IP International Journal of Medical Microbiology and Tropical Diseases, 7*, 62–64.

Kumar, M., & Akhtar, M. S. (2019). Application of microbial biotechnology in improving salt stress and crop productivity. *Salt Stress, Microbes, and Plant Interactions: Mechanisms and Molecular Approaches, 2*, 133–159.

Kumar, M., Kumar, H., Topno, R. K., & Kumar, J. (2019). Analysis of impact of anaerobic condition on the aflatoxin production in *Aspergillus parasiticus* Speare. *Agricultural Science Digest-A Research Journal, 39*(1), 75–78.

Kumar, M., Topno, R. K., Kumar, H., & Kumar, J. (2018). Biological conversion of Carcinogenic Aflatoxin B1 to AFB1-8, 9-exo-epoxide and its potential prevention. *International Journal of Research and Analytical Reviews, 5*(4), 935–946.

Kumar, H., Kumar, M., Kumar, B., Bharti, S., Kumar, P., Kumar, M., & Kumar, J. (2017). Effect of acetone extract of Zea mays L. leaves on the growth of Aspergillus flavus and aflatoxin production. International. *Journal of Advanced Research, 5*(8), 1667–1671.

Kumari, A., Kumar, M., Sahoo, G. C., & Kumar, J. (2019). In silico prediction of Glabridin potency against human Tyrosinase in hyperpigmentation condition. *International Journal of Research and Analytical Reviews*. (IJRAR), E-ISSN. 2348–1269.

Lalitha, K., Kalaimurgan, D., Nithya, K., Venkatesan, S., & Shivakumar, M. S. (2020). Antibacterial, antifungal and mosquitocidal efficacy of copper nanoparticles synthesized from entomopathogenic nematode: Insect–host relationship of bacteria in secondary metabolites of Morganella morganii sp.(PMA1). *Arabian Journal for Science and Engineering, 45*, 4489–4501.

Li, Q., Liu, F., Li, M., Chen, C., & Gadd, G. M. (2022). Nanoparticle and nanomineral production by fungi. *Fungal Biology Reviews, 41*, 31–44.

Litwin, A., Nowak, M., & Różalska, S. (2020). Entomopathogenic fungi: unconventional applications. *Reviews in Environmental Science and Bio/Technology, 19*(1), 23–42.

Madkour, L. H. (2018). Ecofriendly green biosynthesized of metallic nanoparticles: Bio-reduction mechanism, characterization and pharmaceutical applications in biotechnology industry. *Global Drugs and Therapeutics, 3*(1), 1–11.

Mandal, A. K., Katuwal, S., Tettey, F., Gupta, A., Bhattarai, S., Jaisi, S., Bhandari, D. P., Shah, A. K., Bhattarai, N., & Parajuli, N. (2022). Current research on zinc oxide nanoparticles: Synthesis, characterization, and biomedical applications. *Nanomaterials, 12*(17), 3066.

Marouzi, S., Sabouri, Z., & Darroudi, M. (2021). Greener synthesis and medical applications of metal oxide nanoparticles. *Ceramics International, 47*(14), 19632–19650.

Mishra, J., Kumar, M., Sharan, A. K., & Kumar, J. (2019). Detoxification assessment of Aflatoxin in *Aspergillus flavus* under the effect of *Calendula officinalis* Linn's methanolic extract. *Agricultural Science Digest—A Research Journal, 39*(1), 21–25.

Mitchell, M. J., Billingsley, M. M., Haley, R. M., Wechsler, M. E., Peppas, N. A., & Langer, R. (2021). Engineering precision nanoparticles for drug delivery. *Nature Reviews Drug Discovery, 20*(2), 101–124.

Muthuchamy, M., Muneeswaran, T., Rajivgandhi, G., Franck, Q., Muthusamy, A., & Ji-Ming, S. (2020). Biologically synthesized copper and zinc oxide nanoparticles for important biomolecules detection and antimicrobial applications. *Materials Today Communications, 22*, 100766.

Naraian, R., & Abhishek, A. K. B. (2020). Green synthesis and characterization of silver NPs using oyster mushroom extract for antibacterial efficacy. *Journal of Chemistry, Environmental Sciences and its Applications, 7*(1), 13–18.

Narayanan, K. B., & Sakthivel, N. (2011). Facile green synthesis of gold nanostructures by NADPH-dependent enzyme from the extract of Sclerotium rolfsii. *Colloids and Surfaces A: Physicochemical and Engineering Aspects, 380*(1–3), 156–161.

Patel, M., Surti, M., Siddiqui, A. J., & Adnan, M. (2021). Fungi and metal nanoparticles. In *Handbook of greener synthesis of nanomaterials and compounds* (pp. 861–890). Elsevier.

Pour, M. F., Mehdinia, M., Kiamahalleh, M. V., Hoseini, K. D., Hatefnia, H., & Dorieh, A. (2022). Biological durability of particleboard: Fungicidal properties of Ag and Cu nanoparticles against Trametes versicolor white-rot fungus. *Wood Material Science & Engineering, 17*(6), 929–936.

Rai, M., Gupta, I., Bonde, S., Ingle, P., Shende, S., Gaikwad, S., Razzaghi-Abyaneh, M., & Gade, A. (2022). Industrial applications of nanomaterials produced from aspergillus species. In *The Genus Aspergillus-pathogenicity, mycotoxin production and industrial applications*. IntechOpen.

Rai, M., Bonde, S., Golinska, P., Trzcińska-Wencel, J., Gade, A., Abd-Elsalam, K. A., Shende, S., Gaikwad, S., & Ingle, A. P. (2021a). Fusarium as a novel fungus for the synthesis of nanoparticles: Mechanism and applications. *Journal of Fungi, 7*(2), 139.

Rai, M., Ingle, A. P., Trzcińska-Wencel, J., Wypij, M., Bonde, S., Yadav, A., Kratošová, G., & Golińska, P. (2021b). Biogenic silver nanoparticles: What we know and what do we need to know? *Nanomaterials, 11*(11), 2901.

Rajeshkumar, S., & Sivapriya, D. (2020). Fungus-mediated nanoparticles: Characterization and biomedical advances. *Nanoparticles in Medicine*, 185–199.

Rajeshkumar, S., & Rinitha, G. (2018). Nanostructural characterization of antimicrobial and antioxidant copper nanoparticles synthesized using novel Persea americana seeds. *OpenNano, 3*, 18–27.

Raj, N. B., Swamy, M. K., Purushotham, B., & Sukrutha, S. K. (2021). *Applications of microbe-based nanoparticles in agriculture: Present state and future challenges* (pp. 343–382). Microbial Nanobiotechnology: Principles and Applications.

Rana, A., Yadav, K., & Jagadevan, S. (2020). A comprehensive review on green synthesis of nature-inspired metal nanoparticles: Mechanism, application and toxicity. *Journal of Cleaner Production, 272*, 122880.

Ratan, Z. A., Haidere, M. F., Nurunnabi, M. D., Shahriar, S. M., Ahammad, A. S., Shim, Y. Y., Reaney, M. J., & Cho, J. Y. (2020). Green chemistry synthesis of silver nanoparticles and their potential anticancer effects. *Cancers, 12*(4), 855.

Rehman, S., Ansari, M. A., Al-Dossary, H. A., Fatima, Z., Hameed, S., Ahmad, W., & Ali, A. (2021). Current perspectives on mycosynthesis of nanoparticles and their biomedical application. *Modeling and control of drug delivery systems*, 301–311.

Roy, A., Elzaki, A., Tirth, V., Kajoak, S., Osman, H., Algahtani, A., Islam, S., Faizo, N. L., Khandaker, M. U., Islam, M. N., & Emran, T. B. (2021). Biological synthesis of nanocatalysts and their applications. *Catalysts, 11*(12), 1494.

Rizwana, H., Alwhibi, M. S., Aldarsone, H. A., Awad, M. A., Soliman, D. A., & Bhat, R. S. (2021). Green synthesis, characterization, and antimicrobial activity of silver nanoparticles prepared using Trigonella foenum-graecum L. leaves grown in Saudi Arabia. *Green Processing and Synthesis, 10*(1), 421–429.

Sagar, S. S., & Naraian, R. (2018). Biosynthesis of nanoparticles by penicillium and their medical applications. In *New and future developments in microbial biotechnology and bioengineering* (pp. 235–246). Elsevier.

Said, Z., Arora, S., Farooq, S., Sundar, L. S., Li, C., & Allouhi, A. (2022). Recent advances on improved optical, thermal, and radiative characteristics of plasmonic nanofluids: Academic insights and perspectives. *Solar Energy Materials and Solar Cells, 236*, 111504.

Salem, S. S., & Fouda, A. (2021). Green synthesis of metallic nanoparticles and their prospective biotechnological applications: An overview. *Biological Trace Element Research, 199*, 344–370.

Sanusi, I. A., Faloye, F. D., & Gueguim Kana, E. B. (2019). Impact of various metallic oxide nanoparticles on ethanol production by Saccharomyces cerevisiae BY4743: Screening, kinetic study and validation on potato waste. *Catalysis Letters, 149*, 2015–2031.

Saratale, R. G., Saratale, G. D., Shin, H. S., Jacob, J. M., Pugazhendhi, A., Bhaisare, M., & Kumar, G. (2018). New insights on the green synthesis of metallic nanoparticles using plant and waste biomaterials: Current knowledge, their agricultural and environmental applications. *Environmental Science and Pollution Research, 25*, 10164–10183.

Saravanan, A., Kumar, P. S., Karishma, S., Vo, D. V. N., Jeevanantham, S., Yaashikaa, P. R., & George, C. S. (2021). A review on biosynthesis of metal nanoparticles and its environmental applications. *Chemosphere, 264*, 128580.

Sardar, M., Ahmed, W., Al Ayoubi, S., Nisa, S., Bibi, Y., Sabir, M., Khan, M. M., Ahmed, W., & Qayyum, A. (2022). Fungicidal synergistic effect of biogenically synthesized zinc oxide and copper oxide nanoparticles against Alternaria citri causing citrus black rot disease. *Saudi Journal of Biological Sciences, 29*(1), 88–95.

Shah, I. H., Ashraf, M., Sabir, I. A., Manzoor, M. A., Malik, M. S., Gulzar, S., Ashraf, F., Iqbal, J., Niu, Q., & Zhang, Y. (2022). Green synthesis and characterization of copper oxide nanoparticles using Calotropis procera leaf extract and their different biological potentials. *Journal of Molecular Structure, 1259*, 132696.

Sharma, R., & Tripathi, A. (2022). Green synthesis of nanoparticles and its key applications in various sectors. *Materials Today: Proceedings, 48*, 1626–1632.

Siddique, S., & Chow, J. C. (2020). Gold nanoparticles for drug delivery and cancer therapy. *Applied Sciences, 10*(11), 3824.

Shinde, B. H., Inamdar, S. N., Nalawade, S. A., & Chaudhari, S. B. (2023). A systematic review on antifungal and insecticidal applications of biosynthesized metal nanoparticles. *Materials Today: Proceedings, 73*, 412–417.

Singh, A., Gautam, P. K., Verma, A., Singh, V., Shivapriya, P. M., Shivalkar, S., Sahoo, A. K., & Samanta, S. K. (2020). Green synthesis of metallic nanoparticles as effective alternatives to treat antibiotics resistant bacterial infections: A review. *Biotechnology Reports, 25*, e00427.

Singh, J., Dutta, T., Kim, K. H., Rawat, M., Samddar, P., & Kumar, P. (2018). 'Green' synthesis of metals and their oxide nanoparticles: Applications for environmental remediation. *Journal of Nanobiotechnology, 16*(1), 1–24.

Subashini, G., & Bhuvaneswari, S. (2018). Nanoparticles from fungi (myconanoparticles). In *Fungi and their role in sustainable development: current perspectives* (pp. 753–779). Springer.

Surendran, G., & Sherje, A. P. (2022). Cellulose nanofibers and composites: An insight on basics and biomedical applications. *Journal of Drug Delivery Science and Technology*, 103601.

Tamboli, Q. Y., Patange, S. M., Mohanta, Y. K., Sharma, R., & Zakde, K. R. (2023). Green synthesis of cobalt ferrite nanoparticles: An emerging material for environmental and biomedical applications. *Journal of Nanomaterials, 2023*, 1–15.

Teijido, R., Rubio, L. R., Zhang, Q., Lanceros, S., & Vilela, J. L. V. (2023). Biopolymers and their application-oriented composites into a circular economy context. *New Materials for a Circular Economy, 149*, 105–141.

Tyagi, P. K., Arya, A., Mazumder, A. M., & Tyagi, S. (2023). Development of copper nanoparticles and their prospective uses as antioxidants, antimicrobials, anticancer agents in the pharmaceutical sector. *Precision Nanomedicine, 6*(2), 1048–1065.

Vazhacharickal, P. J., Joseph, N. M., Mathew, J. J., & Sajeshkumar, N. K. (2022). Green synthesis of copper and zinc nanoparticles from *Chromolaena odorata* and *Justicia adhatoda* and evaluation of their antifungal activity against *Fusarium oxysporum* cubense: An overview. *Journal of Medicinal Plants, 10*(2), 13–29.

Wang, L., Wu, W. M., Bolan, N. S., Tsang, D. C., Li, Y., Qin, M., & Hou, D. (2021). Environmental fate, toxicity and risk management strategies of nanoplastics in the environment: Current status and future perspectives. *Journal of Hazardous Materials, 401*, 123415.

Wonoputri, V., Wijaya, J., Primayudha, V. A., Khairunnisa, S., & Samadhi, T. W. (2023). February. Performance of antimicrobial ZnO nanoparticle synthesized by biological method: Literature review. In *AIP Conference Proceedings* (Vol. 2667, p. 050007). AIP Publishing LLC.

Yu, G. H., Chi, Z. L., Kappler, A., Sun, F. S., Liu, C. Q., Teng, H. H., & Gadd, G. M. (2020). Fungal nanophase particles catalyze iron transformation for oxidative stress removal and iron acquisition. *Current Biology, 30*(15), 2943–2950.

Yadav, V. K., Khan, S. H., Malik, P., Thappa, A., Suriyaprabha, R., Ravi, R. K., Choudhary, N., Kalasariya, H., & Gnanamoorthy, G. (2020). Microbial synthesis of nanoparticles and their applications for wastewater treatment. In *Microbial biotechnology: Basic research and applications* (pp. 147–187). Springer.

Green Synthesis of Organic Nanomaterials and Their Applications

Abhilekha Sharma, Amit Kumar, Rahul Singh, Richa Saxena, and Dipti Bharti

Organic nanoparticles are solid particles primarily constructed from organic constituents, such as lipid or macromolecules, and they are typically in the range of 10 nanometers to 1 micrometer in size. These nanoparticles (NPs) are constructed using either natural or synthetic organic molecules (Anderson & Deniel, 2003). In the natural world, organic nanoparticles can be found in a variety of forms, including polypeptide clusters, droplets of lipid, dairy suspensions, and more complex structures such as pathogens. While inorganic materials have undergone extensive research and commercial investment, organic nanoparticles have not received comparable attention. One notable advantage of organic nanoparticles is their eco-friendliness due to their limited environmental persistence. Additionally, they offer relatively straightforward methods for incorporating various substances (Michael et al., 2022a). Furthermore, the biodegradable nature of the molecules used to build these organic NPs enhances their attractiveness for applications for pharmaceutical transport and biomedicine (Michael et al., 2022b). The comparative study of organic and inorganic nanoparticles is reported in Table 1.

A. Sharma
Department of Chemistry, School of Sciences, Noida International University, Greater Noida, Uttar Pradesh, India

A. Kumar
Department of Mechanical Engineering, Indian Institute of Technology, Ropar, Punjab, India

R. Singh · D. Bharti (✉)
Department of Applied Science & Humanities, Darbhanga College of Engineering, Darbhanga, Bihar, India

R. Saxena
Department of Biotechnology, Invertis University, Bareilly, UP, India

A. K. Bhardwaj et al. (eds.), *Biogenic Wastes-Enabled Nanomaterial Synthesis*,
https://doi.org/10.1007/978-3-031-59083-2_7

Table 1 Organic versus inorganic nanoparticles

	Inorganic nanoparticles/vehicle	Organic nanoparticles/vehicle
Advantages	Biological activity/metals	Better stability along with multi-structures/poly(lactic-co-glycolic acid)
	Good biocompatibility/carbon nanotubes	Biologically specific outcomes and adequate supply/extracellular vesicle
	Target orientated as well as bioimaging ability/quantum dots	Readily available, robust persistence, environmentally degradable/lipid vesicle
Disadvantages	Instability	Enhancement of biological compatibility required
	Limited target efficiency	Suboptimal targeting precision
	Biotoxicity	Biohazard potential

1 Classification of Nanoparticles

Multiple categories are used for classifying nanomaterials. One way to categorize them is by their size, specifically nanoparticles as one, two, or three dimensions (Fig. 1).

1.1 Commonly Recognized Classes of NPS

1.1.1 Dendrimers

Dendrimers are extensively branched macromolecules with monodisperse characteristics. They exhibit radial dispersion from a central core (Klajnert et al., 2001). Their abundant branching provides distinctive chemical and physical attributes. These qualities set them apart from linear polymers. Dendrimers hold great significance for applications in biomedicine and industry.

Dendrimers are synthesized using two methods: divergent and convergent (Hodge, 1993). Divergent growth starts from a core with a reactive monomer but faces by-product and purification challenges. Convergent synthesis proceeds stepwise from end groups to the core, yielding well-defined structures with easier purification. However, high-generation dendrimers face limitations due to steric hindrance.

1.1.2 Nanocapsules

Nanocapsules, inspired by viral capsids, are hollow polymer shells for substance encapsulation. Natural capsids, found in viruses, serve as protective shells (Kim et al., 2010). Scientists have innovated synthetic nanocapsules with versatile applications in encapsulation, delivery, imaging, and catalysis. Two formation methods exist: interfacial polymerization and direct deposition of preformed polymer.

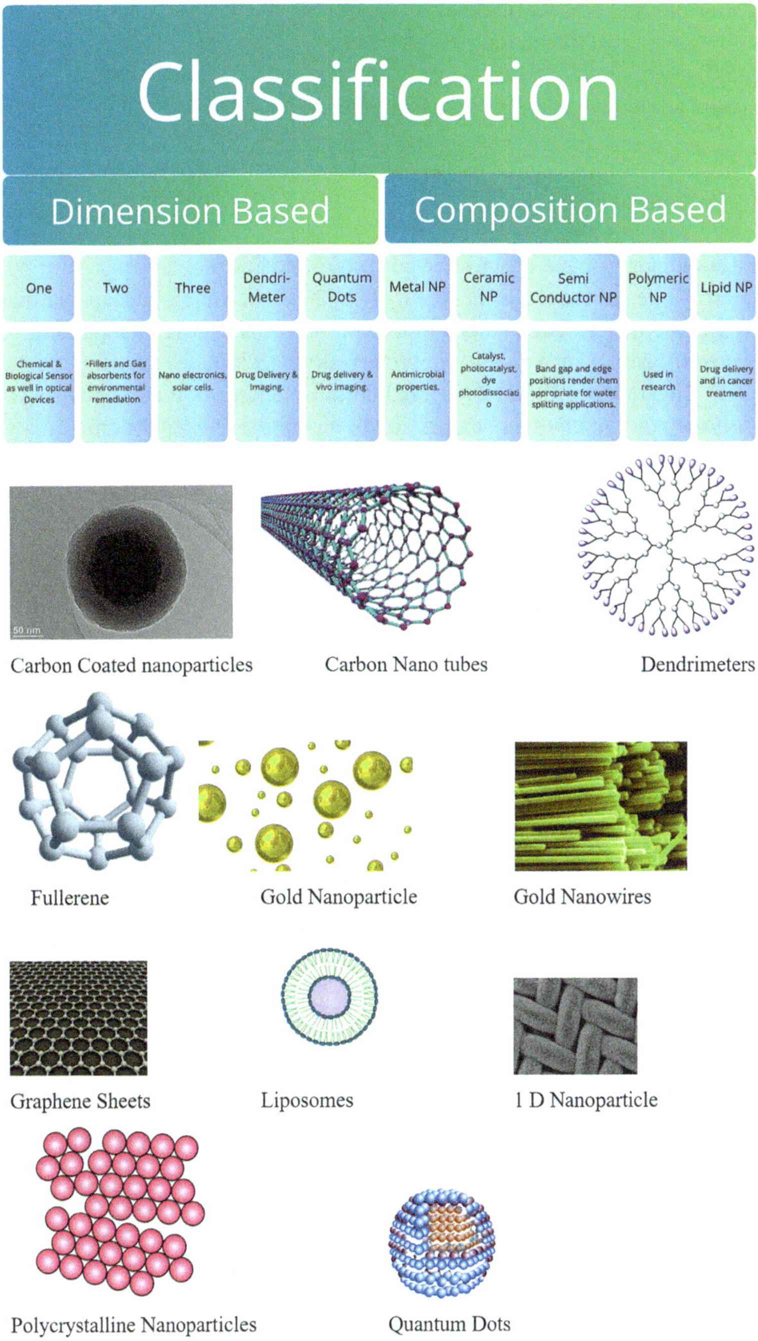

Fig. 1 Various nanomaterials

Interfacial polymerization involves a reaction at an interface, starting with emulsions containing reactive monomers (Reis et al., 2006). An example is (Bouchemal et al., 2004) polyurethane nanocapsules loaded with the antioxidant a-tocopherol, beneficial for medical and cosmetic formulations.

1.1.3 Polymeric NPs

Polymeric nanoparticles, often referred to as nanospheres, are tiny solid particles with matrix-like structures (Bouchemal et al., 2004) capable of encapsulating cargo within the polymer or adsorbing it onto their surfaces. These nanoparticles can be synthesized through in real-time polymerization techniques or using preexisting polymers (Elzatahry & Eldin, 2008). A familiar method for their synthesis is emulsion polymerization, involving dispersion of an emulsionable monomer in a solvent phase. Free radical initiation drives the polymerization until all available monomers react. When encapsulating cargo, it is first dissolved in the reaction solution. The result is a stable scattering of polymer particle colloids in a water-based media known as "latex" (Bhardwaj & Naraian, 2021).

1.2 Fullerene and Carbon Nanotubes (CNTs)

Carbon nanotubes are nanosized tubular structures with two nanoscale dimensions, featuring chemically distinct regions, namely, sidewalls and tips. The tip region exhibits high reactivity, while the sidewall reactivity is mainly influenced by defects (Prato et al., 2008). Carbon nanotubes (CNTs) can serve as molecule containers, depending on their diameter and chemical properties (Koshino et al., 2007). However, solubility and cytotoxicity issues have hindered their practical applications (Prato et al., 2008). To address this, modification approaches include covalent attachment of chemical groups, non-covalent adsorption or encapsulation of molecules, and improved solubility, enabling CNTs to act as carriers for bioactive compounds (Tasis et al., 2006). This synergy of CNT conductivity and bioactive compounds holds potential in bioelectric systems. For instance, water-soluble CNT-BSA conjugates with preserved bioactivity have been demonstrated, showing enhanced solubility and reduced cytotoxicity with increased CNT modification (Huang et al., 2002).

1.3 Nano-cochleates

Viable lipid-derived vesicles primarily consist of anionic lipids as well of bivalent cations. These have a spindle-shaped, multilayered structure with a continual phospholipid double-layer sheet spirally turn over and over, offering minimal hydrophilic area. These carriers are suitable for delivering various molecules, including hydrophobic, amphiphilic, and charged ones. Nano-cochleates are highly resistive to breakdown in the digestive system, making them ideal for administration by mouth. However, challenges arise in food applications due to manufacturing procedures, which often entail the use of other than food agents and surfactants that can adversely affect the structure and stability of encapsulated compounds, raising concerns about toxicity and overall structural integrity.

1.4 Liposomes

Liposomes are sphere entities made up of bilayers of lipid or phospholipid., featuring spacing between the layers to accommodate lipophilic, hydrophobic, and amphiphilic molecules, even with different solubilities (Immordino et al., 2006). They form due to phospholipids' interactions with aqueous media, where polar headgroups face the media and hydrophobic tails cluster into bilayers. Liposomes are classified by size and lamellarity, with variations such as small liposomes and large unilamellar vesicles. Common phospholipids, including phosphatidyl choline, are used for bilayer formation, impacting properties such as charge, acyl chain length, and elasticity (Gangula et al., 2011). Cholesterol is often added to enhance stability, though alternative lipids are explored due to cholesterol consumption concerns. Liposomes are versatile for encapsulating, protecting, and releasing bioactives based on factors such as enzymes, pH, thermo-sensitivity, or antigen-binding.

1.5 Micelles

Micelles are structures formed by assembling monomers, driven by their bipolarity. In micelles, blocks that repel water align with the surrounding polar medium, while hydrophobic blocks cluster in the core, establishing a macromolecular structure (Grover & Maynard, 2010). They can be spherical or tubular and categorized as normal phase (hydrophilic outside) or reversed phase (hydrophilic inside).

1.6 Generating Nanoparticles: Techniques and Approaches

There are several approaches available for synthesizing nanoparticles, among these two main approaches are (1) top-down techniques and (2) bottom-up techniques (Fig. 2).

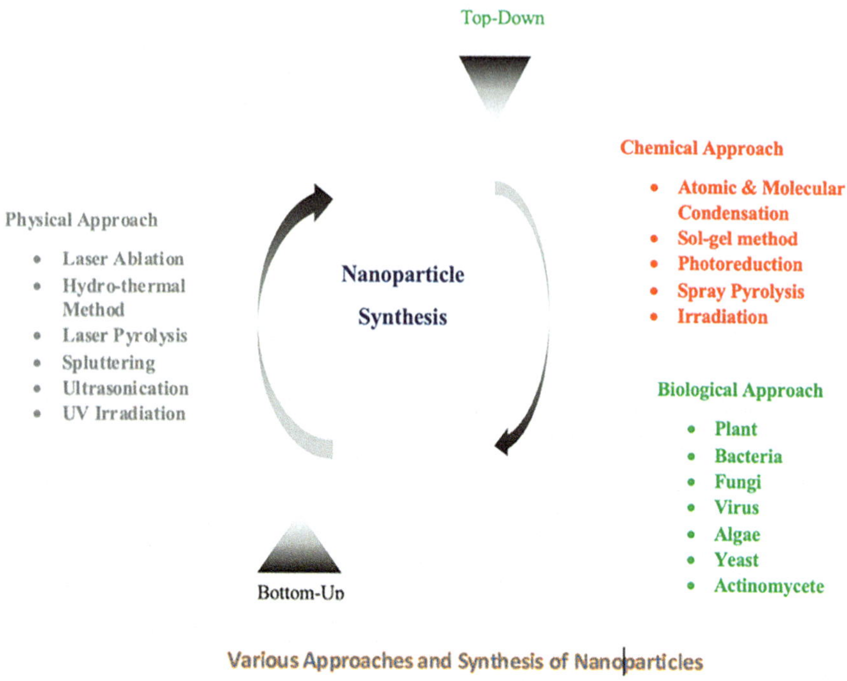

Fig. 2 Top-down and bottom-up approach of nanoparticle synthesis

1.7 Top-Down Approach

This synthesis fragments larger units into small molecules for nanoparticle creation using techniques such as grinding, vacuum deposition, and thin film coating. For example, researchers milled coconut shell particles with a high-energy ball mill, discovering that longer milling times decreased nanoparticle size. Analysis showed reduced crystal size and faded color with increased milling, supported by SEM and X-ray scattering (Bello et al., 2015).

For instance, the nanocrystal TM technology, originally developed by Liversidge et al. utilizes milling to produce drug nanoparticles (NPs) from poorly water-soluble compounds (Haddon, 2002). An alternate approach is wet milling, as exemplified by Sigfridsson et al. (2011), who created ticagrelor drug NPs with a diameter of approximately 230 nm. Additional "top-down" methods encompass microfluidics and lithography, which provide precise control over NP properties and enable the encapsulation of biomolecules such as DNA and proteins (Rolland et al., 2005; Hett, 2004).

1.8 Bottom-Up Approach

The building-up approach creates nanoparticles from simpler substances using methods such as settling, reduction, sol-gel, centrifugation, and eco-friendly techniques (Iravani, 2011).

Through a bottom-up approach, green synthesis involves the amalgamation of metal atoms in their elemental state, culminating in the generation of nanoparticles. These methods yield various nanomaterials, including micelles (Kedar et al., 2010), vesicles and liposomes (Camelo et al., 2009), polymersomes (Christian et al., 2009), polymer conjugates (Canalle et al., 2010), dendrimers (Oliveira et al., 2010), capsules (Cuomo et al., 2010, Moraes et al., 2011, Liu et al., 2010), and polymeric NPs (Rahman et al., 2010; Romero et al., 2010; Schubert et al., 2011).

1.9 Methods of Preparation of Nanoparticles

Organic nanoparticles are capable of being made using methods as follows:

(a) Emulsification deploy procedure
(b) Nanoprecipitation
(c) Dry processes (green synthesis)

Creating organic and bioorganic nanoparticles involves two primary methods. The first follows a two-step process, commencing with the creation of an emulsion system to generate nanodroplets containing solubilized organic compounds. This emulsification can use either high- or low-energy stir techniques. NPs are formed in the next stage by methods such as precipitation, coagulation, or polymeric process (Jing et al., 2015a).

The second approach employs a single-stage method where emulsification is not necessary before nanoparticle formation. These methods depend on the gathering of chemical molecules in liquid by mechanism such as solvent displacement nanoprecipitation or self-assembly produced by ion gelation or polyelectrolyte complex formation (Jing et al., 2015b).

1.10 Emulsification-Based Two-Step Technique

Emulsions can produce microemulsions with stable droplets (10–100 nm) or less stable miniemulsions and macroemulsions with larger droplets (100 nm to 1 μm or more) (50–52). Innovations in emulsification techniques, driven by environmental concerns, have enabled the creation of nanoscale droplets, crucial for designing organic nanoparticles. These advances encompass both low- and high-power emulsifying methods.

The low-power emulsifying procedure for crafting nanoemulsions is classified into three categories. Spontaneously emulsifying process involves the fast dispersion of an aqueous solvent from oil to water (El-Aasser et al., 1988), resulting in emulsions either directly oil dispersed in water (O/W) or inversely water dispersed in oil (W/O) because of interfacial turbulence. Inversion point methods (Sole et al., 2006) gradually dilute microemulsions or liquid crystals with water or oil, leading to their destabilization into nanoemulsions. Phase inversion temperature (PIT) methods, employing polyethylene oxide (PEO)-based surfactants, leverage temperature-induced changes in surfactant affinity for water and oil, resulting in a "transitional phase inversion" of emulsions. These low-energy approaches reduce toxicity and are well-suited for applications such as nanomedicine, pharmaceuticals, and cosmetics, guided by phase diagrams for optimization (Mozafari et al., 2006).

High-energy emulsification primarily relies on mechanical processes, often using devices such as rotor-stator apparatus to apply shear stress, breaking pre-emulsion droplets into smaller, uniform-sized ones. The final droplet size depends mainly on the applied stress and is less affected by the ratio of viscosity between the continuous phase and the dispersed phase (Bengoechea et al., 2007; Lizarraga et al., 2008). Sonication is another widely used method for nanoemulsification and for generating polymer or lipid nanoparticles, including those made from materials such as PLGA-PEG, PCL, or alginate.

1.11 Generation of Nanoparticles from Emulsion

Below are the methods employed to create nano-organic particles from emulsions:

(a) Solvent-driven precipitation
(b) Evaporation of solvent
(c) Diffusion of solvent
(d) Salt-induced precipitation
(e) Emulsion droplet gelation
(f) Emulsion polymerization with polymer formation
(g) Traditional emulsion polymerization
(h) Surfactantless emulsion polymerization
(i) Polymerization at interfaces
(j) Controlled or living radical polymerization

1.12 *One-Step Procedures*

1.12.1 Nanoprecipitation

The nanoprecipitation or solvent exchange technique, pioneered by Fessi et al. (1983) in the late 1980s, is a cost-effective technique for creating nanospheres by utilizing polymers that have already been formed. It consists of four key elements: a polymer, a polymer, a solvent, and a non-solvent. Commonly acetone, ethanol, hexane, methylene chloride, or dioxane is used, with acetone being a frequent choice due to its ability to reduce interfacial tension (Dalpiaz et al., 2009). This reduction leads to an increased surface space and the swift formation of polymer nanoparticles.

A variety of polymers, such as poly(e-caprolactone), polylactide (Legrand et al., 2007), poly(lactide-co-glycolide) (Yallapu et al., 2010), and poly(hydroxyl butyrate) (Deepak et al., 2009) can be synthesized using this method.

1.12.2 Dialysis

Dialysis, a variation of solvent displacement, utilizes dialysis tubing or semipermeable films with appropriate molecular weight cutoffs to physically segregate the polymer (Kostag et al., 2010). This procedure takes place in the presence of a polymer-miscible nonsolvent. As solvent flows via film, the polymer progressively loses its solubility and attains uniform nanoparticle suspensions. The choice of solvent influences both particle size and structure. An array of polymers, including poly(lactide)-b-poly(ethylene oxide) (Lee et al., 2004), polystyrene (Chronopoulou et al., 2009), and poly(L-lactic acid)-b-poly(ethylene glycol) is employed to produce nanoparticles through the dialysis method.

1.12.3 Desolvation

Desolvation is a technique that gradually introduces desolvation agents such as salts, alcohols, or solvents into a macromolecule solution, leading to polymer precipitation. This technique is commonly utilized for creating biopolymeric nanoparticles as human albumin (Langer et al., 2008), bovine serum albumin (Zhao et al., 2010), gliadin, or gelatin (Weber et al., 2000). To enhance nanoparticle stability, cross-linking agents such as glutaraldehyde are often incorporated. Moreover, sodium sulfate can serve as a precipitating agent to facilitate the formation of polymer molecules such as chitin or hyaluronan.

1.13 Green Synthesis of NP

Nanoparticle production via physical and chemical methods raises toxicity and environmental concerns. Physical methods are resource-intensive, while chemical methods involve harmful solvents. Green nanotechnology, using biogenic materials and biotechnology, emerged as an eco-friendly, cost-effective, and sustainable alternative (Khan et al., 2019). Researchers have used eco-friendly ionic liquids to produce purified chitin and nanofibers (Ding et al., 2013; Qin et al., 2010). Recently, environmental considerations have prompted research into solvent-free methods for nanoparticle synthesis as well as using biological sources, an approach toward green synthesis (Vemavarapu et al., 2005).

Various green and biological routes for nanoparticle synthesis are discussed as follows:

1.14 Dry Processes or Supercritical Drying

Supercritical and spray drying processes provide alternative approaches that address the limitations of traditional methods. Two methods harness supercritical fluids to generate nanoparticles. The first dissolves organic macromolecules in a supercritical fluid and then rapidly expands via a sprayer into the atmosphere. This creates uniformly scattered particles due to high supersaturation and rapid pressure reduction, often utilizing CO_2. Particle properties are influenced by factors such as polymer concentration, saturation, processing conditions, molecular mass, and melting point. This method creates nanoparticles of poly(heptadecafluorodecyl acrylate) and poly(L-lactic acid) (Sun et al., 2002).

The second procedure involves converting the supercritical fluid into a fluid, not the air (Sun et al., 2002). It incorporates a CO_2-cosolvent to enhance polymer solubility. Nanosized particles develop as a result of the presence of NaCl in the water solution (Meziani et al., 2005). However, inability of polymers to dissolve in supercritical fluids remains a problem..

1.14.1 Spray Drying

It has applications in producing microsized organic particles and converting nanoparticle suspensions into dry powders, particularly in the biomedical and pharmaceutical fields. This process involves atomizing a liquid into fine droplets and exposing them to hot drying gas for moisture evaporation, resulting in a solid product typically collected with a cyclone unit. Innovative technologies such as the "nano-spray dryer" by BüchiÒ have enabled the preparation of various polymeric nanoparticles, including those made from acacia, whey, PVOH, starch derivative, and glucose polymer (Sun et al., 2002).

1.15 Biological Method for Synthesis of NP

1.15.1 By Bacteria

Bacillus licheniformis to synthesize chitosan-based nanoparticles. These nanoparticles exhibit bactericidal characteristics and may have uses in wound healing and medical textiles, contributing to a safer healthcare environment.

Cyanobacteria are seen as a nano-biofactory to generate complex materials, which will be investigated in a multitude of uses including identification of medical assessment treatments, medications, optical science, and catalytic processes.

Archaeosomes, distinct from liposomes formed with ester phospholipids, are crafted from polar ether lipids sourced from Archaeobacteria. These unique structures exhibit enhanced thermal stability and heightened resilience against oxidative stress, chemical agents, and enzymatic breakdown. Moreover, they demonstrate resistance to the acidic environment and gallstone acid found in the gastrointestinal tract (Patel et al., 2000).

1.15.2 By Plant

Plant-based synthesis of NPs has gained popularity owing to its swift, environmentally benign, and nonhazardous characteristics, offering a one-step biosynthesis approach (Table 2). Plant extracts contain various biomolecules such as polypeptides, carbohydrates, terpenoids, alkaloids, and phenolics.

Green tea antioxidants encompass a collection of polyphenolic group of molecules derived from *Camellia sinensis*, constituting as much as 30% of the dried weight of the leaves of the plant. These compounds are commonly known as catechins. The positive health impacts of TP, either through consumption of green tea or direct ingestion, have been evidenced in lowering the susceptibility to chronic and degenerative illnesses. Additionally, they have been observed to inhibit the formation of human stones.

Green tea, for instance, contains polyphenols that serve the dual roles of reductive agents and stabilizers. Researchers used successfully green tea extract to synthesize polymeric nanoparticles with applications ranging from drug delivery to cancer therapy (Xiang et al., 2017).

Researchers create biocompatible spheres by using natural green tea polyphenols dissolved in water. These spheres can be smooth, porous, or hollow, achieved by controlling temperature and Cu^{2+} levels in tea polyphenol solutions. Their ability to transport substances such as fluorescein and anticancer drugs is evaluated. These spheres emit light, react to stimuli (including GSH), withstand acidity, and respond to alkaline conditions—making them well-suited for biomedical applications (Chen et al., 2013).

Scientist achieved polyphenol-based core at shell nanocomposites via a single-step process involving microwave-assisted oxidation-induced self-assembly and

Table 2 Synthesis of some gold and silver metal-based nanoparticles from various plants

Type	Plant	Morphology/size	Application	Used precursor	Reference
Gold	*Cymbopogon citratus*	Triangular nanoplatelets/50–1800	R radiation blocker	HAuCl4	Shankar et al. (2004)
	Menta piperita	Spherical/up to 150	Antibacterial		Kim et al. (2017)
	Cinnamomum zeylanicum	Spherical/25	Cell labeling and biosensors		Smitha et al. (2009)
	Scutellaria barbata	Spheroid and trigonal/20–35	p-Nitrophenol electrocatalytic breakdown		Wang et al. (2009)
	Breynia rhamnoides	Spherical/5–60	Reduction of 4-nitrophenol		Gangulaet al. (2011)
Silver	*Chelidonium majus* L. plant extract	Spherical/90 mm	Antioxidant and antimicrobial activities	AgNO3	Barbinta-Patrascu et al. (2016)
	Prunus serotina Ehrh. var. capuli	Spherical/20–100 nm	Antioxidant activity		Paulkumar et al. (2017)
	Potato (*Solanum tuberosum*) starch	Spherical/18–52 nm	Biomedical, antibiotic activity, and detectors		Cheviron et al. (2014)
	Peanut (*Arachis hypogaea*) shell extract	Spherical/10–50 nm	Antifungal activity		Velmurugan et al. (2015)
	Anacardium othonianum Rizz nutshell extract	Spherical/5–15 nm	Optoelectronics as well as antifungal properties		Bonatto and Silva (2014)

controlled aggregation within 1 min only (Fei et al., 2014). Agricultural trash, rice hulls, coconut shell, jatropha oil, banana peels, tea tree extract, leftover edible, etc. all have been used to create several nanomaterials and graphene non-biological sources, such polymers.

Polymers, waste rubber or tire, and industrial waste have also been investigated to produce different nanomaterials including grapheme (Table 3) (Singh et al., 2021).

Utilizing the aerial part infusion of *Saccharum officinarum*, researchers have created bio-nanocomposites of chitosan with silver.

The antimicrobial efficacy of chitosan bio-nanocomposites containing silver was assessed against *Bacillus subtilis*, *Klebsiella planticola*, *Streptococcus faecalis*, *Pseudomonas aeruginosa*, and *Escherichia coli*, as determined by Kim et al. (2017) and Paulkumar et al. (2017).

Table 3 Various agricultural sources of application of nanoparticles

Source	Layers	Significant findings	Uses	References
Rice husk ash	2–3	Specific conductance (85 f/g at 500 mV/s)	As electrode	Singh et al. (2017)
Rice husk	2–3	Specific conductance (81 f/g)	Power storage and utilization systems, hydrogen containment	Muramatsu et al. (2014)
Peanut shell	6–7	Higher surface area(2060 m^2/g) and void space volume (1.31 cm^3g^{-1})	Supercapacitor	Purkait et al. (2017)
Tea tree plant	3–4	High-grade, expansive, and flawless	Data storage systems	Jacob et al. (2015)

2 Applications in Pharmaceuticals and Therapeutics

Nanomaterials (NPs) have gained prominence due to their potential to enhance medication delivery, improving treatment efficacy, minimizing side effects, and increasing patient compliance. Various techniques for the characterization of nanomaterials are reported in Fig. 3.

Usage of metronidazole (MZ), an antibiotic, frequently leads to adverse reactions such as queasiness, emesis, epigastric discomfort, and oral dryness, linked to high MZ levels in saliva (Loh et al., 2010). To prevent MZ from dissolving in saliva, it was integrated into chitosan nanoparticles (sized 200–300 nm), enabling controlled release.

Liposomes offer advantages such as drug protection and targeted delivery but face limitations in stability and efficiency. In contrast, polymeric NPs surpass liposomes with enhanced solubility, controlled release, and efficacy (Khan et al., 2019).

Rod-shaped Cu_2O nanoparticles exhibit a high bactericidal ability (MIC 118 l/mL and MBC 138 l/mL) against *S. aureus* bacterium. The excellent antibiotic characteristics of synthesized firm Cu_2O nanoparticles can be used in clinical, medicinal, and environmental settings (Bhardwaj et al., 2019).

Dendrimers are used in biomedical applications, including cardiac testing, MRI contrast agents, drug delivery, gene therapy (Wu et al., 2004), and regenerative medicine. Wang and colleagues have recently designed a polyamidoamine dendrimer targeted with the tripeptide Arg-Gly-Asp (RGD) for delivering bortezomib (BTZ) to bone tumors (Wang et al., 2018).

Cholesterol cochleates designed for addressing hemophilia A have shown promising in vivo results in mouse trials (Miclea et al., 2007).

Gros et al. investigated utilization of polymeric micelles in cancer therapy. These micelles are particularly beneficial for anticancer drugs due to their ability to permeate cell membranes efficiently. However, when administered intravenously, these drugs can lead to embolism. Encapsulation within micelles addresses this issue by increasing drug solubility, thus preventing embolism and allowing for higher drug concentrations (Blanco et al., 2009). As an example, paclitaxel is having 0.15% solubility in water, which can be increased up to 2% when encapsulated within a

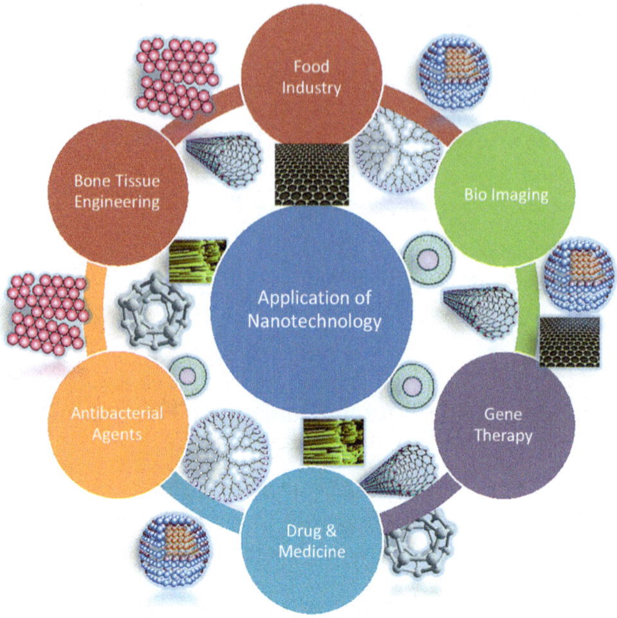

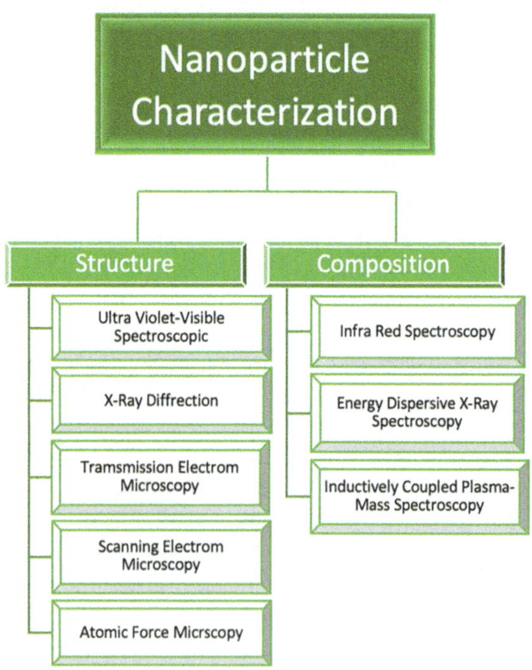

Fig. 3 Various techniques for the characterization of nanomaterials

micelle. Additionally, this encapsulation strategy enhances medicinal robustness by preventing enzymatic breakdown and loss of activity. In cancer treatment, polymeric micelles are frequently employed, typically ranging in size from 10 to 100 nanometer. These micelles are created when biodegradable hydrophilic blocked copolymers self-assemble. The hydrophobic portion of the block copolymer forms a partly solid core when exposed to water, acting as a container for hydrophobic medicines. Simultaneously, the hydrophilic segment creates a protective corona layer around the micelle, preventing rapid clearance from the body.

Carbon nanotubes (CNTs) hold promise in tumour diagnostics and treatment due to their distinctive features. They can detect tumor cells as well as deliver drugs for various treatments, including focused chemo, heat therapy, and light-based treatment and genetic treatment. CNTs exploit tumor characteristics such as enhanced permeability, retention, and the tumor microenvironment for precise drug delivery. Active targeting using ligands or antibodies further enhances their precision. CNTs may also help in overcoming medication resistance, a major difficulty in the treatment of cancer, by accumulating within cells independently of drug resistance mechanisms. The development of multifunctional nanoparticles represents a future frontier in personalized cancer therapy.

Silver nanoparticles produced from oyster mushroom were efficient against *E. coli*, with a 100% deactivation rate achieved at a low concentration of 45–50 g/L. Water disinfection, safe medical equipment, wound healing, preservatives, protective masks, and hand sanitization are all possible applications for these nanoparticles (Bhardwaj et al., 2020).

3 Applications in Food Industry

Liposomes play a diverse role in the food industry. They are used to encapsulate proteolytic enzymes, resulting in faster cheese production with reduced enzyme usage. Additionally, liposomes are employed for fortifying dairy products with vitamins, thereby improving their nutritional content and aiding in digestion. Notably, archaeosomes, which originate from polar ether lipids found in Archaeobacteria, exhibit exceptional thermal stability and resistance to environmental factors, making them valuable for preserving antioxidants during food processing better than synthetic archaeosomes, synthesized with ester phosphor lipids. Moreover, liposomes serve as carriers for delivering functional components such as nutraceuticals, antimicrobials, and flavor additives in various food products (Were et al., 2003).

Biocompatibility and biodegradability are crucial factors in both the food and pharmaceutical industries. Block copolymers employed in these sectors often consist of biocompatible substances such as lipids, proteins, and carbohydrates. Recent safety evaluations of these particular copolymer micelles have confirmed their non-toxic properties, opening up possibilities for their application in pharmaceuticals and potentially in the food industry (Moribe et al., 2010).

4 Applications in Bioimaging

Medical imaging has revolutionized healthcare by enabling visualization, measurement, and monitoring of vital biological processes, playing a pivotal role in disease detection and diagnosis (Cheng et al., 2020; Zhang et al., 2020; Zhou et al., 2020). Despite the limitations of many imaging contrast agents, such as weak signals and cytotoxicity, recent advances in organic nanomaterials show promise in addressing these issues. These nanomaterials, with their nanodimensions, enhance cell permeability and retention and can be densely functionalized for precise receptor targeting, thereby improving signal intensity and imaging resolution.

Semiconducting nanocrystals, commonly known as quantum dots (QDs), provide a versatile range of emission colors depending on their size and can be combined with organic ligands (Jin et al., 2020), offering significant potential for multicolor imaging. QD-based nanocomposites offer several notable advantages compared to traditional organic dyes, including exceptional photostability, a wide range of absorbance spectra, superior photon values, and an extended duration of luminescence (Palui et al., 2015). The Mattoussi research team revealed a notable use of QD-ligand nanocomposites in biological imaging in 2018. They designed a collection of versatile QD-polymer proteins tailored for the targeted internalization of cancer cells. These QD-based components emit a vivid green fluorescence signal. Additionally, QD's surface alteration using SVS-1 peptide with positive charge enables these nanocomposite to specifically adhere to cancer cells and facilitate their escape from endosomes (Kapur et al., 2018).

5 Applications in Gene Therapy

Gene therapy offers immense potential for treating various diseases, involving the delivery of therapeutic nucleic acids such as iRNA or plasmid DNA agents such as CRISPR to alter the expression of a gene (Qu et al., 2020). However, it faces challenges such as susceptibility to enzymatic degradation, poor cellular uptake due to nucleic acid negativity, and the need for effective carriers to escape endosomes.

Cationic polymers are common carriers for nucleic acids due to their opposite charge (Liu et al., 2019). Beyond polymers, diverse nanostructures have emerged as promising carriers. Cheng and team used high in boronic acid dendrimers for intracellular biomolecule delivery, including Cas9 protein for CRISPR-Cas9 genome editing (Liu et al., 2019).

The use of exterior stimulus is to control gene therapy within organic nanosystems. Pu et al. created a near-infrared-absorbing, dendronized semiconducting polymer for efficient delivery of CRISPR/Cas9 plasmids, incorporating a heat-inducible promoter for precise spatial and temporal regulation of gene delivery (Van Bruggen et al., 2019).

6 Applications as Antibacterial Agents

Chitosan nanoparticles, with their potent antibacterial charge, have garnered attention in dental endodontics, particularly in root canal treatments. Their nanoscale form exhibits heightened reactivity compared to bulk chitosan, making them effective against *Enterococcus faecalis* (Kishen et al., 2008). This antibacterial effectiveness remains intact even when dentin or lipids are present.

Incorporating dental canal sealants with nanoparticles of chitosan presents multiple advantages (Del Carpio-Perochena et al., 2015). It prevents microbial infiltration and decreases the development of biofilm at the dentin-root filling junction. As a result, scientists have created endodontic sealers by incorporating chitosan nanoparticles into materials as calcium silicate-based sealers or zinc oxide-eugenol epoxy resin., resulting in sealers with enhanced antibacterial properties.

7 Applications in Bone Cell Synthesis

Composite polyglucosamine nanoparticles are applied in bone cell synthesis. Recent research aims to enhance bone regeneration using poly(ε-caprolactone) (PCL) nanofiber scaffolds containing chitosan nanoparticles or chitosan-PCL nanofibers. Studies with these scaffolds have confirmed their biocompatibility with bone cells such as MG63 cells. Additionally, nanofiber matrices, combining polycaprolactone and chitosan and enhanced with collagen I, have shown mineralization and enhanced basic phosphatase activity of rat bone marrow-derived stromal cells (Cheng et al., 2014). Furthermore, nanoparticles made from chondroitin sulfate, tripolyphosphate, and chitosan show promise for the controlled release of future bone graft applications and osteogenic proteins such as Nel-like molecule-1 (Hou et al., 2012).

8 Outlook for the Future

Nanoparticle (NPs) shows promise in drug delivery, bioimaging, food technology, and industries. Yet, challenges remain in controlling their size and shape during synthesis due to harsh conditions and scalability issues. Because of environmental concerns, NP synthesis necessitates research on the effects and greener synthesis methods. Nanoparticles have the potential to revolutionize energy technology, catalysis, medicine delivery, and cancer therapy.

9 Conclusion and Summary of the Chapter

This chapter provides a comprehensive overview of organic nanoparticles (NPs), covering their classification, synthesis methods, and wide-ranging applications. The small size and extensive surface area of NPs make them adaptable for diverse purposes. The examples emphasized in this article highlight the significant potential of organic nanoparticles. Through the utilization of natural resources, scientists are creating innovative solutions that minimize environmental impact and enhance biocompatibility. The production of nanoparticles utilizing plant extracts, and by other biosources, is reshaping the field of nanotechnology, opening doors for sustainable progress in various industries. Further research and advancement in green nanotechnology are crucial for deepening our comprehension of its principles and optimizing the use of NPs across different domains.

References

Anderson, C. D., & Daniels, E. S. (2003). *Emulsion polymerisation and latex applications* (Vol. 14). iSmithers Rapra Publishing.

Barbinta-Patrascu, M. E., Badea, N., Ungureanu, C., Constantin, M., Pirvu, C., & Rau, I. (2016). Silver-based biohybrids "green" synthesized from Chelidonium majus L. *Optical Materials, 56*, 94–99.

Bello, S. A., Agunsoye, J. O., & Hassan, S. B. (2015). Synthesis of coconut shell nanoparticles via a top down approach: Assessment of milling duration on the particle sizes and morphologies of coconut shell nanoparticles. *Materials Letters, 159*, 514–519.

Bengoechea, C., Cordobés Carmona, F., Puppo, M.C. & Guerrero, A. (2007). *Viscoelasticidad lineal y tamaños de gota de emulsiones o/w estabilizadas por proteínas vegetales.*

Bhardwaj, A. K., & Naraian, R. (2021). Cyanobacteria as biochemical energy source for the synthesis of inorganic nanoparticles, mechanism and potential applications: A review. *3 Biotech, 11*(10), 445.

Bhardwaj, A. K., Kumar, V., Pandey, V., Naraian, R., & Gopal, R. (2019). Bacterial killing efficacy of synthesized rod shaped cuprous oxide nanoparticles using laser ablation technique. *SN Applied Sciences, 1*, 1–8.

Bhardwaj, K., Sharma, A., Tejwan, N., Bhardwaj, S., Bhardwaj, P., Nepovimova, E., Shami, A., Kalia, A., Kumar, A., Abd-Elsalam, K. A., & Kuča, K. (2020). Pleurotus macrofungi-assisted nanoparticle synthesis and its potential applications: A review. *Journal of Fungi, 6*(4), 351.

Blanco, E., Kessinger, C. W., Sumer, B. D., & Gao, J. (2009). Multifunctional micellar nanomedicine for cancer therapy. *Experimental Biology and Medicine, 234*(2), 123–131.

Bonatto, C. C., & Silva, L. P. (2014). Higher temperatures speed up the growth and control the size and optoelectrical properties of silver nanoparticles greenly synthesized by cashew nutshells. *Industrial Crops and Products, 58*, 46–54.

Bouchemal, K. S. E. H. I. N., Briançon, S., Perrier, E., Fessi, H., Bonnet, I., & Zydowicz, N. (2004). Synthesis and characterization of polyurethane and poly (ether urethane) nanocapsules using a new technique of interfacial polycondensation combined to spontaneous emulsification. *International Journal of Pharmaceutics, 269*(1), 89–100.

Camelo, S., Lajavardi, L., Bochot, A., Goldenberg, B., Naud, M. C., Brunel, N., Lescure, B., Klein, C., Fattal, E., Behar-Cohen, F., & de Kozak, Y. (2009). Protective effect of Intravitreal injection

of vasoactive intestinal peptide–loaded liposomes on experimental autoimmune Uveoretinitis. *Journal of Ocular Pharmacology and Therapeutics, 25*(1), 9–22.
Canalle, L. A., Löwik, D. W., & van Hest, J. C. (2010). Polypeptide–polymer bioconjugates. *Chemical Society Reviews, 39*(1), 329–353.
Chen, Z., Wang, C., Chen, J., & Li, X. (2013). Biocompatible, functional spheres based on oxidative coupling assembly of green tea polyphenols. *Journal of the American Chemical Society, 135*(11), 4179–4182.
Cheng, Y., Ramos, D., Lee, P., Liang, D., Yu, X., & Kumbar, S. G. (2014). Collagen functionalized bioactive nanofiber matrices for osteogenic differentiation of mesenchymal stem cells: bone tissue engineering. *Journal of Biomedical Nanotechnology, 10*(2), 287–298.
Cheng, H. B., Li, Y., Tang, B. Z., & Yoon, J. (2020). Assembly strategies of organic-based imaging agents for fluorescence and photoacoustic bioimaging applications. *Chemical Society Reviews, 49*(1), 21–31.
Cheviron, P., Gouanvé, F., & Espuche, E. (2014). Green synthesis of colloid silver nanoparticles and resulting biodegradable starch/silver nanocomposites. *Carbohydrate Polymers, 108*, 291–298.
Christian, D. A., Cai, S., Bowen, D. M., Kim, Y., Pajerowski, J. D., & Discher, D. E. (2009). Polymersome carriers: from self-assembly to siRNA and protein therapeutics. *European Journal of Pharmaceutics and Biopharmaceutics, 71*(3), 463–474.
Chronopoulou, L., Fratoddi, I., Palocci, C., Venditti, I., & Russo, M. V. (2009). Osmosis based method drives the self-assembly of polymeric chains into micro-and nanostructures. *Langmuir, 25*(19), 11940–11946.
Cuomo, F., Lopez, F., Miguel, M. G., & Lindman, B. (2010). Vesicle-templated layer-by-layer assembly for the production of nanocapsules. *Langmuir, 26*(13), 10555–10560.
Dalpiaz, A., Vighi, E., Pavan, B., & Leo, E. (2009). Fabrication via a nonaqueous nanoprecipitation method, characterization and in vitro biological behavior of N6-cyclopentyladenosine-loaded nanoparticles. *Journal of Pharmaceutical Sciences, 98*(11), 4272–4284.
Deepak, V., Kalishwaralal, K., & Gurunathan, S. (2009). Purification, immobilization, and characterization of nattokinase on PHB nanoparticles. *Bioresource Technology, 100*(24), 6644–6646.
del Carpio-Perochena, A., Bramante, C. M., Duarte, M. A. H., de Moura, M. R., Aouada, F. A., & Kishen, A. (2015). Chelating and antibacterial properties of chitosan nanoparticles on dentin. *Restorative Dentistry & Endodontics, 40*(3), 195–201.
Ding, F., Shi, X., Jiang, Z., Liu, L., Cai, J., Li, Z., Chen, S., & Du, Y. (2013). Electrochemically stimulated drug release from dual stimuli responsive chitin hydrogel. *Journal of Materials Chemistry B, 1*(12), 1729–1737.
El-Aasser, M. S., Lack, C. D., Vanderhoff, J. W., & Fowkes, F. M. (1988). The miniemulsification process—different form of spontaneous emulsification. *Colloids and Surfaces, 29*(1), 103–118.
Elzatahry, A. A., & Eldin, M. M. (2008). Preparation and characterization of metronidazole-loaded chitosan nanoparticles for drug delivery application. *Polymers for Advanced Technologies, 19*(12), 1787–1791.
Fei, J., Zhao, J., Du, C., Wang, A., Zhang, H., Dai, L., & Li, J. (2014). One-pot ultrafast self-assembly of autofluorescent polyphenol-based core@ shell nanostructures and their selective antibacterial applications. *ACS Nano, 8*(8), 8529–8536.
Gangula, A., Podila, R., Karanam, L., Janardhana, C., & Rao, A. M. (2011). Catalytic reduction of 4-nitrophenol using biogenic gold and silver nanoparticles derived from Breynia rhamnoides. *Langmuir, 27*(24), 15268–15274.
Grover, G. N., & Maynard, H. D. (2010). Protein–polymer conjugates: synthetic approaches by controlled radical polymerizations and interesting applications. *Current Opinion in Chemical Biology, 14*(6), 818–827.
Haddon, R. C. (2002). Carbon nanotubes. *Accounts of Chemical Research, 35*(12), 997–997.
Hett, A. (2004). *Nanotechnology: Small matter, many unknowns*. Swiss Reinsurance Company.
Hodge, P. (1993). Polymer science branches out. *Nature, 362*(6415), 18–19.

Hou, Y., Hu, J., Park, H., & Lee, M. (2012). Chitosan-based nanoparticles as a sustained protein release carrier for tissue engineering applications. *Journal of Biomedical Materials Research Part A, 100*(4), 939–947.

Huang, W., Taylor, S., Fu, K., Lin, Y., Zhang, D., Hanks, T. W., Rao, A. M., & Sun, Y. P. (2002). Attaching proteins to carbon nanotubes via diimide-activated amidation. *Nano Letters, 2*(4), 311–314.

Immordino, M. L., Dosio, F., & Cattel, L. (2006). Stealth liposomes: Review of the basic science, rationale, and clinical applications, existing and potential. *International Journal of Nanomedicine, 1*(3), 297.

Iravani, S. (2011). Green synthesis of metal nanoparticles using plants. *Green Chemistry, 13*(10), 2638–2650.

Jacob, M. V., Rawat, R. S., Ouyang, B., Bazaka, K., Kumar, D. S., Taguchi, D., Iwamoto, M., Neupane, R., & Varghese, O. K. (2015). Catalyst-free plasma enhanced growth of graphene from sustainable sources. *Nano Letters, 15*(9), 5702–5708.

Jin, Z., Sugiyama, Y., Zhang, C., Palui, G., Xin, Y., Du, L., Wang, S., Dridi, N., & Mattoussi, H. (2020). Rapid photoligation of gold nanocolloids with lipoic acid-based ligands. *Chemistry of Materials, 32*(17), 7469–7483.

Jing, X., Mi, H. Y., Peng, J., Peng, X. F., & Turng, L. S. (2015a). Electrospun aligned poly (propylene carbonate) microfibers with chitosan nanofibers as tissue engineering scaffolds. *Carbohydrate Polymers, 117*, 941–949.

Jing, X., Mi, H. Y., Wang, X. C., Peng, X. F., & Turng, L. S. (2015b). Shish-kebab-structured poly (ε-caprolactone) nanofibers hierarchically decorated with chitosan–poly (ε-caprolactone) copolymers for bone tissue engineering. *ACS Applied Materials & Interfaces, 7*(12), 6955–6965.

Kapur, A., Medina, S. H., Wang, W., Palui, G., Ji, X., Schneider, J. P., & Mattoussi, H. (2018). Enhanced uptake of luminescent quantum dots by live cells mediated by a membrane-active peptide. *ACS Omega, 3*(12), 17164–17172.

Kedar, U., Phutane, P., Shidhaye, S., & Kadam, V. (2010). Advances in polymeric micelles for drug delivery and tumor targeting. *Nanomedicine: Nanotechnology, Biology and Medicine, 6*(6), 714–729.

Khan, I., Khalid, S., & Khan, I. (2019). Nanoparticles: Properties, Application and Toxicities. *Arabian Journal of Chemistry, 12*, 908–993.

Kim, D., Kim, E., Lee, J., Hong, S., Sung, W., Lim, N., Park, C. G., & Kim, K. (2010). Direct synthesis of polymer nanocapsules: self-assembly of polymer hollow spheres through irreversible covalent bond formation. *Journal of the American Chemical Society, 132*(28), 9908–9919.

Kim, J. H., Yu, D., Eom, S. H., Kim, S. H., Oh, J., Jung, W. K., & Kim, Y. M. (2017). Synergistic antibacterial effects of chitosan-caffeic acid conjugate against antibiotic-resistant acne-related bacteria. *Marine Drugs, 15*(6), 167.

Kishen, A., Shi, Z., Shrestha, A., & Neoh, K. G. (2008). An investigation on the antibacterial and antibiofilm efficacy of cationic nanoparticulates for root canal disinfection. *Journal of Endodontics, 34*(12), 1515–1520.

Klajnert, B., & Bryszewska, M. (2001). Dendrimers: properties and applications. *Acta Biochimica Polonica, 48*(1), 199–208.

Koshino, M., Tanaka, T., Solin, N., Suenaga, K., Isobe, H., & Nakamura, E. (2007). Imaging of single organic molecules in motion. *Science, 316*(5826), 853–853.

Kostag, M., Köhler, S., Liebert, T., & Heinze, T. (2010). Pure cellulose nanoparticles from trimethylsilyl cellulose. In *Macromolecular symposia* (Vol. 294, No. 2, pp. 96–106). WILEY-VCH Verlag.

Langer, K., Anhorn, M. G., Steinhauser, I., Dreis, S., Celebi, D., Schrickel, N., Faust, S., & Vogel, V. (2008). Human serum albumin (HSA) nanoparticles: reproducibility of preparation process and kinetics of enzymatic degradation. *International Journal of Pharmaceutics, 347*(1–2), 109–117.

Lee, J., Cho, E. C., & Cho, K. (2004). Incorporation and release behavior of hydrophobic drug in functionalized poly (D, L-lactide)-block–poly (ethylene oxide) micelles. *Journal of Controlled Release, 94*(2–3), 323–335.

Legrand, P., Lesieur, S., Bochot, A., Gref, R., Raatjes, W., Barratt, G., & Vauthier, C. (2007). Influence of polymer behaviour in organic solution on the production of polylactide nanoparticles by nanoprecipitation. *International Journal of Pharmaceutics, 344*(1–2), 33–43.

Liu, J., Qiu, Z., Wang, S., Zhou, L., & Zhang, S. (2010). A modified double-emulsion method for the preparation of daunorubicin-loaded polymeric nanoparticle with enhanced in vitro anti-tumor activity. *Biomedical Materials, 5*(6), 065002.

Liu, C., Wan, T., Wang, H., Zhang, S., Ping, Y., & Cheng, Y. (2019). A boronic acid–rich dendrimer with robust and unprecedented efficiency for cytosolic protein delivery and CRISPR-Cas9 gene editing. *Science Advances, 5*(6), eaaw8922.

Lizarraga, M. S., Pan, L. G., Añon, M. C., & Santiago, L. G. (2008). Stability of concentrated emulsions measured by optical and rheological methods. Effect of processing conditions—I. Whey protein concentrate. *Food Hydrocolloids, 22*(5), 868–878.

Loh, J. W., Yeoh, G., Saunders, M., & Lim, L. Y. (2010). Uptake and cytotoxicity of chitosan nanoparticles in human liver cells. *Toxicology and Applied Pharmacology, 249*(2), 148–157.

Meziani, M. J., Pathak, P., Wang, W., Desai, T., Patil, A., & Sun, Y. P. (2005). Polymeric nanofibers from rapid expansion of supercritical solution. *Industrial & Engineering Chemistry Research, 44*(13), 4594–4598.

Michael, A., Singh, A., Roy, A., & Islam, M. R. (2022a). Fungal- and algal-derived synthesis of various nanoparticles and their applications. *Hindawi Bioinorganic Chemistry and Applications, 2022*, Article ID 3142674, 14.

Michael, A., Singh, A., Roy, A., & Islam, M. R. (2022b). Fungal-and algal-derived synthesis of various nanoparticles and their applications. *Bioinorganic Chemistry and Applications, 2022*, 2022.

Miclea, R. D., Varma, P. R., Peng, A., & Balu-Iyer, S. V. (2007). Development and characterization of lipidic cochleate containing recombinant factor VIII. *Biochimica et Biophysica Acta (BBA)-Biomembranes, 1768*(11), 2890–2898.

Moraes, C. M., de Matos, A. P., Grillo, R., de Melo, N. F., de Paula, E., Dias Filho, N. L., Rosa, A. H., & Fraceto, L. F. (2011). Screening of formulation variables for the preparation of poly (epsilon-caprolactone) nanocapsules containing the local anesthetic benzocaine. *Journal of Nanoscience and Nanotechnology, 11*(3), 2450–2457.

Moribe, K., Maruyama, S., Inoue, Y., Suzuki, T., Fukami, T., Tomono, K., Higashi, K., Tozuka, Y., & Yamamoto, K. (2010). Ascorbyl dipalmitate/PEG-lipid nanoparticles as a novel carrier for hydrophobic drugs. *International Journal of Pharmaceutics, 387*(1–2), 236–243.

Mozafari, M. R., Flanagan, J., Matia-Merino, L., Awati, A., Omri, A., Suntres, Z. E., & Singh, H. (2006). Recent trends in the lipid-based nanoencapsulation of antioxidants and their role in foods. *Journal of the Science of Food and Agriculture, 86*(13), 2038–2045.

Muramatsu, H., Kim, Y. A., Yang, K. S., Cruz-Silva, R., Toda, I., Yamada, T., Terrones, M., Endo, M., Hayashi, T., & Saitoh, H. (2014). Rice husk-derived graphene with nano-sized domains and clean edges. *Small, 10*(14), 2766–2770.

Oliveira, J. M., Salgado, A. J., Sousa, N., Mano, J. F., & Reis, R. L. (2010). Dendrimers and derivatives as a potential therapeutic tool in regenerative medicine strategies—A review. *Progress in Polymer Science, 35*(9), 1163–1194.

Palui, G., Aldeek, F., Wang, W., & Mattoussi, H. (2015). Strategies for interfacing inorganic nanocrystals with biological systems based on polymer-coating. *Chemical Society Reviews, 44*(1), 193–227.

Patel, G. B., Agnew, B. J., Deschatelets, L., Fleming, L. P., & Sprott, G. D. (2000). In vitro assessment of archaeosome stability for developing oral delivery systems. *International Journal of Pharmaceutics, 194*(1), 39–49.

Paulkumar, K., Gnanajobitha, G., Vanaja, M., Pavunraj, M., & Annadurai, G. (2017). Green synthesis of silver nanoparticle and silver based chitosan bionanocomposite using stem extract of Saccharum officinarum and assessment of its antibacterial activity. *Advances in Natural Sciences: Nanoscience and Nanotechnology, 8*(3), 035019.

Prato, M., Kostarelos, K., & Bianco, A. (2008). Functionalized carbon nanotubes in drug design and discovery. *Accounts of Chemical Research, 41*(1), 60–68.

Purkait, T., Singh, G., Singh, M., Kumar, D., & Dey, R. S. (2017). Large area few-layer graphene with scalable preparation from waste biomass for high-performance supercapacitor. *Scientific Reports, 7*(1), 15239.

Qin, Y., Lu, X., Sun, N., & Rogers, R. D. (2010). Dissolution or extraction of crustacean shells using ionic liquids to obtain high molecular weight purified chitin and direct production of chitin films and fibers. *Green Chemistry, 12*(6), 968–971.

Qu, Y., Ju, Y., Cortez-Jugo, C., Lin, Z., Li, S., Zhou, J., Ma, Y., Glab, A., Kent, S. J., Cavalieri, F., & Caruso, F. (2020). Template-mediated assembly of DNA into microcapsules for immunological modulation. *Small, 16*(37), 2002750.

Rahman, M. M., Islam, M. S., Sharmin, N., Chowdhury, J. A., & Jalil, R. U. (2010). Preparation and evaluation of cellulose acetate phthalate and ethyl cellulose based microcapsules of diclofenac sodium using emulsification and solvent-evaporation method. *Dhaka University Journal of Pharmaceutical Sciences, 9*(1), 39–46.

Reis, C. P., Neufeld, R. J., Ribeiro, A. J., & Veiga, F. (2006). Nanoencapsulation I. Methods for preparation of drug-loaded polymeric nanoparticles. *Nanomedicine: Nanotechnology, Biology and Medicine, 2*(1), 8–21.

Rolland, J. P., Maynor, B. W., Euliss, L. E., Exner, A. E., Denison, G. M., & DeSimone, J. M. (2005). Direct fabrication and harvesting of monodisperse, shape-specific nanobiomaterials. *Journal of the American Chemical Society, 127*(28), 10096–10100.

Romero, G., Estrela-Lopis, I., Zhou, J., Rojas, E., Franco, A., Espinel, C. S., Fernández, A. G., Gao, C., Donath, E., & Moya, S. E. (2010). Surface engineered poly (lactide-co-glycolide) nanoparticles for intracellular delivery: Uptake and cytotoxicity—A confocal raman microscopic study. *Biomacromolecules, 11*(11), 2993–2999.

Schubert, S., Delaney, J. T., Jr., & Schubert, U. S. (2011). Nanoprecipitation and nanoformulation of polymers: from history to powerful possibilities beyond poly (lactic acid). *Soft Matter, 7*(5), 1581–1588.

Shankar, S. S., Rai, A., Ankamwar, B., Singh, A., Ahmad, A., & Sastry, M. (2004). Biological synthesis of triangular gold nanoprisms. *Nature Materials, 3*(7), 482–488.

Sigfridsson, K., Björkman, J. A., Skantze, P., & Zachrisson, H. (2011). Usefulness of a nanoparticle formulation to investigate some hemodynamic parameters of a poorly soluble compound. *Journal of Pharmaceutical Sciences, 100*(6), 2194–2202.

Singh, P., Bahadur, J., & Pal, K. (2017). One-step one chemical synthesis process of graphene from rice husk for energy storage applications. *Graphene, 6*(3), 61–71.

Singh, M. P., Bhardwaj, A. K., Bharati, K., Singh, R. P., Chaurasia, S. K., Kumar, S., Singh, R. P., Shukla, A., Naraian, R., & Vikram, K. (2021). Biogenic and non-biogenic waste utilization in the synthesis of 2D materials (graphene, h-BN, g-C2N) and their applications. *Frontiers in Nanotechnology, 3*, 685427.

Smitha, S. L., Philip, D., & Gopchandran, K. G. (2009). Green synthesis of gold nanoparticles using Cinnamomum zeylanicum leaf broth. *Spectrochimica acta Part A: Molecular and biomolecular spectroscopy, 74*(3), 735–739.

Sole, I., Maestro, A., González, C., Solans, C., & Gutiérrez, J. M. (2006). Optimization of nanoemulsion preparation by low-energy methods in an ionic surfactant system. *Langmuir, 22*(20), 8326–8332.

Sun, Y. P., Rolling, H. W., Bandara, J., Meziani, J. M., & Bunker, C. E. (2002). Preparation and processing of nanoscale materials by supercritical fluid technology. In *Supercritical fluid technology in materials science and engineering: Synthesis, properties, and applications* (pp. 491–576). Marcel Dekker.

Tasis, D., Tagmatarchis, N., Bianco, A., & Prato, M. (2006). Chemistry of carbon nanotubes. *Chemical Reviews, 106*, 1105–1136.

Van Bruggen, C., Hexum, J. K., Tan, Z., Dalal, R. J., & Reineke, T. M. (2019). Nonviral gene delivery with cationic glycopolymers. *Accounts of Chemical Research, 52*(5), 1347–1358.

Velmurugan, P., Sivakumar, S., Young-Chae, S., Seong-Ho, J., Pyoung-In, Y., Jeong-Min, S., & Sung-Chul, H. (2015). Synthesis and characterization comparison of peanut shell extract silver nanoparticles with commercial silver nanoparticles and their antifungal activity. *Journal of Industrial and Engineering Chemistry, 31*, 51–54.

Vemavarapu, C., Mollan, M. J., Lodaya, M., & Needham, T. E. (2005). Design and process aspects of laboratory scale SCF particle formation systems. *International Journal of Pharmaceutics, 292*(1–2), 1–16.

Wang, Y., He, X., Wang, K., Zhang, X., & Tan, W. (2009). Barbated Skullcup herb extract-mediated biosynthesis of gold nanoparticles and its primary application in electrochemistry. *Colloids and Surfaces B: Biointerfaces, 73*(1), 75–79.

Wang, M., Cai, X., Yang, J., Wang, C., Tong, L., Xiao, J., & Li, L. (2018). A targeted and pH-responsive bortezomib nanomedicine in the treatment of metastatic bone tumors. *ACS Applied Materials & Interfaces, 10*(48), 41003–41011.

Weber, C., Coester, C., Kreuter, J., & Langer, K. (2000). Desolvation process and surface characterisation of protein nanoparticles. *International Journal of Pharmaceutics, 194*(1), 91–102.

Were, L. M., Bruce, B. D., Davidson, P. M., & Weiss, J. (2003). Size, stability, and entrapment efficiency of phospholipid nanocapsules containing polypeptide antimicrobials. *Journal of Agricultural and Food Chemistry, 51*(27), 8073–8079.

Wu, G., Barth, R. F., Yang, W., Chatterjee, M., Tjarks, W., Ciesielski, M. J., & Fenstermaker, R. A. (2004). Site-specific conjugation of boron-containing dendrimers to anti-EGF receptor monoclonal antibody cetuximab (IMC-C225) and its evaluation as a potential delivery agent for neutron capture therapy. *Bioconjugate Chemistry, 15*(1), 185–194.

Xiang, S., Yang, P., Guo, H., Zhang, S., Zhang, X., Zhu, F., & Li, Y. (2017). Green tea makes polyphenol nanoparticles with radical-scavenging activities. *Macromolecular Rapid Communications, 38*(23), 1700446.

Yallapu, M. M., Gupta, B. K., Jaggi, M., & Chauhan, S. C. (2010). Fabrication of curcumin encapsulated PLGA nanoparticles for improved therapeutic effects in metastatic cancer cells. *Journal of Colloid and Interface Science, 351*(1), 19–29.

Zhang, Y., Fang, F., Li, L., & Zhang, J. (2020). Self-assembled organic nanomaterials for drug delivery, bioimaging, and cancer therapy. *ACS Biomaterials Science & Engineering, 6*(9), 4816–4833.

Zhao, D., Zhao, X., Zu, Y., Li, J., Zhang, Y., Jiang, R., & Zhang, Z. (2010). Preparation, characterization, and in vitro targeted delivery of folate-decorated paclitaxel-loaded bovine serum albumin nanoparticles. *International Journal of Nanomedicine, 5*, 669–677.

Zhou, N., Huo, F., Yue, Y., & Yin, C. (2020). Specific fluorescent probe based on "protect–deprotect" to visualize the norepinephrine signaling pathway and drug intervention tracers. *Journal of the American Chemical Society, 142*(41), 17751–17755.

Synthesis of Metal NPs Using Plant Flowers and Fruits

Milan Hait, Nand Kumar Kashyap, and Arnab Kanti Giri

Abbreviations

ABTS	2,2′-azino-bis (3-ethylbenzothiazoline-6-sulfonic acid)
AFM	Atomic force microscopy
AgNPs	Silver nanoparticles
AuNPs	Gold nanoparticles
CBPE	Cavandis banana peel extract
CdNPs	Cadmium nanoparticles
Cu_2ONPs	Copper oxide nanoparticles
CuNPs	Copper nanoparticles
CVH	Cardiovascular health
DLS	Dynamic light scattering
DPPH	2,2 diphenyl-1picrylhydrazyl
FAO	Food and Agriculture Organization
FDA	Food and Drug Administration
FeNPs	Iron nanoparticles
FTIR	Fourier transform infrared spectroscopy
GID	Gastrointestinal disorder
GRAS	Generally recognized as safe
HPLC	High-performance liquid chromatography
LDL	Low-density lipoprotein
MB	Methylene blue
MNPs	Metal nanoparticles

M. Hait (✉) · N. K. Kashyap
Department of Chemistry, Dr. C. V. Raman University, Bilaspur, Chhattisgarh, India

A. K. Giri
Research Associate, NME Division, IIT Khargapur, Khargapur, WB, India

A. K. Bhardwaj et al. (eds.), *Biogenic Wastes-Enabled Nanomaterial Synthesis*,
https://doi.org/10.1007/978-3-031-59083-2_8

NPs	Nanoparticles
RhB	Rhodamine B
ROS	Reactive oxygen species
SEM	Scanning electron microscopy
TEM	Transmission electron microscopy
TiO_2NPs	Titanium oxide nanoparticles
USDA	United States Department of Agriculture
UV	Ultraviolet
Vis	Visible
WHO	World Health Organization
XRD	X-ray diffraction
ZnNPs	Zinc nanoparticles
ZnO	Zinc oxide
ZnONPs	Zinc oxide nanoparticles

1 Introduction

In contemporary nanotechnology, the creation and implementation of nanomaterials take centre stage. Herbs, plants, weeds, roots, flowers, and fruits include secondary metabolites like phenolic compounds, terpenoids, essential oils, flavonoids, and other natural products that aid in reducing metal ions and the creation of nanoparticles (NPs) (Sebesta et al., 2021). Many plants have synthesized substances that help maintain health. Among them are aromatic molecules, phenolic compounds, and tannins. There are likely more than 12,000 different secondary metabolites, but scientists have only been able to identify a few thousand. Humans have discovered many of the plants and herbs used in cooking and therapeutics (Jain et al., 2009). There has been a discussion on using natural chemicals derived from plants, especially higher plants, as a potential new source for antibiotics. Each species has its own unique chemical characteristics. Plants, especially higher plants, are a potential new source of antibiotics (Materon et al., 2021). Each species has its own unique chemical characteristics. This strategy piques interest since it offers a possible supply of bioactive molecules that have been proclaimed relatively harmless by the general population and typically act at many unique target locations, minimizing the likelihood of resistance (Ahmed et al., 2020). Ayurveda, an ancient Indian medical system, prescribes a wide variety of medications derived from locally available plants and animals for the treatment of a wide range of illnesses. Ayurveda, the traditional Indian system of medicine, relies heavily on the use of plant substances as therapeutic agents (Kumar et al., 2020a, b). Herbal remedies are growing in popularity in developed nations, but there are rising questions about their quality, safety, and effectiveness. The use of herbal medicine in treating serious illnesses is as ancient as humanity itself. Although there has been significant progress in the drug industry and the field of antibiotics, therapeutic plants remain an important raw material for drugs used to treat various ailments. Significant changes have occurred in the medical system on a global scale during the last two decades (Barhoum et al.,

2022). Medication derived from natural sources is becoming popular in the West because people are becoming more aware of the toxicity and adverse consequences associated with the prolonged use of synthetic medications and antibiotics (Sangeetha et al., 2020).

Nanotechnology has broad applications in the agricultural sector, healthcare, catalytic processes, skincare products, electronics, optical science, the drug industry, sensing instruments, and fabric materials (Bachheti et al., 2020). These many uses have prompted the development of new fields of study (Kalia et al., 2020). Scientists are increasingly interested in developing methods to create green nanoparticles from various plant components due to the sector's rapid expansion. The key benefits of synthesizing metallic nanoparticles from various plant parts are their low cost, low environmental impact, long life span, and ease of production (Kumar et al., 2020a, b). Metal and metal oxide NPs are useful MNPs because they have excellent surface-to-volume ratios and biomedical characteristics. In the bottom-up method used to create biogenic MNPs, individual atoms and compounds are employed as building blocks that then combine into larger structures. The biomimetic biological system synthesizes pure metals and their MNPs (Bhardwaj et al., 2018). Nanoparticles (NPs) are the most widely manufactured nanomaterials and have several uses. Synthesizing MNPs via conventional methods is not safe for the environment or health. Alternatives to this include using living organisms or biological extracts to synthesize MNPs in a green, affordable, and energy-efficient way. The physically reducing agent concentration, beginning precursor salt concentration, agitation, reaction duration, pH, temperature, and light impact biologically derived MNPs. The fundamental obstacle to industrializing the biological technique was the unexplored dependency between various reaction parameters (Miu & Dinischiotu, 2022). Economic forecasters predict that the worldwide demand for nanomaterials will continue to rise in the coming years. The production of MNPs far exceeds that of any other nanomaterial class (Inshakova et al., 2017). Nanocomposites are used in the aerospace and automotive sectors, food packaging, additives, sports equipment, textiles, beauty products, and information technology. Given the current state of research, it is possible that MNPs used in biomedicine and agriculture will flood the global market in the near future (Yadid et al., 2019; Hofmann et al., 2020). Because of the positive effects that natural products have on the economy, developing nations could be interested in MNPs (Salamanca-Buentello et al., 2005). Although certain MNPs may be dangerous, numerous goods with MNPs are currently on the market (Tang et al., 2015). Traditional synthesis techniques release hazardous byproducts and require greater energy. Biological MNP synthesis may be a viable long-term option. Its environmental, cost-effectiveness, and energy advantages outweigh those of traditional approaches (Jamkhande et al., 2019). Synthesizing MNPs in a sustainable manner, scientists tap into the reducing power of chemicals found in living beings. Microorganisms and plants are global, easily available, and safe to handle. Also, incorporating biomass waste into green synthesis can help speed up the process of shifting metal nanoparticle creation to a circular economy (Parveen et al., 2016). During green synthesis, MNPs have bio-compounds coated on their surfaces. Thus, this extra layer helps explain why green MNPs have better biological properties than their chemically reduced counterparts. The involvement of bio-compounds also safeguards against hazardous by-products contaminating green MNPs (Javed et al., 2020). During the chemical and physical synthesis of MNPs, potentially harmful compounds may bind to them. Bio-assisted synthesis has mostly stayed in the theoretical stage because of the complexity of the interconnected factors involved (Samuel et al., 2022). These variables impact MNPs' characteristics after formulation. Optimizing a biological process takes time and resources. The interior environments of organisms vary from species to species and even from individual to individual within the same species, making it difficult to standardize and repeat techniques. Reviews have focused on green MNP synthesis due to the decade-long interest in this field (Dikshit et al., 2021). The more extensive evaluations of the green reaction parameters were only performed on a particular kind of nanoparticle, such as silver (Ag) (Srikar et al., 2016), copper (Cu) (Din and Rehan, 2017), zinc oxide (Zno) (Xu et al., 2021), gold (Au) (Rani et al., 2022), palladium (Pd) (Tognacchini et al., 2020), and platinum (Pt) (Chandra et al., 2020; Rana et al., 2020; Yan et al., 2020). Currently, no in-depth research has been conducted on the general parameters that govern the synthesis of MNPs in a manner that is beneficial to the environment. Thus, this book chapter clarifies the procedures that can be used to regulate the attributes of MNPs while performing green synthesis. We highlighted the recently obtained information regarding the parameters that affect the biological synthesis of MNPs, and we emphasized the stunningly diverse range of biological agents that participate in nanobiotechnology. Along with that, it is possible that this chapter may help frame future proposals for how to go about building unique steps for the synthesis of ecological MNPs.

2 Importance of Flowers and Fruits in Daily Life

Plants are crucial to human survival; without vegetation, life on Earth would have ceased to exist long ago. The number of known plant species is in the millions, and it is continually growing (Salem & Fouda, 2021). They provide us with a wide range of resources that help us meet our basic needs, such as food, air, flowers, roots, leaves, fruits, wood, medicine, shelter, and many more items. All other forms of life on Earth rely on plants as their major food source. Humans, flowers, and fruits have a unique bond. Even in times of famine or scarcity, humans have always had a peculiar affinity for flowers and fruits. Humans have physically and emotionally invested in the cultivation of flowers for their aesthetic values (Haviland-Jones et al., 2005). Although flowers and fruits can serve as a basic source of medicine and an indicator of the land's fertility, the main motivation for growing flowers appears to be aesthetic rather than utilitarian (Jadoun et al., 2021). But this is more than skin deep; smelling, touching, and looking at flowers is an embodied, multi-sensory experience that requires physical movement. Flowers and fruits need gentle handling, watering, and nurturing from man, adding another relational dimension to the experience (Berger, 2009).

2.1 Flowers

A flower is a component of a plant that has a colourful, attractive, and comforting aroma. Certain flowers may not have appealing colours, but their scent is quite pleasing. A bud will eventually grow into a flower. The size and shape of the buds on various flowers may vary greatly. Typically, the bud of a plant will have some leaves covering it (Hait et al., 2018). These leaves shield the bud from harm and promote its development into a beautiful flower. Flowers create fruit, and the majority of the fruits include seed pods. These seeds have the potential to develop into more plants. Some of these flowers have a scent that is extremely enticing. These are the ingredients that go into manufacturing fragrances. Examples of this category of flowers are roses, jasmine, sirish (Hait et al., 2020) keora, hibiscus, kumhi (Kashyap et al., 2022b), and lavender. Flowers of kachnar, kela, munga, and kaddu can be cooked and consumed like vegetables. Some other flowers, like kumhi, hibiscus, Kali haldi (Bhardwaj et al., 2023), and Aami haldi (Hait et al., 2019), are used as medicines (Kashyap et al., 2022a, b). Cloves are dried flower buds that come from clove plants. The essential oil obtained by pressing cloves is put to medicinal use. A similar lotion, prepared from the arnica flower, is applied to wounds and injuries in order to heal them. Rose water has a long history of usage in the medical field as a flavouring component in confections and beverages. Flowers have a long history of application in the cosmetics and dyeing industries (Huss et al., 2018). People have a special connection with flowers, and it's probable that their beauty inspired individuals to cultivate and pass on their genes (Comba et al., 1999). Flowers stimulate the visual, viscera-motor, sensory-motor, and affective cerebral circuits, resulting in a positive emotional response. Flowers provide visual impact that activates multiple areas of the brain (Sidhu & Zafar, 2018). Ayurveda and Siddha claim that certain flowers are therapeutic agents. The chemical drug remedy list has more than a thousand entries for flowers (Sarid & Zaccai, 2016; Shubhashree et al., 2015). "Kaiyadevanighantu" primarily highlights the medicinal properties of the flowers of numerous therapeutic plants. Research into the phytochemical elements contained in flowers and their pharmacological effects has the potential to address modern medical difficulties because of the wealth of old medicinal knowledge. Saponin, quercetin, cyanide, alkaloids, flavonoids, tannins, and terpenoids are present in the *Hibiscus rosa-sinensis* flower (Nishteswar, 2015; Khan et al., 2010). Isolating the flavonoids, alkaloids, phenolics, and tannins in *Mimusops elengi*'s bloom requires a number of distinct processes (Khani et al., 2018; Koppula, 2013). There is evidence that *Mimusops elengi* methanolic extract may stop the spread of harmful germs (Arullappan et al., 2009). Cyclohexyl ethanoic acid, derived from *Nyctanthes arbortristis* flower extract, is efficient against *Plasmodium falciparum* (Ruban and Gajalakshmi, 2012; Meena et al., 2014; Sundari et al., 2012). Additional evidence suggests that the *Nyctanthes arbortristis* flower extract has greater antidiabetic potential than their leaf extract (Reddy & Jose, 2013; Tuntiwachwuttikul et al., 2003). An NBT (nitro blue tetrazolium) superoxide scavenging assay showed that two flavonoids isolated from *Tussilago farfara* flower buds are superior antioxidants to quercetin (Khatune et al., 2005; Kim et al., 2006). The flower and fruit extracts have a wide range of active components that could act as oxidizing or reducing agents. This could be useful in the green synthesis of NPs, especially MNPs or MONPs (Nanu et al., 2008). Numerous scientific investigations have established that flowers are not only edible but also a rich resource of beneficial substances. Flowers provide phytochemicals that combat inflammation, cholesterol, cancer, obesity, fungus, diabetes, fever, colds, coughs, and act as analgesics (Manisha et al., 2021). Flowers are the most visually striking and aromatic parts of plants. Nature provides us with a variety of food, like fruits, roots, rhizomes, spices, leaves, nuts, stems, and flowers. The distinctive aromas, tastes, and visual appeal of many flowers have made them a trendy new addition to the world of haute cuisine (Ahmed et al., 2016). The flowers boost the food's visual appeal and acceptance by giving it colour and a pleasant scent. Beverages such as herbal teas and wines can benefit from the addition of flowers for their aroma and flavour. Spreads like butter and fruit preserves benefit from their addition, as do vinegar, marinades, and salad dressings. There are many health advantages and therapeutic uses of flowers beyond their aesthetic value. Flowers are a rich source of phytochemicals, such as flavonoids, carotenoids, anthocyanins, and phenolic compounds. The phytochemicals in flowers have been studied for their antioxidant, anti-inflammatory, antibacterial, immune-stimulating, and other health benefits (Dzubak et al., 2006).

2.2 *Fruits*

Humans have been aware that flowers may produce fruits. Fruits often include nourishment as well as seeds inside them. We consume fruits because they are a source of nourishment for our bodies. The taste and flavour of fruits are often rather pleasant. In most cases, fruits include seeds, which may later develop into new plants. There is just one seed in several fruits, including mangoes, peaches, and plums. There are a lot of seeds in certain other fruits as well, including lemons, papayas, oranges, and watermelons. There are cultivars of bananas, grapes, and papayas that produce fruits that do not have seeds (Bellavia et al., 2013). Fruits are the edible portion of a plant, which includes the seeds and surrounding tissues. Fruits are essential parts of a healthy diet and aid in the prevention of major illnesses due to their high vitamin and dietary fibre content. Dietary guidelines benefit from the inclusion of fruits because of their health-giving qualities. Fruits are the ripened ovaries of plants or the succulent edible part of woody plants and are the sweet and fleshy product of a tree or plant that contains seed and can be eaten as food. Fruits are nutrient-dense, energy-rich food that provides a wide variety of health benefits (Mintah et al., 2012). Fruits are the edible and fleshy seed-related structures of some plants. According to FAO (2003), raising fruit and vegetable intake is a critical public health concern at the present time. Nutritional diseases, including impaired immunity, congenital malformations, and mental and physical retardation, are only some of the consequences of widespread micro-nutrient deficiencies, which prompted the declaration. These deficits arise from a lack of information about the nutritional benefits of fruits, which leads to a lack of their intake (Conner et al., 2017; Hart et al., 2005). Fruit eating has been linked to several health advantages, including increased life expectancy, enhanced mental health, enhanced CVH (cardiovascular health), decreased risk of some malignancies, and improved ability to maintain a healthy weight. Fruits provide a sufficient concentration of potassium, which is required to reduce bone loss and kidney stone formation (Boffetta et al., 2010). Fruits help the body and mind work together because they provide essential nutrients like fibre, which aids in digestion and boosts memory. Potassium, sodium, vitamins, protein, fibre, and folic acid are essential elements found in fruits. Fruit eating ensures maximum health, delivers quick energy, and supplies essential vitamins and minerals for proper bodily functioning (Amao, 2018). Fruit extracts include naturally occurring bioactive compounds with proven medicinal value. The synthesis of MNPs has historically included chemical processes, with the chemical serving as a reducing agent. When it became clear that these conventional MNPs posed a biological risk, researchers looked into greener alternatives. There are a variety of green approaches to synthesizing these MNPs, including the use of bacteria and plants, most often flower and fruit extracts. Consuming fruit and fruit-derived products not only makes an individual's health better, but it also lowers the risk of developing a variety of illnesses, like ageing, cardiovascular disease (CVD), cancer, vision disorder, immunity, gastrointestinal disorders (GID), tension, lipidemia, and a reduction in low-density lipoprotein (LDL) (Sidhu & Zafar, 2018). The United States Department of Agriculture (USDA) has suggested that one should fill up at least half of their plate with fruits and vegetables in order to promote a healthy way of life. This is due to their high levels of beneficial phytochemicals and other nutrients, including fibre, minerals, vitamins, proteins, and fats. FAO and the World Health Organization (WHO) recommend that eating a minimum of 400 g of vegetables and fruits every day reduces the risk of developing chronic illnesses such as cancer, diabetes, heart disease, and obesity and improves levels of micro-nutrients in the body (Hemmati et al., 2020).

3 Green Synthesis of Nanoparticles

Nanomaterials such as hybrids, MNPs, MONPs (metal oxide nanoparticles), and bio-inspired materials have all benefited from the increased focus on green synthesis as sustainable, dependable, and environmentally friendly. MNPs are often cited as the "medical miracle" that sparked the "Industrial Revolution". The use of MNPs loaded with cosmeceuticals has emerged as a promising new option among the many methods for skin renewal (Nadagouda & Varma, 2008). MNPs derived from fruits and flowers are rich in several secondary metabolites, like vitamins, flavonoids, tannins, phenols, alkaloids, and terpenoids. They reduce direct contact between molecules, making them useful as both a bio-reductant of metal ions and a capping agent for sterically stabilizing MNPs (Yugandhar et al., 2018). Flowers contain a wide variety of bioactive chemicals that have significant ethnobotanical value, especially in regard to the treatment of illness via "Pushpa Ayurveda", or floral therapy (Variya et al., 2016). Gold, silver, copper, zinc, iron (Fe), and cadmium (Cd), among others, may be synthesized using these chemicals since they are effective reducing agents. Flower-active molecules have several functions, including reducing agents, stabilizing agents, and precursor molecules for the synthesis of MNPs. Furthermore, no toxic derivatives are formed during the synthesis, making it safe for the environment. The produced MNPs have interesting and varied characteristics and potential use in many different areas (Kumar et al., 2020a, b). The use of plants in the synthesis of nanoparticles is attracting greater interest since the procedure is fast and non-pathogenic, requires only one step, and is environmentally friendly and cost-effective. MNP biomolecules include enzymes, proteins, amino acids, terpenoids, phenolics, vitamins, and polysaccharides (Ahmed et al., 2016). Bark, flowers, fruits, fruit peels, and roots are some components used in the creation of MNPs of varying sizes and shapes, including platinum, silver, titanium (Ti), and gold (Ibrahim, 2015). Synthesizing AgNPs using numerous fruit peel extracts was a rapid, cheap, and environmentally friendly alternative to the traditional procedure (Ndikau et al., 2017; Naganathan and Thirunavukkarasu, 2017). Besides acting as a bio-reductant, stabilizing agent, and capping agent, fruit extract also played these roles (Ahmed et al., 2018). The colour change reveals that flower, fruit, or fruit waste extract forms AgNPs, and the extract itself serves as a capping agent for the particles (Alaqad & Saleh, 2016; Patel & Mehata, 2019).

Visual inspection may validate the occurrence of green synthesis since the reduction of metallic ions often results in a change in the colour of the reaction mixture. Ultraviolet-visible (UV-Vis) spectroscopy is the most simple and necessary tool for investigations involving green nano-chemistry, and it allows for a more precise examination of MNPs (Olfati et al., 2021; Chakravarty et al., 2022). The formation of biologically synthesized MNPs, whose principal characterization often involves UV-Vis spectroscopy, depends on the particle components. The optical extinction spectra of colloidal solutions of green NPs display an absorption peak within a certain range. Direct observation of NP form and size using electron microscopy is commonplace, but hydrodynamic diameter estimation via dynamic light scattering (DLS) is useful. Through XRD (X-ray diffraction) analysis, explore the particles' crystalline structure. Another prominent technique is Fourier transform infrared spectroscopy (FTIR), which may reveal which functional groups remain in MNPs after environmentally friendly production (Mourdikoudis et al., 2018). The fundamental concept of FTIR is to expose the sample to infrared light, which the sample's structural-functional groups subsequently absorb. The spectrum's depiction of the absorbance varies according to the wavenumbers, which are inversely proportional to the wavelength. The wavenumber provides details of the functional group present in the active molecules. Zinc oxide nanoparticles (ZnONPs) obtained from *Deverra tortuosa* showed a large peak at 3436 cm^{-1} in their FTIR spectra, indicating the presence of hydroxyl groups. The amine functional groups found in proteins accounted for the other peaks (Selim et al., 2020). HPLC is a useful technique because it can separate plant extracts from other components of the mixture. HPLC gives information on the component parts of the biological agent that were employed in the green synthesis. Many studies have found evidence to support the idea that cell-free stock solutions for microbial cultures have the potential to act as coating agents (Liu et al., 2022). Electrophoresis, immunoblotting, and sequencing are viable methods for identifying and characterizing the proteins and play a key role in MNPs'functionalization. Active biomolecules serve as capping agents and are significant because they attach to particle surfaces, increasing their stability (Chowdhury et al., 2014). The zeta potential, which determines the amount of electrical charge that is present on the surface of NPs, is an important indicator of the stability of nanoparticle solutions. In general, suspensions of NPs with a zeta potential value either below −25 or above +25 eV are stable because electrostatic repulsion between surfaces has a significant positive or negative charge. The aggregation of particles is possible because of the mild repulsion that exists between lower-charged NPs and the present van der Waals attraction force (Guria et al., 2016). The absorption peaks of various MNPs synthesized via the green approach are given in Table 1.

4 Green Synthesis of Nanoparticles Mediated by Flowers and Fruits

Flowers and fruits have their own unique chemical characteristics, some of which might be useful in the synthesis of MNPs. The synthesis of MNPs using fruit and flower medium has advantages over other biological MNP synthesis procedures, especially those that include microorganisms as mediators. This is because the synthesis of MNPs necessitates the cultivation of microorganisms under sterile and pure culture conditions. It may be difficult and time-consuming to separate NPs while processing bacterial broth cultures. It also takes more time to transform soluble metallic salts into MNPs or MONPs (Kumar et al., 2020a). Flower- and fruit-based MNP synthesis offers an extra benefit over biological MNP synthesis. Microbes play an important role in the biological process of MNP synthesis; hence, it is important to keep them in a sterile setting and to only use pure strains. MNPs may be difficult to isolate from microbial broth cultures during post-production processing (Kumar et al., 2020b). The preparation of nanoparticles might be considered "green synthesis", which refers to the use of biological materials like plants or microorganisms. Plant-mediated green synthesis procedures are preferable and are primarily focused on research and fabrication breakthroughs. This is because employing microorganisms poses a biohazard, requires particular isolation techniques, and is more likely to result in contamination (Singh et al., 2016; Zhang et al., 2016). Green synthesis creates NPs from plant extracts and metal salt solutions. Biochemical reduction can lower a metal ion's electrochemical potential by using different plant extracts and other parameters like pH, temperature, concentration, pressure, and response time. This process reduces metal ions to zero, allowing

Table 1 Absorption peak of various MNPs synthesized via green chemistry

S.N.	MNP types	Absorbance range (nm)	Absorption peaks (nm)	References
1	Au	530–550	534	Veeramani et al. (2022)
			540	Yang et al. (2022)
			544	Guo et al. (2020)
			549	Alikhani et al. (2022)
2	Ag	400–450	402	Farshori et al. (2022)
			415	Narayanan et al. (2021)
			432	Alsammarraie et al. (2018)
			443	Chakravarty et al. (2022)
3	Pd	<300	268	Xiong et al. (2020)
			293	Sonbol et al. (2021)
			<300	Momeni and Nabipour (2015)
4	Pt	260–295	262	Fanoro et al. (2021)
			263	Aygun et al. (2020)
			269	Fahmy et al. (2021)
5	Mg	250–290	250	Abdallah et al. (2019)
			260	Fouda et al. (2021)
			270	Saman et al. (2021)
			282	Hassan et al. (2021)
			295	Eltaweil et al. (2022)
6	TiO_2	310–360	315	Sunny et al. (2022)
			320	Devikala et al. (2021)
			322	Anbumani et al. (2022)
			355	Ansari et al. (2022)
7	NiO	300–350	300	Louafi et al. (2021)
			319	Sabouri et al. (2018)
			328	Hussein and Mohammed, (2021)
			348	Al-Zaqri et al. (2022)
8	Cu	550–590	535	Yallappa et al. (2013)
			555	Nasrollahzadeh et al. (2017)
			580	Keihan et al. (2017)
			581	Shubhashree et al. (2022)
9	Fe_3O_4	360–410	365	Shafaei et al. (2019)
			405	Shalaby et al. (2021)
			410	Subhashini et al. (2018)

them to nucleate and form nanoparticles. Recent research has demonstrated that numerous flower and fruit phytochemicals and enzymes with critical functional groups like hydroxyl and carboxyl may act as reducing, capping, and stabilizing agents (Kuppusamy et al., 2016; Kandiah & Chandrasekaran, 2021). Figure 1 shows the mechanism of MNP synthesis from flowers and fruits.

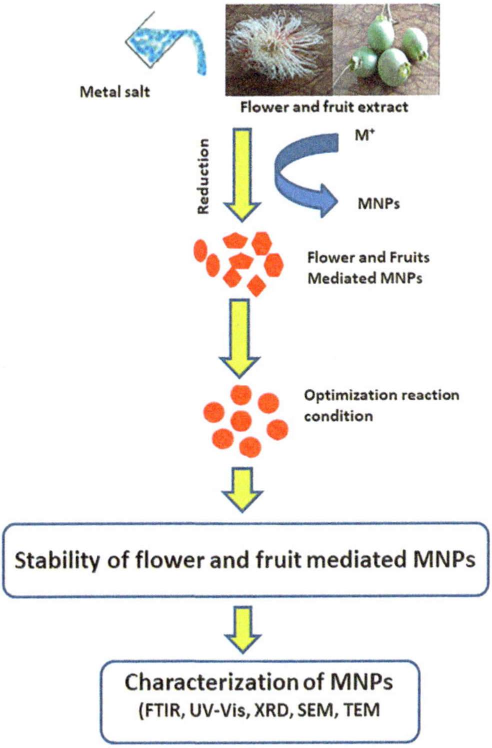

Fig. 1 Mechanism of synthesis of MNPs from flowers and fruits

4.1 Silver Nanoparticles (AgNPs)

When compared to larger particles with the same chemical structure, silver nanoparticles (AgNPs) with larger surface areas have better biological activity, catalytic activity, and atomic behaviour. The larger surface area of AgNPs compared to particles with similar chemical structures accounts for their increased interest in biochemical and catalytic applications. In the first step of making AgNPs, Ag^+ ions are turned into AgO. In the second step, colloidal AgNPs are clumped together to make oligomeric clusters. In the third step, these clusters are made more stable. Various microorganisms, plant extracts, flower and fruit extracts, and animal products contain enzymes necessary for the reduction of Ag^+ ions (Kumar et al., 2020a; Xu et al., 2006; Mashwani et al., 2016). The applications of flower- and fruit-derived AgNPs are displayed in Table 2.

Table 2 Application of Application of Ag-MNPs using different flowers and fruits

MNPs	Scientific name	Family	Application	References
Flowers	*Lablab purpureus*	Fabaceae	Antibacterial effect against *Escherichia coli* and *Staphylococcus aureus*	Muruganantham et al. (2018)
	Plumeria rubra	Apocynaceae	Antibacterial effect against *E. coli* and *Bacillus* sp.	Mandal (2018)
	Catharanthus roseus	Apocynaceae	Antibacterial effect against *E. coli*, *Pseudomonas putida*, *S. aureus*, *Klebsiella pneumoniae*, and *B. subtilis*	Manisha et al. (2014)
	Cassia angustifolia	Fabaceae	Antioxidant and anti-cancer potential	Bharathi and Bhuvaneshwari (2018)
	Allamanda cathartica	Apocynaceae	Antioxidant and antibacterial potential against *Salmonella typhimurium*, *S. aureus*, *E. coli*, and *K. pneumoniae*	Karnuakaran et al. (2016)
	Malva sylvestris	Malvaceae	Antibacterial efficacy against *E. coli*, *S. aureus*, and *Streptococcus pyogenes*	Esfanddarani et al. (2018)
	Caesalpinia pulcherrima	Fabaceae	Anti-fungal potential against *Candida glabrata*, antioxidant and cytotoxic effect, anti-cancer potential	Moteriya and Chanda (2016)
	Tussilago farfara	Asteraceae	Antibacterial action against *Enterococcus faecium* and cytotoxic potential	Lee et al. (2019)
	Tagetes erecta	Asteraceae	Antibacterial effect against *E. coli* and *P. aeruginosa*; anti-fungal action against *C. albicans*	Padalia et al. (2014)
	Madhuca longifolia	Sapotaceae	Antibacterial action against *B. cereus* and *S. saprophyticus*	Patil et al. (2018)
	Hibiscus rosa-sinensis	Malvaceae	Antibacterial action against *Aeromonas hydrophila*	Surya et al. (2016)
	Ipomoea digitata Linn	Convolvulaceae	Antibacterial action against *S. epidermidis*, catalytic impact against MB	Varadavenkatesan et al. (2019)
	Chrysanthemum indicum L.	Asteraceae	Larvicidal and pupicidal potential against *Anopheles stephensi*	Arokiyaraj et al. (2015)
Fruits	*Terminalia chebula*	Combretaceae	Catalytic action against MB	Edison and Sethuraman (2012)
	Vitis vinifera	Vitaceae	Antibacterial potential against *B. subtilis* and *K. planticola*	Gnanajobitha et al. (2013)

(continued)

Table 2 (continued)

MNPs	Scientific name	Family	Application	References
	Emblica officinalis	Phyllanthaceae	Antibacterial action against *S. aureus*, *B. subtilis*, *E. coli*, and *K. pneumoniae*	Ramesh et al. (2015)
	Phyllanthus emblica	Phyllanthaceae	Antibacterial effect against *Aspergillus oryzae*	Masum et al. (2019)
	Phyllanthus emblica	Phyllanthaceae	Antibacterial impact against *S. aureus* and *K. pneumoniae*	Renuka et al. (2020)
	Ficus carica	Moraceae	Anti-cancer potential against human breast cancer cell line MCF-7	Jacob et al. (2017)
	Crataegus pentagyna	Rosaceae	Antibacterial effect against *E. coli*, *S. aureus*, *E. faecalis*, *P. aeruginosa*, and *Acinetobacter baumannii*; photocatalytic potential against rhodamine B, eosin, and MB	Ebrahimzadeh et al. (2020)
	Phoenix dactylifera	Arecaceae	Antibacterial action against *E. coli*, *S. aureus*, *E. faecalis*, and *P. aeruginosa*; anti-fungal effect against *C. albicans*; and anti-cancer potential	Zafar and Zafar (2019)
	Phoenix dactylifera	Arecaceae	Antibacterial action against *S. aureus*, *S. epidermidis*, *K. pneumoniae*, and *E. coli*; catalytic potential against 4-nitrophenol	Farhadi et al. (2017)
	Malus pumila	Rosaceae	Antibacterial effect against *E. coli*, *S. aureus*, and *P. aeruginosa*	Ali et al. (2016)
	Punica granatum	Lythraceae	Antibacterial potential against *E. coli*, *S. aureus*, and *P. aeruginosa*	Shanmugavadivu et al. (2014)
	Musa paradisiaca	Musaceae	Antibacterial impact against *E. coli*, *S. aureus*, *P. aeruginosa*, and *B. subtilis*; anti-fungal potential against *C. albicans*	Ibrahim (2015)
	Musa paradisiaca	Musaceae	Anti-microbial action against *E. coli*, *K. pneumoniae*, *S. aureus*, and *B. subtilis*; antioxidant potential	Kokila et al. (2015)
	Prunus armeniaca	Rosaceae	Anti-microbial action against *E. coli*, *S. aureus*, *P. aeruginosa*, and *B. subtilis*	Ajmal et al. (2016)

(continued)

Table 2 (continued)

MNPs	Scientific name	Family	Application	References
	Punica granatum	Lythraceae	Anti-microbial action against *E. coli*, *S. aureus*, *S. epidermidis*, *P. aeruginosa*, *Proteus vulgaris*, *S. typhi*, and *K. pneumoniae*; anti-cancer effect against colon cancer cell lines	Devanesan et al. (2018)
	Ananas comosus	Bromeliaceae	Antioxidant, anti-cancer effect against HepG2 (liver cancer cell lines), anti-diabetic potential, and antibacterial action against *B. cereus*, *Listeria monocytogenes*, *E. faecium*, and *S. aureus*	Das et al. (2019)
	Dimocarpus longan Lour	Sapindaceae	Anti-microbial action against *E. coli*, *S. aureus*, *P. aeruginosa*, *B. subtilis*, and *C. albicans*; anti-cancer effect against prostate cancer cell lines	He et al. (2016)

4.2 Gold Nanoparticles (AuNPs)

AuNPs of a variety of sizes and morphologies have garnered a large amount of interest in regard to their applications in the area of medicine. These applications include functioning as carriers for drugs, such as tumour detectors and radiation dose boosters, and there are other possible uses as well (Gibson et al., 2007; Qian et al., 2008; Hainfeld et al., 2010; McMahon et al., 2011; Elia et al., 2014; Hainfeld et al., 2004). In current times, AuNPs have attracted a considerable amount of attention owing to the biocompatibility, optic density, shape, and size characteristics that they exhibit. The area of therapeutic drugs makes extensive use of AuNPs with a variety of sizes and morphologies for a broad range of applications, such as the detection of tumours and the administration of medications (Pechyen et al., 2021). The applications of flower- and fruit-derived AuNPs are displayed in Table 3.

4.3 Other Nanoparticles

Nanoparticles composed of various metals, such as Ti, Cd, Cu, Fe, Zn, and Mg, have recently emerged as a new class in research due to their revolutionary applications. Under a stirring condition at 70 degrees Celsius for 4 h, the herb rosemary, scientifically known as *Rosmarinus officinalis* L., was applied in the synthesis of MgO nano-flowers (Abdallah et al., 2019). The manufacture of CdNPs included the use

Table 3 Application of Au-MNPs using different flowers and fruits

MNPs	Scientific name	Family	Application	References
Flowers	*Plumeria alba* Linn	Apocynaceae	Antibacterial action against *E. coli*	Nagaraj et al. (2012)
	Gnidia glauca	Thymelaeaceae	Chemocatalytic action against 4-nitrophenol	Ghosh et al. (2012)
	Mangifera indica	Anacardiaceae	Catalytic action against 4-nitrophenol	Nayan et al. (2018)
	Tussilago farfara	Asteraceae	Antibacterial activity against *E. faecium*, cytotoxic effect, and anti-cancer action	Lee et al. (2019)
Fruits	*Punica granatum*	Lythraceae	Catalytic activity against 4-nitrophenol	Dash and Bag (2014)
	Citrus macroptera	Rutaceae	Anti-biofilm action against *P. aeruginosa*, anti-cancer effect	Majumdar et al. (2019)
	Euphoria longana Lam.	Sapindaceae	Anti-cancer and antioxidant potential	Khan et al. (2016)
	Punica granatum	Lythraceae	Antioxidant potential	Gubitosa et al. (2018)
	Citrus maxima	Rutaceae	Catalytic properties	Yu et al. (2016)
	Citrullus lanatus	Cucurbitaceae	Anti-microbial effect against *B. cereus*, *E. coli*, *L. monocytogenes*, *S. aureus*, and *S. typhimurium*; antioxidant and anti-proteas inhibition potential	Patra and Baek (2015)
	Prunus domestica	Rosaceae	Catalytic properties	Dauthal and Mukhopadhyay (2012)
	Vitis vinifera	Vitaceae	Apoptotic action	Nirmala et al. (2016) and Nirmala et al. (2017)

of extracts from the petals of marigold flowers (Hajra et al., 2016). In brief, when $CdCl_2$ solution and marigold flower extract were combined, a yellow nanoparticle solution had the appearance of spheres. *Syzygium aromaticum* bud extract, flower extract and zinc nitrate (Zn $(NO_3)_2.6H_2O$) were employed synthesized ZnONPs (Lakshmeesha et al., 2019). A fresh $FeCl_2$ solution was used with *Piliostigma thonningii* flower extract and synthesized FeNPs. Reductants were already present in the flower extract and served the dual purpose of agents (reducing and stabilizing) (Igwe & Nwamezie, 2018). The manufacture of titanium dioxide nanoparticles (TiO_2NPs) included the use of extract from the flower of *Calotropis gigantean* (Marimuthu et al., 2013). The CuNPs synthesized from *Mimusops elengi* flower powder exhibited potent antimicrobial, anticoagulant, antibacterial, anti-pupae, and anti-larva activities (Sarah & Iyer, 2019). Table 4 displays various other types of nanoparticles synthesized from flowers and fruits.

Table 4 Application of other types of MNPs using different flowers and fruits

Types of MNPs	Plant parts	Scientific name	Family	Application	References
Cu	Flower	*Mimusops elengi*	Sapotaceae	Antibacterial effect against *E. coli*, *S. pseudomonas*, *B. subtilis*, *A. flavus*, *C. albicans*, *Penicillium*, and *A. fumigates*; antioxidant; thrombolytic; anti-larva; and anti-cancer	Sarah and Iyer (2019)
Fe	Flower	*Piliostigma thonningii*	Fabaceae	Anti-microbial potential against *E. coli* and *S. aureus*	Igwe and Nwamezie (2018)
Zn	Flower	*Nyctanthes arbor-tristis*	Oleaceae	Antibacterial action against *A. niger*, *Botrytis cinerea*, *F. oxysporum*, and *P. expansum*	Jamdagni et al. (2016)
Zn	Flower	*Syzygium aromaticum*	Myrtaceae	Antibacterial effect against *F. graminearum*	Lakshmeesha et al. (2019)
Zn	Flower	*Jacaranda mimosifolia*	Bignoniaceae	Antibacterial action against *E. faecium*	Sharma et al. (2016)
Cd	Flower	*Tagetes* sp.	Asteraceae	Larvicidal action against *Aedes albopictus*	Hajra et al. (2016)
Ti	Flower	*Calotropis gigantean*	Apocynaceae	Acaricidal action against *Rhipicephalus microplus* and *Haemaphysalis bispinosa*	Marimuthu et al. (2013)
Mg	Flower	*Rosmarinus officinalis* L.	Lamiaceae	Antibacterial action against *Xanthomonas oryzae*	Abdallah et al. (2019)
Zno	Fruit	*Citrus maxima*	Rutaceae	Photocatalytic action against MB; anti-microbial effect against *K. aerogenes*, *E. coli*, and *S. aureus*; sensory properties	Pavithra et al. (2017)
Zno	Fruit	*Garcinia mangostana*	Clusiaceae	Photocatalytic action against MG	Aminuzzaman et al. (2018)
Zno	Fruit	*Punica granatum*	Lythraceae	Anti-cancer effect against HCT-116 (colorectal cancer cell line)	Sukri et al. (2019)
Zno	Fruit	*Ananas comosus*	Bromeliaceae	Antibacterial potential against *E. coli*	Ahmad et al. (2019)
CuO	Fruit	*Syzygium alternifolium*	Myrtaceae	Antiviral action	Yugandhar et al. (2018)

(continued)

Table 4 (continued)

Types of MNPs	Plant parts	Scientific name	Family	Application	References
CuO	Fruit	*Ziziphus spina-christi* L.	Rhamnaceae	Anti-microbial effect against *E. coli* and *S. aureus*	Khani et al. (2018)
CuO	Fruit	*Capparis spinosa*	Capparaceae	Anti-microbial potential against *S. aureus* and *Bacillus cereus*	Ebrahimi et al. (2017)
CuO	Fruit	*Citrus medica* Linn.	Rutaceae	Anti-microbial potential against *P. aeruginosa*, *Propionibacterium acnes*, *S. typhi*, *F. culmorum*, *F. oxysporum*, and *F. graminearum*	Shende et al. (2015)
CuO	Fruit	*Fragaria ananassa*	Rosaceae	Anti-microbial action against *S. aureus*, *S. saprophyticus*, *B. subtilis*, *S. pneumoniae*, *E. coli*, *S. typhimurium*, *P. mirabilis*, *P. aeruginosa*, *C. guilliermondii*, *C. parapsilosis*, *C. albicans*, *C. krusei*, and *C. glabrata*; antioxidant	Hemmati et al. (2020)
CuO	Fruit	*Psidium guajava* L.	Myrtaceae	Antibacterial effect against *E. coli* and *S. aureus*	Carloling et al. (2015)
CuO	Fruit	*Punica granatum*	Lythraceae	Antibacterial effect against *Micrococcus luteus*, *P. aeruginosa*, *S. enterica*, and *Enterobacter aerogenes*	Kaur et al. (2016)

4.3.1 Copper Oxide Nanoparticles (Cu_2ONPs)

There are several characteristics of copper oxide nanoparticles (Cu_2O-NPs) that are different from other transition metal oxides, including its low band gap (2.0 eV), large surface area, and exceptional stability in solutions. Anti-fouling protective coatings, biocatalysis, biosensors, electrical chemistry, and storage of energy are some of the many disciplines in which it has found practical uses. As a result, nanotechnology experts consider it the second option after noble MNPs. They have found broad application in the non-enzymatic detection of clinical analytes because of their capacity to stimulate electron transfer processes even while operating at low voltage (Verma & Kumar, 2019).

4.3.2 Zinc Oxide Nanoparticles (ZnONPs)

ZnONPs offer a broad variety of uses in a number of industries, including the biological, electrical, and optical sectors, which have currently piqued the attention of scientists working in the area of nanotechnology (Agarwal et al., 2017). The synthesis of ZnONPs is simple, risk-free, and cost-effective. ZnO-NPs are categorized by the Food and Drug Administration (FDA) as a chemical that is generally regarded as safe (GRAS) (Jayaseelan et al., 2012). In the field of medicine, ZnONPs are most often understood for their anti-inflammatory and wound-healing capabilities. As a result of the inherent UV-filtering capacity that ZnONPs possess, these nanoparticles are also increasingly finding their way into cosmetic goods like sunscreen creams. ZnONPs are used in the pharmaceutical industry as drug delivery agents because of their anti-diabetic, anti-cancer, anti-fungal, and antibacterial capabilities (Pulit-prociak et al., 2016).

5 Characterization of NPs

Both the synthesis and characterization of MNPs are crucial in order to identify nanoparticles and understand their uses. For the purpose of the characterization of MNPs, several different types of microscopy are used, especially scanning electron microscopy (SEM), transmission electron microscopy (TEM), atomic force microscopy (AFM), and spectroscopy, including X-ray photoelectron spectroscopy (XRP), UV-Vis spectroscopy, Fourier transform infrared spectroscopy (FTIR), dynamic light scattering (DLS), and X-ray diffraction (XRD). Many characteristics, like pore size, shape, particle size, crystalline structure, surface area, and fractal dimensions, are used. Another techniques, like dispersion, orientation, or intercalation, may also be used to determine the characterization of MNPs (El-Nour et al., 2010).

In scanning electron microscopy (SEM), a cylindrical stub made of copper is used. This stub has one side covered with a material that is made of double-sided carbon. After a sample has been placed inside the stub using a pipette, the stub is then secured using a holder. The scanning electron microscope reveals information about the shape and morphology of MNPs (Ibrahim, 2015). The SEM method generates an extremely high-resolution picture, allowing for the determination of the shape and size of the synthesized MNPs. The SEM micro-graphs help in deducing the specimens' three-dimensional structure. Because the electron beam in SEM is so narrow, it produces a tremendously deep field (Joshia et al., 2008). In the transmission electron microscopy (TEM) method, carbon-coated copper girds are loaded with the synthesized MNPs. The micro-graphs of an NP drop are created using an accelerating voltage and a transmission electron microscope (Heera & Shanmugam, 2015). The electrons transmitting through the material interact, creating a picture. The shape and particle size for a precise profile of the MNPs are clearly visible in a magnified and focused picture on the imaging equipment. The resolution of TEM is better than that of the more common scanning electron microscope (SEM) for characterizing MNPs. At a resolution of 0.2 nm, it is able to reveal details about the cellular structure, membrane interaction, and NP accumulation state. By providing a sharp picture through diffraction and providing insight into the specimen's investigation, morphology, and structure, TEM is one of the most reliable methods for nanoparticle characterization (Ahmed et al., 2015). MNPs like Au, Ag, Pd, Cu, and Pt acquire their colours due to their absorption. Incidental light absorbs EMR (electromagnetic radiation) and conduction electrons because of oscillations. Dilution of the AgNPs solution with purified water brings the absorption range to between 0.5 and 0.7. The wavelength of the absorption maximum in the solvent provides an indication of the particle size (Hurley et al., 2015). Nano-sized silver atoms emit light in the 400 nm range. The FTIR spectroscopy of MNPs reveals their functional group. In FTIR, a powder sample is prepared using powder diffraction calculation. Firstly, place the prepared sample between two NaCl cells, then place the holder of KBr pellet cells on top of the sample. The computer software uses the signal to detect functional groups (Ibrahim, 2015). X-ray diffractometry (XRD) verifies the crystalline nature of nanoparticles. To acquire and verify the structure and size of nanoparticles, a small quantity of material is required. MNPs synthesized with n-butyl alcohol exhibit a peak at a concentration indicative of a putative mono-crystal. The peak's clarity demonstrates its nanometric dimensions (Alaqad & Saleh, 2016). If one takes into account the rapid consumption of reactants at elevated temperatures, the resulting nanoparticles are smaller. Increasing temperatures and reaction rates cause a drop in the absorbance peak. As the temperature rises, the time required to react decreases as well. High temperatures enhance the kinetic energy of molecules, leading to a greater demand for silver ions, which in turn leads to a more uniform particle size during the synthesis of MNPs. As the reaction temperature rises, the UV-Vis spectrum exhibits increasingly fine and narrow peaks in the shorter wavelength region, indicating the formation of smaller MNPs (Sharma et al., 2021). When compared to the optical diffraction limit, AFM provides a clearer, higher-resolution image by a factor of 100. To determine how much water and other transparent or opaque substances absorb or transmit light, researchers use UV-Vis spectroscopy (Begum et al., 2018). It is crucial for studying the catalytic behaviour of MNPs and their applications in a phototactic and sensory nature (Boddolla & Thodeti, 2018). XRD analyzes the sample to identify the lattice arrangements, including the chemical composition, bond angles, unit cell array, crystalline character, and lattice plane arrangement of MNPs. DLS is a method that is widely used to determine the size and nature of MNPs in colloid solutions within these ranges. The FTIR method reveals the vibrational stretching frequency of metal-oxygen bonding (Srivastava, 2012; Bhardwaj et al., 2022a, b). Figure 2 shows the characterization tool for MNPs synthesized by flowers and fruits.

5.1 Biological Activities of Flower- and Fruit-Derived MNPs

The biological function and pharmacological effectiveness of MNPs have been the subject of significant research. These effects include antioxidant, antibacterial, anti-fungal, insecticidal, catalytic, anti-inflammatory, and anti-cancer properties.

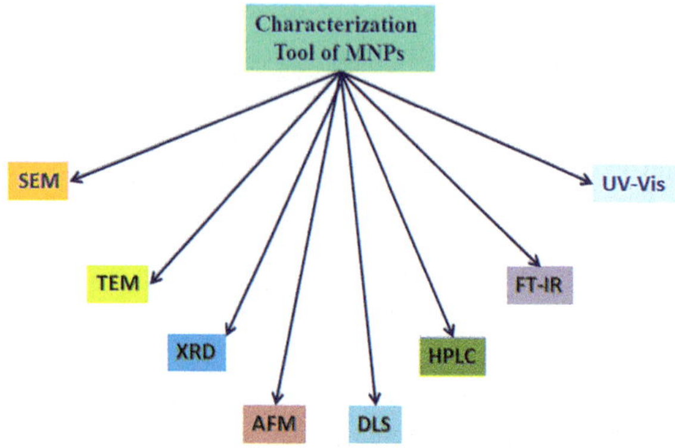

Fig. 2 Characterization tool of MNPs synthesized by flowers and fruits

Fig. 3 Various biological activities of MNPs synthesized from flowers and fruits

MNPs are finding more widespread use in a variety of sectors, including electronics, catalysis, medicine, and biotechnology. Metal nanoparticles (MNPs) have found extensive application due to their antibacterial, antiviral, anti-fungal, anti-inflammatory, and anti-cancer properties, as well as their ability to mediate the creation of other metal nanoparticles (Chand et al., 2020, 2021; Khan et al., 2019). The diverse biological activities of MNPs synthesized from flowers and fruits are shown in Fig. 3.

5.2 Antibacterial Activity

The antibacterial activity of MNPs is very effective against both gram-positive and gram-negative bacteria. Inconsistent data exist on whether an antibiotic treatment is effective against gram-positive or gram-negative bacteria. Some studies have found that gram-negative bacteria are more susceptible to MNPs than gram-positive bacteria. MNPs' antibacterial effect may originate from their interactions with the bacterial cell wall, which leads to the synthesis of pores in the cell wall at the sites of MNP deposition and alters the membrane's permeability. Possible causes for the reported differences in sensitivity between the two bacterial species include differences in bacterial structure, MNP shape and size, bacterial inoculum, nanoparticle size, exposure time, and nutrient medium (Raut et al., 2014). Silver is non-reactive while it is in its ionic state, but when it comes into contact with water, it begins to emit silver ions. It has been shown that NPs aggregate within the membrane and have the potential to eventually penetrate cells, which may cause harm to the cell wall and cell membrane (Klueh et al., 2000). It is believed that metal atoms attach to the thiol group of enzymes and synthesize stable S-M bonds with the thiol-containing molecules. These bindings cause the enzymes in the cell membrane to become inactive. The fact that MNPs are known to have an impact on germs is one of their many benefits (Haqq & Chattree, 2018). For MNPs' antibacterial properties to be evident, they must interact with bacterial cells. MNPs are able to enter the bacterial cell, accumulate along the metabolic route, and ultimately affect cellular function. The binding of MNPs to deoxyribonucleic acid, lysosomes, ribosomes, and enzymes results in oxygen species, heterogeneous modifications, changes in the membrane of a cell's permeability, electrolyte equilibrium, enzyme inhibition, protein deactivation, and abnormalities in gene expression (Wang & Thanou, 2010). When combined with medications like imipenem, vancomycin, and norfloxacin, synthesized AuNPs show enhanced antibacterial efficacy against *Escherichia coli*. When combined with vancomycin and norfloxacin, however, AuNPs show even greater anti-fungal efficacy against *Aspergillus flavus*. *E. coli* shows the largest zone of inhibition from FeNPs, followed by *Staphylococcus aureus* (Igwe & Nwamezie, 2018). The synthesis of AgNPs from the *Catharanthus roseus* flower shows promising antibacterial action against *Bacillus subtilis*, *E. coli*, *Klebsiella pneumoniae*, *Pseudomonas putida*, and *S. aureus* (Manisha et al., 2014). Padalia et al. (2014) reported that the antibacterial action of AgNPs synthesized from the flower extract of *Tagetes erecta* was higher against *S. aureus* than against *Bacillus cereus*. Sharma et al. (2016) found that *Tussilago farfara* flower bud extract, which contains sesquiterpenoids, reduced AgNP formation. These AgNPs outperformed gram-negative and gram-positive bacteria, with the highest antibacterial effectiveness against *Enterococcus faecium*. Abdallah et al. (2019) found that MgONPs generated from the floral extract of *Rosmarinus officinalis* L. showed strong inhibitory efficiency against *Xanthomonas oryzae* pv. oryzae Biofilms.

5.3 Antioxidant Activity

Flowers and fruits have antioxidant capabilities because their extracts include a wide variety of phytochemicals. Studies have shown that many fruit and flower extracts have potent antioxidant properties. *Careya arborea* Roxb., *Vaccinium cyanococcus*, *Rubus* subg., *Cornus mas* L., *Citrullus lanatus*, *Vitis vinifera*, *Terminalia arjuna*, and *Punica granatum* L. are fruits and flowers that are rich in various phytochemicals, vitamins, and minerals (Timoszyk, 2018). *Cassia angustifolia* flower extract has several aromatic active components, including carbonyls, phenols, nitro, alkane, and alkyl halides, which serve as reducing, capping, and stabilizing agents during the synthesis of AgNPs (Bharathi & Bhuvaneshwari, 2018). Using 2,2'-azinobis(3-ethylbenzothiazoline-6-sulfonic acid) (ABTS) radical cation-scavenging; 2,2-diphenyl-1-picrylhydrazyl free radical scavenging (DPPH-FRS); ferric reducing antioxidant power (FRAP); superoxide anion radical; and reducing power assessments, the antioxidant potential of AgNPs made from *Caesalpinia pulcherrima* flower extract was found to be more effective at an antioxidant potential capacity (Moteriya & Chanda, 2016). The viability analysis of AgNPs' genotoxicity in organisms using the HeLa cell line revealed that particles had no noticeable detrimental effects at doses below the detection threshold (Kumar et al., 2020a, b). *Cavendish* banana peel extract (CBPE) was used to synthesize AgNPs, which were tested for their antioxidant capacity against DPPH (Kokila et al., 2015). Researchers have shown that strawberry-derived Cu_2ONPs have potent DPPH-FRS action, and this action increases with increasing concentration. Cu_2ONPs and DPPH react with strawberry extract by transferring electrons and hydrogen ions to DPPH. Typically, DPPH has high absorbance at a 517 nm wavelength. However, it undergoes a transformation after gaining an electron or hydrogen atom from an active molecule, resulting in a decrease in wavelength at 517 nm. The colour change (purple to yellow) indicates the antioxidant's ability to neutralize free radicals. Antioxidant compounds extracted from strawberries include anthocyanins, flavonols, flavan-3-ols, and tannins (Hemmati et al., 2020). These molecules have been responsible for the maintenance of redox homeostasis in numerous phases of antioxidant reactions. AuNPs obtained from longan fruit extract exhibit dose-dependent antioxidant capacity against DPPH and rise in proportion to the amount of AuNPs present. However, oxidative degradation can change the catalytic potential of protein cytochrome-C, despite the fact that it is highly sensitive to hydrogen peroxide. Cyt-C degradation may be greatly slowed down by increasing the amount of AuNPs. The presence of AuNPs slows down the process of cytochrome C degradation (Khan et al., 2016).

5.4 Insecticidal Activity

Researchers have discovered that MNPs generated from flowers and fruits have the ability to kill insects. Cadmium nanoparticles (CdNPs) made from the extract of *Tagetes erecta* petals killed 69% of *Aedes albopictus* when they were given in low doses (10 ppm). CdNPs, on the other hand, demonstrated a death rate of 100% after 3 days of therapy against *A. albopictus* at the same dose (Hajra et al., 2016). The AgNPs obtained from *Chrysanthemum indicum* L. flower extract had lethal capabilities (Arokiyaraj et al., 2015). *Anopheles stephensi* mosquito larvae and pupae were tested with varying concentrations of *C. indicum*-synthesized AgNPs. The results indicated that AgNPs were the most effective in killing *A. stephensi* mosquitoes, which are the vectors of the disease. When tested on the larvae of *Haemaphysalis bispinosa* and *Rhipicephalus microplus*, TiO_2 nanoparticles made from the flowers of *Calotropis gigantea* had a range of LC_{50} values (Marimuthu et al., 2013).

5.5 Catalytic Activity

The widespread organic waste pollutants may cause significant harm to ecosystems due to herbicides, insecticides, and synthetic colours that all use 4-nitrophenol or its derivatives (Singh et al., 2018). Since 4-nitrophenol is both toxic and inhibitive, it poses a significant threat to the natural world. With its usage as a precursor in antipyretics and analgesics, the 4-nitrophenol reduction product has found to be employed as a mediator for dyes, as an antioxidant in rubber, in the formation of film developers, as a corrosion inhibitor, and in the manufacture of biofilms (Panigrahi et al., 2007; Woo and Lai, 2012). Reducing 4-nitrophenol using AuNPs (Sharma et al., 2015), AgNPs (Lim et al., 2016), CuONPs (Rostami-Vartooni et al., 2016), or PdNPs using $NaBH_4$ as a reductant and metal catalyst is the simplest and most efficient technique (Gopalakrishnan et al., 2017). Nayan et al. (2018) created AuNPs by synthesizing *Mangifera indica* flower extract. In the presence of $NaBH_4$ and at room temperature, these AuNPs demonstrated significant nano-catalysis in the aqueous phase by converting 4-nitrophenol into 4-aminophenol. Many researchers employed $NaBH_4$ as a model-reducing agent in the synthesis of AgNPs from *Ipomoea digitata* flower extract to catalyze the degradation of methylene blue (MB) dye (Varadavenkatesan et al., 2019). According to Khani et al. (2018), Cu_2ONPs synthesized from *Ziziphus spina-christi* (L.) fruit extract were able to absorb 90% of crystal violet under regulated conditions. A study by Yu et al. (2016) found that using *Punica granatum* fruit extracts to make colloidal AuNPs can change 4-nitrophenol to 4-aminophenol in as little as 10 min. In contrast, the synthesis of AuNPs from *Citrus maxima* required a few minutes for the same amount (Dash & Bag, 2014). Edison and Sethuraman (2012) also reported that AgNPs generated from *Terminalia chebula* may achieve a maximal reduction of MB to leucomethylene blue (LMB), which is colourless, in 25 min. Ebrahimzadeh et al. (2020) found that AgNPs made from *Crataegus pentagyna* fruit extract broke eosin, MB, and rhodamine B (Rh-B) dyes when they were exposed to sunlight for 90 min, with degradation rates of 70%, 75%, and 80%, respectively. Many reports show that ZnONPs made from the pericarp of the *Garcinia mangostana* fruit effectively oxidize malachite green dye by forming hydroxyl (·OH) radicals during photocatalytic processes. This causes 99% of the dye to break down in 3 h under sunlight in a water solution (Aminuzzaman et al., 2018).

5.6 *Other Activities*

5.6.1 Anti-cancer Activity

Cancer is one of the leading causes of death nowadays. It was responsible for more than 9.5 million deaths throughout the globe in 2018, with 70% of those fatalities occurring in low- and middle-income nations (Bray et al., 2018). The management and treatment of cancer often include four main procedures: chemotherapy, hormone therapy, radiation therapy, and surgery. It is known that many different medicinal herbs have anti-cancer and cytotoxic capabilities (Cragg & Newman, 2005). It is believed that the biological and pharmacological activity of plants is due to polyphenols such as alkaloids, flavonoids, phenolic acids, and terpenes (Balunas & Kinghorn, 2005; Ren et al., 2003; Hu, 2011). We know that triterpenoids exhibit cytotoxic efficacy and are also claimed to have anti-cancer potential. The flavonoids are rutin, myricetin, kaempferol, and quercetin. The use of nanotechnology in the fight against cancer has opened up new lines of inquiry involving disciplines as diverse as biology, chemistry, engineering, and medicine in the pursuit of better protocols for diagnosis, detection, and therapy (Dzubak et al., 2006). Majumdar et al. (2019) synthesized AuNPs from *Citrus macroptera* fruit extract and tested them in vitro for their anti-cancer effects on $HepG_2$ liver cancer cells. The data demonstrated that AuNPs had a higher IC_{50} value. Using longan fruit extract, Khan et al. (2016) synthesized AuNPs and tested them for their anti-cancer effects on MCF-7 (human breast cancer cells). The experiment's findings showed that cell viability was significantly reduced when AuNPs were present at concentrations of 6.20–100 g/mL. Significant anti-cancer potential is also associated with a spherical shape, a small size, the capping of phytochemicals, and uniform distribution. In addition, the phytochemicals act as a cap to aid in controlling the balance between anti-apoptotic proteins. Jacob et al. (2017) made AgNPs from the dried fruit extract of *Ficus carica* and tested their ability to kill MCF-7 cells. The outcome demonstrated that, in contrast to fruit extract, AgNPs' LD_{50} was variable. The amount of reactive oxygen species (ROS) was considered while analyzing the potential for cytotoxicity. AgNPs synthesized from the peel extract of *Punica granatum* have been shown to exhibit increased toxicity against colon cancer cell lines when administered at a dose of 12.50 mg (Devanesan et al., 2018). There is evidence that AgNPs made from *Dimocarpus longan* Lour fruit extract have dose-dependent cytotoxic effects against a prostate cancer cell line (He et al., 2016). The cytotoxic potential of AuNPs was synthesized using grape peel and seeds and has been investigated. The activity of AuNPs caused the concentration of ROS to rise, sped up apoptosis, and changed the shape of A431 cells, which is typical of apoptosis. It was believed that the interaction of AuNPs with the mitochondrial membrane's potential led to these changes (Nirmala et al., 2016; Nirmala et al., 2017).

5.6.2 Anti-microbial Activity

Due to their ability to cross bacterial membranes and affect cellular activity and metabolic processes, MNPs may exhibit antibacterial action. MNPs penetrate bacterial cells, where they disrupt metabolic processes by binding to metabolic building blocks. Heterogeneous changes occur when MNPs bind to simple molecules; they include oxidative stress, permeability changes, protein inactivation, enzymatic suppression, electrolyte balancing, and gene variations (Wang & Thanou, 2010). The barriers provided by the membranes and cell walls of the bacteria's cells allow them to withstand the hostile conditions found outside. A major function of the bacterial cell wall is to give bacteria their characteristic shapes. The uptake of MNPs via the cell membrane is a multi-step process that differs between gram-positive and gram-negative bacteria (Lesniak et al., 2013). Cu_2ONPs made from *Ziziphus spina-christi* (L.) fruit extract can kill bacteria, but they do not work on *E. coli*. Instead, they kill *S. aureus* (Khani et al., 2018). Many studies discussed Cu_2ONPs made from *Capparis spinosa* fruit extract, which demonstrated robust anti-microbial effectiveness against *B. cereus*, *S. aureus*, *E. coli*, and *K. pneumoniae*. Reports of AuNPs mediated by *Citrus macroptera* fruit extract suppressing the growth of *Pseudomonas aeruginosa*, a bacterium responsible for biofilm formation, are also available. Many particles at low concentrations were able to prevent biofilm formation (Singaravelu et al., 2007). *Emblica officinalis* extracts are used to synthesize AgNPs. *E. officinalis* fruit extract is less anti-microbial than synthesized AgNPs. Biological AgNPs are strong antibacterials against *B. subtilis*, *E. coli*, *K. pneumoniae*, and *S. aureus* (Ramesh et al., 2015). ZnONPs made from *Citrus maxima* extract also killed bacteria like *K. aerogenes*, *S. aureus*, and *E. coli* (Pavithra et al., 2017).

6 Future Scope in the Use of Flower- and Fruit-Mediated MNPs

Researchers are under pressure to find innovative, environmentally responsible ways to manufacture the things that society requires. Despite having high expectations for MNPs in the coming years, worries about their impact on the natural world cast a shadow over their promising future. Conventional methods for MNP synthesis have drawbacks and are not sustainable for the public's attention. The increasing prevalence of MNPs in consumer items, however, necessitates that we consider viable substitutes. In light of the variety of life on Earth, scientists are continually making strides towards their goal of discovering new MNPs capable of synthesis (Miu & Dinischiotu, 2022). Learning how to regulate MNP properties is crucial if green MNPs are to gain traction in the commercial sector and mitigate the unfavourable effects of expanding nanotechnological activities on the environment. We believe the next step in green MNP synthesis research should be to optimize the reaction itself rather than find additional species amenable to this process. We may customize MNP activities throughout their design to match application demands. MNP synthesis using a single biomolecule removes natural extract variability and enhances biological process repeatability and homogeneity. These approaches improve our understanding of how molecules engage in synthesis. Single-molecule techniques preclude biomolecule collaboration in living systems. The strength of the bio-compound layer employed to stabilize MNPs requires additional investigation. The centrifugation technique may destroy its coating. However the use of computational modelling will soon help us comprehend the various elements of MNP synthesis. Artificial neural networks that predict the bio-inspired MNP attributes may help scale up green MNP synthesis (Natsuki et al., 2015).

In nanotechnology, AgNPs have garnered much attention due to the exceptional qualities they possess. Applications for AgNPs include electrical devices, antibacterial potential used in the medical sector, food storage, textile coatings, and applications related to the environment. In industrial settings, AgNPs perform admirably as catalysts due to their high surface energy and their ability to stimulate surface reactivity (Biswas & Dey, 2015). AgNPs have better antibacterial properties and are used in the sterilization of medical equipment, household appliances, the purification of drinking water, and the textile coating business (El-Nour et al., 2010). AgNP-infused fibres were very effective against *E. coli*. AgNPs have many uses as anti-microbial, germicidal, and bactericidal like air sanitisers, sprays, shampoos, soaps, detergents, wet wipes, face masks, socks, slippers, pillows, toothpaste, and coatings for refrigerators, washing machines, vacuum cleaners, food storage containers, air filters, and mobile phones (Natsuki et al., 2015). AgNPs have antibacterial, anti-HIV, antiviral, anti-fungal, anti-microbial, fungicide, wound-dressing, ointment, lotion, and burn infection properties, making them useful for post-operative infection prevention (Khatoon et al., 2017). The most significant disadvantage of microbe-mediated MNP synthesis is the possibility that microorganisms would lose their capacity to synthesize MNPs as a result of mutations that occur over an extended period of time (Naraian & Bhardwaj, 2020). In addition, not all microorganisms are capable of ZnONPs synthesis; hence, screening processes need to identify their capability (Begum et al., 2019; Chhikara et al., 2018; Dulta et al., 2021; El-Belely et al., 2021; Nazir et al., 2021). Due to the environment and specific maintenance requirements, microbe-derived MNPs are not suitable for large-scale manufacture. In order to increase on a larger scale, thorough optimization studies are required to determine the effectiveness of all relevant variables. Coordination between the fundamental sciences, chemical engineering, and industry is required to economically use microbe-mediated ZnONPs (Park et al., 2021; Ratan et al., 2020; Sultan et al., 2019; Wang et al., 2017). Flower- and fruit-mediated MNP synthesis faces technical obstacles, and while green nanoscience has garnered attention, efforts to standardize NP synthesis protocols are ongoing (Bhardwaj et al., 2022a). The development of industrial-scale MNP synthesis technology, tools, and protocols for green synthesis must be improved (Singh et al., 2021). Nanotoxicity is another crucial concern for the large-scale application of MNP synthesis (Naraian & Bhardwaj, 2020). Commercially and environmentally safe nano-based goods, regulatory authorities, and legislation must be recognized (Bhardwaj et al., 2022b).

Due to the scarcity of commercial-grade items, they cannot compete with traditional materials. This book chapter has focused on the synthesis of MNPs using flowers and fruits. The development of MNPs such as Au, Ag, Fe, Pd, Pt, Zn, Cd, ZnO, TiO_2, and magnetic iron oxide (Fe_3O_4) is crucial to the progress of scientific fields.

7 Conclusion

In recent years, several methods rooted in tradition and convention have facilitated the synthesis of a diverse range of MNPs. However, it is worth noting that only a limited proportion of these MNPs have gained widespread use in the context of toxic substances and environmentally hazardous waste. Moreover, the limited comprehension of health-related issues and toxicity hinders the application of MNP synthesis through this approach, particularly within clinical domains. Metal nanoparticles (MNPs) obtained from flowers and fruits have garnered significant attention as a potentially impactful alternative. Biological MNPs are costly and pose potential hazards. The process of MNP synthesis involves the use of plant extracts. The process of synthesizing MNPs using renewable resources places greater emphasis on harnessing plants and their multiple parts, including leaves, roots, stems, flowers, and fruits, as agents for both reduction and stabilization. The distinctive characteristics of MNPs have garnered significant scientific interest across multiple academic fields. For many years, scientists have continually adhered to an established protocol for the synthesis of MNPs. Their synthesis involves costly, time-consuming, and various toxic chemicals, so the more preferred alternative is biological synthesis. Studies have shown that flower- and fruit-mediated MNP synthesis is more stable than microorganism and other synthesis, as well as being more eco-friendly, cheaper, and easier to maintain in terms of cell cultures. MNP synthesis is necessary since conventional methods involve high energy, high temperature, and toxic chemicals. Flower and fruit waste synthesis is cost-effective, energy-efficient, and expanding. Flowers and fruits contain biomolecules that can stabilize and decrease MNPs derived from metal ions. These responses confer unique features on the MNPs. MNPs are used in a variety of ecological uses given their low cost, low effect on the environment, and rapid output. Some of these uses include textile covering, food storage, bio-therapeutic areas, water purification, and many others. Flower- and fruit-mediated MNP synthesis is a secure innovation that does not need isolation and maintenance, unlike algae, bacteria, and fungi-based MNP synthesis. This chapter focuses on the possible use of MNPs obtained from flowers and fruits. Considering a variety of therapeutic attributes, those with the greatest MNP potential remained the gold and silver nanoparticles (AuNPs and AgNPs). The bio-accumulation and toxicology of green MNPs prohibit their use as safe drugs for human consumption, which necessitates additional research into the subject in scientific circles. The making of flower and fruit MNPs not only is without waste and economical but also is secure, renewable, sustainable, and non-toxic. However, extracts of fruits and flowers may be able to synthesize MNPs quicker than biological methods, which require microbial colonies to maintain their efficacy. Thus, flower and fruit extract MNPs may have considerable significance and new applications.

References

Abdallah, Y., Ogunyemi, S. O., Abdelazez, A., Zhang, M., Hong, X., Ibrahim, E., Hossain, A., Fouad, H., Li, B., & Chen, J. (2019). The green synthesis of MgO nano-flowers using *Rosmarinus officinalis* L. (Rosemary) and the antibacterial activities against Xanthomonas oryzae pv. Oryzae. *BioMed Research International, 2019*, 5620989. https://doi.org/10.1155/2019/5620989

Agarwal, H., Kumar, S. V., & Rajeshkumar, S. A. (2017). Review on green synthesis of zinc oxide nanoparticles—An eco-friendly approach. *Resource-Efficient Technologies, 3*(2), 406–413. https://doi.org/10.1016/j.reffit.2017.03.002

Ahmad, R. A., Harun, Z., Othman, M. H. D., Basri, H., Yunos, M. Z., Ahmad, A., Akhair, S. H. M., Rashid, A. Q. A., Azhar, F. H., & Alias, S. S. (2019). Biosynthesis of zinc oxide nanoparticles by using fruits extracts of *Ananas comosus* and its antibacterial activity. *Malaysian Journal of Fundamental and Applied Sciences, 15*(2), 268–273. https://doi.org/10.11113/mjfas.v15n2.1217

Ahmed, S. S., Messali, Z., Ouahabi, A., Trepout, S., Messaoudi, C., & Marco, S. (2015). Nonparametric denoising methods based on contourlet transform with sharp frequency localization: Application to low exposure time electron microscopy images. *Entropy, 17*(5), 3461–3478.

Ahmed, S., Ahmad, M., Swami, B. L., & Ikram, S. (2016). A review on plants extract mediated synthesis of silver nanoparticles for antimicrobial applications: A green expertise. *Journal of Advanced Research, 7*(1), 17–28. https://doi.org/10.1016/j.jare.2015.02.007

Ahmed, S., Kaur, G., Sharma, P., Singh, S., & Ikram, S. (2018). Fruit waste (peel) as bioreductant to synthesize silver nanoparticles with antimicrobial, antioxidant and cytotoxic activities. *Journal of Applied Biomedicine, 16*, 221–231. https://doi.org/10.1016/j.jab.2018.02.002

Ahmed, A., Zangeneh, A., & Zangeneh, M. M. (2020). Green synthesis and chemical characterization of gold nanoparticle synthesized using *Camellia sinensis* leaf aqueous extract for the treatment of acute myeloid leukemia in comparison to daunorubicin in a leukemic mouse model. *Applied Organometallic Chemistry, 34*(3), e5290. https://doi.org/10.1002/aoc.5290

Ajmal, N., Saraswat, K., Sharma, V., & Zafar, M. E. (2016). Synthesis and antibacterial activity of silver nanoparticles from *Prunus armeniaca* (Apricot) fruit peel extract. *Bulletin of Environment, Pharmacology and Life Sciences, 5*(8), 91–94.

Alaqad, K., & Saleh, T. A. (2016). Gold and silver nanoparticles: Synthesis methods, characterization routes and applications towards drugs. *Journal of Environmental & Analytical Toxicology, 6*(4), 1000384. https://doi.org/10.4172/2161-0525.1000384

Ali, Z. A., Yahya, R., Sekaran, S. D., & Puteh, R. (2016). Green synthesis of silver nanoparticles using apple extract and its antibacterial properties. *Advances in Materials Science and Engineering, 2016*, 4102196. https://doi.org/10.1155/2016/4102196

Alikhani, N., Hekmati, M., Karmakar, B., & Veisi, H. (2022). Green synthesis of gold nanoparticles (Au NPs) using *Rosa canina* fruit extract and evaluation of its catalytic activity in the degradation of organic dye pollutants of water. *Inorganic Chemistry Communications, 139*, 109351. https://doi.org/10.1016/j.inoche.2022.109351

Alsammarraie, F. K., Wang, W., Zhou, P., Mustapha, A., & Lin, M. (2018). Green synthesis of silver nanoparticles using turmeric extracts and investigation of their antibacterial activities. *Colloids and Surfaces. B, Biointerfaces, 171*, 398–405. https://doi.org/10.1016/j.colsurfb.2018.07.059

Al-Zaqri, N., Umamakeshvari, K., Mohana, V., Muthuvel, A., & Boshaala, A. (2022). Green synthesis of nickel oxide nanoparticles and its photocatalytic degradation and antibacterial activity. *Journal of Materials Science: Materials in Electronics, 33*(15), 11864–11880. https://doi.org/10.1007/s10854-022-08149-1

Amao, I. (2018) *Health benefits of fruits and vegetables: Review from sub-Saharan Africa*. pp. 33–53. https://doi.org/10.5772/intechopen.74472.

Aminuzzaman, M., Ying, L. P., Goh, W. S., & Watanabe, A. (2018). Green synthesis of zinc oxide nanoparticles using aqueous extract of *Garcinia mangostana* fruit pericarp and their photocatalytic activity. *Bulletin of Materials Science, 41*(2), 50. https://doi.org/10.1007/s12034-018-1568-4

Anbumani, D., Dhandapani, K. V., Manoharan, J., Babujanarthanam, R., Bashir, A. K. H., Muthusamy, K., Alfarhan, A., & Kanimozhi, K. (2022). Green synthesis and antimicrobial efficacy of titanium dioxide nanoparticles using *Luffa acutangula* leaf extract. *Journal of King Saud University, Science, 34*, 101896. https://doi.org/10.1016/j.jksus.2022.101896

Ansari, A., Siddiqui, V. U., Rehman, W. U., Akram, M. K., Siddiqi, W. A., Alosaimi, A. M., Hussein, M. A., & Rafatullah, M. (2022). Green synthesis of TiO_2 nanoparticles using *Acorus calamus* leaf extract and evaluating its photocatalytic and in vitro antimicrobial activity. *Catalysts, 12*(2), 181. https://doi.org/10.3390/catal12020181

Arokiyaraj, S., Kumar, V. D., Elakya, V., Kamala, T., Park, S. K., Saravanan, M., Bououdina, M., Arasu, M. V., Kovendan, K., & Vincent, S. (2015). Biosynthesized silver nanoparticles using floral extract of *Chrysanthemum indicum* L.-potential for malaria vector control. *Environmental Science and Pollution Research International, 22*(13), 9759–9765. https://doi.org/10.1007/s11356-015-4148-9

Arullappan, S., Zakaria, Z., & Basri, D. F. (2009). Preliminary screening of antibacterial activity using crude extracts of *Hibiscus rosa-sinensis*. *International Journal of Life Sciences Research, 20*(2), 109–118.

Aygun, A., Gulbagca, F., Ozer, L. Y., Ustaoglu, B., Altunoglu, Y. C., Baloglu, M. C., Atalar, M. N., Alma, M. H., & Sen, F. (2020). Biogenic platinum nanoparticles using black cumin seed and their potential usage as antimicrobial and anticancer agent. *Journal of Pharmaceutical and Biomedical Analysis, 179*, 112961. https://doi.org/10.1016/j.jpba.2019.112961

Bachheti, R. K., Fikadu, A., Bachheti, A., & Husen, A. (2020). Biogenic fabrication of nanomaterials from flower-based chemical compounds, characterization and their various applications: A review. *Saudi Journal of Biological Sciences, 27*(10), 2551–2562. https://doi.org/10.1016/j.sjbs.2020.05.012

Balunas, M. J., & Kinghorn, A. D. (2005). Drug discovery from medicinal plants. *Life Sciences, 78*(2), 431–441. https://doi.org/10.1016/j.lfs.2005.09.012

Barhoum, A., García-Betancourt, M. L., Jeevanandam, J., Hussien, E. A., Mekkawy, S. A., Mostafa, M., Omran, M. M., Abdalla, M. S., & Bechelany, M. (2022). Review on natural, incidental, bioinspired, and engineered nanomaterials: History, definitions, classifications, synthesis, properties, market, toxicities, risks, and regulations. *Nanomaterials, 12*(2), 177. https://doi.org/10.3390/nano12020177

Begum, R., Farooqi, H., & Naseem, K. (2018). Applications of UV/Vis spectroscopy in characterization and catalytic activity of Noble metal nanoparticles fabricated in responsive polymer microgels: A review. *Critical Reviews in Analytical Chemistry, 48*(6), 503–516. https://doi.org/10.1080/10408347.2018.1451299

Begum, R., Najeeb, J., Sattar, A., Naseem, K., Irfan, A., Al-Sehemi, A. G., & Farooqi, Z. H. (2019). Chemical reduction of methylene blue in the presence of nanocatalysts: A critical review. *Reviews in Chemical Engineering, 36*(6), 567–576. https://doi.org/10.1515/revce-2018-0047

Bellavia, A., Larsson, S. C., Bottai, M., Wolk, A., & Orsini, N. (2013). Fruit and vegetable consumption and all-cause mortality: A dose-response analysis. *The American Journal of Clinical Nutrition, 98*(2), 454–459. https://doi.org/10.3945/ajcn.112.056119

Berger, R. (2009). Nature therapy: Developing a framework for practice. In *School of Health and Social Sciences*. University of Abertay, Dundee Abertay of University Press: Dundee UK.

Bharathi, D., & Bhuvaneshwari, V. (2018). Evaluation of the cytotoxic and antioxidant activity of phyto-synthesized silver nanoparticles using *Cassia angustifolia* flowers. *BioNanoScience, 9*(2), 155–163. https://doi.org/10.1007/s12668-018-0577-5

Bhardwaj, A. K., & Naraian, R. (2021). Cyanobacteria as biochemical energy source for the synthesis of inorganic nanoparticles, mechanism and potential applications: A review. *3 Biotech, 11*, 445. https://doi.org/10.1007/s13205-021-02992-5

Bhardwaj, A. K., Shukla, A., Maurya, S., Singh, S. C., Uttam, K. N., Sundaram, S., Singh, M. P., & Gopal, R. (2018). Direct sunlight enabled photo-biochemical synthesis of silver nanoparticles and their bactericidal efficacy: Photon energy as key for size and distribution control. *Journal of Photochemistry and Photobiology B: Biology, 188*, 42–49.

Bhardwaj, A. K., Naraian, R., Sundaram, S., & Kaur, R. (2022a). Biogenic and non-biogenic waste for the synthesis of nanoparticles and their applications. In *Bioremediation: Green approaches for a clean and sustainable environment* (pp. 207–218). CRC Press.

Bhardwaj, K., Chopra, C., Bhardwaj, P., Dhanjal, D. S., Singh, R., Najda, A., Cruz-Martins, N., Singh, S., Sharma, R., Kuca, K., & Manickam, S. K. (2022b). Biogenic metallic nanoparticles from seed extracts: Characteristics, properties, and applications. *Journal of Nanomaterials, 2022*, 2271278. https://doi.org/10.1155/2022/2271278

Bhardwaj, A. K., Kashyap, N. K., Hait, M., Bera, S. K., & Dewangan, H. (2023). Physicochemical characterization of rhizome of *Curcuma caesia* Roxb. *ES Food and Agroforestry, 11*, 813. https://doi.org/10.30919/esfaf813

Biswas, P. K., & Dey, S. (2015). Effects and applications of silver nanoparticles in different fields. *The International Journal of Recent Scientific Research, 6*(8), 5880–5884.

Boddolla, S., & Thodeti, S. (2018). A review on characterization techniques of nanomaterials. *International Journal of Engineering Mathematics, 7*(1), 169–175.

Boffetta, P., Couto, E., Wichmann, J., Ferrari, P., Trichopoulos, D., & Bueno-de-Mesquita, H. B. (2010). Fruit and vegetable intake and overall cancer risk in the European prospective investigation into cancer and nutrition. *Journal of the National Cancer Institute, 102*(8), 529–537. https://doi.org/10.1093/jnci/djq072

Bray, F., Ferlay, J., Soerjomataram, I., Siegel, R. L., Torre, L. A., & Jemal, A. (2018). Global cancer statistics 2018: GLOBOCAN estimates of incidence and mortality worldwide for 36 cancers in 185 countries. *CA: a Cancer Journal for Clinicians, 68*(6), 394–424. https://doi.org/10.3322/caac.21492

Carloling, G., Priyadharshini, M. N., Vinodhini, E., Ranjitham, A. M., & Shanthi, P. (2015). Biosynthesis of copper nanoparticles using aqueous guava extract-characterisation and study of antibacterial effects. *International Journal of Pharma and Bio Sciences, 5*(2), 25–43.

Chakravarty, A., Ahmad, I., Singh, P., Ud Din Sheikh, M., Aalam, G., Sagadevan, S., & Ikram, S. (2022). Green synthesis of silver nanoparticles using fruits extracts of Syzygium cumini and their bioactivity. *Chemical Physics Letters, 795*, 139493. https://doi.org/10.1016/j.cplett.2022.139493

Chand, K., Cao, D., & Eldin, F. D. (2020). Green synthesis, characterization and photocatalytic application of silver nanoparticles synthesized by various plant extracts. *Arabian Journal of Chemistry, 13*(11), 8248–8261. https://doi.org/10.1016/j.arabjc.2020.01.009

Chand, K., Jiao, C., & Lakhan, M. N. (2021). Green synthesis, characterization and photocatalytic activity of silver nanoparticles synthesized with Nigella sativa seed extract. *Chemical Physics Letters, 763*, 138218. https://doi.org/10.1016/j.cplett.2020.138218

Chandra, H., Kumari, P., Bontempi, E., & Yadav, S. (2020). Medicinal plants: Treasure trove for green synthesis of metallic nanoparticles and their biomedical applications. *Biocatalysis and Agricultural Biotechnology, 24*(1), 101518. https://doi.org/10.1016/j.bcab.2020.101518

Chhikara, N., Kour, R., Jaglan, S., Gupta, P., Gat, Y., & Panghal, A. (2018). *Citrus medica*: Nutritional, phytochemical composition and health benefits—A review. *Food & Function, 9*(4), 1978–1992. https://doi.org/10.1039/C7FO02035J

Chowdhury, S., Basu, A., & Kundu, S. (2014). Green synthesis of protein capped silver nanoparticles from phytopathogenic fungus *Macrophomina phaseolina* (Tassi) Goid with antimicrobial properties against multidrug-resistant bacteria. *Nanoscale Research Letters, 9*, 365. https://doi.org/10.1186/1556-276X-9-365

Comba, L., Corbet, S. A., Barron, A., Bird, A., Collinge, S., Miyazaki, C., & Powell, M. (1999). Garden flowers: Insect visits and the floral reward of horticulturally-modified variants. *Annals of Botany, 83*(1), 73–86. https://doi.org/10.1006/anbo.1998.0798

Conner, T. S., Brookie, K. L., Carr, A. C., Mainvil, L. A., & Vissers, M. C. M. (2017). Let them eat fruit! The effect of fruit and vegetable consumption on psychological well-being in young adults: A randomized controlled trial. *PLoS One, 12*, e0171206. https://doi.org/10.1371/journal.pone.0171206

Cragg, G. M., & Newman, D. J. (2005). Plants as a source of anticancer agents. *Journal of Ethnopharmacology, 100*, 72–79. https://doi.org/10.1016/j.jep.2005.05.011

Das, G., Patra, J. Y., Debnath, D., Ansari, A., & Shin, H. S. (2019). Investigation of antioxidant, antibacterial, antidiabetic, and cytotoxicity potential of silver nanoparticles synthesized using the outer peel extract of *Ananas comosus* (L.). *PLoS One, 14*(8), e0220950. https://doi.org/10.1371/journal.pone.0220950

Dash, S. S., & Bag, B. G. (2014). Synthesis of gold nanoparticles using renewable *Punica granatum* juice and study of its catalytic activity. *Applied Nanoscience, 4*(1), 55–59. https://doi.org/10.1007/s13204-012-0179-4

Dauthal, P., & Mukhopadhyay, M. (2012). Prunus domestica fruit extract-mediated synthesis of gold nanoparticles and its catalytic activity for 4-nitrophenol reduction. *Industrial and Engineering Chemistry Research, 51*(40), 13014–13020. https://doi.org/10.1021/ie300369g

Devanesan, S., AlSalh, M. S., Balaji, R. V., Ranjitsingh, A. J. A., Ahamed, A., Alfuraydi, A. A., AlQahtani, F. Y., Aleanizy, F. S., & Othman, A. H. (2018). Antimicrobial and cytotoxicity effects of synthesized silver nanoparticles from *Punica granatum* peel extract. *Nanoscale Research Letters, 13*(1), 315. https://doi.org/10.1186/s11671-018-2731-y

Devikala, S., Abisharani, J. M., & Bharath, M. (2021). Biosynthesis of TiO_2 nanoparticles from *Caesalpinia pulcherrima* flower extracts. *Materials Today Proceedings, 40*, 185–188. https://doi.org/10.1016/j.matpr.2020.08.578

Dikshit, P. K., Kumar, J., Das, A. K., Sadhu, S., Sharma, S., Singh, S., Gupta, P. K., & Kim, B. S. (2021). Green synthesis of metallic nanoparticles: Applications and limitations. *Catalysts, 11*(8), 902. https://doi.org/10.3390/catal11080902

Din, M. I., & Rehan, R. (2017). Synthesis, characterization, and applications of copper nanoparticles. *Analytical Letters, 50*(1), 50–62. https://doi.org/10.1080/00032719.2016.1172081

Dulta, K., Kosarsoy, A., Gçeli, G., Chauhan, P., Jasrotia, R., & Chauhan, P. K. (2021). Ecofriendly synthesis of zinc oxide nanoparticles by *Carica papaya* leaf extract and their applications. *Journal of Cluster Science, 33*(1), 603–617. https://doi.org/10.1007/s10876-020-01962-w

Dzubak, P., Hajduch, M., Vydra, D., Hustova, A., Kvasnica, M., Biedermann, D., Markova, L., Urban, M., & Sarek, J. (2006). Pharmacological activities of natural triterpenoids and their therapeutic implications. *Natural Product Reports, 23*(3), 394–411. https://doi.org/10.1039/B515312N

Ebrahimi, K., Shiravand, S., & Mahmoudvand, H. (2017). Biosynthesis of copper nanoparticles using aqueous extract of *Capparis spinosa* fruit and investigation of its antibacterial activity. *Journal of Pharmaceutical Research, 21*(4), 866–871. https://doi.org/10.12991/mpj.2017.31

Ebrahimzadeh, M. A., Naghizadeh, A., Amiri, O., Shirzadi-Ahodashti, M., & Mortazavi-Derazkola, S. (2020). Green and facile synthesis of Ag nanoparticles using *Crataegus pentagyna* fruit extract (CP-AgNPs) for organic pollution dyes degradation and antibacterial application. *Bioorganic Chemistry, 94*, 103425. https://doi.org/10.1016/j.bioorg.2019.103425

Edison, T. J. I., & Sethuraman, M. G. (2012). Instant green synthesis of silver nanoparticles using *Terminalia chebula* fruit extract and evaluation of their catalytic activity on reduction of methylene blue. *Process Biochemistry, 47*, 1351–1357. https://doi.org/10.1016/j.procbio.2012.04.025

El-Belely, E. F., Farag, M. M. S., Said, H. A., Amin, A. S., Azab, E., Gobouri, A. A., & Fouda, A. (2021). Green synthesis of zinc oxide nanoparticles (ZnO-NPs) using Arthrospira platensis (class: Cyanophyceae) and evaluation of their biomedical activities. *Nanomaterials, 11*(1), 95. https://doi.org/10.3390/nano11010095

Elia, P., Zach, R., Hazan, S., Kolusheva, S., Porat, Z., & Zeiri, Y. (2014). Green synthesis of gold nanoparticles using plant extracts as reducing agents. *International Journal of Nanomedicine, 9*, 4007–4021. https://doi.org/10.2147/IJN.S57343

El-Nour, K. M. A., Eftaiha, A. A., Al-Warthan, A., & Ammar, R. A. (2010). Synthesis and applications of silver nanoparticles. *Arabian Journal of Chemistry, 3*(3), 135–140.

Eltaweil, A. S., Fawzy, M., Hosny, M., Abd El-Monaem, E. M., Tamer, T. M., & Omer, A. M. (2022). Green synthesis of platinum nanoparticles using *Atriplex halimus* leaves for potential antimicrobial, antioxidant, and catalytic applications. *Arabian Journal of Chemistry, 15*(1), 103517. https://doi.org/10.1016/j.arabjc.2021.103517

Esfanddarani, H. M., Kajani, A. A., & Bordbar, A. K. (2018). Green synthesis of silver nanoparticles using flower extract of *Malva sylvestris* and investigation of their antibacterial activity. *IET Nanobiotechnology, 12*(4), 412–416. https://doi.org/10.1049/iet-nbt.2017.0166

Fahmy, S. A., Fawzy, I. M., Saleh, B. M., Issa, M. Y., Bakowsky, U., & Azzazy, H. M. E. (2021). Green synthesis of platinum and palladium nanoparticles using *Peganum harmala* L. seed alkaloids: Biological and computational studies. *Nanomaterials, 11*, 965. https://doi.org/10.3390/nano11040965

Fanoro, O. T., Parani, S., Maluleke, R., Lebepe, T. C., Varghese, R. J., Mgedle, N., Mavumengwana, V., & Oluwafemi, O. S. (2021). Biosynthesis of smaller-sized platinum nanoparticles using the leaf extract of *Combretum erythrophyllum* and its antibacterial activities. *Antibiotics, 10*, 1275. https://doi.org/10.3390/antibiotics10111275

FAO. (2003). *Increasing fruit and vegetable consumption becomes a global priority*. FAO News Room Focus.

Farhadi, S., Ajerloo, B., & Mohammadi, A. (2017). Green biosynthesis of spherical silver nanoparticles by using date palm (*Phoenix dactylifera*) fruit extract and study of their antibacterial and catalytic activities. *Acta Chimica Slovenica, 64*(1), 129–143. https://doi.org/10.17344/acsi.2016.2956

Farshori, N. N., Al-Oqail, M. M., Al-Sheddi, E. S., Al-Massarani, S. M., Saquib, Q., Siddiqui, M. A., Wahab, R., & Al-Khedhairy, A. A. (2022). Green synthesis of silver nanoparticles using *Phoenix dactylifera* seed extract and its anticancer effect against human lung adenocarcinoma cells. *Journal of Drug Delivery Science and Technology, 70*, 103260. https://doi.org/10.1016/j.jddst.2022.103260

Fouda, A., Hassan, S. E. D., Saied, E., & Hamza, M. F. (2021). Photocatalytic degradation of real textile and tannery effluent using biosynthesized magnesium oxide nanoparticles (MgO-NPs), heavy metal adsorption, phytotoxicity, and antimicrobial activity. *Journal of Environmental Chemical Engineering, 9*(4), 105346. https://doi.org/10.1016/j.jece.2021.105346

Ghosh, S., Patil, S., Ahire, M., Kitture, R., Gurav, D. D., Jabgunde, A. M., Kale, S., Pardesi, K., Shinde, V., & Bellare, J. (2012). *Gnidia glauca* flower extract mediated synthesis of gold nanoparticles and evaluation of its chemocatalytic potential. *Journal of Nanbiotechnology, 10*(17), 2012. https://doi.org/10.1186/1477-3155-10-17

Gibson, J. D., Khanal, B. P., & Zubarev, E. R. (2007). Paclitaxel-functionalized gold nanoparticles. *Journal of the American Chemical Society, 129*(37), 11653–11661. https://doi.org/10.1021/ja075181k

Gnanajobitha, G., Paulkumar, K., Vanaja, M., Rajeshkumar, S., Malarkodi, C., Annadurai, G., & Kannan, C. (2013). Fruit-mediated synthesis of silver nanoparticles using *Vitis vinifera* and evaluation of their antimicrobial efficacy. *Journal of Nanostructure in Chemistry, 3*(1), 67. https://doi.org/10.1186/2193-8865-3-67

Gopalakrishnan, R., Loganathan, B., Dinesh, S., & Raghu, K. (2017). Strategic green synthesis, characterization and catalytic application to 4-nitrophenol reduction of palladium nanoparticles. *Journal of Cluster Science, 28*(2), 2123–2131. https://doi.org/10.1007/s10876-017-1207-z

Gubitosa, J., Rizzi, V., Lopedota, A., Fini, P., Laurenzana, A., Fibbi, G., Fanelli, F., Petrella, A., Laquintana, V., & Denora, N. (2018). One pot environmental friendly synthesis of gold nanoparticles using *Punica Granatum* juice: A novel antioxidant agent for future dermatological and cosmetic applications. *Journal of Colloid and Interface Science, 521*, 50–61. https://doi.org/10.1016/j.jcis.2018.02.069

Guo, Y., Jiang, N., Zhang, L., & Yin, M. (2020). Green synthesis of gold nanoparticles from *Fritillaria cirrhosa* and its anti-diabetic activity on Streptozotocin induced rats. *Arabian Journal of Chemistry, 13*(4), 5096–5106. https://doi.org/10.1016/j.arabjc.2020.02.009

Guria, M. K., Majumdar, M., & Bhattacharyya, M. (2016). Green synthesis of protein capped nano-gold particle: An excellent recyclable nano-catalyst for the reduction of nitro-aromatic pollutants at higher concentration. *Journal of Molecular Liquids, 222*, 549–557. https://doi.org/10.1016/j.molliq.2016.07.087

Hainfeld, J. F., Slatkin, D. N., & Smilowitz, H. M. (2004). The use of gold nanoparticles to enhance radiotherapy in mice. *Physics in Medicine and Biology, 49*(1), 309–315. https://doi.org/10.1088/0031-9155/49/18/N03

Hainfeld, J. F., Dilmanian, F. A., Zhong, Z., Slatkin, D. N., Kalef-Ezra, J. A., & Smilowitz, H. M. (2010). Gold nanoparticles enhance the radiation therapy of a murine squamous cell carcinoma. *Physics in Medicine and Biology, 55*(11), 3045–3059. https://doi.org/10.1088/0031-9155/55/11/004

Hait, M., Nemu, S. C., Kashyap, N. K., & Chaturwedi, A. (2018). Physicochemical and phytochemical exploration on flower of *Delonex regia*. *Journal of Medicinal Plants Studies, 6*(3), 15–18.

Hait, M., Bhardwaj, A. K., Kashyap, N. K., & Vaishnav, M. M. (2019). Physicochemical and phytochemical evaluation on non- areal part of *Curcuma caesia*. *Journal of Pharmaceutical Innovation, 8*(5), 514–517.

Hait, M., Kashyap, N. K., Deepak, J., & Patel, A. (2020). Seclusion of bioactive components from flower of *Samanea saman*. *Plant Archives, 20*(2), 4581–4584.

Hajra, A., Dutta, S., & Mondal, N. K. (2016). Mosquito larvicidal activity of cadmium nanoparticles synthesized from petal extracts of marigold (*Tagetes* sp.) and rose (*Rosa* sp.) flower. *Journal of Parasitic Diseases, 40*(4), 1519–1527. https://doi.org/10.1007/s12639-015-0719-4

Haqq, S. M., & Chattree, A. (2018). A review: A green approach for the synthesis of silver nanoparticles and its antibacterial applications. *Asian Journal of Pharmaceutical and Clinical Research, 11*(8), 74–78.

Hart, A. D., Azubuike, C. U., & Barimala, S. C. (2005). Vegetable consumption patterns of households in selected areas of the old rivers state of Nigeria. *African Journal of Food Agriculture Nutrition and Development, 5*(1), 587–593. https://doi.org/10.18697/ajfand.8.1365

Hassan, S. E., Fouda, A., Saied, E., Farag, M. M. S., Eid, A. M., Barghoth, M. G., Awad, M. A., Hamza, M. F., & Awad, M. F. (2021). Rhizopus oryzae-mediated green synthesis of magnesium oxide nanoparticles (MgO-NPs): A promising tool for antimicrobial, mosquitocidal action, and tanning effluent treatment. *Journal of Fungi, 7*(5), 372. https://doi.org/10.3390/jof7050372

Haviland-Jones, J., Rosario, H. H., Wilson, P., & Mc Guire, T. R. (2005). An environmental approach to positive emotion: Flowers. *Evolutionary Psychology, 3*(1), 104–132. https://doi.org/10.1177/147470490500300109

He, Y., Du, Z., Ma, S., Cheng, S., Jiang, S., Liu, Y., Li, D., Huang, H., Zhang, K., & Zheng, X. (2016). Biosynthesis, antibacterial activity and anticancer effects against prostate cancer (pc-3) cells of silver nanoparticles using *Dimocarpus longan* Lour. peel extract. *Nanoscale Research Letters, 11*(1), 300. https://doi.org/10.1186/s11671-016-1511-9

Heera, P., & Shanmugam, S. (2015). Review article nanoparticle characterization and application: An overview. *International Journal of Current Microbiology and Applied Sciences, 4*(2), 379–386.

Hemmati, S., Ahmed, A., Salehabadi, Y., Zangeneh, A., & Zangeneh, M. M. (2020). Synthesis, characterization, and evaluation of cytotoxicity, antioxidant, antifungal, antibacterial, and cutaneous wound healing effects of copper nanoparticles using the aqueous extract of strawberry fruit and L-ascorbic acid. *Polyhedron, 180*, 114425. https://doi.org/10.1016/j.poly.2020.114425

Hofmann, T., Lowry, G. V., Ghoshal, S., Tufenkji, N., Brambilla, D., Dutcher, J. R., Gilbertson, L. M., Giraldo, J. P., Kinsella, J. M., Landry, M. P., Lovell, W., Naccache, R., Paret, M., Pedersen, J. A., Unrine, J. M., White, J. C., & Wilkinsen, K. J. (2020). Technology readiness and overcoming barriers to sustainably implement nanotechnology-enabled plant agriculture. *Nature Food, 1*(7), 416–425. https://doi.org/10.1038/s43016-020-0110-1

Hu, M. L. (2011). Dietary polyphenols as antioxidants and anticancer agents: More questions than answers. *Chang Gung Medical Journal, 34*(5), 449–460.

Hurley, K. R., Ring, H. L., Kang, H., Klein, N. D., & Haynes, C. L. (2015). Characterization of magnetic nanoparticles in biological matrices. *Analytical Chemistry, 87*(23), 11611–11619. https://doi.org/10.1021/acs.analchem.5b02229

Huss, E., Yosef, K. B., & Zaccai, Z. (2018). Humans' relationship to flowers as an example of the multiple components of embodied aesthetics. *Behavioral Science, 8*(3), 32. https://doi.org/10.3390/bs8030032

Hussein, B. Y., & Mohammed, A. M. (2021). Biosynthesis and characterization of nickel oxide nanoparticles by using aqueous grape extract and evaluation of their biological applications. *Results in Chemistry, 3*(11), 100142. https://doi.org/10.1016/j.rechem.2021.100142

Ibrahim, H. M. (2015). Green synthesis and characterization of silver nanoparticles using banana peel extract and their antimicrobial activity against representative microorganisms. *Journal of Radiation Research and Applied Science, 8*(2015), 265–275. https://doi.org/10.1016/j.jrras.2015.01.007

Igwe, O. U., & Nwamezie, F. (2018). Green synthesis of iron nanoparticles using flower extract of *Piliostigma thonningii* and their antibacterial activity evaluation. *Chemistry International, 4*(2), 60–66.

Inshakova, E., & Inshakov, O. (2017). World market for nanomaterials: Structure and trends. In *MATEC Web of Conferences* (Vol. 129, p. 02013). EDP Sciences. https://doi.org/10.1051/matecconf/201712902013

Jacob, S. J. P., Prasad, V. L. S., Sivasankar, S., & Muralidharan, P. (2017). Biosynthesis of silver nanoparticles using dried fruit extract of *Ficus carica* screening for its anticancer activity and toxicity in animal models. *Food and Chemical Toxicology, 109*, 951–956. https://doi.org/10.1016/j.fct.2017.03.066

Jadoun, S., Arif, R., Jangid, N. K., & Meena, R. K. (2021). Green synthesis of nanoparticles using plant extracts: A review. *Environmental Chemistry Letters, 19*(1), 355–374. https://doi.org/10.1007/s10311-020-01074-x

Jain, D., Daima, H. K., Kachhwaha, S., & Kothari, S. L. (2009). Synthesis of plant-mediated zinc nanoparticle using papaya fruit extract and evaluation of their anti-microbial activities. *Digest Journal of Nanomaterials and Biostructures, 4*(3), 557–563.

Jamdagni, P., Khatri, P., & Rana, J. S. (2016). Green synthesis of zinc oxide nanoparticles using flower extract of *Nyctanthes arbor-tristis* and their antifungal activity. *Journal of King Saud University, Science, 30*(2), 168–175. https://doi.org/10.1016/j.jksus.2016.10.002

Jamkhande, P. G., Ghule, N. W., Bamer, A. H., & Kalaskar, M. G. (2019). Metal nanoparticles synthesis: An overview on methods of preparation, advantages and disadvantages, and applications. *Journal of Drug Delivery Science and Technology, 53*, 101174. https://doi.org/10.1016/j.jddst.2019.101174

Javed, R., Zia, M., Naz, S., Aisida, S. O., Ain, N., & Ao, Q. (2020). Role of capping agents in the application of nanoparticles in biomedicine and environmental remediation: Recent trends and future prospects. *Journal of Nanbiotechnology, 18*, 172. https://doi.org/10.1186/s12951-020-00704-4

Jayaseelan, C., Rahuman, A. A., Kirthi, A. V., Marimuthu, S., Santhoshkumar, T., Bagavan, A., Guarav, K., Karthik, L., & Rao, K. V. (2012). Novel microbial route to synthesize ZnO nanoparticles using Aeromonas hydrophila and their activity against pathogenic bacteria and fungi. *Spectrochimica Acta. Part A, Molecular and Biomolecular Spectroscopy, 90*, 78–84. https://doi.org/10.1016/j.saa.2012.01.006

Joshia, M., Bhattacharyya, A., & Ali, S. W. (2008). Characterization techniques for nanotechnology applications in textiles. *Indian Journal of Fibre & Textile Research, 33*(1), 304–317.

Kalia, A., & Singh, S. (2020). Myco-decontamination of azo dyes: Nano-augmentation technologies. *3Biotech, 10*(9), 384. https://doi.org/10.1007/s13205-020-02378-z

Kandiah, M., & Chandrasekaran, K. N. (2021). Green synthesis of silver nanoparticles using *Catharanthus Roseus* flower extracts and the determination of their antioxidant, antimicrobial, and photocatalytic activity. *Journal of Nanotechnology, 2021*, 5512786. https://doi.org/10.1155/2021/5512786

Karnuakaran, G., Jagathambal, M., Gusev, A., Kloesnikov, E., Mandal, A. R., & Kuznestov, D. (2016). *Allamanda cathartica* flower's aqueous extract-mediated green synthesis of silver nanoparticles with excellent antioxidant and antibacterial potential for biomedical application. *MRS Communications, 6*(1), 41–46. https://doi.org/10.1557/mrc.2016.2

Kashyap, N. K., Deepak, J., Bhardwaj, A. K., Hait, M., & Pal, D. K. (2022a). *In-vitro* antibacterial and antifungal activity of *Curcuma amada* Roxb. against human pathogens. *Trends in Sciences, 19*(22), 4839. https://doi.org/10.48048/tis.2022.4839

Kashyap, N. K., Hait, M., Roy Mahapatra, G. S., & Vaishnav, M. M. (2022b). Proximate and elemental analysis of *Careya Arborea* Roxb plant's root. *ES Food & Agroforestry, 7*, 41–47. https://doi.org/10.30919/esfaf620

Kaur, P., Thakur, A., & Chaudhary, A. (2016). Biogenesis of copper nanoparticles using peel extract of *Punica granatum* and their antimicrobial activity against opportunistic pathogens. *Green Chemistry Letters and Reviews, 9*(1), 33–38. https://doi.org/10.1080/17518253.2016.1141238

Keihan, A. H., Veisi, H., & Veasi, H. (2017). Green synthesis and characterization of spherical copper nanoparticles as organometallic antibacterial agent. *Applied Organometallic Chemistry, 31*(7), e3642. https://doi.org/10.1002/aoc.3642

Khan, Z. S., Shinde, V. N., Bhosle, N. O., & Nasreen, S. (2010). Chemical composition and antimicrobial activity of angiospermic plants. *Middle-East Journal of Scientific Research, 6*(1), 56–61.

Khan, A. F., Yuan, Q., Wei, Y., Khan, S. U., Tahir, K., Khna, Z. U. H., Ahmad, A., Ali, F., Ali, S., & Nazir, S. (2016). Longan fruit juice mediated synthesis of uniformly dispersed spherical AuNPs: Cytotoxicity against human breast cancer cell line MCF-7, antioxidant and fluorescent properties. *RSC Advances, 6*, 23775–23782. https://doi.org/10.1039/C5RA27100B

Khan, M. S. J., Kamal, T., Ali, F., Asiri, A. M., & Khan, S. B. (2019). Chitosan-coated polyurethane sponge supported metal nanoparticles for catalytic reduction of organic pollutants. *International Journal of Biological Macromolecules, 132*, 772–783. https://doi.org/10.1016/j.ijbiomac.2019.03.205

Khani, R., Roostaei, B., Bagherzade, G., & Moudi, M. (2018). Green synthesis of copper nanoparticles by fruit extract of *Ziziphus spina-christi* (L.) Willd: Application for adsorption of triphenylmethane dye and antibacterial assay. *Journal of Molecular Liquids, 255*, 541–549. https://doi.org/10.1016/j.molliq.2018.02.010

Khatoon, N., Mazumder, J. A., & Sardar, M. (2017). Biotechnological applications of green synthesized silver nanoparticles. *Journal of Nanosciences: Current Research, 2*(1), 1000107. https://doi.org/10.4172/2572-0813.1000107

Khatune, N. A., Mossadik, M. A., Rahman, M. M., Khondkar, P., Haque, M. E., & Gray, A. I. (2005). A benzofuranone from the flowers of *Nyctanthes arbortristis* and its antibacterial and cytotoxic activities. *Dhaka University Journal of Pharmaceutical Sciences, 4*(1), 33–37. https://doi.org/10.3329/dujps.v4i1.191

Kim, M. R., Lee, J. Y., Lee, H. H., Aryal, D. K., Kim, Y. G., Kim, S. K., Woo, E. R., & Kang, K. W. (2006). Antioxidative effects of quercetin-glycosides isolated from the flower buds of *Tussilago farfara* L. *Food and Chemical Toxicology, 44*(8), 1299–1307. https://doi.org/10.1016/j.fct.2006.02.007

Klueh, U., Wagner, V., Kelly, S., Johnson, A., & Bryers, J. D. (2000). Efficacy of the silver coated fabric to prevent bacterial colonization and subsequent device-based biofilm formation. *Journal of Biomedical Materials Research Part B: Applied Biomaterials, 53*(6), 621–631. https://doi.org/10.1002/1097-4636(2000)53:6<621::aid-jbm2>3.0.co;2-q

Kokila, T., Ramesh, P. S., & Geetha, D. (2015). Biosynthesis of silver nanoparticles from Cavendish banana peel extract and its antibacterial and free radical scavenging assay: A novel biological approach. *Applied Nanoscience, 5*(8), 911–920. https://doi.org/10.1007/s13204-015-0401-2

Koppula, S. B. (2013). Antimicrobial activity of floral extracts on selected human pathogens. *The International Journal of Bio-Pharma Research, 2*(6), 141–143.

Kumar, H., Bhardwaj, K., Kuca, K., Kalia, A., Nepovimova, E., Verma, R., & Kumar, D. (2020a). Flower-based green synthesis of metallic nanoparticles: Applications beyond fragrance. *Nanomaterials, 10*(4), 766. https://doi.org/10.3390/nano10040766

Kumar, H., Bhardwaj, K., Sharma, R., Nepovimova, E., Kuca, K., Dhanjal, D. S., Verma, R., Bhardwaj, P., Sharma, S., & Kumar, D. (2020b). Fruit and vegetable peels: Utilization of high value horticultural waste in novel industrial applications. *Molecules, 25*(12), 2812. https://doi.org/10.3390/molecules25122812

Kuppusamy, P., Yusoff, M., Maniam, G., & Govindan, N. (2016). Biosynthesis of metallic nanoparticles using plant derivatives and their new avenues in pharmacological applications—An updated report. *Saudi Pharmaceutical Journal, 24*(4), 473–484. https://doi.org/10.1016/j.jsps.2014.11.013

Lakshmeesha, T. R., Kalagatur, N. K., Mudili, V., Mohan, C. D., Rangappa, S., Prasad, B. D., Ashwini, B. S., Hashem, A., Alqarawi, A. A., & Malik, J. A. (2019). Biofabrication of zinc oxide nanoparticles with *Syzygium aromaticum* flower buds extract and finding its novel application in controlling the growth and mycotoxins of *Fusarium graminearum. Frontiers in Microbiology, 10*, 1244. https://doi.org/10.3389/fmicb.2019.01244

Lee, Y. J., Song, K., Cha, S. H., Cho, S., Kim, Y. S., & Park, Y. (2019). Sesquiterpenoids from *Tussilago farfara* flower bud extract for the eco-friendly synthesis of silver and gold nanoparticles possessing antibacterial and anticancer activities. *Nanomaterials, 9*(6), 819. https://doi.org/10.3390/nano9060819

Lesniak, A., Salvati, A., Santos-Martinez, M. J., Radomski, M. W., Dawson, K. A., & Aberg, C. (2013). Nanoparticle adhesion to the cell membrane and its effect on nano particle uptake efficiency. *Journal of the American Chemical Society, 135*(1), 1438–1444. https://doi.org/10.1021/ja309812z

Lim, S. H., Ahn, E. Y., & Park, Y. (2016). Green synthesis and catalytic activity of gold nanoparticles synthesized by Artemisia capillaries water extract. *Nanoscale Research Letters, 11*(1), 474. https://doi.org/10.1186/s11671-016-1694-0

Liu, P., Zhao, W. H., Song, Y. P., Wang, Q., Kan, Y. F., Wang, S. Y., Xia, J. L., Bilal, M., Zhu, X. Y., & Wang, Z. X. (2022). Characterization, antimicrobial, and antioxidant potentialities of first-time isolated silver nanoparticles synthesizing protein secreted by *Lysinibacillus sphaericus. Process Biochemistry, 122*(1), 230–237. https://doi.org/10.1016/j.procbio.2022.08.032

Louafi, O., Khelef, A., Zeroual, S., Laouini, S. E., & Tedjani, M. L. (2021). Effect of nickel nitrate concentration on the size of nickel oxide nanoparticles bio-synthesized by *Artemisia herba-alba* aqueous leaves extract and improving their antioxidant activities. *Journal of Inorganic and Organometallic Polymers, 32*(6), 1116–1128. https://doi.org/10.1007/s10904-021-02152-5

Majumdar, M., Biswas, S. C., Choudhury, R., Upadhyay, P., Adhikary, A., Roy, D. N., & Misra, T. K. (2019). Synthesis of gold nanoparticles using *Citrus macroptera* fruit extract: Anti-biofilm and anticancer activity. *Chemistry Select, 4*(19), 5714–5723. https://doi.org/10.1002/slct.201804021

Mandal, P. (2018). Biosynthesis of silver nanoparticles by *Plumeria rubra* flower extract: Characterization and their antimicrobial activities. *International Journal Engineering Science, 7*(1), 1–6.

Manisha, D. R., Alwala, J., Kudle, K. R., & Rudra, M. P. P. (2014). Biosynthesis of silver nanoparticles using flower extracts of *Catharanthus roseus* and evaluation of its antibacterial efficacy. *World Journal of Pharmaceutical Sciences, 3*(5), 877–885.

Manisha, B., Mbiri, R. S., Senanayake, H. A., & Sakshi. (2021). Health benefits of flowers and their value-added products. *The International Journal of Creative Research Thoughts, 9*(5), 527–540.

Marimuthu, S., Rahuman, A. A., Jayaseelan, C., Kirthi, A. V., Santhoshkumar, T., Velayutham, K., Bagavan, A., Kamaraj, C., Elango, G., & Iyappan, M. (2013). Acaricidal activity of synthesized titanium dioxide nanoparticles using Calotropisgigantea against *Rhipicephalus microplus* and

Haemaphysalis bispinosa. *Asian Pacific Journal of Tropical Medicine, 6*(9), 682–688. https://doi.org/10.1016/S1995-7645(13)60118-2

Mashwani, Z. R., Khan, M. A., Khan, T., & Nadhman, A. (2016). Applications of plant terpenoids in the synthesis of colloidal silver nanoparticles. *Advances in Colloid and Interface Science, 234*, 132–141. https://doi.org/10.1016/j.cis.2016.04.008

Masum, M. M. I., Siddiqa, M. M., Ali, K. A., Zhang, Y., Abdallah, Y., Ibrahim, E., Qiu, W., Yan, C., & Li, B. (2019). Biogenic synthesis of silver nanoparticles using *Phyllanthus emblica* fruit extract and its inhibitory action against the pathogen Acidovorax oryzae strain RS-2 of rice bacterial brown stripe. *Frontiers in Microbiology, 10*, 820. https://doi.org/10.3389/fmicb.2019.00820

Materon, E. M., Miyazaki, C. M., Carr, O., Joshi, N., Picciani, P. H. S., Dalmaschio, C. J., Davis, F., & Shimizu, F. M. (2021). Magnetic nanoparticles in biomedical applications: A review. *Applied Surface Science Advances, 6*, 100163. https://doi.org/10.1016/j.apsadv.2021.100163

McMahon, S. J., Hyland, W. B., Muir, M. F., Coulter, J. A., Jain, S., Butterworth, K. T., Schettino, G., Dickson, G. R., Hounsell, A. R., & O'Sullivan, J. M. (2011). Biological consequences of nanoscale energy deposition near irradiated heavy atom nanoparticles. *Scientific Reports, 1*, 18. https://doi.org/10.1038/srep00018

Meena, A. K., Patidar, D., & Singh, R. K. (2014). Ameliorative effect of *Hibiscus rosa-sinensis* on phenylhydrazine induced haematotoxicity. *International Journal of Innovative Research in Science, Engineering and Technology, 3*(2), 8678–8683.

Mintah, B. K., Eliason, A. E., Nsiah, M., Baah, E. M., Hagan, E., & Ofosu, D. B. (2012). Consumption of fruits among students: A case of public university in Ghana. *African Journal of Food Agriculture Nutrition and Development, 12*(2), 5979–5993. https://doi.org/10.18697/ajfand.50.9785

Miu, B. A., & Dinischiotu, A. (2022). New green approaches in nanoparticles synthesis: An overview. *Molecules, 27*(19), 6472. https://doi.org/10.3390/molecules27196472

Momeni, S., & Nabipour, I. (2015). A simple green synthesis of palladium nanoparticles with Sargassum alga and their electrocatalytic activities towards hydrogen peroxide. *Applied Biochemistry and Biotechnology, 176*(7), 1937–1949. https://doi.org/10.1007/s12010-015-1690-3

Moteriya, P., & Chanda, S. (2016). Synthesis and characterization of silver nanoparticles using *Caesalpinia pulcherrima* flower extract and assessment of their in vitro antimicrobial, antioxidant, cytotoxic, and genotoxic activities. *Artificial Cells, Nanomedicine, and Biotechnology, 45*(8), 1556–1567. https://doi.org/10.1080/21691401.2016.1261871

Mourdikoudis, S., Pallares, R. M., & Thanh, N. T. K. (2018). Characterization techniques for nanoparticles: Comparison and complementarity upon studying nanoparticle properties. *Nanoscale, 10*, 12871–12934. https://doi.org/10.1039/C8NR02278J

Muruganantham, N., Govindharaju, R., Anitha, P., & Anusuya, V. (2018). Synthesis and characterization of silver nanoparticles using *Lablab purpureus* flowers (purple colour) and its anti-microbial activities. *International Journal of Biological Sciences, 5*(6), 1–7. https://doi.org/10.26438/ijsrbs/v5i6.17

Nadagouda, M. N., & Varma, R. S. (2008). Green synthesis of silver and palladium nanoparticles at room temperature using coffee and tea extract. *Green Chemistry, 10*, 859–862. https://doi.org/10.1039/B804703K

Naganathan, K., & Thirunavukkarasu, S. (2017). Green way genesis of silver nanoparticles using multiple fruit peels waste and its antimicrobial, anti-oxidant and anti-tumor cell line studies. *Materials Science and Engineering, 191*(1), 012009. https://doi.org/10.1088/1757-899X/191/1/012009

Nagaraj, B., Malakar, B., Divya, T. K., Krishnamurthy, N. B., Liny, P., & Dinesh, R. (2012). Environmental benign synthesis of gold nanoparticles from the flower extracts of *Plumeria alba* Linn. (Frangipani) and evaluation of their biological activities. *International Journal of Drug Development & Research, 4*(1), 144–150.

Nanu, R., Raghuveer, I., Chitme, H., & Chandra, R. (2008). Antidiabetic activity of *Nyctanthes arbortristis*. *Pharmacognosy Magazine, 4*(16), 335–340.

Naraian, R., & Bhardwaj, A. K. (2020). Green synthesis and characterization of silver NPs using oyster mushroom extract for antibacterial efficacy. *Journal of Chemistry, Environmental Sciences and its Applications, 7*(1), 13–18. https://doi.org/10.15415/jce.2020.71003

Narayanan, M., Divya, S., Natarajan, D., Senthil-Nathan, S., Kandasamy, S., Chinnathambi, A., Alahmadi, T. A., & Pugazhendhi, A. (2021). Green synthesis of silver nanoparticles from aqueous extract of *Ctenolepis garcini* L. and assess their possible biological applications. *Process Biochemistry, 107*(1), 91–99. https://doi.org/10.1016/j.procbio.2021.05.008

Nasrollahzadeh, M., Momeni, S. S., & Sajadi, S. M. (2017). Green synthesis of copper nanoparticles using *Plantago asiatica* leaf extract and their application for the cyanation of aldehydes using $K_4Fe(CN)_6$. *Journal of Colloid and Interface Science, 506*(2), 471–477. https://doi.org/10.1016/j.jcis.2017.07.072

Natsuki, J., Natsuki, T., & Hashimoto, Y. (2015). A review of silver nanoparticles: Synthesis methods, properties and applications. *International Journal of Materials Science and Applications, 4*(5), 325–332.

Nayan, V., Onteru, S. K., & Singh, D. (2018). *Mangifera indica* flower extract mediated biogenic green gold nanoparticles: Efficient nanocatalyst for reduction of 4-nitrophenol. *Environmental Progress & Sustainable Energy, 37*(1), 283–294. https://doi.org/10.1002/ep.12669

Nazir, A., Akbar, A., Baghdadi, H. B., Ur Rehman, S., Al-Abbad, E., Fatima, M., Iqbal, M., Tamam, N., Alwadai, N., & Abbas, M. (2021). Zinc oxide nanoparticles fabrication using *Eriobotrya japonica* leaves extract: Photocatalytic performance and antibacterial activity evaluation. *Arabian Journal of Chemistry, 14*(8), 103251. https://doi.org/10.1016/j.arabjc.2021.103251

Ndikau, M., Noah, N. M., Andala, D. M., & Masika, E. (2017). Green synthesis and characterization of silver nanoparticles using *Citrullus lanatus* fruit rind extract. *International Journal of Analytical Chemistry, 2017*, 8108504. https://doi.org/10.1155/2017/8108504

Nirmala, J. G., Akila, S., Nadar, M. S., Narendhirakannan, R. T., & Chatterjee, S. (2016). Biosynthesized Vitis vinifera seed gold nanoparticles induce apoptotic cell death in A431 skin cancer cells. *RSC Advances, 6*, 82205–82218. https://doi.org/10.1039/C6RA16310F

Nirmala, J. G., Akila, S., Nadar, M. S., Narendhirakannan, R. T., & Chatterjee, S. (2017). *Vitis vinifera* peel polyphenols stabilized gold nanoparticles induce cytotoxicity and apoptotic cell death in A431 skin cancer cell lines. *Advanced Powder Technology, 28*(4), 1170–1184. https://doi.org/10.1016/j.apt.2017.02.003

Nishteswar, K. (2015). Pushpayurveda (flowers of medicinal plants) delineated in Kaiyadevanighantu. *Punarna, 2*(1), 1–10.

Olfati, A., Kahrizi, D., Balaky, S. T. J., Sharifi, R., Tahir, M. B., & Darvishi, E. (2021). Green synthesis of nanoparticles using *Calendula officinalis* extract from silver sulfate and their antibacterial effects on Pectobacterium caratovorum. *Inorganic Chemistry Communications, 125*, 108439. https://doi.org/10.1016/j.inoche.2020.108439

Padalia, H., Moteriya, P., & Chanda, S. (2014). Green synthesis of silver nanoparticles from marigold flower and its synergistic antimicrobial potential. *Arabian Journal of Chemistry, 8*(5), 732–741. https://doi.org/10.1016/j.arabjc.2014.11.015

Panigrahi, S., Basu, S., Praharaj, S., Pande, S., Jana, S., Pal, A., Ghosh, S. K., & Pal, T. (2007). Synthesis and size-selective catalysis by supported gold nanoparticles: Study on heterogeneous and homogeneous catalytic process. *The Journal of Physical Chemistry, C 111*(12), 4596–4605. https://doi.org/10.1021/jp067554u

Park, J. K., Rupa, E. J., Arif, M. H., Li, J. F., Anandapadmanaban, G., Kang, J. P., Kim, M., Ahn, J. C., Akter, R., & Yang, D. C. (2021). Synthesis of zinc oxide nanoparticles from Gynostemma pentaphyllum extracts and assessment of photocatalytic properties through malachite green dye decolorization under UV illumination—A green approach. *Optik, 239*, 166249. https://doi.org/10.1016/j.ijleo.2020.166249

Parveen, K., Banse, V., & Ledwani, L. (2016). Green synthesis of nanoparticles: Their advantages and disadvantages. *AIP Conference Proceedings, 1724*, 020048. https://doi.org/10.1063/1.4945168

Patel, R., & Mehta, M. (2019). Green synthesis of silver nanoparticles by using fruit and vegetable waste: A review. *International Journal of Applied Engineering Research, 9*(3), 78–85.

Patil, M. P., Singh, R. D., Koli, P. B., Patil, K. T., Jagdale, B. S., Tipare, A. R., & Kim, G. D. (2018). Antibacterial potential of silver nanoparticles synthesized using *Madhuca longifolia* flower extract as a green resource. *Microbial Pathogenesis, 121*, 184–189. https://doi.org/10.1016/j.micpath.2018.05.040

Patra, J. K., & Baek, K. H. (2015). Novel green synthesis of gold nanoparticles using *Citrullus lanatus rind* and investigation of proteasome inhibitory activity, antibacterial, and antioxidant potential. *International Journal of Nanomedicine, 10*, 7253–7264. https://doi.org/10.2147/IJN.S95483

Pavithra, N. S., Lingaraju, K., Raghu, G. K., & Nagaraju, G. (2017). *Citrus maxima* (Pomelo) juice mediated eco-friendly synthesis of ZnO nanoparticles: Applications to photocatalytic, electrochemical sensor and antibacterial activities. *Spectrochimica Acta. Part A, Molecular and Biomolecular Spectroscopy, 185*, 11–19. https://doi.org/10.1016/j.saa.2017.05.032

Pechyen, C., Ponsanti, K., Tangnorawich, B., & Ngernyuang, N. (2021). Waste fruit peel-mediated green synthesis of biocompatible gold nanoparticles. *Journal of Materials Research and Technology, 14*(5), 2982–2991. https://doi.org/10.1016/j.jmrt.2021.08.111

Pulit-prociak, J., Chwastowski, J., Kucharski, A., & Banach, M. (2016). Applied surface science functionalization of textiles with silver and zinc oxide nanoparticles. *Applied Surface Science, 385*, 543–553. https://doi.org/10.1016/j.apsusc.2016.05.167

Qian, X., Peng, X. H., Ansari, D. O., Yin-Goen, Q., Chen, G. Z., Shin, D. M., Yang, L., Young, A. N., Wang, M. D., & Nie, S. (2008). *In vivo* tumor targeting and spectroscopic detection with surface-enhanced Raman nanoparticle tags. *Nature Biotechnology, 26*(1), 83–90. https://doi.org/10.1038/nbt1377

Ramesh, P. S., Kokila, T., & Geetha, D. (2015). Plant mediated green synthesis and antibacterial activity of silver nanoparticles using *Emblica Officinalis* fruit extract. *Spectrochimica Acta. Part A, Molecular and Biomolecular Spectroscopy, 142*, 339–343. https://doi.org/10.1016/j.saa.2015.01.062

Rana, A., Yadav, K., & Jagadevan, S. A. (2020). Comprehensive review on green synthesis of nature-inspired metal nanoparticles: Mechanism, application and toxicity. *Journal of Cleaner Production, 272*(5), 122880. https://doi.org/10.1016/j.jclepro.2020.122880

Rani, S., Kumar, P., Dahiya, P., Dang, A. S., & Suneja, P. (2022). Biogenic synthesis of zinc nanoparticles, their applications, and toxicity prospects. *Frontiers in Microbiology, 13*, 824427. https://doi.org/10.3389/fmicb.2022.824427

Ratan, Z. A., Haidere, M. F., Nurunnabi, M., Shahriar, S. M., Ahammad, A. J. S., Shim, Y. Y., Reaney, M. J. T., & Cho, J. Y. (2020). Green chemistry synthesis of silver nanoparticles and their potential anticancer effects. *Cancer, 12*(4), 855. https://doi.org/10.3390/cancers12040855

Raut, R. W., Mendhulkar, V. D., & Kashid, S. B. (2014). Photosensitized synthesis of silver nanoparticles using Withania Somnifera leaf powder and silver nitrate. *Journal of Photochemistry and Photobiology B: Biology, 132*, 45–55.

Reddy, L. J., & Jose, B. (2013). Evaluation of antibacterial activity of *Mimusops elengi* L. flowers and *Trichosanthes cucumerina* L. fruits from South India. *International Journal of Pharmacy and Pharmaceutical Sciences, 5*(2), 362–364.

Ren, W., Qiao, Z., Wang, H., Zhu, L., & Zhang, L. (2003). Flavonoids: Promising anticancer agents. *Medicinal Research Reviews, 23*(3), 519–534. https://doi.org/10.1002/med.10033

Renuka, R., Devi, K. R., Sivakami, M., Thilagavathi, T., & Uthrakumar, R. (2020). Biosynthesis of silver nanoparticles using *Phyllanthus emblica* fruit extract for antimicrobial application. *Biocatalysis and Agricultural Biotechnology, 24*, 101567. https://doi.org/10.1016/j.bcab.2020.101567

Rostami-Vartooni, A., Nasrollahzadeh, M., & Alizadeh, M. (2016). Green synthesis of perlite supported silver nanoparticles using *Hamamelis virginiana* leaf extract and investigation of its catalytic activity for the reduction of 4-nitrophenol and Congo red. *Journal of Alloys and Compounds, 680*, 309–314. https://doi.org/10.1016/j.jallcom.2016.04.008

Rout, P. K., Sahoo, D., & Misra, L. N. (2010). Comparison of extraction methods of *Mimusops elengi* L. flowers. *Industrial Crops and Products, 32*(2), 678–680. https://doi.org/10.1016/j.indcrop.2010.05.019

Ruban, P., & Gajalakshmi, K. (2012). In vitro antibacterial activity of *Hibiscus rosa-sinensis* flower extract against human pathogens. *Asian Pacific Journal of Tropical Biomedicine, 2*(5), 399–403. https://doi.org/10.1016/S2221-1691(12)60064-1

Sabouri, Z., Akbari, A., Hosseini, H. A., & Darroudi, M. (2018). Facile green synthesis of NiO nanoparticles and investigation of dye degradation and cytotoxicity effects. *Journal of Molecular Structure, 1173*, 931–936. https://doi.org/10.1016/j.molstruc.2018.07.063

Salamanca-Buentello, F., Persad, D. L., Court, E. B., Martin, D. K., Daar, A. S., & Singer, P. A. (2005). Nanotechnology and the developing world. *PLoS Medicine, 2*(5), e97. https://doi.org/10.1371/journal.pmed.0020097

Salem, S. S., & Fouda, A. (2021). Green synthesis of metallic nanoparticles and their prospective biotechnological applications: An overview. *Biological Trace Element Research, 199*(1), 344–370. https://doi.org/10.1007/s12011-020-02138-3

Saman, S., Balouch, A., Talpur, F. N., Memon, A. A., Mousavi, B. M., & Verpoort, F. (2021). Green synthesis of MgO nanocatalyst by using *Ziziphus mauritiana* leaves and seeds for biodiesel production. *Applied Organometallic Chemistry, 35*(5), e6199. https://doi.org/10.1002/aoc.6199

Samuel, M. S., Ravikumar, M., John, J. A., Selvarajan, E., Patel, H., Chander, P. S., Soundarya, J., Vuppala, S., Balaji, R., & Chandrasekar, N. (2022). A review on green synthesis of nanoparticles and their diverse biomedical and environmental applications. *Catalysts, 12*(5), 459. https://doi.org/10.3390/catal12050459

Sangeetha, V. S., Vanitha, A., Devi, P., & Bhakyajothi, V. (2020). *Green synthesis of zinc oxide nanoparticles using flower extract of Phyla nodiflora*. Dhanalakshmi College of Arts and Science for Women (Autonomous).

Sarah, S. L. R., & Iyer, P. R. (2019). Green synthesis of copper nanoparticles from the flowers of *Mimusops elengi*. *International Journal of Scientific Research, 10*(6), 32956–32963.

Sarid, O., & Zaccai, M. (2016). Changes in mood states are induced by smelling familiar and exotic fragrances. *Frontiers in Psychology, 7*(1), 1724. https://doi.org/10.3389/fpsyg.2016.01724

Sebesta, M., Kolencik, M., Sunil, B. R., Illa, R., Mosnacek, J., Ingle, A. P., & Urik, M. (2021). Field application of ZnO and TiO_2 nanoparticles on agricultural plants. *Agronomy, 11*, 2281. https://doi.org/10.3390/agronomy11112281

Selim, Y. A., Azb, M. A., Ragab, I., & Abd-El-Azim, M. H. M. (2020). Green synthesis of zinc oxide nanoparticles using aqueous extract of *Deverra tortuosa* and their cytotoxic activities. *Scientific Reports, 10*(1), 3445. https://doi.org/10.1038/s41598-020-60541-1

Shafaei, F., Babaei, S. E., Shahvelayati, A. S., & Janatabadi, F. H. (2019). Biosynthesis of Fe_3O_4-magnetic nanoparticles using clover leaf aqueous extract: Green synthesis of 1, 3-benzoxazole derivatives. *Journal of the Chinese Chemical Society, 67*(4), 891–897. https://doi.org/10.1002/jccs.201800489

Shalaby, S. M., Madkour, F. F., El-Kassas, H. Y., Mohamed, A. A., & Elgarahy, A. M. (2021). Green synthesis of recyclable iron oxide nanoparticles using Spirulina platensis microalgae for adsorptive removal of cationic and anionic dyes. *Environmental Science and Pollution Research International, 28*(46), 65549–65572. https://doi.org/10.1007/s11356-021-15544-4

Shanmugavadivu, M., Kuppusamy, S., & Ranjithkumar, R. (2014). Synthesis of pomegranate peel extract mediated silver nanoparticles and its antibacterial activity. *American Journal of Advanced Drug Delivery, 2*(2), 174–182.

Sharma, J. K., Akhtar, M. S., Ameen, S., Srivastva, P., & Singh, G. (2015). Green synthesis of CuO nanoparticles with leaf extract of *Calotropis gigantea* and its dye-sensitized solar cells applications. *Journal of Alloys and Compounds, 632*(2), 321–325. https://doi.org/10.1016/j.jallcom.2015.01.172

Sharma, D., Sabela, M. I., Kanchi, S., Mdluli, P. S., Singh, G., Stenstrom, T. A., & Bisetty, K. (2016). Biosynthesis of ZnO nanoparticles using *Jacaranda mimosifolia* flowers extract:

Synergistic antibacterial activity and molecular simulated facet specific adsorption studies. *Journal of Photochemistry and Photobiology B: Biology, 162*, 199–207. https://doi.org/10.1016/j.jphotobiol.2016.06.043

Sharma, D., Shandilya, P., & Saini, N. K. (2021). Insights into the synthesis and mechanism of green synthesized antimicrobial nanoparticles, answer to the multidrug resistance. *Materials Today Chemistry, 19*(3), 100391. https://doi.org/10.1016/j.mtchem.2020.100391

Shende, S., Ingle, A. P., Gade, A., & Rai, M. (2015). Green synthesis of copper nanoparticles by *Citrus medica* Linn. (Idilimbu) juice and its antimicrobial activity. *World Journal of Microbiology and Biotechnology, 31*(6), 865–873. https://doi.org/10.1007/s11274-015-1840-3

Shubhashree, M. N., Shantha, T. R., Ramarao, V., Reddy, M. P., & Venkateshawarulu, G. (2015). A review on therapeutic uses of flowers as depicted in classical texts of Ayurveda and Siddha. *Journal of Research and Education in Indian Medicine, 21*(1), 1–14.

Shubhashree, K. R., Reddy, R., Gangula, A. K., Nagananda, G. S., Badiya, P. K., Ramamurthy, S. S., Aramwit, P., & Reddy, N. (2022). Green synthesis of copper nanoparticles using aqueous extracts from *Hyptis suaveolens* (L.). *Materials Chemistry and Physics, 280*, 125795. https://doi.org/10.1016/j.matchemphys.2022.125795

Sidhu, J. S., & Zafar, T. A. (2018). Bioactive compounds in banana fruits and their health benefits. *Food Quality and Safety, 2*(4), 183–188. https://doi.org/10.1093/fqsafe/fyy019

Singaravelu, G., Arockiamary, J. S., Kumar, V. G., & Govindaraju, K. (2007). A novel extracellular synthesis of monodisperse gold nanoparticles using marine algae, *Sargassum wightii Greville. Colloids and Surfaces. B, Biointerfaces, 57*(1), 97–101. https://doi.org/10.1016/j.colsurfb.2007.01.010

Singh, P., Kim, Y., Zhang, D., & Yang, D. (2016). Biological synthesis of nanoparticles from plants and microorganisms. *Trends in Biotechnology, 34*(7), 588–599. https://doi.org/10.1016/j.tibtech.2016.02.006

Singh, J., Dutta, T., Kim, K. H., Rawat, M., Samddar, P., & Kumar, P. (2018). Green synthesis of metals and their oxide nanoparticles: Applications for environmental remediation. *Journal of Nanbiotechnology, 16*(84), 2018. https://doi.org/10.1186/s12951-018-0408-4

Singh, M. P., Bhardwaj, A. K., Bharati, K., Singh, R. P., Chaurasia, S. K., Kumar, S., Singh, R. P., Shukla, A., Naraian, R., & Vikram, K. (2021). Biogenic and non-biogenic waste utilization in the synthesis of 2D materials (graphene, h-BN, g-C_2N) and their applications. *Frontiers in Nanotechnology, 3*, 2021. https://doi.org/10.3389/fnano.2021.685427

Sonbol, H., Ameen, F., Alyahya, S., Almansob, A., & Alwakeel, S. (2021). Padina boryana mediated green synthesis of crystalline palladium nanoparticles as potential nanodrug against multidrug resistant bacteria and cancer cells. *Scientific Reports, 11*, 5444. https://doi.org/10.1038/s41598-021-84794-6

Srikar, S. K., Giri, D. D., Pal, D. B., Mishra, P. K., & Upadhyay, S. N. (2016). Green synthesis of silver nanoparticles: A review. *Green and Sustainable Chemistry, 6*(1), 34–56. https://doi.org/10.4236/gsc.2016.61004

Srivastava, R. (2012). Synthesis and characterization techniques of nanomaterials. *International Journal of Green Nanotechnology, 4*(1), 17–27. https://doi.org/10.1080/19430892.2012.654738

Subhashini, G., Ruban, P., & Daniel, T. (2018). Biosynthesis and characterization of magnetic (Fe_3O_4) iron oxide nanoparticles from a red seaweed *Gracilaria edulis* and its antimicrobial activity. *International Journal of Advanced Science and Research Manag, 3*(10), 184–189.

Sukri, S. N., Shameli, K., Wong, M. M. T., Teow, S. Y., Chew, J., & Ismail, N. F. (2019). Cytotoxicity and antibacterial activities of plant-mediated synthesized zinc oxide (ZnO) nanoparticles using *Punica granatum* (pomegranate) fruit peels extract. *Journal of Molecular Structure, 1189*, 57–65. https://doi.org/10.1016/j.molstruc.2019.04.026

Sultan, M., Waheed, A., Bibi, I., & Islam, A. (2019). Ecofriendly reduction of methylene blue with polyurethane catalyst. *International Journal of Polymeric Science, 2019*, 3168618. https://doi.org/10.1155/2019/3168618

Sundari, U. T., Rekha, S., & Parvathi, A. (2012). Phytochemical analysis of some therapeutic medicinal flowers. *International Journal of Pharmaceutics, 2*(3), 583–585.

Sunny, N. E., Mathew, S. S., Kumar, S. V., Saravanan, P., Rajeshkannan, R., Rajasimman, M., & Vasseghian, Y. (2022). Effect of green synthesized nano-titanium synthesized from *Trachyspermum ammi* extract on seed germination of Vigna radiate. *Chemosphere, 300*, 134600. https://doi.org/10.1016/j.chemosphere.2022.134600

Surya, S., Kumar, G. D., & Rajakumar, R. (2016). Green synthesis of silver nanoparticles from flower extract of *Hibiscus rosa-sinensis* and its antibacterial activity. *International Journal of Innovative Research in Science, Engineering and Technology, 5*(4), 5242–5247.

Tang, S., Wang, M., Germ, K. E., Du, H. M., Sun, W. J., Gao, W. M., & Mayer, G. D. (2015). Health implications of engineered nanoparticles in infants and children. *World Journal of Pediatrics, 11*(3), 197–206. https://doi.org/10.1007/s12519-015-0028-0

Timoszyk, A. (2018). A review of the biological synthesis of gold nanoparticles using fruit extracts: Scientific potential and application. *Bulletin of Materials Science, 41*(6), 154. https://doi.org/10.1007/s12034-018-1673-4

Tognacchini, A., Rosenkranz, T., Van der, E. A., Machinet, G. E., Echevarria, G., & Puschenreiter, M. (2020). Nickel phytomining from industrial wastes: Growing nickel hyperaccumulator plants on galvanic sludges. *Journal of Environmental Management, 254*, 109798. https://doi.org/10.1016/j.jenvman.2019.109798

Tuntiwachwuttikul, P., Rayanil, K., & Taylor, W. C. (2003). Chemical constituents from the flowers of *Nyctanthes arbortristis. Science Asia, 29*(1), 21–30. https://doi.org/10.2306/scienceasia1513-1874.2003.29.021

Varadavenkatesan, T., Selvaraj, R., & Vinayagam, R. (2019). Dye degradation and antibacterial activity of green synthesized silver nanoparticles using *Ipomoea digitata* Linn. flower extract. *International journal of Environmental Science and Technology, 16*(5), 2395–2404. https://doi.org/10.1007/s13762-018-1850-4

Variya, B. C., Bakrania, A. K., & Patel, S. S. (2016). *Emblica officinalis* (Amla): A review for its phytochemistry, ethnomedicinal uses and medicinal potentials with respect to molecular mechanisms. *Pharmacological Research, 111*, 180–200. https://doi.org/10.1016/j.phrs.2016.06.013

Veeramani, S., Narayanan, A. P., Yuvaraj, K., Sivaramakrishnan, R., Pugazhendhi, A., Rishivarathan, I., Jose, S. P., & Ilangovan, R. (2022). *Nigella sativa* flavonoids surface coated gold NPs (Au-NPs) enhancing antioxidant and anti-diabetic activity. *Process Biochemistry, 114*, 193–202. https://doi.org/10.1016/j.procbio.2021.01.004

Verma, N., & Kumar, N. (2019). Synthesis and biomedical applications of copper oxide nanoparticles: An expanding horizon. *ACS Biomaterials Science & Engineering, 5*(3), 1170–1188. https://doi.org/10.1021/acsbiomaterials.8b01092

Wang, M., & Thanou, M. (2010). Targeting nanoparticles to cancer. *Pharmacological Research, 62*(1), 90–99. https://doi.org/10.1016/j.phrs.2010.03.005

Wang, L., Hu, C., & Shao, L. (2017). The antimicrobial activity of nanoparticles: Present situation and prospects for the future. *International Journal of Nanomedicine, 12*(4), 1227–1249. https://doi.org/10.2147/IJN.S121956

Woo, Y., & Lai, D. Y. (2012). Aromatic amino and nitro-amino compounds and their halogenated derivatives. In B. Cohrssen & C. H. Powell (Eds.), *Patty's toxicology; Bingham E.* Wiley.

Xiong, Y., Huang, L., Mahmud, S., Yang, F., & Liu, H. (2020). Bio-synthesized palladium nanoparticles using alginate for catalytic degradation of azo dyes. *Chinese Journal of Chemical Engineering, 28*(5), 1334–1343. https://doi.org/10.1016/j.cjche.2020.02.014

Xu, Z. P., Zeng, Q. H., Lu, G. Q., & Yu, A. B. (2006). Inorganic nanoparticles as carriers for efficient cellular delivery. *Chemical Engineering Science, 61*(3), 1027–1040. https://doi.org/10.1016/j.ces.2005.06.019

Xu, J., Huang, Y., Zhu, S., Abbes, N., Jing, X., & Zhang, L. (2021). A review of the green synthesis of ZnO nanoparticles using plant extracts and their prospects for application in antibacterial textiles. *Journal of Engineered Fibers and Fabrics, 16*(2021), 1–14. https://doi.org/10.1177/15589250211046242

Yadid, M., Feiner, R., & Dvir, T. (2019). Gold nanoparticle-integrated scaffolds for tissue engineering and regenerative medicine. *Nano Letters, 19*(4), 2198–2206. https://doi.org/10.1021/acs.nanolett.9b00472

Yallappa, S., Manjanna, J., Sindhe, M. A., Satyanarayan, N. D., Pramod, S. N., & Nagaraja, K. (2013). Microwave assisted rapid synthesis and biological evaluation of stable copper nanoparticles using *T. Arjuna* bark extract. *Spectrochimica Acta. Part A, Molecular and Biomolecular Spectroscopy, 110*, 108–115. https://doi.org/10.1016/j.saa.2013.03.005

Yan, C., Wang, F., Liu, H., Liu, H., Pu, S., Lin, F., Geng, H., Ma, S., Zhang, Y., & Tian, Z. (2020). Deciphering the toxic effects of metals in gold mining area: Microbial community tolerance mechanism and change of antibiotic resistance genes. *Environmental Research, 189*, 109869. https://doi.org/10.1016/j.envres.2020.109869

Yang, Z., Liu, Z., Zhu, J., Xu, J., Pu, Y., & Bao, Y. (2022). Green synthesis and characterization of gold nanoparticles from Pholiota adiposa and their anticancer effects on hepatic carcinoma. *Drug Delivery, 29*(1), 997–1006. https://doi.org/10.1080/10717544.2022.2056664

Yu, J., Xu, D., Guan, H. N., Wang, C., Huang, L. K., & Chi, D. F. (2016). Facile one-step green synthesis of gold nanoparticles using *Citrus maxima* aqueous extracts and its catalytic activity. *Materials Letters, 166*(1), 110–112. https://doi.org/10.1016/j.matlet.2015.12.031

Yugandhar, P., Vasavi, T., Rao, Y. J., Devi, P. U. M., Narasimha, G., & Savithramma, N. (2018). Cost effective, green synthesis of copper oxide nanoparticles using fruit extract of *Syzygium alternifolium* (Wt.) Walp., characterization and evaluation of antiviral activity. *Journal of Cluster Science, 29*(4), 743–755. https://doi.org/10.1007/s10876-018-1395-1

Zafar, S., & Zafar, A. (2019). Biosynthesis and characterization of silver nanoparticles using *Phoenix dactylifera* fruits extract and there in vitro antimicrobial and cytotoxic effects. *Open Biotechnology Journal, 13*(1), 37–46. https://doi.org/10.2174/1874070701913010037

Zhang, X. F., Liu, Z. G., Shen, W., & Gurunathan, S. (2016). Silver nanoparticles: Synthesis, characterization, properties, applications, and therapeutic approaches. *International Journal of Molecular Sciences, 17*(9), 1534. https://doi.org/10.3390/ijms17091534

Recent Advancements in the Green Synthesis of Bioactive Metallic Nanoparticles from Biological Entities and Their Biomedical Applications

Parthasarathy Panchatcharam

1 Introduction

Nanotechnology, a promising branch of science, presents a vast array of possibilities in agriculture and medicine. Due to their potential for innovation in a variety of commercial sectors, research into metal nanoparticles (NPs) has drawn the attention of scientists increasingly. Nanoparticles are solid particles in a particulate dispersion that measure between 1 and 100 nm in at least one dimension, with unique characteristics and functions that differ from bulk metals. The significant surface area of metallic nanoparticles provides more opportunities for interaction with other molecules, which is a remarkable feature. As a result of these distinct and fascinating properties, nanoparticles have garnered significant attention in a diverse range of applications, including electronic devices, biochemical sensors, catalysts, tumour imaging, bioassays, pharmaceutical treatment, and drug delivery procedures (Koul et al., 2021). There are two main methods used for creating and stabilizing nanoparticles: the "bottom-up" and "top-down" approaches. The "top-down" method involves reducing bulk material into smaller particles using physical and chemical methods like milling, grinding, and thermal ablation. On the other hand, the "bottom-up" method involves the self-assembly of atoms into nuclei, which then develop into nanoscale particles using chemical and biological techniques. However, these physical methods demand a substantial amount of energy and expenditure, resulting in a lower yield of nanoscale materials (Ovais et al., 2018). Chemical techniques are used for the synthesis of nanoparticles, such as electrochemistry, chemical reduction, and photochemical reduction. Chemical processes have gradually gained popularity in recent years due to their capacity to create precise particles in terms of size and shape while utilizing less energy during the reduction process. However, the use

P. Panchatcharam (✉)
Department of Electronics and Communication Engineering, CMR Institute of Technology, Bengaluru, India

A. K. Bhardwaj et al. (eds.), *Biogenic Wastes-Enabled Nanomaterial Synthesis*,
https://doi.org/10.1007/978-3-031-59083-2_9

of dangerous chemicals like hydrazine or potassium bitartrate in these methods can pose environmental hazards, resulting in carcinogenicity, genotoxicity, and cytotoxicity, limiting their use in biomedical applications. As a result, researchers are now focusing on developing an environmentally friendly approach that can efficiently control the morphology, size, stability, and characteristics of nanoparticles. Microbe-mediated bio-mediated synthesis of nanoparticles is a potential substitute for chemical synthesis. Utilizing various biological entities, including bacteria, actinomycetes, fungi, algae, viruses, and yeast to produce nanoparticles is a reliable and eco-friendly method. This energy-efficient, cost-effective, and readily scalable "green" process creates particles with a variety of sizes, shapes, and compositional characteristics (Gahlawat & Choudhury, 2019). The main focus of this article is on recent advancements and developments in the biosynthesis of nanoparticles, despite the existence of prior reviews on the topic. The paper explores the underlying mechanisms involved in fabricating metal nanoparticles at the nanoscale, with a specific emphasis on the use of microbial agents for the biosynthesis of these particles. Furthermore, recent achievements in large-scale production are outlined, and the article concludes by providing some insight into the future of nanoparticle biosynthesis.

2 Synthesis of Nanoparticles

Nanoparticle synthesis can be produced economically, more sustainably, and without the use of harmful substances by using natural resources like plants and microbes. This strategy involves using reducing and stabilizing agents to generate stable nanoparticles, which has gained a lot of interest recently due to the high stability characteristics of the resulting nanoparticles. Various studies have successfully produced nanoparticles using plants and microorganisms. There are different green synthesis methods, but generally, they involve reacting metallic salts with organisms or their extracts and then using biological reduction to convert the metal into nanoparticles. After proper characterization, the nanoparticles can be readily used. The microbial synthesis of nanoparticles involves using fungi, bacteria, viruses, and their products to produce well-defined, structured nanoparticles. The biological synthesis of NPs using microbes has numerous advantages over chemical and physical methods. Microbial synthesis is a one-pot sustainable method that eliminates the use of harmful reducing agents and organic solvents. Microbial synthesis is also highly efficient, and the process can be scaled up to meet industrial production demands. Additionally, microbial synthesis allows for the synthesis of highly stable and uniform NPs that exhibit enhanced biological properties. The use of living organisms for NP synthesis has also gained significant interest due to their unique ability to synthesize NPs of diverse sizes, shapes, and compositions. However, the microbial synthesis of NPs has some limitations. The process can be affected by various factors such as the type of microbe used, growth conditions, and media composition. The purity of the microbial culture is also essential to obtain reproducible results. Moreover, the downstream processing of the synthesized NPs can be challenging and time-consuming. Thus, the microbial synthesis of NPs is a promising green technology that offers several benefits over conventional methods. With further development and optimization, microbial synthesis has the potential to become a viable alternative to chemical and physical methods for the production of NPs with various applications in biomedicine, catalysis, and energy production.

In nanoparticle synthesis, bacteria have commonly been utilized due to their remarkable capacity for rapid growth. This attribute makes them an attractive option for researchers who seek to produce nanoparticles in a cost-effective manner. By harnessing the unique properties of bacteria, scientists can create nanoparticles with a high degree of precision and control over their physical and chemical properties. Moreover, bacteria offer other benefits in nanoparticle synthesis, such as their ability to produce nanoparticles in large quantities and their compatibility with a range of materials. Moreover, the bacterial synthesis of nanoparticles is often regarded as a more environmentally friendly alternative to traditional chemical methods, which can produce hazardous byproducts. In addition to bacteria, other living sources such as actinobacteria, algae, yeast, fungi, and viruses have also been explored for the fabrication of nanoparticles (NPs). These living sources have unique properties that make them attractive for NP synthesis, such as their ability to survive in harsh environmental conditions, their adaptability to different substrates, and their diverse metabolic capabilities. Yeast, for example, is a eukaryotic microorganism that can produce extracellular enzymes capable of reducing metallic ions into NPs. Yeast-mediated synthesis of NPs has been demonstrated for the synthesis of gold, silver, copper, and iron oxide NPs. Fungi can secrete various enzymes, including reductases, that can reduce metallic ions into NPs. Fungal-mediated synthesis of NPs has been reported for the synthesis of palladium, silver, selenium, and gold NPs. Algae are photosynthetic microorganisms that are used for the synthesis of NPs; they can survive in diverse environments and have unique metabolic capabilities. Algae can absorb metal ions present in the media and reduce them into NPs using their metabolic enzymes. Viruses are yet another potential source of NP synthesis. Viruses are intracellular parasites that can only reproduce within host cells. Viruses can be used to synthesize NPs by using their capsid proteins to template the formation of NPs. Virus-mediated synthesis of NPs has been reported for the synthesis of gold, silver, platinum, and palladium NPs. Figure 1 depicts the process of synthesis of microbe-based nanoparticles. Overall, using living sources for NP synthesis has several benefits over conventional physical and chemical methods, including environmental friendliness, low cost, and the ability to create NPs with distinct morphologies and properties. The type of metal ion, the intended size and shape of the NPs, and the scalability of the synthesis process all influence the choice of living source for NP synthesis (Ali et al., 2020).

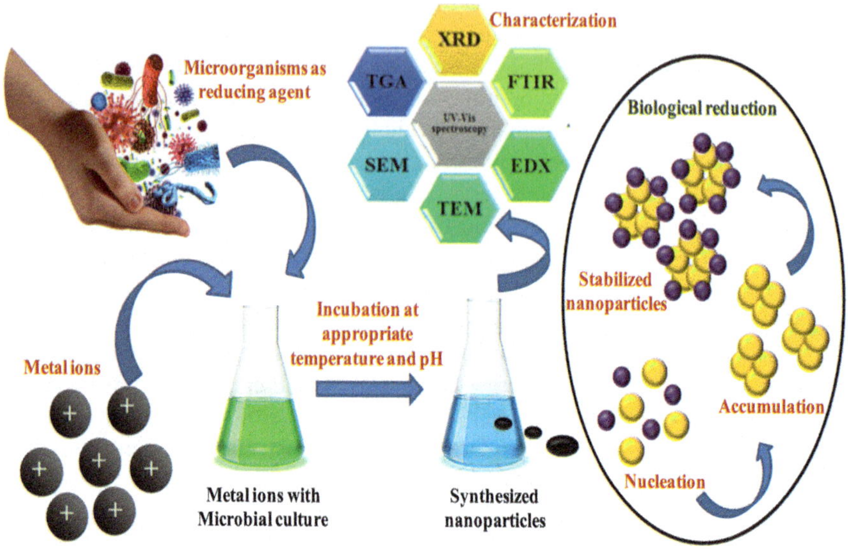

Fig. 1 Flow for the synthesis of biometallic nanoparticles using microorganisms

2.1 Silver Nanoparticles

Silver nanoparticles (AgNPs) have been used in various industries, resulting in an increase in their production. There are various methods available for synthesizing AgNPs, and several microorganisms have been employed for this purpose. An example is the use of the CO strain of *Pseudomonas poae*, which was isolated from garlic leaves to synthesize AgNPs. The resulting nanoparticles ranged in size from 19.8 to 44.9 nm and showed high stability and anti-fungal properties against the *Fusariumgraminearum* pathogen (Ibrahim et al., 2020).

Another research involved the use of *Bacillus siamensis* strain C1, which is an endophytic bacterium extracted from *Coriandrum sativum* to produce AgNPs with a size range of 25–50 nm. These nanoparticles demonstrated strong antibacterial properties against rice pathogens like bacterial brown stripe and bacterial leaf blight. Additionally, synthesized AgNPs showed inhibitory effects on biofilm formation, bacterial growth, and the swimming motility of Ao strain RS-1 and Xoo strain LND0005 (Ibrahim et al., 2019). In another experiment, AgNPs were successfully produced using the strain SZT1 of *Bacillus cereus*, which was identified by gene sequence analysis. The resultant nanoparticles were spherical in shape and had a size range of 18–39 nm. These nanoparticles also demonstrated chemical disease control benefits for rice plants (Ahmed et al., 2020).

Biosynthesized AgNPs from *Penicillium duclauxii, a seed-borne fungus, were prepared to analyze anti-microbial effect against Bipolaris sorghicola, which is responsible for the target leaf spot disease on sorghum plants. The synthesized AgNPs exhibited approximately 54.2 nm range of size with good anti-fungal properties* (Almaary et al., 2020). The silver nanoparticles were also synthesized using various marine algal extracts as capping and reducing agents. It includes *Cystoseira myrica* (brown algae), *Gracilaria foliifera* (red algae), and *Ulva rigida* (green algae) species. It was observed that the developed Ag nanoparticles had nanoparticles in the range of 17 nm, 24 nm, and 12 nm with excellent anti-cancer properties against normal skin and breast cancer cell lines. In addition to these properties, it also exhibited better anti-fungal and antibacterial properties against pathogenic dermatophyte moulds and food-borne bacteria (Algotiml et al., 2022). Using the SNPGA-8 strain of *Streptomyces hirsutus,* AgNPs were synthesized. It showed excellent antibacterial activity against *Alternaria alternata, Candida albicans, Fusarium oxysporum, Candida glabrata, Enterococcus faecalis, Pseudomonas aeruginosa, Staphylococcus aureus*, and *Escherichia coli*. It also exhibited anti-tumour activity in the MTT cytotoxicity assay (Pallavi et al., 2022).

AgNPs of rod and spherical shapes were synthesized using *Bacillus, Brevundimonas*, and *Rhodococcus* bacterial strains extracted from a marine ciliate *Euplotes focardii*. These nanoparticles ranged in size from 20 to 50 nm and exhibited potent anti-microbial activity against nasal pathogens (John et al., 2022). Similarly, AgNPs were produced using the microalgal species *Sargassum tenerrimum*, ranging from 13 to 46 nm in size. These nanoparticles demonstrated a significant impact on pathogenic bacteria and showed promise in treating various fatal diseases (Solanki & Patel, 2023). *Sphingobium* sp. was used for the biological synthesis of silver nanoparticles (AgNPs) ranging in size from 7 to 22 nm. These nanoparticles exhibited excellent antibacterial activity against *S. aureus* and *E. coli*, with a minimal inhibitory concentration (MIC) of 50 lg/mL and 6.25 lg/mL, respectively (Akter & Huq, 2020).

The study utilized the fungal species *Fusarium oxysporum* to synthesize spherical silver nanoparticles ranging in size from 1 to 50 nm. In silico studies were conducted to identify the antibacterial activity of these nanoparticles against microorganisms such as *E. coli* and *P. aeruginosa*, and metal docking was carried out using SYBYL X 1.1.1 software to identify the binding area of the nanoparticles in the bacterial outer membrane for their antibacterial activity. Furthermore, in vitro studies demonstrated the nanoparticles' significant antibacterial potential against the selected pathogens (Srivastava et al., 2019). *Aeromonas schubertii*, an epiphytic bacterium, and *Shewanella chilikensis*, an endophytic bacterium, were isolated from the seaweed *Stoechospermum marginatum* using imprint and serial dilution methods for the production of AgNPs. The nanoparticles developed from the strains of the seaweed were analyzed for biological activities like biofilm production, hemolytic activity, angiogenesis, and biofilm inhibition assay. The results showed promising results to the analysis, especially in the development of new blood vessels in the chicken embryo (Pradeep et al., 2023). The anti-microbial effectiveness of biosynthesized silver nanoparticles (AgNPs) using a yeast strain, *Saccharomyces cerevisiae*, was investigated against various selected pathogens. The spherical nanoparticles ranged in size from 3 to 60 nm and demonstrated excellent anti-microbial action against *P. aeruginosa, E. coli, C. albicans, B. subtilis, and S. aureus*. The minimum inhibitory concentration was found to be 25 μg/mL for *B. subtilis*,

E. coli, *P. aeruginosa*, and *C. albicans* and 12.5 μg/mL for *S. aureus* (Nagarajan & Panchatcharam, 2023a).

Actinomycetes strains were recovered from mangrove sediment (Canv1-58) and soil from the Brazilian Caatinga biome (Caat5-35) and identified by sequencing the 16S rRNA gene. These strains were used to create AgNPs with typical sizes of 30 and 60 nm, respectively, and were referred to as AgNPs 35 and AgNPs 58. As phytopathogens like *Colletotrichum gloeosporioides*, *Alternaria solani*, and *Alternaria alternata* were tested for AgNPs' anti-fungal effects, they showed good inhibition of these pathogens (Parthasarathy & Vivekanandan, 2020). Additionally, secondary metabolites were extracted from the actinomycetes strains *Microbacterium proteolyticum* LA2(R) and *S. rochei* LA2(O) to synthesize AgNPs, which were tested against meningitis-causing bacteria. The minimal inhibitory concentration (MIC) test demonstrated the effectiveness of these nanoparticles against *Neisseria meningitidis*, *Streptococcus pneumoniae*, and *Haemophilus influenzae*, with RAgNPs showing MICs of 138.80 μg/mL, 109.4 μg/mL, and 120.60 μg/mL, respectively, and OAgNPs showing MICs of 129.06 μg/mL, 105.80 μg/mL, and 114.40 μg/mL, respectively (Bano et al., 2023).

2.2 Gold Nanoparticles

Gold nanoparticles (AuNPs) possess distinctive optical and biological features and are extensively used across various fields. In a recent investigation, marine fungal strains, which were identified by sequencing the 16S rRNA gene, were utilized to synthesize gold nanoparticles (AuNPs) as well as AgNPs. The synthesis of silver nanoparticles was accomplished using *Aspergillus* sp., while *Alternaria* sp. was employed for the development of biogenic AuNPs. Chitosan was also used to generate chitosan-AuNP and chitosan-AgNP conjugates, which displayed robust antimicrobial activity against both gram-positive and negative bacteria such as *E. coli* and *S. aureus*. The findings indicate that the conjugates of silver nanoparticles exhibited greater antibacterial activity than those of gold nanoparticles. Furthermore, the conjugates prepared were tested against biofilm-forming bacteria, such as *P. aeruginosa*, *S. aureus*, *E. coli*, and *B. subtilis*. Similar to the antibacterial effect of the strains, AuNP conjugates showed minimal activities on the stains than AgNPs (Mostafa et al., 2022). AuNPs were developed using bacteria from hyper-alkaline spring (pH 11). Through the sequencing study, the isolated bacteria were found to be phylogenetically related to *Pseudomonas stutzeri* and *Lysinibacillus* sp. Biosynthesized AuNPs using the isolated bacteria showed an irregular spherical shape with an excellent inhibitory effect on the pyocyanin production of the *Pseudomonas aeruginosa* PA01 (San Diego et al., 2018).

The utilization of *Penicillium rubens* cell-free filtrate has enabled the synthesis of AuNPs, which displayed an average size of 14.92 ± 5 nm. These nanoparticles have demonstrated notable anti-microbial effects against various microbial species, including *Penicillium verrucosum* MTCC 4935, *B. subtilis* NCIM-2010, and *E. coli* NCIM-2065, as well as an inhibitory effect on the ochratoxin A (OTA) biosynthesis. Moreover, these nanoparticles possess the potential to degrade organic pollutants such as bromothymol blue, methylene blue, phenol red, 4-nitrophenol, and Congo red (Parthasarathy & Vivekanandan, 2018). Biosynthesized AuNPs by extracellular metabolites from the marine bacterial species *Panna microdon*, *Selachimorpha* sp., and *Rastrelliger kanagurta* have been tested for antibacterial and anti-mycobacterial activities. The synthesized AuNPs from the bacterial strains showed good anti-microbial activities. Along with these, it has shown good biocompatibility and less toxicity when tested using the zebrafish larvae as an experimental organism (Rajasekar et al., 2020). AuNPs synthesized using cyanobacteria *Spirulina platensis* and *Oscillatoria* sp. were used for the demonstration of anti-microbial properties. The developed AuNPs were found to be in numerous shapes, such as spherical, triangular, and pentagonal dimensions, with a mean size of 15.49–55.08 nm. It exhibited an excellent minimal inhibitory effect against the bacterial species *Bacillus subtilis* ATCC 19659, *Candida tropicalis* ATCC 1380, and *Salmonella typhi* ATCC14028 with a concentration of about 1.95 μg/ml, 1.1 μg/ml, 3.90 μg/ml, respectively (Mathew et al., 2021).

2.3 Copper Nanoparticles

Due to their exceptional physical and chemical characteristics, copper nanoparticles (CuNPs) are used extensively in numerous fields. In recent research, CuNPs have been generated using bacterial strains that were isolated from an Antarctic consortium. A significant amount of copper oxide nanoparticles (CuONPs) with an average size of 30 nm were produced after the bacterial culture was incubated with a copper ion solution. Against pathogenic microorganisms like *C. albicans*, *S. aureus*, *and E. coli*, these CuONPs showed strong antibacterial action (John et al., 2021). Another investigation utilized copper-resistant bacteria *Pseudomonas fluorescens* MAL2 cell-free supernatant to synthesize copper nanoparticles. The bacteria were isolated from metal-contaminated soil in Egypt, and under optimized conditions of 35 °C and pH 7, the copper sulphate solution was added to the bacterial supernatant for the development of CuNPs. The formed CuNPs displayed toxicity against the pest *Tribolium castaneum*, with a 37 ppm LC50 after 5 days of treatment (El-Saadony et al., 2020).

Aspergillus niger strain STA9 was used under optimal circumstances for the fungus-based synthesis of copper nanoparticles. The produced nanoparticles, with an average size of 500 nm, significantly inhibited human hepatocellular carcinoma cell lines (IC50 value of 3.09 g/ml). Additionally, by inhibiting alpha-glucosidase and in agar well diffusion tests against *Klebsiella pneumoniae*, *S. aureus*, *B. subtilis*, *Micrococcus luteus*, and *E. coli*, these nanoparticles demonstrated anti-diabetic and antibacterial action (Noor et al., 2020). Copper and silver nanoparticles were synthesized using the fungus *Schizophyllum commune* and used for the analysis of anti-microbial activity. The nanoparticles found to be spherically shaped possess excellent inhibition against the microbial species *Salmonella abony*, *Staphylococcus aureus*, *Klebsiella pneumoniae*, and *Escherichia coli* (Fatima & Wahid, 2022).

In a study conducted in southern India, various biological agents, including actinomycetes, fungi, and bacteria, were isolated from tea soils and used for the biosynthesis of nanoparticles. Copper and silica nanoparticles were developed using *Streptomyces griseus*, *Pseudomonas fluorescens*, and *Trichoderma atroviride.* Effective growth inhibition was achieved by the nanoparticles against *Pestalotiopsis theae* and *Poria hypolateritia*, the main causative agents of red root rot in tea plants. Further, these nanoparticles were formulated and tested for toxicity using zebrafish as a model organism with no toxic effects observed (Natesan et al., 2021). The fungal species *Fusarium oxysporum* was used for the synthesis of copper nanoparticles via incubation with copper acetate solution. The developed nanoparticles showed effective anti-microbial activity against *P. aeruginosa*, *E. coli*, *K. pneumoniae*, and *S. aureus* (Nagarajan & Panchatcharam, 2023b). In another study, copper nanoparticles were biosynthesized from the cell-free supernatant of *Aspergillus niger* and were tested for their antibacterial activity in wound healing against *Staphylococcus aureus*. The nanoparticles with 20 nm in size showed significant inhibition against the bacterium when used alone compared to their combination with ciprofloxacin (Suood et al., 2021).

The study evaluated the effect of biosynthesized copper nanoparticles on fluconazole-resistant *Candida albicans*. The nanoparticles' minimum inhibitory concentration (MIC) and minimum bactericidal concentration (MBC) against *C. albicans* were found to be 2.5 mg/ml and 5 mg/ml, respectively, with a tolerance level of 2. The anti-fungal activity of the nanoparticles was tested using absorbance measurements and the well-plate method, which demonstrated that the nanoparticles inhibited the growth of both *C. albicans* and the fluconazole-resistant strain. Additionally, the nanoparticles were found to prevent *C. albicans* from forming biofilms. The results suggest that copper nanoparticles could be used as an alternative to traditional anti-fungal medications (Rasool et al., 2019). The study found that the best method for creating CuONPs was by using the cell-free supernatant of marine *Streptomyces* sp. with 5 mM of $CuSO_4$. After 1 h of incubation, well-defined CuONPs were formed. The CuONPs showed anti-microbial activity against various microorganisms, including *Enterococcus faecalis* ATCC 29212, *Salmonella typhimurium* ATCC 14028, *P. aeruginosa* ATCC 9027, *E. coli* ATCC 8939, yeast (*C. albicans* ATCC 10237), and fungi (*Rhizoctonia solani*, *Aspergillus niger*, and *Fusarium solani*), with the highest activity against *Candida albicans*. The CuONP groups also showed lower levels of liver enzymes and higher levels of antioxidants in comparison to the paracetamol group, as revealed by biochemical results. Additionally, histopathological examination demonstrated the potential of CuONPs in improving liver damage caused by paracetamol (Nagarajan & Panchatcharam, 2023c). The potential utilization of fungi to synthesize copper nanoparticles (CuNPs) by utilizing reducing agents such as proteins and enzymes was investigated. The fungus *Aspergillus flavus* was used to produce small, spherical particles of CuNPs with a diameter ranging from 2 to 60 nm. The reduction of copper ions occurred on the mycelia surface due to the action of extracellularly secreted enzymes (Nagarajan & Panchatcharam, 2023d). The study utilized an endophytic fungus, *Aspergillus terreus*, obtained from the medicinal tree *Aegle marmelosa* in the Indian Western Ghats, to create CuONPs. Among the 16 different endophytic fungal strains isolated, FCBY1 demonstrated the highest antioxidant and antagonistic activities. The CuONPs created from this strain showed reactive oxygen species (ROS)-scavenging, anti-fungal, and antibacterial properties against clinical pathogens at low concentrations. Additionally, the study evaluated the angiogenesis inhibition in tumour cells using in vivo HET-CAM testing and assessed the anti-cancer activity of CuONPs on colon cancer cell lines (HT-29) using MTT and FACS analyses. The results demonstrated that the CuONPs inhibited blood vessel growth by 31.36% and 81.81% at the highest concentration (60 L) and showed concentration-dependent anti-cancer activities. The findings suggest that CuONPs produced by endophytic fungi, particularly *Aspergillus terreus*, may have promising potential as cancer therapeutics (Mani et al., 2021). Copper oxide nanoparticles (CuONP) were synthesized using *Purpureocillium lilacinum*, yielding spherical nanoparticles with sizes varying from 4.03 nm to 8.83 nm. A comparative study was done to evaluate the photocatalytic degradation of safranin and naval blue using CuONP. The results showed that CuONPs had possible catalytic activity, with respective degradation rates for navy blue and safranin of 57.5% and 63%, respectively (Hammad et al., 2022).

The significant inhibition of biofilm formation by the CuNPs synthesized through actinomycetes-mediated biosynthesis is a promising result as biofilm formation can be a challenge in various industries, including medicine and agriculture. The observed anti-microbial activity and the potential for catalytic activity of wound healing suggest that the CuNPs synthesized using actinomycete isolates from soil samples could have a vast range of applications in numerous fields (Navashree & Parthasarathy, 2023). A biological pathway was utilized to examine the creation of CuONPs, which were lessened and stabilized by marine endophytic actinomycetes. The results indicated that the created CuONPs were spherical, and pure, with an average size of 20 nm. At a concentration of 750 mg/mL, the CuONPs displayed effective antibacterial properties against *P. mirabilis* and *E. coli*, both of which generate biofilms. The efficacy of removing biofilms was verified using the 24-well polystyrene method with a 24-h time frame, revealing a 92% and 94% reduction against *P. mirabilis* and *E. coli* that form biofilms, respectively. Furthermore, CuONPs exhibited a 54% inhibitory range in a cytotoxicity test against A549 lung cancer cells at 500 mg/mL concentration, suggesting their potential as a cancer treatment (Zhao et al., 2022).

2.4 Zinc Nanoparticles

Due to their remarkable thermal, electrical, photocatalytic, and optical properties, zinc nanoparticles have been extensively synthesized and utilized in various fields. Using the biomass filtrate of *P. aeruginosa*, isolated from rhizosphere sediment, ZnONPs were biosynthesized. During the characterization process, the spherical, crystalline ZnONPs exhibited surface plasmon resonance (SPR) with a maximum value of 380 nm. ZnONP bioactivities, such as their anti-microbial, anti-candidal, and larvicidal properties, were also additionally investigated in the study. The biosynthesized ZnONPs demonstrated concentration-dependent activity against pathogenic bacteria and fungi, including *C. albicans*, *E. coli*, *B. subtilis*, *S. aureus*, and *P. aeruginosa.* ZnONPs had inhibition zones ranging from 11.7 ± 0.3 mm to 29.3 ± 0.3 mm at 200 ppm against the tested pathogens, with MIC values of 50 ppm for *C. albicans*, *B. subtilis*, and *E. coli* and 200 ppm for *P. aeruginosa* and *S. aureus*. In addition, the research discovered that *Culex pipiens* had a high mortality rate in the biosynthesized ZnONPs, with percentages of 100 ± 0.0% at 200 ppm after 24 h (Abdo et al., 2021). ZnONPs were developed using a culture supernatant of *Acinetobacter schindleri* SIZ7, isolated from the Sivakasi waste-filling region. The anti-microbial efficacy of the synthesized ZnONPs against four food-borne pathogens was evaluated against *Escherichia coli* (MTCC 739), *Staphylococcus aureus* (MTCC 96), *Salmonella enterica* (MTCC 9844), and *Vibrio parahaemolyticus* (MTCC 451). The results showed that the MIC of 100 g ml^{-1} demonstrated that the ZnONPs had potent antibacterial activity against *E. coli* and *S. enterica* (Busi et al., 2016).

A study found that *Streptomyces* sp. isolated from soil in India could synthesize ZnONPs that demonstrated a significant concentration of biofilm inhibition (BIC) against multidrug-resistant *K. pneumoniae*. The ZnONPs had a BIC of 200 g/mL, with a 92% hydrophobicity index and 88% biofilm degradation potential. This research suggests that actinomycetes have the potential to synthesize ZnONPs for biomedical applications such as antibacterials and anti-biofilms against pathogenic microbes (Rajivgandhi et al., 2022). In another study, a bacterial strain from agricultural soil in Saudi Arabia, identified as *Bacillus foraminis*, was used to produce ZnONPs. The synthesis of ZnONPs was confirmed by the formation of a white precipitate at the end of the incubation period, and the particles were characterized as polydisperse, spherical, and ranging in size from 5.40 to 6.79 nm. The ZnONPs have potential applications in various fields due to their antibacterial properties and environmentally friendly nature (El-Ghwas, 2022). The *Bacillus cereus* RNT6 native strain was utilized to produce ZnONPs to combat bacterial panicle blight (BPB) caused by *Burkholderia glumae* and *B. gladioli*. The ZnONPs demonstrated significant antibacterial activity with 2.83 cm and 2.18 cm zone of inhibition, respectively, at 50 g/mL concentration. Additionally, the cell counts of the two pathogens in the broth culture were reduced by 71.2% and 68.1%, respectively. Ultrastructure studies on both cells treated with ZnONPs showed morphological damage compared to the control group. The ZnONPs were characterized by their functional groups, crystalline structure, and spherical shape, with sizes ranging from 21 to 35 nm (Ahmed et al., 2021).

2.5 Other Metal Nanoparticles

Due to their exceptional physical and chemical characteristics, numerous other nanoparticles have been applied in a variety of disciplines. In a recent study, stable tin oxide nanoparticles (SnO_2NPs) were produced using an ethanolic extract of red algae *Galaxaura elongata*, and their cytotoxic and antibacterial effects were investigated. The biosynthesized SnO_2NPs had a typical size of 35 nm with nanocrystallite structures. The zone of inhibition against bacterial strains (*E. coli*, *S. aureus*, *B. subtilis*, and *P. aeruginosa*) ranged from 16 to 24 mm when the NPs'antibacterial activity was tested. Moreover, the extract exhibited a modest MIC value of 0.31–9.47 g/mL against the tested organisms, compared to 15.6 g/mL and 62.5 g/mL for tin oxide nanoparticles. The SnO_2NPs were also found to be cytotoxic against A549 human lung cancer cells, with 28.08 g/mL IC50 (Al-Enazi et al., 2021).

In another study, the potential antibacterial activity of selenium nanoparticles (SeNPs) produced through mycosynthesis using *Penicillium chrysogenum* PTCC 5031 was investigated. The SeNPs had an average hydrodynamic size of 24.65 nm, a PdI of 0.392, and a zeta potential of −34 mV at pH 7, indicating their uniformity and good monodispersity. The findings revealed that the SeNPs exhibited antibacterial activity against gram-positive bacteria such as *Listeria monocytogenes* and *S. aureus*, with a zone of inhibition (ZOI) of 13 and 10 mm, respectively (Nagarajan & Panchatcharam, 2023e).

The study investigated the potential use of *Agaricus bisporus* in creating palladium nanoparticles (PdNPs) for biological purposes. The biosynthesis process resulted in the formation of triangular and spherical PdNPs ranging in size from 13 to 18 nm, with a surface charge of −24.3 mV. The PdNPs exhibited strong antioxidant activity against 1,1-diphenyl-2-picrylhydrazyl (DPPH) free radicals, with a maximum radical scavenging activity of 77% observed at 50 g/mL. Antibacterial assays demonstrated that the PdNPs had a greater inhibitory effect on *S. pyrogens*, *B. subtilis*, and *S. aureus*, compared to *Proteus vulgaris*, *K. pneumoniae*, and *Enterobacter aerogenes*. The PdNPs also exhibited strong cytotoxicity against PK13 cell lines and anti-inflammatory activity, with a maximum growth inhibition of 79% observed at the highest dose (50 g/mL) and an IC50 value of 26.1 g/mL. These findings suggest that PdNPs have the potential to be used in biomedical applications, including as antioxidants, antibacterials, and anti-inflammatory agents (Parthasarathy & Vivekanandan, 2019).

In vitro and in vivo experiments were conducted to evaluate the effectiveness of SeNPs against *R. solani*. SeNPs were produced by *Bacillus megaterium*, and their impact on plant growth was assessed, specifically on the yield, morphological, and metabolic markers of *Vicia faba* seeds. SeNPs displayed strong anti-fungal properties against *R. solani*, with minimum inhibitory and minimum fungicidal concentrations of 0.0625 and 1 mM, respectively. Additionally, SeNPs reduced the severity of root rot disease and the pre-post emergence of *R. solani* in plants (Hashem et al., 2021). SeNPs were biosynthesized using *Streptomyces* sp., which was isolated from magnesite soil in Tamil Nadu, India. The SeNPs displayed antibacterial, synergistic, larvicidal, and anthelminthic activity, indicating their potential for a wide range of biological applications (Ramya et al., 2020). The extracellular synthesis of SeNPs was carried out in an alternate study using the probiotic bacteria *Lactobacillus acidophilus*. The SeNPs had a zeta potential of +37.86 mV, a hydrodynamic radius of 34.13 nm, a surface plasmon resonance at 385 nm, and a typical diameter of 2–15 nm. With minimum inhibitory concentrations (MIC90) varying from 1 to 10 g/ml, the synthesized SeNPs demonstrated antibacterial activity against distinct bacterial strains. Additionally, the SeNPs showed their capacity to prevent and destroy bacterial biofilms. Human HEK-293 normal cell lines were used to test the SeNPs' cytocompatibility, and the MTT assay findings revealed that the SeNPs were cytocompatible (Parthasarathy, 2018).

In order to biosynthesize iron oxide nanoparticles (IONPs), *Bacillus coagulans* bacterium supernatant was used as a bio-reducing and stabilizing agent. The IONPs produced had tiny, average particle sizes of 15.2 nm and showed fine agglomeration. Using the good diffusion technique, the IONPs' antibacterial and anti-biofilm activity was evaluated against uropathogenic *E. coli* bacteria isolated from patients with urinary tract infections (UTIs). The IONPs showed potent anti-biofilm activity at a minimum inhibitory concentration (MIC) of 150 g/ml in both Congo red and tube method tests. This activity was concentration-dependent and strongly antibacterial for the IONPs (Al-Maliki & Taj-Aldeen, 2021).

An endophytic bacterium from *Terminalia chebula* is used to synthesize zirconia nanoparticles with potential applications in dental restorative materials. The nanoparticles were found to exhibit anti-inflammatory, antibacterial, antioxidant, and anti-biofilm activities against oral pathogens. The biocompatibility of the nanoparticles was also tested on human primary gingival fibroblast cell lines (Balaji & Shanmugam, 2021).

3 Factors That Influence the Structure and Morphology of Nanoparticles

Microorganism-based nanoparticle synthesis is gaining popularity because it is beneficial to the environment. Numerous variables influence the properties of the nanoparticles created by microbial synthesis. The size, form, and stability of the nanoparticles can be affected by the microorganism strain used in the synthesis. The reduction of metal ions and the consequent formation of nanoparticles can also be influenced by the metabolic pathways and enzymes of various microbial species. The size, shape, and stability of nanoparticles can be affected by the pH level of the reaction mixture, with alkaline conditions favouring bigger particles and acidic conditions smaller particles. The temperature can have an impact on nanoparticles as well because it can encourage the nucleation and aggregation of particles, which can result in bigger sizes. Additionally, the concentration of metal ions in the reaction mixture can affect the size, shape, and stability of the nanoparticles, with high concentrations leading to larger particles and low concentrations leading to smaller particles. The length of time taken for synthesis is also an important factor, with longer reaction times leading to larger particles due to continuous nucleation and growth. Finally, the availability of nutrients such as carbon and nitrogen sources can impact nanoparticle synthesis as excess nutrients can promote cell growth and affect nanoparticle synthesis (Patil & Kim, 2018).

4 Application of Bioactive Metallic Nanoparticles

The exceptional characteristics and potential uses of bioactive metallic nanoparticles that are synthesized by microorganisms have made them highly sought-after. Microorganisms, including algae, actinomycetes fungi and bacteria, produce these nanoparticles using a process called biomineralization. In biomineralization, microorganisms employ a range of enzymes and biomolecules to convert metallic ions found in their environment into nanoparticles that are subsequently coated and stabilized with organic molecules. Various applications of biosynthesized nanoparticles have been shown in Fig. 2.

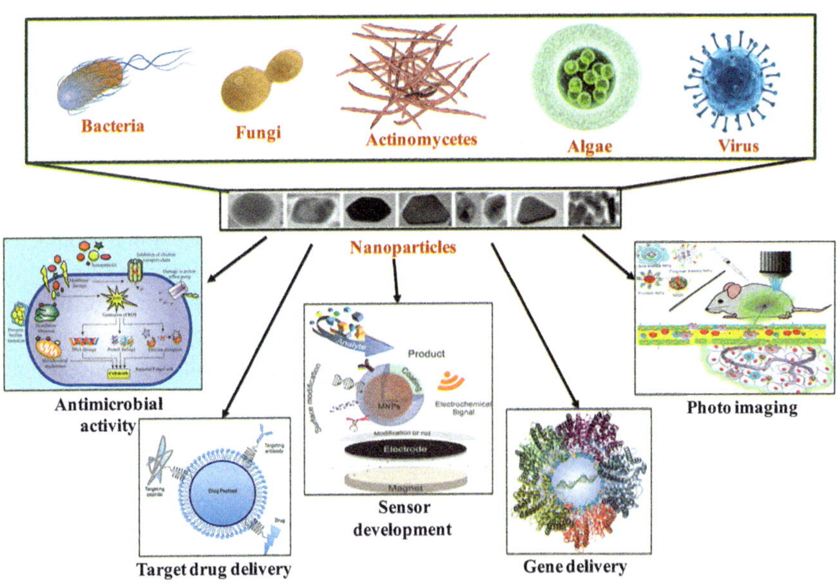

Fig. 2 Applications of biosynthesized nanoparticles

One of the most promising areas of application for bioactive metallic nanoparticles synthesized by microorganisms is in medicine. In particular, silver nanoparticles have been extensively studied for their anti-microbial properties. They can be used as a disinfectant for wound healing and can also be incorporated into medical devices such as catheters to prevent bacterial infections. AuNPs have also been investigated for use in cancer therapy as they have been shown to have potential as a drug delivery system and can also enhance the effectiveness of radiation therapy. In 2021, researchers from the Indian Institute of Technology Bombay developed a new technique for synthesizing silver nanoparticles using a bacterium called *Bacillus* sp. The resulting nanoparticles were found to have potent anti-viral activity and to be antibacterial and could be used as an effective disinfectant for medical applications (Jain et al., 2021).

In agriculture, bioactive metallic nanoparticles synthesized by microorganisms have potential applications in the development of nanofertilizers. For example, iron nanoparticles synthesized by bacteria can be used as a slow-release fertilizer, improving plant growth and crop yield. Other applications in agriculture include the use of nanoparticles for seed treatment and crop protection. In 2020, researchers from the University of California, Riverside, developed a method for synthesizing iron nanoparticles using a fungus called *Trichoderma reesei*. The resulting nanoparticles were found to be an effective fertilizer for tomato plants, improving plant growth and increasing crop yield (Parthasarathy, 2023).

Bioactive metallic nanoparticles synthesized by microorganisms can also be used for environmental remediation. Nanoparticles of zero-valent iron (nZVI) synthesized by bacteria can be used to remove heavy metals from contaminated water and soil. Other applications in environmental remediation include the use of nanoparticles for the degradation of pollutants and the treatment of wastewater. In 2022, researchers from Van Yüzüncü Yıl University developed a method for synthesizing palladium nanoparticles using a bacterium called *Shewanella oneidensis*. The resulting nanoparticles were found to be effective and could be used for environmental remediation (Seckin et al., 2022).

Finally, bioactive metallic nanoparticles synthesized by microorganisms have potential applications in various industrial sectors. For example, copper nanoparticles synthesized by bacteria can be used in the production of conductive inks and coatings. Other applications include the use of nanoparticles for catalysis and the production of nanocomposites. In 2020, researchers from the Indian Institute of Technology Hyderabad developed a method for synthesizing copper nanoparticles using a bacterium called *Pseudomonas aeruginosa*. The resulting nanoparticles were found to be an effective anti-microbial agent and could be used in the production of medical devices and other industrial applications (Hamad et al., 2023).

Consequently, the use of bioactive metallic nanoparticles produced by microorganisms is a fascinating field of study with numerous potential applications across multiple disciplines. These nanoparticles have the potential to become more crucial in industry, environmental remediation, agriculture, and health with more research and development. It is possible to anticipate that as research in this field advances, even more cutting-edge and intriguing applications will be discovered.

5 Conclusion

The application of microbial cells to produce nanoparticles (NPs) in a short time is beneficial, but there are challenges in achieving consistent size, shape, composition, and symmetry due to environmental factors. Microbial synthesis has been potentially used in biomedicine and cancer treatment; further research is needed to modify and scale up the methods for commercial production and to investigate the toxicity and biocompatibility of NPs on various organs and systems. Furthermore, the potential effect of metallic nanoparticles on agriculture and the environment must be investigated before they are used in plant disease management. So in this review, the application of various metallic nanoparticles in biomedicine and agriculture, especially in plant disease management, has been discussed. Further studies on metallic NP application in health and bioengineering will help understand the promising risks for humans and the environment.

References

Abdo, A. M., Fouda, A., Eid, A. M., Fahmy, N. M., Elsayed, A. M., Khalil, A. M. A., Alzahrani, O. M., Ahmed, A. F., & Soliman, A. M. (2021). Green synthesis of Zinc Oxide Nanoparticles (ZnO-NPs) by Pseudomonas aeruginosa and their activity against pathogenic microbes and common house mosquito, Culex pipiens. *Materials, 14*(22), 6983.

Ahmed, T., Shahid, M., Noman, M., Niazi, M. B. K., Mahmood, F., Manzoor, I., Zhang, Y., Li, B., Yang, Y., Yan, C., & Chen, J. (2020). Silver nanoparticles synthesized by using Bacillus cereus SZT1 ameliorated the damage of bacterial leaf blight pathogen in rice. *Pathogens, 9*(3), 160.

Ahmed, T., Wu, Z., Jiang, H., Luo, J., Noman, M., Shahid, M., Manzoor, I., Allemailem, K. S., Alrumaihi, F., & Li, B. (2021). Bioinspired green synthesis of zinc oxide nanoparticles from a native Bacillus cereus strain RNT6: Characterization and antibacterial activity against rice panicle blight pathogens Burkholderia glumae and B. gladioli. *Nanomaterials, 11*(4), 884.

Akter, S., & Huq, M. A. (2020). Biologically rapid synthesis of silver nanoparticles by Sphingobium sp. MAH-11T and their antibacterial activity and mechanisms investigation against drug-resistant pathogenic microbes. *Artificial Cells, Nanomedicine, and Biotechnology, 48*(1), 672–682.

Al-Enazi, N. M., Ameen, F., Alsamhary, K., Dawoud, T., Al-Khattaf, F., & AlNadhari, S. (2021). Tin oxide nanoparticles (SnO2-NPs) synthesis using Galaxaura elongata and its anti-microbial and cytotoxicity study: A greenery approach. *Applied Nanoscience, 13*, 1–9.

Algotiml, R., Gab-Alla, A., Seoudi, R., Abulreesh, H. H., El-Readi, M. Z., & Elbanna, K. (2022). Anticancer and antimicrobial activity of biosynthesized Red Sea marine algal silver nanoparticles. *Scientific Reports, 12*(1), 2421.

Ali, M. A., Ahmed, T., Wu, W., Hossain, A., Hafeez, R., Islam Masum, M. M., Wang, Y., An, Q., Sun, G., & Li, B. (2020). Advancements in plant and microbe-based synthesis of metallic nanoparticles and their antimicrobial activity against plant pathogens. *Nanomaterials, 10*(6), 1146.

Almaary, K. S., Sayed, S. R., Abd-Elkader, O. H., Dawoud, T. M., El Orabi, N. F., & Elgorban, A. M. (2020). Complete green synthesis of silver-nanoparticles applying seed-borne Penicillium duclauxii. *Saudi Journal of Biological Sciences, 27*(5), 1333–1339.

Al-Maliki, Q. A., & Taj-Aldeen, W. R. (2021). Antibacterial and antibiofilm activity of bacteria mediated synthesized Fe3O4 nanoparticles using bacillus Coagulans. *Journal of Nanostructures, 11*(4), 782–789.

Balaji, S., & Shanmugam, V. K. (2021). Enhanced antibiofilm activity of endophytic bacteria mediated zirconium nanoparticles. *Biointerface Research in Applied Chemistry, 12*(1), 74–82.

Bano, N., Iqbal, D., Al Othaim, A., Kamal, M., Albadrani, H. M., Algehainy, N. A., Alyenbaawi, H., Alghofaili, F., Amir, M., & Roohi. (2023). Antibacterial efficacy of synthesized silver nanoparticles of Microbacterium proteolyticum LA2 (R) and Streptomyces rochei LA2 (O) against biofilm forming meningitis causing microbes. *Scientific Reports, 13*(1), 4150.

Busi, S., Rajkumari, J., Pattnaik, S., Parasuraman, P., & Hnamte, S. (2016). Extracellular synthesis of zinc oxide nanoparticles using Acinetobacter schindleri SIZ7 and its antimicrobial property against foodborne pathogens. *Journal of Microbiology, Biotechnology and Food Sciences, 5*(5), 407.

El-Ghwas, D. E. (2022). Characterization and biological synthesis of zinc oxide nanoparticles by new strain of Bacillus foraminis. *Biodiversitas Journal of Biological Diversity, 23*(1). https://doi.org/10.13057/biodiv/d230159

El-Saadony, M. T., Abd El-Hack, M. E., Taha, A. E., Fouda, M. M., Ajarem, J. S., Maodaa, N. S., Allam, A. A., & Elshaer, N. (2020). Ecofriendly synthesis and insecticidal application of copper nanoparticles against the storage pest Tribolium castaneum. *Nanomaterials, 10*(3), 587.

Fatima, F., & Wahid, I. (2022). Eco-friendly synthesis of silver and copper nanoparticles by Shizophyllum commune fungus and its biomedical applications. *International journal of Environmental Science and Technology, 19*(8), 7915–7926.

Gahlawat, G., & Choudhury, A. R. (2019). A review on the biosynthesis of metal and metal salt nanoparticles by microbes. *RSC Advances, 9*(23), 12944–12967.

Hamad, A. M., Atiyea, Q. M., Hameed, D. N. A., & Dalaf, A. H. (2023). Green synthesis of copper nanoparticles using strawberry leaves and study of properties, anti-cancer action, and activity against bacteria isolated from Covid-19 patients. *Karbala International Journal of Modern Science, 9*(1), 12.

Hammad, E. N., Salem, S. S., Zohair, M. M., Mohamed, A. A., & El-Dougdoug, W. (2022). Purpureocillium lilacinum mediated biosynthesis copper oxide nanoparticles with promising removal of dyes. *Biointerface Research in Applied Chemistry, 12*(2), 1397–1404.

Hashem, A. H., Abdelaziz, A. M., Askar, A. A., Fouda, H. M., Khalil, A. M., Abd-Elsalam, K. A., & Khaleil, M. M. (2021). Bacillus megaterium-mediated synthesis of selenium nanoparticles and their antifungal activity against Rhizoctonia solani in faba bean plants. *Journal of Fungi, 7*(3), 195.

Ibrahim, E., Fouad, H., Zhang, M., Zhang, Y., Qiu, W., Yan, C., Li, B., Mo, J., & Chen, J. (2019). Biosynthesis of silver nanoparticles using endophytic bacteria and their role in inhibition of rice pathogenic bacteria and plant growth promotion. *RSC Advances, 9*(50), 29293–29299.

Ibrahim, E., Zhang, M., Zhang, Y., Hossain, A., Qiu, W., Chen, Y., Wang, Y., Wu, W., Sun, G., & Li, B. (2020). Green-synthesization of silver nanoparticles using endophytic bacteria isolated from garlic and its antifungal activity against wheat Fusarium head blight pathogen Fusarium graminearum. *Nanomaterials, 10*(2), 219.

Jain, A. S., Pawar, P. S., Sarkar, A., Junnuthula, V., & Dyawanapelly, S. (2021). Bionanofactories for green synthesis of silver nanoparticles: Toward antimicrobial applications. *International Journal of Molecular Sciences, 22*(21), 11993.

John, M. S., Nagoth, J. A., Zannotti, M., Giovannetti, R., Mancini, A., Ramasamy, K. P., Miceli, C., & Pucciarelli, S. (2021). Biogenic synthesis of copper nanoparticles using bacterial strains isolated from an antarctic consortium associated to a psychrophilic marine ciliate: Characterization and potential application as antimicrobial agents. *Marine Drugs, 19*(5), 263.

John, M. S., Nagoth, J. A., Ramasamy, K. P., Mancini, A., Giuli, G., Miceli, C., & Pucciarelli, S. (2022). Synthesis of bioactive silver nanoparticles using new bacterial strains from an antarctic consortium. *Marine Drugs, 20*(9), 558.

Koul, B., Poonia, A. K., Yadav, D., & Jin, J. O. (2021). Microbe-mediated biosynthesis of nanoparticles: Applications and future prospects. *Biomolecules, 11*(6), 886.

Mani, V. M., Kalaivani, S., Sabarathinam, S., Vasuki, M., Soundari, A. J. P. G., Das, M. A., Elfasakhany, A., & Pugazhendhi, A. (2021). Copper oxide nanoparticles synthesized from an endophytic fungus Aspergillus terreus: Bioactivity and anti-cancer evaluations. *Environmental Research, 201*, 111502.

Mathew, A. A., Parthasarathy, P., & Vivekanandan, S. (2021). Development of copper nanoparticles from E-waste for biomedical applications. In *Advances in automation, signal processing, instrumentation, and control: Select proceedings of i-CASIC 2020* (pp. 703–715). Springer Singapore.

Mostafa, E. M., Abdelgawad, M. A., Musa, A., Alotaibi, N. H., Elkomy, M. H., Ghoneim, M. M., Badawy, M. S. E., Taha, M. N., Hassan, H. M., & Hamed, A. A. (2022). Chitosan silver and gold nanoparticle formation using endophytic fungi as powerful antimicrobial and anti-biofilm potentialities. *Antibiotics, 11*(5), 668.

Nagarajan, N., & Panchatcharam, P. (2023a). Biosensor nanoarchitectonics of Cu–Fe-nanoparticles/Zeolite-A/Graphene nanocomposite for enhanced electrooxidation and dopamine detection. *Heliyon, 9*(9), e19741.

Nagarajan, N., & Panchatcharam, P. (2023b). Fabrication of polyvinyl alcohol doped CuO thin films for improved Amperometric-non enzymatic hydrogen peroxide sensing. *Journal of Biomedical Photonics & Engineering, 9*(3), 030309.

Nagarajan, N., & Panchatcharam, P. (2023c). Cost-effective and eco-friendly copper recovery from waste printed circuit boards using organic chemical leaching. *Heliyon, 9*(3), e13806.

Nagarajan, N., & Panchatcharam, P. (2023d, February). Development of highly sensitive and stable electrochemical sensor for H_2 O_2 detection by facile synthesized copper nanoparticle thin film. In *2023 fifth International Conference on Electrical, Computer and Communication Technologies (ICECCT)* (pp. 1–9). IEEE.

Nagarajan, N., & Panchatcharam, P. (2023e, February). A review on functionalized biosensors and its mechanism behind the clinical diagnosis of dopamine. In *2023 fifth International Conference on Electrical, Computer and Communication Technologies (ICECCT)* (pp. 1–7). IEEE.

Natesan, K., Ponmurugan, P., Gnanamangai, B. M., Manigandan, V., Joy, S. P. J., Jayakumar, C., & Amsaveni, G. (2021). Biosynthesis of silica and copper nanoparticles from Trichoderma, Streptomyces and Pseudomonas spp. evaluated against collar canker and red root-rot disease of tea plants. *Archives of Phytopathology and Plant Protection, 54*(1–2), 56–85.

Navashree, N., & Parthasarathy, P. (2023). A comparative study on electrochemical behaviour of various electrolytes by cyclic voltammetry: GCE as electrode material. *Materials Today: Proceedings*. https://doi.org/10.1016/j.matpr.2023.05.175

Noor, S., Shah, Z., Javed, A., Ali, A., Hussain, S. B., Zafar, S., Ali, H., & Muhammad, S. A. (2020). A fungal based synthesis method for copper nanoparticles with the determination of anticancer, antidiabetic and antibacterial activities. *Journal of Microbiological Methods, 174*, 105966.

Ovais, M., Khalil, A. T., Ayaz, M., Ahmad, I., Nethi, S. K., & Mukherjee, S. (2018). Biosynthesis of metal nanoparticles via microbial enzymes: A mechanistic approach. *International Journal of Molecular Sciences, 19*(12), 4100.

Pallavi, S. S., Rudayni, H. A., Bepari, A., Niazi, S. K., & Nayaka, S. (2022). Green synthesis of Silver nanoparticles using Streptomyces hirsutus strain SNPGA-8 and their characterization, antimicrobial activity, and anticancer activity against human lung carcinoma cell line A549. *Saudi Journal of Biological Sciences, 29*(1), 228–238.

Parthasarathy, P. (2018, September). Synthesis and UV detection characteristics of TiO2 thin film prepared through sol gel route. In *IOP conference series: Materials science and engineering* (Vol. 360, No. 1, p. 012056). IOP Publishing.

Parthasarathy, P. (2023). Electrochemical detection of Uropathogenic Escherichia Coli using PVA/R-GO/PEI modified nanocomposite electrode. *Engineered Science, 24*, 903.

Parthasarathy, P., & Vivekanandan, S. (2018). A comprehensive review on thin film-based nano-biosensor for uric acid determination: Arthritis diagnosis. *World Review of Science, Technology and Sustainable Development, 14*(1), 52–71.

Parthasarathy, P., & Vivekanandan, S. (2019, March). Structural, optical and electrochemical response studies of TiO2–ZrO2 nanocomposite for uric acid detection. In *2019 Innovations in Power and Advanced Computing Technologies (i-PACT)* (Vol. 1, pp. 1–6). IEEE.

Parthasarathy, P., & Vivekanandan, S. (2020). Biocompatible TiO2-CeO2 nano-composite synthesis, characterization and analysis on electrochemical performance for uric acid determination. *Ain Shams Engineering Journal, 11*(3), 777–785.

Patil, M. P., & Kim, G. D. (2018). Marine microorganisms for synthesis of metallic nanoparticles and their biomedical applications. *Colloids and Surfaces B: Biointerfaces, 172*, 487–495.

Pradeep, M. A., Janani, M. S., & Kaviraj, M. (2023). Characterization of Silver nanoparticles synthesized by epiphytic and endophytic bacteria isolated from Stoechospermum marginatum and evaluation of its angiogenic property.

Rajasekar, T., Karthika, K., Muralitharan, G., Maryshamya, A., Sabarika, S., Anbarasu, S., Revathy, K., Prasannabalaji, N., & Kumaran, S. (2020). Green synthesis of gold nanoparticles using extracellular metabolites of fish gut microbes and their antimicrobial properties. *Brazilian Journal of Microbiology, 51*, 957–967.

Rajivgandhi, G., Gnanamangai, B. M., Prabha, T. H., Poornima, S., Maruthupandy, M., Alharbi, N. S., Kadaikunnan, S., & Li, W. J. (2022). Biosynthesized zinc oxide nanoparticles (ZnO NPs) using actinomycetes enhance the anti-bacterial efficacy against K. Pneumoniae. *Journal of King Saud University-Science, 34*(1), 101731.

Ramya, S., Shanmugasundaram, T., & Balagurunathan, R. (2020). Actinobacterial enzyme mediated synthesis of selenium nanoparticles for antibacterial, mosquito larvicidal and anthelminthic applications. *Particulate Science and Technology, 38*(1), 63–72.

Rasool, U., Sah, S. K., & Hemalatha, S. (2019). Effect of biosynthesized copper nanoparticles (CUNPS) on the growth and biofilm formation of fluconazole-resistant Candida albicans. *The Journal of Microbiology, Biotechnology and Food Sciences, 9*(1), 21.

San Diego, K. D., Alindayu, J. I. A., & Baculi, R. Q. (2018). Biosynthesis of gold nanoparticles by bacteria from hyperalkaline spring and evaluation of their inhibitory activity against pyocyanin production. *Journal of Microbiology, Biotechnology & Food Sciences, 8*(2), 781–787.

Seckin, H., Tiri, R. N. E., Meydan, I., Aygun, A., Gunduz, M. K., & Sen, F. (2022). An environmental approach for the photodegradation of toxic pollutants from wastewater using Pt–Pd nanoparticles: Antioxidant, antibacterial and lipid peroxidation inhibition applications. *Environmental Research, 208*, 112708.

Solanki, A. D., & Patel, I. C. (2023). Sargassum tenerrimum-mediated green synthesis of silver nanoparticles along with antimicrobial activity. *Applied Nanoscience, 13*(6), 4415–4425.

Srivastava, S., Bhargava, A., Pathak, N., & Srivastava, P. (2019). Production, characterization and antibacterial activity of silver nanoparticles produced by Fusarium oxysporum and monitoring of protein-ligand interaction through in-silico approaches. *Microbial Pathogenesis, 129*, 136–145.

Suood, A. M., Saleh, M. K., & Thalij, K. M. (2021, November). Synthesis of copper nanoparticles using aspergillus Niger and their efficacy against pathogenic Staphylococcus aureus. In *IOP conference series: Earth and environmental science* (Vol. 910, No. 1, p. 012083). IOP Publishing.

Zhao, H., Maruthupandy, M., Al-mekhlafi, F. A., Chackaravarthi, G., Ramachandran, G., & Chelliah, C. K. (2022). Biological synthesis of copper oxide nanoparticles using marine endophytic actinomycetes and evaluation of biofilm producing bacteria and A549 lung cancer cells. *Journal of King Saud University-Science, 34*(3), 101866.

Green Synthesis of Algal Nanoparticles: Harnessing Nature's Biofactories for Sustainable Nanomaterials

Shailendra Kumar Singh, Abhijeet Sharma, Lokesh Sharma, and Shanthy Sundaram

1 Introduction

In recent decades, there has been a noticeable surge in the development of nanotechnology using inorganic nanoparticles and nanomaterials in various industries. These advancements hold great promise in fields like targeted drug delivery, bio- and chemical sensing, catalysis, antimicrobial coatings, and optoelectronic devices (Panwar et al., 2019). However, concerns have arisen regarding the sustainability of traditional methods for synthesizing nanomaterials, which often involve energy-intensive processes, high temperatures, toxic solvents, or the creation of harmful chemical by-products (Clarke et al., 2018). To address these concerns, scientists worldwide have been exploring environmentally friendly and sustainable alternatives to chemical synthesis. One such approach that has garnered significant attention is the use of living organisms, their components, extracts, or biomolecules as catalysts for the eco-friendly production of nanomaterials, commonly known as the "biosynthesis of nanomaterials."

Within this context, algae, photosynthetic organisms capable of converting sunlight and carbon dioxide into energy, have emerged as particularly promising candidates for nanoparticle synthesis due to their unique biochemical properties (Dahoumane et al., 2017). By harnessing algae as live cell factories or utilizing algal extracts as natural reducing agents, it is possible to achieve a green and clean method for synthesizing various metallic nanomaterials. The vast diversity of algal strains offers opportunities for applications in biomaterials and green chemistry. These organisms can be harnessed through various methods, leading to the production of

S. K. Singh
Department of Biochemistry, S. S. Khanna Girls' Degree College, Prayagraj, Uttar Pradesh, India

A. Sharma · L. Sharma · S. Sundaram (✉)
Centre of Biotechnology, University of Allahabad, Prayagraj, Uttar Pradesh, India

A. K. Bhardwaj et al. (eds.), *Biogenic Wastes-Enabled Nanomaterial Synthesis*,
https://doi.org/10.1007/978-3-031-59083-2_10

nanomaterials composed of noble metals, oxides, and chalcogenides (Wang et al., 2021). The wide range of sizes and shapes attainable with these nanomaterials further enhances their versatility and the associated benefits they offer. The recent interest in the synthesis of metal nanoparticles using algae can be attributed to their simplicity and cost-effectiveness in terms of mass production and downstream processing (Dahoumane et al., 2017; Mishra et al., 2020). Moreover, their use significantly reduces the reliance on hazardous chemicals, underscoring their environmental friendliness. This chapter aims to spotlight the accomplishments and current status of metallic nanoparticle biosynthesis mediated by algae, as well as their applications. The article provides valuable insights into improving the production yield of algae-based nanomaterials and explores their potential for environmental applications.

2 Nanomaterials: Unveiling Synthesis Routes for Sustainable Fabrication

Although the term "nanomaterials" emerged in the late nineteenth century, the production and use of such materials can be traced back to ancient civilizations. Nanomaterials are defined as materials that possess one or more peripheral nanoscale dimensions (range 1–100 nm) (Borm et al., 2006). The nanomaterials can be synthesized using two prominent approaches depicted in Fig. 1: top-down and bottom-up approaches (Abed et al., 2022).

1. *Top-down fabrication*: In the top-down approach, bulk materials are subjected to physical or chemical modifications to obtain nanoscale structures. The common techniques employed in this approach include ball milling and lithography. These methods involve mechanically machining the bulk materials to reduce them into fine particles with nanodimensions. However, it is important to note that top-down methods do not offer precise control over the particle size and

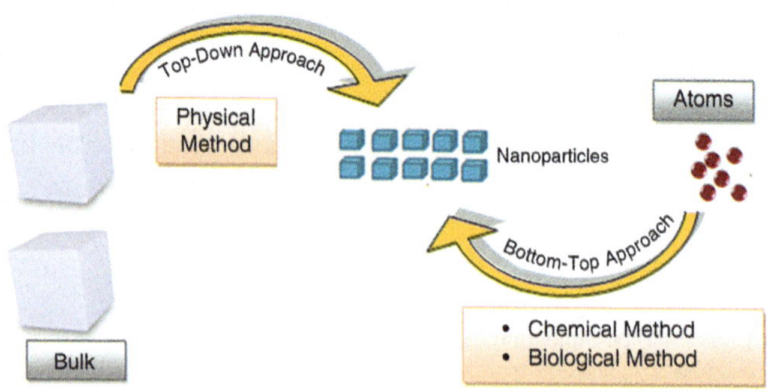

Fig. 1 Mechanism of approaches involved in nanoparticle synthesis

structure. The resulting nanoparticles may exhibit variations in size and structure due to the limitations of the mechanical processes involved (Tiwari et al., 2008).
2. *Bottom-up assembly*: The bottom-up method involves the construction of nanoscale structures by assembling atoms, molecules, or fine particles either through self-assembly or coprecipitation methods. Scientists commonly rely on this approach for the synthesis of nanoparticles as it allows for the gradual building of a material from the bottom up (Belusso et al., 2019). This method encompasses various techniques, including atom-by-atom, molecule-by-molecule, and cluster-by-cluster assembly. Bottom-up methods, such as chemical vapor deposition (CVD), sol-gel, spinning, pyrolysis, and biological synthesis, are extensively utilized in nanoparticle synthesis. These methods offer advantages over top-down approaches as they provide greater control over particle size, shape, and composition. The ability to precisely manipulate the assembly process from the atomic or molecular level allows for the creation of well-defined nanomaterials with desired properties. Consequently, bottom-up methods have gained popularity among researchers in the field of nanoparticle synthesis (Sau & Rogach, 2012).

Both approaches are further categorized into physical, chemical, and biological techniques for synthesizing nanoparticles (Fig. 2):

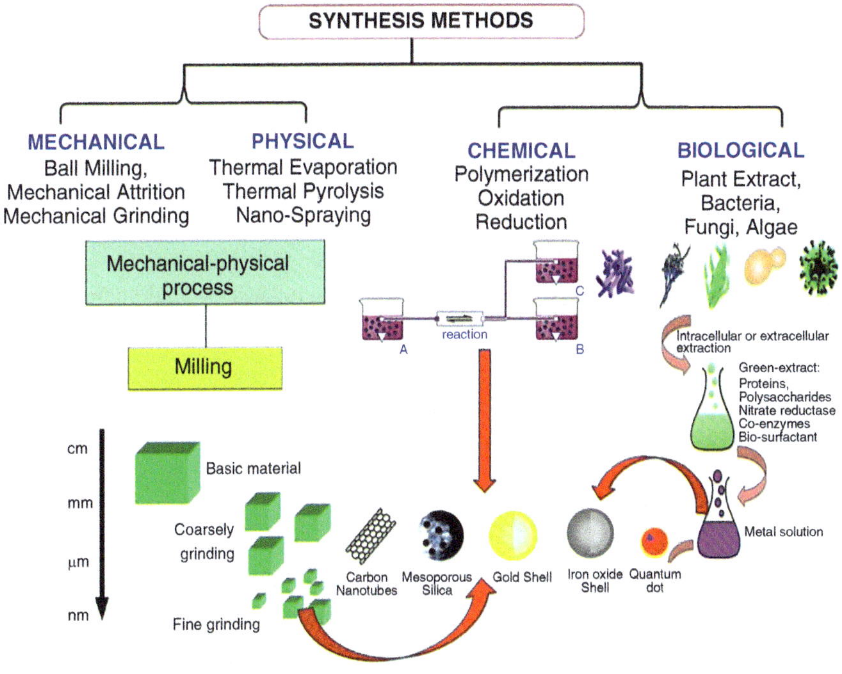

Fig. 2 Represents the nanoparticle synthesis methods and their events

1. *Chemical synthesis*: this involves controlled oxidation-reduction, precipitation hydrolysis, thermolysis, polymerization, and condensation of precursor chemicals to form nanoparticles. It often utilizes chemical reagents, solvents, and specific reaction conditions (Hasan, 2015).
2. *Physical synthesis*: physical methods include techniques to produce nanostructures, like milling, evaporation, vapor deposition, sputtering, or laser ablation, where nanomaterials are formed through physical processes, such as condensation or fragmentation (Iravani et al., 2014).
3. *Biological synthesis*: also known as biosynthesis, this approach utilizes biological agents such as microorganisms, plants, or enzymes to synthesize nanomaterials. It offers eco-friendly and sustainable alternatives to chemical methods (Singh et al., 2016a, b).

The choice of synthesis method depends on various factors, including the desired properties of the nanomaterial, the scalability of the process, cost-effectiveness, and the intended application. Different synthesis methods offer different levels of control over particle size, shape, composition, and dispersion (Ealia & Saravanakumar, 2017). Researchers must consider these factors and select an appropriate synthesis method accordingly to achieve the desired characteristics of the nanoparticles. In order to achieve the desired characteristics of nanoparticles, precise control over the processing conditions is necessary. The resulting nanoparticles should exhibit the following characteristics (Rane et al., 2018):

(i) *Identical size of all particles*: the synthesis method should ensure that the nanoparticles have a uniform size distribution, with minimal variation in particle size.
(ii) *Identical shape*: the nanoparticles should possess a consistent shape, whether it is spherical, rod-like, or any other desired morphology.
(iii) *Identical chemical composition and crystal structure*: the synthesis process should yield nanoparticles with a consistent chemical composition and crystal structure throughout the material.
(iv) *Individually dispersed with no agglomeration*: the nanoparticles should be well-dispersed, with no tendency to form aggregates or agglomerates. This ensures their stability and prevents undesired changes in their properties.

3 Bio-Based Green Approach for Nanomaterial Synthesis

In recent years, there has been increasing concern about the sustainability of nanomaterial synthesis using physicochemical methods. Both top-down and bottom-up approaches often involve energy-intensive processes, high temperatures, the use of toxic solvents, and the generation of undesirable toxic by-products. These factors have raised environmental and human health risks associated with conventional synthesis methods. As a result, there is a growing focus on developing sustainable and environmentally friendly approaches for the production of nanostructures. The aim is to minimize or eliminate the use of toxic substances and reduce the overall impact on the environment and human health (Hussain et al., 2016). Within this context, biological routes have gained significant attention and have made notable contributions to the field of nanotechnology. The biological green methods of nanomaterials involve living organisms or utilizing materials derived from biological sources, such as plants, microorganisms, and various biowastes like vegetable waste, fruit peel waste, eggshells, and agricultural waste. Bioactive compounds in biological materials, such as carbonyl groups, terpenoids, phenolics, flavonones, amines, amides, proteins, pigments, alkaloids, and other reducing agents, can facilitate the in vivo or in vitro bio-based synthesis of metallic nanoparticles. This process leads to the production of metallic nanoparticles in the size range of 1 to 100 nm (Ying et al., 2022). By harnessing the capabilities of biological systems, researchers are able to achieve the synthesis of nanomaterials using minimal or negligible toxic substances, thereby minimizing the impact on the environment and human health. These environmentally friendly methods contribute to the advancement of nanotechnology while addressing the sustainability concerns associated with conventional physicochemical synthesis approaches (Nadaroglu et al., 2017).

Biological methods offer several advantages as they do not employ harmful chemicals that can harm the environment. Consequently, these methods are considered more efficient, simpler, and economically viable for nanoparticle synthesis. Additionally, they can be easily scaled up for larger operations (Parveen et al., 2016). Among biological materials, microorganisms such as bacteria, algae, and fungi, as well as their extracts, have been successfully employed as biocatalysts for nanomaterial synthesis. These microorganisms possess cost-effective and environmentally benign characteristics, making them ideal for nanoparticle production. However, this chapter specifically focuses on the methods involved in synthesizing nanoparticles using algae and highlights their applications (Shukla & Iravani, 2018).

4 Algae-Mediated Nanoparticle Synthesis: Sustainable Approach for Nanoparticle Production

Algae are an attractive medium for the processing of diverse nanomaterials, owing to the inclusion of bioactive compounds in their cell extracts, such as pigments and antioxidants that serve as biocompatible reductants. The biosynthesis of nanoparticles using algae is a simple process that involves the following steps:

(i) *Selection of algae*: algae is a term that encompasses a diverse range of organisms, comprising over 30,000 species. These species can be harnessed for nanoparticle biosynthesis, including both microalgae and macroalgae. Various families of algae, such as Cyanophyceae, Chlorophyceae, Phaeophyceae, and Rhodophyceae, have been utilized for the intracellular and extracellular synthesis of nanoparticles (NPs) (Khan et al., 2022). Popular examples include *Chlorella*, *Spirulina*, *Dunaliella*, and diatoms. The selection of algae for nanoparticle synthesis depends on factors such as the desired properties of the nanoparticles and the availability of the specific algae species.

(ii) *Isolation and cultivation*: the chosen algae are isolated and cultivated under controlled conditions, providing them with the necessary nutrients and optimal growth conditions. This step ensures sufficient biomass for nanoparticle production.

(iii) *Harvesting and extraction*: the algae biomass is harvested, typically through centrifugation or filtration, to separate it from the culture medium. The biomass is then processed to extract the intracellular or extracellular components that serve as the precursors for nanoparticle synthesis (Mukherjee et al., 2021).

(iv) *Nanoparticle synthesis*: algal components, such as proteins, pigments, polysaccharides, or lipids, can be extracted and utilized for nanoparticle synthesis through various physical, chemical, or biological methods. There are two main types of nanoparticles that can be synthesized using algae: organic nanoparticles and inorganic nanoparticles (Alprol et al., 2023). Organic nanoparticles are formed using compounds such as poly-ε-lysine, chitosan, cationic quaternary polyelectrolytes, quaternary ammonium compounds, and other similar materials. These organic nanoparticles have the advantage of being biocompatible and can be used in applications such as drug delivery systems. However, they are generally less stable at high temperatures compared to inorganic nanoparticles.

    Inorganic nanoparticles, on the other hand, are synthesized using salts of metals such as silver (Ag), gold (Au), platinum (Pt), copper (Cu), zinc (Zn), titanium (Ti), magnesium (Mg), and others. These inorganic nanoparticles often exhibit excellent stability at high temperatures and are particularly useful as antimicrobial agents due to their inherent properties. Inorganic nanoparticles can also possess unique optical, electrical, and catalytic properties, making them valuable in a wide range of applications.

    The choice between organic and inorganic nanoparticles depends on the specific requirements of the application and the desired properties of the nanoparticles. Inorganic nanoparticles are typically favored for their stability at high temperatures and antimicrobial properties, while organic nanoparticles may be preferred for their biocompatibility and suitability for certain drug delivery systems. Some algae-mediated nanoparticles have been summarized in Table 1.

(v) *Characterization and purification*: the synthesized nanoparticles are characterized to determine their size, shape, composition, stability, and other physicochemical properties. Purification techniques, such as centrifugation or filtration, are employed to remove any impurities or unreacted components.

Table 1 Types of algae-mediated NPs with their conditions

Algae	NPs	Synthesis conditions	Shape and size	Reducing agent	References
Red algae					
Pterocladia capillacea	Cu	Incubation at 70 °C for 3 h	Spherical, 62 nm	Extracellular polysaccharide	Aboeita et al. (2022)
Portieria hornemannii	Ag	Incubation on an orbital shaker at 150 rpm under 25 °C for 48 h	Spherical, 70–75 nm	Extracellular compounds	Fatima et al. (2020)
Gelidium corneum	Ag	Incubation at 100 °C for 20 min	Spherical, 20–50 nm	Extracellular compounds	Öztürk et al. (2020)
Gracilaria edulis	Ag	Incubation on an orbital shaker at 150 rpm under 40 °C	Spherical, 12.5–100 nm	Proteins and terpenoids	Pugazhendhi et al. (2018)
Galaxaura elongata	Au	Incubation for 10 min to 12 h at the speed of 120 RPM	Rod, truncated and triangular shaped, particle size from 3.85 nm to 77.13 nm	Palmitic acid	Abdel-Raouf et al. (2017)
Gracilaria sp.	Ag-Au	Overnight incubation at 60 °C, pH 6	Crystalline	Functional groups	Al-Naamani et al. (2017)
Brown algae					
Spatoglossum asperum	Fe3O4	Incubation at 25 °C for 24 h	Spherical, 16 nm	Extracellular compounds	Palaniyandi et al. (2023)
Cystoseira myrica	Ag	Incubation at 75 °C for 30 min	Spherical, 8 to 15 nm	Extracellular compounds	Rajeshkumar et al. (2022)
Cystoseira crinita	MgO	Incubation at 25 °C for 24 h	Spherical, 3 to 8 nm	Extracellular compounds	Fouda et al. (2022)
Cystoseira baccata	Au	24 h stirring at room temperature	Poly-crystalline nature with spherical shape, 8.4 nm ± 2.2 nm	Polyphenols and polysaccharides	González-Ballesteros et al. (2017)
Ecklonia cava	Au	1 min incubation at 80 °C on magnetic stirrer	Spherical, 20–50 nm	Hydroxyl and phenolic groups	Venkatesan et al. (2014)

(continued)

Table 1 (continued)

Algae	NPs	Synthesis conditions	Shape and size	Reducing agent	References
Sargassum muticum	ZnO	Reaction temperature at 450 °C for 4 h	Hexagonal, 30 nm to 57 nm	Sulfate and hydroxyl groups	Azizi et al. (2014)
Turbinaria conoides	Au	24 h incubation	Poly-dispersed, rectangular, triangular, 6–10 nm	Carboxylic, fucoidan, and polyphenolic groups	Rajeshkumar et al. (2013)
Padina pavonica	Ag	24 h incubation at room temperature	Poly-dispersed and spherical, 10–72 nm	Terpenoids	Sahayaraj et al. (2012)
Fucus vesiculosus	Au	pH 2 to 9, incubation period 1 h to 8 h	Spherical, 20 nm to 50 nm	Hydroxyl groups present in polysaccharides	Mata et al. (2009)
Sargassum spp.	Au	pH 7, 120 min incubation at 25 °C	Wide edges, hexagon and triangle shape, 300 nm–400 nm	Hydroxyl, amine functional groups	Liu et al. (2005)
Blue green algae					
Oscillatoria sp.	AgO	10.49 and 45.81 nm	30 min at 50 °C	Extracellular compounds	El-Sheekh et al. (2021)
Spirulina platensis	Au	Triangular, pentagonal, and slightly spherical shapes, 15.49 to 55.08 nm	30 min at 50 °C	Extracellular compounds	El-Sheekh et al. (2021)
Chlamydomonas reinhardtii	CdSNPs	–	Incubation at 65 °C, centrifugation	Extracellular proteins	Rao and Pennathur (2017)
Nostoc ellipsosporum	Au	Decahedral and icosahedron, 20–40 nm	3 h incubation, pH 5	Proteins and carboxylate groups	Parial et al. (2016)
Spirulina platensis	Au	Octahedral and cubic	2 day incubation at 25 °C, centrifugation at 10 k rpm	Proteins and peptides	Suganya et al. (2015)
Lyngbya majuscula	Au	Hexagonal and spherical shape, 2 nm to 25 nm	Incubation at 25 °C and centrifugation at 10,000 rpm	Proteins	Parial and Pal (2014)

(continued)

Table 1 (continued)

Algae	NPs	Synthesis conditions	Shape and size	Reducing agent	References
Aphanothece sp. and *Oscillatoria* sp.	Ag	Spherical, 44–79 nm	–	Bioactive compounds	Sudha et al. (2013)
Spirulina platensis	Si	Crystalline	24 h of incubation at 25 °C, pH 7	–	Mahdieh et al. (2012)
Spirulina platensis	Au	Spherical, 20–30 nm	40 h incubation	Biomolecules (amino, carboxylic, phosphate, thiol)	Kalabegishvili et al. (2012)
Oscillatoria willei	Ag	Spherical, 10–25 nm	24 h incubation	Tryptophan	Ali et al. (2011)
Plectonema boryanum	Au	20–25 nm	24 h incubation at 200 °C	–	Lengke et al. (2006)
Microgreen algae					
Oscillatoria sancta	Ag	24–34 nm	20 min at 60 °C	Extracellular components	Elumalai et al. (2021)
Botryococcus braunii	Ag	Spherical, 16 nm	180 min continuous stirring at 25 °C, dried AgNPs at 55 °C for 5 h	Peptides and amines	Arya et al. (2019)
Acutodesmus dimorphus	Ag	Spherical, 2 nm to 20 nm	1 day incubation, centrifugation at 16 k rpm for 20 min at 4 °C	Proteins	Chokshi et al. (2016)
Pithophora oedogonia	Ag	Cubical and hexagonal, 24 nm to 55 nm	Incubation at 25 °C, centrifugation at 15 k rpm for 15 min	Proteins and functional groups	Sinha et al. (2015)
Enteromorpha flexuosa	Ag	Spherical, 2–32 nm	1 h incubation	Functional groups	Yousefzadi et al. (2014)
Chlorococcum humicola	Ag	Spherical, 16 nm	3 day incubation at 28 °C, centrifugation at 12 k rpm	Intracellular protein molecules	Jena et al. (2014)
Chlorella pyrenoidosa	Au	Icosahedral and spherical, 25–30 nm	Incubation at 100 °C at 100 rpm, pH 8	Enzymes	Oza et al. (2012)

(continued)

Table 1 (continued)

Algae	NPs	Synthesis conditions	Shape and size	Reducing agent	References
Tetraselmis cochinensis	Au	Triangular, face-centered lattice, and spherical, 5 nm to 35 nm	2 day incubation at 28 °C to 29 °C on 200 rpm	Cell wall and cytoplasm's enzymes	Senapati et al. (2012)
Chlamydomonas reinhardtii	Ag	Rectangular and spherical, 1 nm to 15 nm	5 h to 10 h of incubation	Cellular proteins like histones	Barwal et al. (2011)
Tetraselmis suecica	Au	Spherical, 120 nm	5 min incubation at 90 °C	Hydroxyl, carbonyl, and nitrate function groups	Shakibaie et al. (2010)
Chlorella vulgaris	Au	9–20 nm	3 h incubation for 45–90 °C	Biomolecules (amino, carboxylic, phosphate, thiol)	Wei and Qian (2008)
Scenedesmus sp.	Ag	15–20 nm	Incubation at 37 °C on 200 rpm shaking	–	Vigneshwaran et al. (2007)
Chlorella vulgaris	Au	Triangular and truncated	48 h of incubation	Proteins	Xie et al. (2007)
Macrogreen algae					
Caulerpa sertularioides	Ag	Spherical, 24 to 57 nm	Incubation at room temperature	Extracellular components	Anjali et al. (2022)
Ulva intestinalis	Ag-Au	–	Incubation at room temperature	Polysaccharides	González-Ballesteros et al. (2019)
Spirogyra varians	Ag	FCC, 17.6 nm	20 min incubation in dark	Amino, carboxylic, and hydroxyl groups	Salari et al. (2016)
Gracilaria edulis	Ag and ZnO	55–99 nm and 66–95 nm	Centrifugation at 10,000 rpm	Alcohol, amides, and nitro groups	Priyadharshini et al. (2014)
Ulva flexuosa	Ag	Circular and face-centered lattice, 2 nm to 32 nm	1 day incubation, centrifugation at 5000 rpm	Peptides	Rahimi et al. (2014)
Codium capitatum	Ag	3–44 nm	Incubation in the dark at room temperature	Amine, peptide, and sulfate groups	Priyadharshini et al. (2014)

(continued)

Table 1 (continued)

Algae	NPs	Synthesis conditions	Shape and size	Reducing agent	References
Ulva reticulata	Ag	Spherical, 40–50 nm	Incubation at room temperature	Carboxylic acids, benzene rings, and fluoroalkane	Kannan et al. (2013)
Enteromorpha compressa4	Ag	Spherical, 0–50 nm	–	Benzene rings and hydrogen bonded alcohols	Kannan et al. (2013)

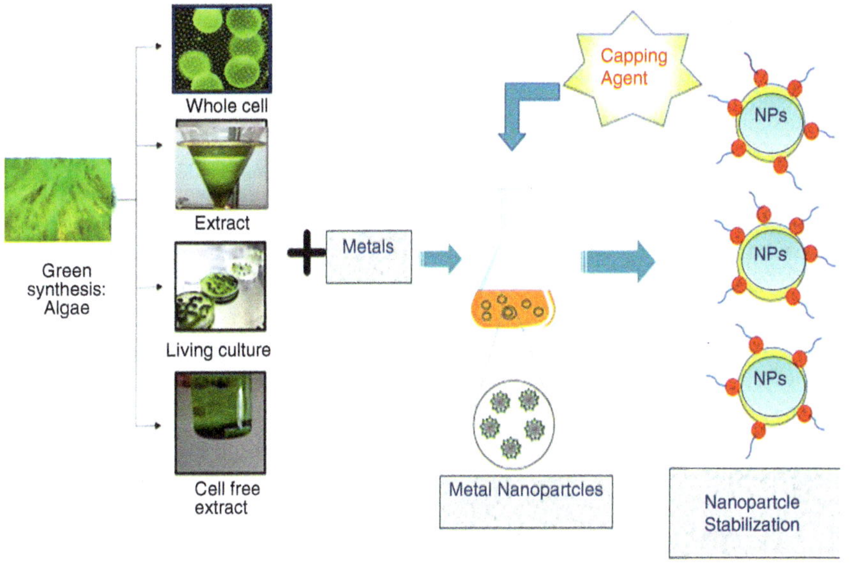

Fig. 3 Schematic illustration of the mechanism involved in the synthesis of NPs through algae-mediated routes

5 Algal Routes for the Biosynthesis of Nanoparticles

Various techniques are commonly employed for the synthesis of nanoparticles (NPs), including chemical reduction, photochemical synthesis, and biological approaches utilizing the metabolic activity of algae (Khanna et al., 2019). The biosynthesis of NPs through algae-mediated routes typically involves the following three steps, as shown in Fig. 3:

(i) The algal extract is mixed with water or an organic solvent and subjected to heating or boiling for a specific duration.
(ii) Molar ionic metallic solutions are prepared separately.
(iii) The algal extract is exposed to the molar solution of the metal ion precursor in a flask, with or without continuous stirring, for a specific duration.

During the reaction, a color change occurs, indicating the nucleation process in which neighboring nucleon particles bind together to form thermodynamically stable NPs of diverse shapes and sizes. Numerous algal routes have been explored for the biosynthesis of nanoparticles, showcasing the potential of algae as efficient and sustainable sources for NP production (Lewis Oscar et al., 2016).

5.1 NP Synthesis from Living Algal Cells

This is the simplest among the algal routes that uses living algal cultures to synthesize NPs. Both algae and metal salt solutions of ionic metallic compounds are incubated for a definite period under controlled conditions, either with regular stirring or without stirring. Sometimes algal cells are expected to continue their metabolism, photosynthesis, and growth along with aqueous salt solutions. Thus, culture stages and the molar concentration of salt solutions need to be considered for sizes and the stability of produced NPs. Based on the characteristics of algae, NP synthesis can be accomplished by either an extracellular or an intracellular method in a precise and quantitative manner. In the extracellular method, secreted extracellular materials (e.g., polysaccharides, reducing carbohydrates, proteins, peptides, etc.) in the algal culture are used to reduce the metal ions to nanoparticles. The algae cells produced very narrow size nanoparticles (NPs) intracellularly. The intracellular formation can be often visualized with the naked eye and further evidenced by optical and electron microscopes. Many authors showed the potential of algal phyla to promote the production of various NPs from live cells. Dağlıoğlu and Yılmaz Öztürk (2019) used Scenedesmaceae to synthesize AgNP, whereas Pytlik et al. (2017) used diatom *Stephanopyxis turris* to synthesize AuNP.

5.2 NP Synthesis Using Extracted Biomolecules

The initial stage in these approaches is to extract biomolecules from live algal cells using vortexing with glass beads, sonication, and lyophilization techniques. Consequently, the aspect of intra- and extracellular synthesis is meaningless in either of these forms. Extracted biomolecules in various forms, such as fine powder, the cell-free filtrate of shattered cells, cell-free supernatants, and so on, are then subjected to nanoparticle production. The biomolecules present in the extract reduce the metal ions and coat the surface of the nanoparticles. Unlike the previous method, the current method requires some energy input owing to the implementation of some harvesting techniques. Somasundaram et al. (2021) showed that AgNP could be synthesized by the extracts of *Sargassum coreanum* extract, whereas Aboelfetoh et al. (2017) used green algae (*Caulerpa serrulata*) to synthesize AgNP.

5.3 NP Synthesis Using Cell-Free Supernatant

A cell-free supernatant (CFS) is a liquid containing the metabolites and enzymes resulting from the growth of algal cells and the residual nutrients of the medium used. Proteins, amino acids, and secondary metabolites such as terpenoids, alkaloids, and phenolic compounds secreted extracellularly by algae are considered main candidates for surface capping. Algal cells are removed by centrifugation and filtration methods, and CFS is used for the synthesis of NPs. Many researchers showed that cell-free supernatants of various species of cyanobacteria and Chlorophyta can promote the synthesis of NPs. Yugay et al. (2020) used *Saccharina cichorioides* cell free extract to synthesize AgNP, and Yousefzadi et al. (2014) used *Enteromorpha flexuosa* to prepare AgNP.

5.4 NP Synthesis Using Whole Algal Cells

This method involves harvesting algal cells and exposing them to precursor salt solutions. Algae possess a variety of extracellular matrices, including cell walls. Most green, red, and brown algae have cell walls consisting of a composite of a fibrillar polysaccharide framework associated with a glycoprotein matrix. Functional groups present in these matrices, such as hydroxyl, methyl, carbonyl, carboxyl, sulfhydryl (thiol), sulfonate, sulfate, phosphate, and amino groups, play pivotal roles in metal reduction processes. However, the metabolic activity of isolated cells can also be reduced considerably within a short time due to cell breakout in the stressful environment. El-Sheekh et al. (2021) used *Spirulina platensis* and *Oscillatoria* sp. whole cells to synthesize AgNP, whereas Dağlıoğlu and Yılmaz Öztürk (2019) used the whole cell of Desmodesmus sp. (*Scenedesmaceace*) to synthesize AgNP.

6 Factors Affecting the Synthesis Process

Many researchers have demonstrated the feasibility of manufacturing nanoparticles with well-defined sizes and shapes tailored for specific applications by adjusting various reaction parameters. These parameters include pH, temperature, concentration of cell extracts, metal ion concentration, activity of catalysts, and incubation time. By carefully controlling these factors, researchers can modulate the nucleation and growth kinetics of nanoparticles, ultimately influencing their size, shape, and properties to meet the requirements of diverse applications. Some dominant factors that affect nanoparticle biosynthesis are described below (Fig. 4).

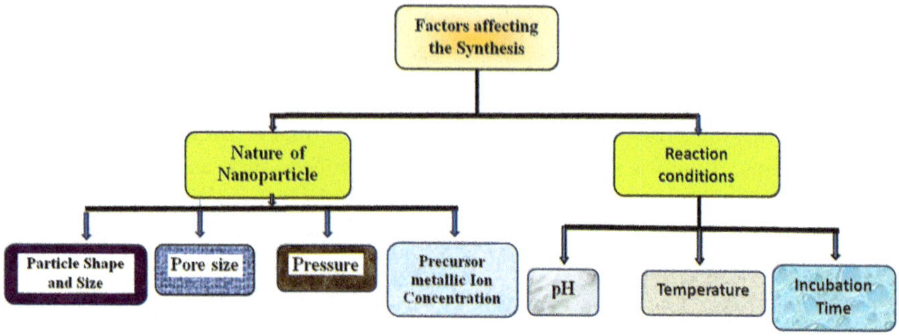

Fig. 4 Factors affecting the nanoparticle synthesis process included the nature of nanoparticles and their reaction condition

6.1 pH of Reaction-Mixture

The pH of the solution medium plays a crucial role in the synthesis of nanoparticles using green technology methods. The pH level affects significantly the size, stability, morphology, and texture of the nanoparticles produced. When the pH of the solution is acidic or lower, indicating a high concentration of H^+ ions, the reducing power potential of the functional groups present on the extracellular matrices decreases. As a result, particle agglomeration tends to occur, leading to the formation of nanoparticles with larger particle sizes. On the other hand, when the pH of the solution is alkaline or higher, it increases the reducing power of the surface functional groups. This increase in reducing power helps prevent particle agglomeration, resulting in the formation of nanoparticles with smaller sizes. Additionally, the hydrophobic and hydrophilic interactions of intramolecular forces also play a role in preventing nanoparticle agglomeration. By altering the pH of the solution media, the size of nanoparticles can be controlled. Several researchers have investigated the influence of pH on the shape and size of synthesized nanoparticles. For example, Rajeshkumar et al. (2014) demonstrated that higher pH levels enhance the synthesis of silver nanoparticles when using *Sargassum longifolium* as a reducing agent. Parial et al. (2012) reported that various algae can synthesize nanoparticles with different shapes at different pH levels.

6.2 Temperature

The temperature of the reaction medium is an important thermodynamic variable and plays a crucial role in determining the size and distribution of the nanoparticle formed. A study showed that an increase in the reaction temperature may not only fasten the rate of synthesis but also produce NPs with a uniform size. With an increase in temperature, the particles gain kinetic energy and move faster. Thus, all three methods for NP synthesis involve a broad temperature range from ambient temperature to very high temperatures, e.g., physical method (>350 °C) and chemical method (<350 °C). However, the green technology route requires temperatures less than 100 °C or room temperature. Li et al. (2021b) showed that the minimum 30 nm size of AgNP can be reached at 50 °C.

6.3 Particle Shape and Size

Particle shape and size analysis plays a crucial role in understanding the properties and behavior of nanoparticles. The surface energy distribution, density, and unique characteristics of NPs can be determined, enabling the prediction of their properties under specific conditions and applications. The size of nanoparticles determines their surface-to-volume ratio, which is significantly higher compared to larger particles. This increased surface area per unit volume strongly influences their properties. For example, the melting point of NPs is reduced with a decreasing size at the nanometer scale. This is because materials composed of NPs have a much greater surface area per unit volume ratio, resulting in enhanced chemical reactivity.

Moreover, changes in the shape of NPs also affect the distribution of energy on their surfaces, leading to variations in their chemical properties and dynamic behavior. NPs with different configurations may possess similar energy, which enables the easy transformation of NPs' shape and behavior. For instance, copper nanoparticles smaller than 50 nm exhibit super hardness and lack the malleability or ductility typically observed in bulk copper. This demonstrates the unique properties exhibited by NPs due to their small size and distinct surface characteristics. In terms of synthesis, researchers have observed that smaller silver nanoparticles can be synthesized at lower temperatures, while higher temperatures tend to result in the formation of agglomerated and larger-sized particles, as reported by Prasad et al. (2013). Additionally, various shapes of nanoparticles, such as rods, spheres, triangles, and diamonds, can be synthesized using different types of algae, as shown by Parial et al. (2012). This emphasizes the versatility of green synthesis methods in controlling the shape and size of nanoparticles.

6.4 Pore Size

The quality and application of nanoparticles are greatly influenced by the porosity of the synthesized nanoparticles. The pore size of NPs needs to address the variety of catalytic reactions and adsorption applications that exist in chemical industries. Researchers achieved to increase the immobilization of biomolecules onto porous NPs for drug delivery and the biomedical field. By controlling the pore size of nanoparticles, researchers can tailor the specific surface area and pore volume, optimizing their performance for various applications. Different pore sizes can accommodate different sizes of molecules or allow for specific interactions, leading to enhanced selectivity and efficiency in adsorption and catalysis processes. This is particularly relevant in the chemical industry, where a variety of catalytic reactions and adsorption applications exist. Moreover, the presence of pores allows for the incorporation and controlled release of therapeutic agents, making porous nanoparticles promising candidates for targeted drug delivery systems.

6.5 Pressure

Pressure can lead to unique phase transitions, resulting in the formation of new crystal structures or modifications in the existing ones. The application of pressure to the reaction medium drives atomic phase transitions in the shape and size of synthesized NPs. These phase transitions can cause alterations in the electronic band structure, optical properties, and mechanical behavior of NPs. In most circumstances, when synthesized NPs come into contact with or near the surface of other NPs, their characteristics change. Thus, pressure serves as an important thermodynamic variable, which not only changes the chemical and crystal structures of NPs but also influences their properties such as electronic, optical, and mechanical stability. By subjecting NPs to pressure, it is possible to induce innovative changes in their chemical and physical properties, which may not be achievable through conventional solution-based chemical methods. As a result, pressure provides a means to tune and control the properties of nanoparticles, enabling the exploration of novel applications.

6.6 Incubation Time

The duration of incubation in the reaction medium plays a crucial role in the synthesis of nanoparticles using green technology methods. The amount of time the reaction medium is incubated can have a significant impact on the quality and type of nanoparticles produced. Increasing the contact duration or incubation time generally leads to an increase in the intensity of the surface plasmon resonance (SPR) band without any significant shift. This indicates a rapid rise in the quantity of nanoparticles synthesized. The incubation time directly influences the nucleation and growth processes, affecting the size, shape, and concentration of the nanoparticles. Furthermore, the properties of the synthesized nanoparticles can also change over time due to various factors such as the synthesis process itself, exposure to light, storage conditions, etc. For instance, nanoparticles may undergo particle aggregation during long-term storage, which can impact their stability and potential applications. Additionally, particles may shrink or expand, and their properties may degrade or evolve during extended storage periods. It is important to consider these time-dependent effects when utilizing nanoparticles in practical applications. The variation in incubation time can occur in different ways, and its influence on nanoparticle synthesis is dependent on the specific system and conditions. For example, Singh et al. (2012) reported that the maximum concentration of gold nanoparticles (AuNPs) using *Padina gymnospora* as a reducing agent could be achieved after 12 days of incubation. Lahr and Vikesland (2014) demonstrated that longer incubation times are required to synthesize AuNPs at initial low gold concentrations.

6.7 Precursor Metallic Ion Concentration

In chemical nanotechnology, a precursor is a chemical compound that is used to form another chemical through various procedures and processes. The concentration of the precursor plays a crucial role in controlling the size of nanoparticles during synthesis. However, there have been conflicting results regarding the relationship between precursor concentration and nanoparticle size. Traditionally, it has been believed that increasing precursor concentration leads to an increase in nanoparticle size. This is because a higher concentration of precursor molecules provides more building blocks for the growth of nanoparticles, resulting in larger sizes. However, recent studies have challenged this notion and reported instances where increasing precursor concentration actually leads to a decrease in nanoparticle size. The conflicting results can be attributed to several factors, including the specific synthesis method, reaction conditions, and properties of the precursor itself. Different synthesis techniques and conditions can influence the rate of nucleation and growth of nanoparticles, ultimately affecting their final size. Additionally, the properties of the precursor, such as its reactivity, solubility, and surface adsorption, can interact with the reaction environment and impact the nanoparticle formation process. Therefore, achieving precise control over nanoparticle size solely through adjustments in precursor concentration may not be straightforward or universally applicable. It requires a comprehensive understanding of the specific system and experimental conditions involved. Other factors, such as the addition of stabilizing agents, temperature, pH, and reaction time, may also need to be considered and optimized to achieve the desired nanoparticle size.

Table 1 summarizes the algae-mediated NPs with their conditions.

7 Perspective on the Green Synthesis of Algae-Mediated Nanomaterial

The green synthesis of algae-mediated nanomaterials offers a promising perspective for sustainable and environmentally friendly manufacturing processes. Algae, as natural and renewable resources, provide a green alternative to traditional methods

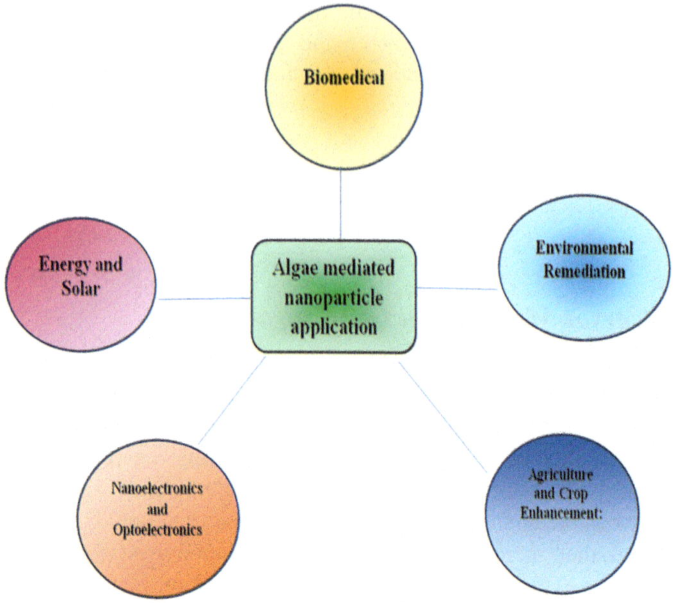

Fig. 5 Applications of nanoparticles

that rely on hazardous chemicals and energy-intensive processes (Singh et al., 2013a, b). Algae-mediated metallic NPs have demonstrated broad applicability in environmental health, food packaging, sensing devices, and the photo-electrochemical and chemical industries (see Fig. 5).

7.1 Biomedical Applications

Algal nanoparticles have emerged as promising candidates for various biomedical applications. They possess several advantageous properties, such as the ability to serve as drug delivery systems, surface functionalization with specific targeting ligands, and the loading of therapeutic agents (Singh et al., 2021). Algal nanoparticles also exhibit low toxicity and high biocompatibility, making them suitable for diverse medical applications (Singh et al., 2020). Additionally, these nanoparticles have demonstrated antimicrobial activity, effectively inhibiting various bacteria and fungi. For instance, gold nanoparticles (AuNPs) synthesized from algal sources like *Lemanea fluviatilis*, *Spirulina platensis*, *Galaxaura elongata*, and *Turbinaria conoides* have exhibited antibacterial activity against different bacterial pathogens. Bhattacharya et al. (2019) investigated the disinfection ability of algae-mediated synthesized copper oxide (CuO) nanoparticles in drinking water, showing their effectiveness against *Escherichia coli*. Furthermore, silver nanoparticles (AgNPs) produced by *Gelidiella acerosa* and *Sargassum longifolium* displayed strong antifungal activity against pathogens such as *Aspergillus fumigatus*, *Humicola insolens*, *Fusarium* sp., *Candida albicans*, and *Trichoderma reesei*.

Algal nanoparticles also have the potential to serve as green antifouling compounds. Vijayan et al. (2014) synthesized AgNPs using *Turbinaria conoides* and found them to be efficient in controlling biofilm formation in bacteria like *E. coli*, *Salmonella* sp., *Serratia liquefaciens*, and *Aeromonas hydrophila*. Moreover, algae-mediated nanoparticles can respond to internal biological stimuli (such as acidic pH, redox environment, and enzyme presence) as well as external stimuli (including temperature, ultrasound, magnetic field, and light). These responsive nanocarriers enable controlled release profiles and find applications in various medical treatments, including cancer therapy. They can act as smart nanocarriers for the delivery of chemotherapeutic and imaging agents to tumor-specific sites. Algal-mediated iron oxide (Fe_3O_4) nanoparticles derived from brown seaweed *Sargassum muticum* have demonstrated cytotoxicity and anticancer activity. Algal nanoparticles also serve as antioxidant agents, neutralizing free radicals in biological cells and playing important roles in preventing chronic diseases such as cardiovascular diseases, aging, heart disease, anemia, and cancer. Studies have shown that nanoparticles exhibit greater antioxidant activity compared to natural antioxidants like vitamin C. For instance, Keshari et al. (2020) demonstrated that AgNPs exhibited 29.55% antioxidant activity, while vitamin C displayed 24.28% antioxidant activity. Green-synthesized nanoparticles also exhibit stronger antioxidant properties due to surface modifications by phytochemicals. Sreelekha et al. (2021) reported that green-synthesized nanoparticles showed 91% DPPH scavenging activity at a concentration of 5 μg/ml, whereas chemically synthesized nanoparticles only displayed 79% activity at the same concentration. Additionally, Sharma et al. (2014) biosynthesized AuNPs using the edible freshwater epilithic red alga *Lemanea fluviatilis*, which exhibited remarkable antioxidant activity in the DPPH assay.

7.2 Environmental Remediation

Algal nanoparticles can be employed for environmental remediation purposes. They can be used to remove heavy metals, organic pollutants, and dyes from wastewater due to their high adsorption capacity. Algae-derived nanoparticles also have the potential to serve as nanocatalysts for the degradation of environmental contaminants. Apart from the medicinal applications, algae-mediated metallic NPs have extensive applicability in environmental health, food packaging, sensing devices, and photo-electrochemical and chemical industries. ZnO, a well-known photocatalyst, has received extensive attention for the efficient removal of toxic pollutants such as methyl orange, methylene blue, and 4-nitrophenol in aquatic environments. In a study by Khalafi et al. (2019), utilized extracts of *Chlorella* sp. as a biotemplate to synthesize safe, stable, and high-quality ZnO nanoparticles. Aziz et al. (2015) demonstrated that biosynthetic silver nanoparticles, produced from *C. pyrenoidosa*, showed outstanding photocatalytic activity in the removal of methylene blue under visible light. The performance of these nanoparticles surpassed that of commercial Ag nanoparticles and bulk silver powder, highlighting their potential for environmental remediation applications. Additionally, Subramaniyam et al. (2015) successfully synthesized Fe_3O_4 nanoparticles using the alga *Chlorococcum* sp. MM11 and demonstrated their applicability in the restoration of Cr^{6+}. The Fe nanoparticles efficiently reduced 92% of 4 $mg.L^{-1}$ Cr^{6+} to Cr^{3+}, showcasing their potential for heavy metal ion removal in water treatment.

Recently, algae-mediated NPs have also applied to selectively detect metal ions in water due to the merits of unique quantum effects. Rojas-Perez et al. (2015) documented the synthesis of gold nanoparticles (Au-NPs) derived from the alga *Osmundaria obtusiloba*. These nanoparticles were specifically employed for the detection of 4-nitrothiophenol. Researchers also revealed the application of algae-mediated AgNPs (Elgamouz et al., 2020) and palladium nanoparticles (PdNPs) (Momeni & Nabipour, 2015) as a colorimetric and electrochemical sensor, respectively for hydrogen peroxide (H_2O_2). These studies demonstrate the diverse applications of algae-mediated nanoparticles in areas such as photocatalysis, heavy metal ion removal, selective detection of compounds, and sensing of hydrogen peroxide. The eco-friendly synthesis approach using algae extracts presents a promising avenue for the production of functional nanoparticles with enhanced properties for various applications.

7.3 Agriculture and Crop Enhancement

Algal nanoparticles can be used to improve agricultural practices. They can be applied as nanofertilizers, nanopesticides, or nanosensors for plant health monitoring. The nanoparticles can enhance nutrient uptake, protect plants against pests and diseases, and provide real-time data on plant physiological parameters.

7.4 Energy and Solar Applications

Algal nanoparticles can be utilized in energy-related applications. For instance, they can serve as efficient catalysts for fuel cells, photocatalysts for water splitting, or light-absorbing materials for solar panels. Algae-derived nanoparticles have the potential to contribute to the development of sustainable energy technologies.

7.5 Nanoelectronics and Optoelectronics

Algal nanoparticles possess unique optical and electronic properties that make them attractive for nanoelectronics and optoelectronics. They can be incorporated into devices such as sensors, displays, and optoelectronic circuits. The natural pigments found in algae, such as chlorophyll, can be harnessed for photovoltaic applications (Li et al., 2021a).

It is worth noting that while algal nanoparticles hold great promise, further research is needed to optimize their synthesis methods, understand their long-term effects, and scale up production for commercial applications. Nonetheless, the field of algal nanoparticle biosynthesis presents exciting opportunities for various industries (Singh et al., 2016a, b).

8 Challenges

Harnessing the potential of algae-mediated nanoparticles (NPs) possesses the capacity to transform the approach in the human-health-related domain, as well as in the agricultural and food industries, by providing innovative tools for rapid disease detection and treatment. Algae, with their abundance of secondary metabolites, are ideal candidates for eco-friendly NP biosynthesis, where they can function as a reducer, capping agent, and stabilizer in every form of NP synthesis, such as in metal, metal oxide, or bimetallic NPs. The integration of nanotechnology with biological systems, such as microalgae and cyanobacteria, is opening up new possibilities for environmental remediation, biotechnological applications, and sustainable resource utilization (Singh et al., 2012).

However, despite the immense potential, there are still several limitations and challenges that need to be addressed to ensure commercial feasibility.

One of the major hurdles in the field of algae-mediated NP synthesis is the lack of a comprehensive understanding of the precise mechanisms involved in NP production. Advanced studies are required, particularly with diverse algal species, to isolate biomolecules that can work efficiently in NP synthesis (Grasso et al., 2019).

The interaction between algae and metal precursors often results in NPs with varying sizes and morphologies. Therefore, a better understanding of the optimization of the process becomes very necessary in which selections of algal species, media composition, inoculum growth phase, NP synthesis mechanisms, and its synthesis mechanism are crucial to exert control over particle size, shape, and monodispersity. Furthermore, while a limited number of metal NPs (such as AuNPs and AgNPs) have been extensively investigated, there is a need to explore the synthesis of new NPs with diverse compositions, including metal oxides (e.g., ZnO), palladium, nitrides, carbides, sulfides, silicon, and carbon-based NPs, using algae as the biofactory (Kumar et al., 2021).

Other challenges include the relatively slow reaction kinetics, which can take days to weeks, as well as issues related to cell viability and colloidal stability, leading to significant agglomeration and limited NP yields. Therefore, it is crucial to develop simple and cost-effective techniques that can make the synthesis process commercially viable. Additionally, the application of emerging characterization technologies holds promise in improving the characteristics of algal-mediated NPs for industrial applications.

Addressing these limitations and challenges will pave the way for the successful translation of algae-mediated NP synthesis into practical and commercially viable processes, further enhancing their potential impact in various industries.

9 Conclusion

Algae-mediated nanoparticle synthesis holds the potential to revolutionize the manufacturing industry by offering greener alternatives to traditional methods while significantly reducing environmental impacts. The synthesis of nanoparticles (NPs) can be achieved through various approaches, including top-down and bottom-up methods, which encompass physical, chemical, and biological techniques. Among these methods, biological routes that utilize materials from biological sources such as plants, bacteria, algae, and fungi have made substantial contributions to the field of nanotechnology. Notably, the use of living cells in algae-mediated NP synthesis stands out for its ability to provide precise control over the size and shape of nanomaterials. A significant advantage of algae-mediated synthesis lies in tapping into the remarkable capabilities of photosynthetic organisms, which can convert sunlight and carbon dioxide into energy. This intrinsic capability paves the way for the sustainable production of metallic nanoparticles, eliminating the need for energy-intensive heating or chemical reactions. Whether by harnessing algae as live cell factories or employing algal extracts as natural reducing agents, this approach ensures a greener and cleaner synthesis process. Furthermore, the production of exopolysaccharides by algae facilitates the formation of colloidally stabilized NPs. By fine-tuning reaction parameters such as pH, temperature, cell-extract concentration, metal ion concentration, catalyst activity, and incubation time, it becomes possible to manufacture nanoparticles with well-defined sizes and shapes tailored for specific applications. These nanoparticles find diverse and essential roles in biotechnology, medicine, cosmetics, and pharmaceuticals. They can interact with the cell membranes of targeted harmful biological entities, causing significant damage to pathogens. Additionally, nanoparticles serve as effective nanocarriers for controlled drug release, with extensive applications in fields like cancer therapy. It is worth noting that green-synthesized nanoparticles often exhibit higher antioxidant activity compared to their chemically synthesized counterparts. This eco-friendly approach not only advances the field of nanotechnology but also contributes to more sustainable and environmentally responsible practices in various industries.

Acknowledgments Dr. Shailendra Kumar Singh gratefully acknowledges the financial support from the University Grants Commission (UGC) through Dr. D. S. Kothari Postdoctoral Scheme (Grant Number: EN/19-20/0020).

References

Abdel-Raouf, N., Al-Enazi, N. M., & Ibraheem, I. B. (2017). Green biosynthesis of gold nanoparticles using *Galaxaura elongata* and characterization of their antibacterial activity. *Arabian Journal of Chemistry, 10*, S3029–S3039.

Abed, A., Derakhshan, M., Karimi, M., Shirazinia, M., Mahjoubin-Tehran, M., Homayonfal, M., ... & Mirzaei, H. (2022). Platinum nanoparticles in biomedicine: Preparation, anti-cancer activity, and drug delivery vehicles. *Frontiers in Pharmacology, 13*, 797804.

Aboeita, N. M., Fahmy, S. A., El-Sayed, M. M., Azzazy, H. M. E. S., & Shoeib, T. (2022). Enhanced anticancer activity of nedaplatin loaded onto copper nanoparticles synthesized using red algae. *Pharmaceutics, 14*(2), 418.

Aboelfetoh, E. F., El-Shenody, R. A., & Ghobara, M. M. (2017). Eco-friendly synthesis of silver nanoparticles using green algae (Caulerpa serrulata): Reaction optimization, catalytic and antibacterial activities. *Environmental Monitoring and Assessment, 189*, 1–15.

Ali, A., Ali, M. A., Ali, M. U., & Mohammad, S. (2011). Hospital outcomes of obstetrical-related acute renal failure in a tertiary care teaching hospital. *Renal Failure, 33*(3), 285–290.

Al-Naamani, L., Dobretsov, S., Dutta, J., & Burgess, J. G. (2017). Chitosan-zinc oxide nanocomposite coatings for the prevention of marine biofouling. *Chemosphere, 168*, 408–417.

Alprol, A. E., Mansour, A. T., El-Beltagi, H. S., & Ashour, M. (2023). Algal extracts for green synthesis of zinc oxide nanoparticles: Promising approach for algae bioremediation. *Materials, 16*(7), 2819.

Anjali, R., Palanisamy, S., Vinosha, M., Selvi, A. M., Sathiyaraj, G., Marudhupandi, T., Mohandoss, S., Prabhu, N. M., & You, S. (2022). Fabrication of silver nanoparticles from marine macro algae Caulerpa sertularioides: Characterization, antioxidant and antimicrobial activity. *Process Biochemistry, 121*, 601–618.

Arya, A., Mishra, V., & Chundawat, T. S. (2019). Green synthesis of silver nanoparticles from green algae (Botryococcus braunii) and its catalytic behavior for the synthesis of benzimidazoles. *Chemical Data Collections, 20*, 100190.

Azizi, S., Ahmad, M. B., Namvar, F., & Mohamad, R. (2014). Green biosynthesis and characterization of zinc oxide nanoparticles using brown marine macroalga Sargassum muticum aqueous extract. *Materials Letters, 116*, 275–277.

Aziz, N., Faraz, M., Pandey, R., Shakir, M., Fatma, T., Varma, A., ... & Prasad, R. (2015). Facile algae-derived route to biogenic silver nanoparticles: synthesis, antibacterial, and photocatalytic properties. *Langmuir, 31*(42), 11605–11612.

Barwal, I., Ranjan, P., Kateriya, S., & Yadav, S. C. (2011). Cellular oxido-reductive proteins of Chlamydomonas reinhardtii control the biosynthesis of silver nanoparticles. *Journal of Nanobiotechnology, 9*(1), 1–12.

Belusso, L. C., Lenz, G. F., Fiorini, E. E., Pereira, A. J., Sequinel, R., Bini, R. A., Felix, J. F., & Schneider, R. (2019). Synthesis of silver nanoparticles from bottom up approach on borophosphate glass and their applications as SERS, antibacterial and glass-based catalyst. *Applied Surface Science, 473*, 303–312.

Bhattacharya, D., Sahu, S. K., Banerjee, S., Das, J. K., Mishra, M., & Sahoo, G. C. (2019). Antibacterial and disinfection activities of CuO nanoparticles: Synthesis, characterization and implication for drinking water treatment. *Journal of Environmental Chemical Engineering, 7*(2), 102997.

Borm, P. J. A., & Müller-Schulte, D. (2006). *Nanoparticles in drug delivery and environmental exposure: same size, same risks*?. 235–249.

Clarke, G., Rogov, A., McCarthy, S., Bonacina, L., Gun'ko, Y., Galez, C., ... & Prina-Mello, A. (2018). Preparation from a revisited wet chemical route of phase-pure, monocrystalline and SHG-efficient BiFeO3 nanoparticles for harmonic bio-imaging. *Scientific Reports, 8*(1), 10473.

Chokshi, K., Pancha, I., Ghosh, T., Paliwal, C., Maurya, R., Ghosh, A., & Mishra, S. (2016). Green synthesis, characterization and antioxidant potential of silver nanoparticles biosynthesized from de-oiled biomass of thermotolerant oleaginous microalgae Acutodesmusdimorphus. *RSC Advances, 6*(76), 72269–72274.

Dahoumane, S. A., Jeffryes, C., Mechouet, M., & Agathos, S. N. (2017). Biosynthesis of inorganic nanoparticles: A fresh look at the control of shape, size and composition. *Bioengineering, 4*(1), 14.

Dağlıoğlu, Y., & Yılmaz Öztürk, B. (2019). A novel intracellular synthesis of silver nanoparticles using Desmodesmussp.(Scenedesmaceae): Different methods of pigment change. *RendicontiLincei. ScienzeFisiche e Naturali, 30*, 611–621.

Ealia, S. A. M., & Saravanakumar, M. P. (2017, November). A review on the classification, characterisation, synthesis of nanoparticles and their application. In *IOP conference series: Materials science and engineering* (Vol. 263, No. 3, p. 032019). IOP Publishing.

El-Sheekh, M. M., Hassan, L. H., & Morsi, H. H. (2021). Evaluation of antimicrobial activities of blue-green algae-mediated silver and gold nanoparticles. *Rendiconti Lincei. Scienze Fisiche e Naturali, 32*(4), 747–759.

Elgamouz, A., Idriss, H., Nassab, C., Bihi, A., Bajou, K., Hasan, K., ... & Patole, S. P. (2020). Green synthesis, characterization, antimicrobial, anti-cancer, and optimization of colorimetric sensing of hydrogen peroxide of algae extract capped silver nanoparticles. *Nanomaterials, 10*(9), 1861.

Elumalai, D., Hemavathi, M., Rekha, G. S., Pushpalatha, M., Leelavathy, R., Vignesh, A., Ashok, K., & Babu, M. (2021). Photochemical synthesizes of silver nanoparticles using Oscillatoria sancta micro algae against mosquito vectors Aedes aegypti and Anopheles stephensi. *Sensing and Bio-Sensing Research, 34*, 100457.

Fatima, R., Priya, M., Indurthi, L., Radhakrishnan, V., & Sudhakaran, R. (2020). Biosynthesis of silver nanoparticles using red algae Portieria hornemannii and its antibacterial activity against fish pathogens. *Microbial Pathogenesis, 138*, 103780.

Fouda, A., Eid, A. M., Abdel-Rahman, M. A., El-Belely, E. F., Awad, M. A., Hassan, S. E. D., Al-Faifi, Z. E., & Hamza, M. F. (2022). Enhanced antimicrobial, cytotoxicity, larvicidal, and repellence activities of brown algae, Cystoseira crinita-mediated green synthesis of magnesium oxide nanoparticles. *Frontiers in Bioengineering and Biotechnology, 10*, 849921.

González-Ballesteros, N., Prado-López, S., Rodríguez-González, J. B., Lastra, M., & Rodríguez-Argüelles, M. (2017). Green synthesis of gold nanoparticles using brown algae Cystoseirabaccata: Its activity in colon cancer cells. *Colloids and Surfaces B: Biointerfaces, 153*, 190–198.

González-Ballesteros, N., Diego-González, L., Lastra-Valdor, M., Rodríguez-Argüelles, M. C., Grimaldi, M., Cavazza, A., Bigi, F., & Simón-Vázquez, R. (2019). Immunostimulant and biocompatible gold and silver nanoparticles synthesized using the Ulva intestinalis L. aqueous extract. *Journal of Materials Chemistry B, 7*(30), 4677–4691.

Grasso, G., Zane, D., & Dragone, R. (2019). Microbial nanotechnology: Challenges and prospects for green biocatalytic synthesis of nanoscale materials for sensoristic and biomedical applications. *Nanomaterials, 10*(1), 11.

Hasan, S. (2015). A review on nanoparticles: Their synthesis and types. *Research Journal of Recent Sciences, 2277*, 2502.

Hussain, I., Singh, N. B., Singh, A., Singh, H., & Singh, S. C. (2016). Green synthesis of nanoparticles and its potential application. *Biotechnology Letters, 38*, 545–560.

Iravani, S., Korbekandi, H., Mirmohammadi, S. V., & Zolfaghari, B. (2014). Synthesis of silver nanoparticles: Chemical, physical and biological methods. *Research in Pharmaceutical Sciences, 9*(6), 385.

Jena, J., Pradhan, N., Nayak, R. R., Dash, B. P., Sukla, L. B., Panda, P. K., & Mishra, B. K. (2014). Microalga Scenedesmus sp.: A potential low-cost green machine for silver nanoparticle synthesis. *Journal of Microbiology and Biotechnology, 24*(4), 522–533.

Kalabegishvili, T. L., Kirkesali, E. I., Rcheulishvili, A. N., Ginturi, E. N., Murusidze, I. G., Pataraya, D. T., Gurielidze, M. A., Tsertsvadze, G. I., Gabunia, V. N., Lomidze, L. G., & Gvarjaladze, D. N. (2012). Synthesis of gold nanoparticles by some strains of Arthrobacter genera. *Journal of Materials Science and Engineering A: Structural Materials: Properties, Microstructure and Processing, 2*(2), 164–173.

Kannan, R. R. R., Stirk, W. A., & Van Staden, J. (2013). Synthesis of silver nanoparticles using the seaweed Codium capitatum PC Silva (Chlorophyceae). *South African Journal of Botany, 86*, 1–4.

Keshari, A. K., Srivastava, R., Singh, P., Yadav, V. B., & Nath, G. (2020). Antioxidant and antibacterial activity of silver nanoparticles synthesized by Cestrum nocturnum. *Journal of Ayurveda and Integrative Medicine, 11*(1), 37–44.

Khan, F., Shahid, A., Zhu, H., Wang, N., Javed, M. R., Ahmad, N., Xu, J., Alam, M. A., & Mehmood, M. A. (2022). Prospects of algae-based green synthesis of nanoparticles for environmental applications. *Chemosphere, 293*, 133571.

Khalafi, T., Buazar, F., & Ghanemi, K. (2019). Phycosynthesis and enhanced photocatalytic activity of zinc oxide nanoparticles toward organosulfur pollutants. *Scientific Reports, 9*(1), 6866.

Khanna, P., Kaur, A., & Goyal, D. (2019). Algae-based metallic nanoparticles: Synthesis, characterization and applications. *Journal of Microbiological Methods, 163*, 105656.

Kumar, B. R., Mathimani, T., Sudhakar, M. P., Rajendran, K., Nizami, A. S., Brindhadevi, K., & Pugazhendhi, A. (2021). A state of the art review on the cultivation of algae for energy and other valuable products: Application, challenges, and opportunities. *Renewable and Sustainable Energy Reviews, 138*, 110649.

Lahr, R. H., & Vikesland, P. J. (2014). Surface-enhanced Raman spectroscopy (SERS) cellular imaging of intracellulary biosynthesized gold nanoparticles. *ACS Sustainable Chemistry & Engineering, 2*(7), 1599–1608.

Lengke, M. F., Fleet, M. E., & Southam, G. (2006). Morphology of gold nanoparticles synthesized by filamentous cyanobacteria from gold (I)– thiosulfate and gold (III)– chloride complexes. *Langmuir, 22*(6), 2780–2787.

Lewis Oscar, F., Vismaya, S., Arunkumar, M., Thajuddin, N., Dhanasekaran, D., & Nithya, C. (2016). Algal nanoparticles: Synthesis and biotechnological potentials. In *Algae-organisms for imminent biotechnology* (Vol. 7, pp. 157–182). IntechOpen.

Li, S. N., Wang, R., & Ho, S. H. (2021a). Algae-mediated biosystems for metallic nanoparticle production: From synthetic mechanisms to aquatic environmental applications. *Journal of Hazardous Materials, 420*, 126625.

Li, A. S. H., Izzati, S., Toemen, S., & Lee, S. L. (2021b). *Chitosan Modified Titanium Dioxide Photocatalysts for Dye Degradation*.

Liu, B., Xie, J., Lee, J. Y., Ting, Y. P., & Chen, J. P. (2005). Optimization of high-yield biological synthesis of single-crystalline gold nanoplates. *The Journal of Physical Chemistry B, 109*(32), 15256–15263.

Mahdieh, M., Zolanvari, A., & Azimee, A. S. (2012). Green biosynthesis of silver nanoparticles by Spirulina platensis. *Scientia Iranica, 19*(3), 926–929.

Mata, Y. N., Torres, E., Blazquez, M. L., Ballester, A., González, F. M. J. A., & Munoz, J. A. (2009). Gold (III) biosorption and bioreduction with the brown alga Fucus vesiculosus. *Journal of Hazardous Materials, 166*(2–3), 612–618.

Mishra, B., Saxena, A., & Tiwari, A. (2020). Biosynthesis of silver nanoparticles from marine diatoms Chaetoceros sp., Skeletonema sp., Thalassiosira sp., and their antibacterial study. *Biotechnology Reports, 28*, e00571.

Momeni, S., & Nabipour, I. (2015). A simple green synthesis of palladium nanoparticles with Sargassum alga and their electrocatalytic activities towards hydrogen peroxide. *Applied Biochemistry and Biotechnology, 176*, 1937–1949.

Mukherjee, A., Sarkar, D., & Sasmal, S. (2021). A review of green synthesis of metal nanoparticles using algae. *Frontiers in Microbiology, 12*, 693899.

Nadaroglu, H., Güngör, A. A., & Selvi, İ. N. C. E. (2017). Synthesis of nanoparticles by green synthesis method. *International Journal of Innovative Research and Reviews, 1*(1), 6–9.

Oza, G., Pandey, S., Mewada, A., Kalita, G., Sharon, M., Phata, J., Ambernath, W., & Sharon, M. (2012). Facile biosynthesis of gold nanoparticles exploiting optimum pH and temperature of fresh water algae Chlorella pyrenoidusa. *Advances in Applied Science Research, 3*(3), 1405–1412.

Öztürk, B. Y., Gürsu, B. Y., & Dağ, İ. (2020). Antibiofilm and antimicrobial activities of green synthesized silver nanoparticles using marine red algae Gelidium corneum. *Process Biochemistry, 89*, 208–219.

Panwar, N., Soehartono, A. M., Chan, K. K., Zeng, S., Xu, G., Qu, J., ... & Chen, X. (2019). Nanocarbons for biology and medicine: sensing, imaging, and drug delivery. *Chemical Reviews, 119*(16), 9559–9656.

Palaniyandi, T., Baskar, G. V. B., Viswanathan, S., Abdul Wahab, M. R., Govindaraj, M. K., Sivaji, A., Rajendran, B. K., & Kaliamoorthy, S. (2023). Biosynthesis of iron nanoparticles using brown algae Spatoglossum asperum and its antioxidant and anticancer activities through in vitro and in silico studies. *Particulate Science and Technology, 41*, 1–14.

Parial, D., & Pal, R. (2014). Green synthesis of gold nanoparticles using cyanobacteria and their characterization. *Indian Journal of Applied Research, 4*(1), 69.

Parial, D., Patra, H. K., Dasgupta, A. K., & Pal, R. (2012). Screening of different algae for green synthesis of gold nanoparticles. *European Journal of Phycology, 47*(1), 22–29.

Parial, D., Gopal, P. K., Paul, S., & Pal, R. (2016). Gold (III) bioreduction by cyanobacteria with special reference to in vitro biosafety assay of gold nanoparticles. *Journal of Applied Phycology, 28*, 3395–3406.

Parveen, K., Banse, V., & Ledwani, L. (2016, April). Green synthesis of nanoparticles: Their advantages and disadvantages. In *AIP conference proceedings* (Vol. 1724, No. 1). AIP Publishing.

Prasad, T. N., Kambala, V. S. R., & Naidu, R. (2013). Phyconanotechnology: Synthesis of silver nanoparticles using brown marine algae Cystophora moniliformis and their characterisation. *Journal of Applied Phycology, 25*, 177–182.

Priyadharshini, R. I., Prasannaraj, G., Geetha, N., & Venkatachalam, P. (2014). Microwave-mediated extracellular synthesis of metallic silver and zinc oxide nanoparticles using macroalgae (Gracilaria edulis) extracts and its anticancer activity against human PC3 cell lines. *Applied Biochemistry and Biotechnology, 174*, 2777–2790.

Pugazhendhi, A., Prabakar, D., Jacob, J. M., Karuppusamy, I., & Saratale, R. G. (2018). Synthesis and characterization of silver nanoparticles using Gelidiumamansii and its antimicrobial property against various pathogenic bacteria. *Microbial Pathogenesis, 114*, 41–45.

Pytlik, N., Kaden, J., Finger, M., Naumann, J., Wanke, S., Machill, S., & Brunner, E. (2017). Biological synthesis of gold nanoparticles by the diatom Stephanopyxisturris and in vivo SERS analyses. *Algal Research, 28*, 9–15.

Rahimi, Z., Yousefzadi, M., Noori, A., & Akbarzadeh, A. (2014). Green synthesis of silver nanoparticles using Ulva flexousa from the Persian Gulf, Iran. *Journal of the Persian Gulf, 5*(15), 9–16.

Rajeshkumar, S., Malarkodi, C., Gnanajobitha, G., Paulkumar, K., Vanaja, M., Kannan, C., & Annadurai, G. (2013). Seaweed-mediated synthesis of gold nanoparticles using Turbinariaconoides and its characterization. *Journal of Nanostructure in Chemistry, 3*, 1–7.

Rajeshkumar, S., Malarkodi, C., Paulkumar, K., Vanaja, M., Gnanajobitha, G., & Annadurai, G. (2014). Algae mediated green fabrication of silver nanoparticles and examination of its antifungal activity against clinical pathogens. *International Journal of Metals, 2014*, 1–8.

Rajeshkumar, S., Sellami, H., Ali, D., Chitra, S., Ponnanikajamideen, M., & Arunachalam, K. (2022). Bioconjugation of fluorescent gold nanoparticles synthesized using marine brown algae *Sargassum longifolium*. *Journal of Nanomaterials, 2022*, 6765227.

Rane, A. V., Kanny, K., Abitha, V. K., & Thomas, S. (2018). Methods for synthesis of nanoparticles and fabrication of nanocomposites. In *Synthesis of inorganic nanomaterials* (pp. 121–139). Woodhead Publishing.

Rao, M. D., & Pennathur, G. (2017). Green synthesis and characterization of cadmium sulphide nanoparticles from Chlamydomonas reinhardtii and their application as photocatalysts. *Materials Research Bulletin, 85*, 64–73.

Rojas-Perez, A., Adorno, L., Cordero, M., Ruiz, A., Mercado-Diaz, Z., Rodriguez, A., ... & Diaz Vazquez, L. M. (2015). Biosynthesis of gold nanoparticles using Osmudaria obtusiloba extract and their potential use in optical sensing application. *Austin Journal of Biosensors & Bioelectronics, 1*(5), 1–9.

Sahayaraj, K., Rajesh, S., & Rathi, J. M. (2012). Silver nanoparticles biosynthesis using marine alga Padina Pavonica (Linn.) and its microbicidal activity. *Digest Journal of Nanomaterials & Biostructures (DJNB), 7*(4), 1557–1567.

Salari, Z., Danafar, F., Dabaghi, S., & Ataei, S. A. (2016). Sustainable synthesis of silver nanoparticles using macroalgae Spirogyra varians and analysis of their antibacterial activity. *Journal of Saudi Chemical Society, 20*(4), 459–464.

Sau, T. K., & Rogach, A. L. (Eds.). (2012). *Complex-shaped metal nanoparticles: Bottom-up syntheses and applications*. Wiley.

Senapati, S., Syed, A., Moeez, S., Kumar, A., & Ahmad, A. (2012). Intracellular synthesis of gold nanoparticles using alga Tetraselmiskochinensis. *Materials Letters, 79*, 116–118.

Shakibaie, M., Forootanfar, H., Mollazadeh-Moghaddam, K., Bagherzadeh, Z., Nafissi-Varcheh, N., Shahverdi, A. R., & Faramarzi, M. A. (2010). Green synthesis of gold nanoparticles by the marine microalga Tetraselmissuecica. *Biotechnology and Applied Biochemistry, 57*(2), 71–75.

Sharma, B., Purkayastha, D. D., Hazra, S., Thajamanbi, M., Bhattacharjee, C. R., Ghosh, N. N., & Rout, J. (2014). Biosynthesis of fluorescent gold nanoparticles using an edible freshwater red alga, Lemanea fluviatilis (L.) C. Ag. and antioxidant activity of biomatrix loaded nanoparticles. *Bioprocess and Biosystems Engineering, 37*, 2559–2565.

Shukla, A. K., & Iravani, S. (Eds.). (2018). *Green synthesis, characterization and applications of nanoparticles*. Elsevier.

Singh, S. K., Bansal, A., Jha, M. K., & Dey, A. (2012). An integrated approach to remove Cr (VI) using immobilized Chlorella minutissima grown in nutrient rich sewage wastewater. *Bioresource Technology, 104*, 257–265.

Singh, M., Kalaivani, R., Manikandan, S., Sangeetha, N., & Kumaraguru, A. K. (2013a). Facile green synthesis of variable metallic gold nanoparticle using Padina gymnospora, a brown marine macroalga. *Applied Nanoscience, 3*, 145–151.

Singh, S. K., Bansal, A., Jha, M. K., & Jain, R. (2013b). Production of biodiesel from wastewater grown Chlorella minutissima. *Indian Journal of Chemical Technology, 20*, 341–345.

Singh, P., Kim, Y. J., Zhang, D., & Yang, D. C. (2016a). Biological synthesis of nanoparticles from plants and microorganisms. *Trends in Biotechnology, 34*(7), 588–599.

Singh, S. K., Sundaram, S., Sinha, S., Rahman, M. A., & Kapur, S. (2016b). Recent advances in CO_2 uptake and fixation mechanism of cyanobacteria and microalgae. *Critical Reviews in Environmental Science and Technology, 46*(16), 1297–1323.

Singh, S. K., Kaur, R., Bansal, A., Kapur, S., & Sundaram, S. (2020). Biotechnological exploitation of cyanobacteria and microalgae for bioactive compounds. In *Biotechnological production of bioactive compounds* (pp. 221–259). Elsevier.

Singh, S. K., Kaur, R., Rahman, M. A., Mishra, M., & Sundaram, S. (2021). Evaluation of potent cyanobacteria species for UV-protecting compound synthesis using bicarbonate-based culture system. *3 Biotech, 11*, 1–15.

Sinha, S. N., Paul, D., Halder, N., Sengupta, D., & Patra, S. K. (2015). Green synthesis of silver nanoparticles using fresh water green alga Pithophoraoedogonia (Mont.) Wittrock and evaluation of their antibacterial activity. *Applied Nanoscience, 5*, 703–709.

Somasundaram, C. K., Atchudan, R., Edison, T. N. J. I., Perumal, S., Vinodh, R., Sundramoorthy, A. K., Babu, R. S., Alagan, M., & Lee, Y. R. (2021). Sustainable synthesis of silver nanoparticles using marine algae for catalytic degradation of methylene blue. *Catalysts, 11*(11), 1377.

Sreelekha, E., George, B., Shyam, A., Sajina, N., & Mathew, B. (2021). A comparative study on the synthesis, characterization, and antioxidant activity of green and chemically synthesized silver nanoparticles. *BioNanoScience, 11*, 489–496.

Subramaniyam, V., Subashchandrabose, S. R., Thavamani, P., Megharaj, M., Chen, Z., & Naidu, R. (2015). Chlorococcum sp. MM11—a novel phyco-nanofactory for the synthesis of iron nanoparticles. *Journal of Applied Phycology, 27*, 1861–1869.

Sudha, S. S., Rajamanickam, K., & Rengaramanujam, J. (2013). Microalgae mediated synthesis of silver nanoparticles and their antibacterial activity against pathogenic bacteria. *Indian Journal of Experimental Biology., 51*(5), 393–399.

Suganya, K. U., Govindaraju, K., Kumar, V. G., Dhas, T. S., Karthick, V., Singaravelu, G., & Elanchezhiyan, M. (2015). Blue green alga mediated synthesis of gold nanoparticles and its antibacterial efficacy against Gram positive organisms. *Materials Science and Engineering: C, 47*, 351–356.

Tiwari, D. K., Behari, J., & Sen, P. (2008). Time and dose-dependent antimicrobial potential of Ag nanoparticles synthesized by top-down approach. *Current Science, 95*, 647–655.

Venkatesan, J., Manivasagan, P., Kim, S. K., Kirthi, A. V., Marimuthu, S., & Rahuman, A. A. (2014). Marine algae-mediated synthesis of gold nanoparticles using a novel Ecklonia cava. *Bioprocess and Biosystems Engineering, 37*, 1591–1597.

Vigneshwaran, N., Ashtaputre, N. M., Varadarajan, P. V., Nachane, R. P., Paralikar, K. M., & Balasubramanya, R. H. (2007). Biological synthesis of silver nanoparticles using the fungus Aspergillus flavus. *Materials Letters, 61*(6), 1413–1418.

Vijayan, S. R., Santhiyagu, P., Singamuthu, M., Kumari Ahila, N., Jayaraman, R., & Ethiraj, K. (2014). Synthesis and characterization of silver and gold nanoparticles using aqueous extract of seaweed, Turbinaria conoides, and their antimicrofouling activity. *The Scientific World Journal.*

Wang, H., Wang, L., Lin, D., Feng, X., Niu, Y., Zhang, B., & Xiao, F. S. (2021). Strong metal–support interactions on gold nanoparticle catalysts achieved through Le Chatelier's principle. *Nature Catalysis, 4*(5), 418–424.

Wei, D., & Qian, W. (2008). Facile synthesis of Ag and Au nanoparticles utilizing chitosan as a mediator agent. *Colloids and Surfaces B: Biointerfaces, 62*(1), 136–142.

Xie, J., Lee, J. Y., Wang, D. I., & Ting, Y. P. (2007). Identification of active biomolecules in the high-yield synthesis of single-crystalline gold nanoplates in algal solutions. *Small, 3*(4), 672–682.

Ying, S., Guan, Z., Ofoegbu, P. C., Clubb, P., Rico, C., He, F., & Hong, J. (2022). Green synthesis of nanoparticles: Current developments and limitations. *Environmental Technology & Innovation, 26*, 102336.

Yousefzadi, M., Rahimi, Z., & Ghafori, V. (2014). The green synthesis, characterization and antimicrobial activities of silver nanoparticles synthesized from green alga Enteromorpha flexuosa (wulfen) J. Agardh. *Materials Letters, 137*, 1–4.

Yugay, Y. A., Usoltseva, R. V., Silant'ev, V. E., Egorova, A. E., Karabtsov, A. A., Kumeiko, V. V., Ermakova, S. P., Bulgakov, V. P., & Shkryl, Y. N. (2020). Synthesis of bioactive silver nanoparticles using alginate, fucoidan and laminaran from brown algae as a reducing and stabilizing agent. *Carbohydrate Polymers, 245*, 116547.

Eco-Friendly Production of Organic Nanoparticles and Their Uses

Vijaya Ilango

1 Introduction

The development of engineered nanomaterials is one of the most significant advances in nanotechnology and material science (Sun et al., 2021). Nanotechnology has entered various fields, such as the delivery of drugs and other biomedical applications (Bhattacharya et al., 2022). Numerous advancements in nanoscale science have resulted in an abundance of nano-dimensioned materials to aid in related research, and as a result, a range of important nano-sized materials are being created on a commercial scale (Kessler, 2011). Many potential uses of nanomaterials have yet to be discovered, and their real-world applications remain underutilized. Nanotechnology has made significant advancements as one of the concepts in science and technology that has recently experienced rapid development. With upcoming bids in a wide range of disciplines, the nanomaterial with unique physicochemical features has the potential to produce novel systems, structures, devices, and nanoplatforms. Nanoparticles, which are very small particles, have better thermal conductivity, chemical stability, catalytic reactivity, and nonlinear optical performance due to their large surface-area-to-volume ratio. Nanomaterials are particles of nanoscale dimensions (Mirzaei & Darroudi, 2017).

The field of nanotechnology has received a great deal of attention recently. The prefix "nano" surfaced in the previous decade and stems from the ancient Greek phrase for "very small" (Furxhi et al., 2020). Materials with a size of between 1 and 100 nanometers fall within the category of nanotechnology (Yap et al., 2020). Although the compound's size is what makes it a nanomaterial, its geometry and shape also significantly affect how it works. Nanoscale materials are used in almost every industry, including but not limited to agriculture, electronics, and medicine.

V. Ilango (✉)
Department of Chemistry, Birla Institute of Technology & Science Pilani, Dubai Campus, Dubai, UAE

A. K. Bhardwaj et al. (eds.), *Biogenic Wastes-Enabled Nanomaterial Synthesis*,
https://doi.org/10.1007/978-3-031-59083-2_11

Nanotechnology enables the use of nanoparticles to transform the constituents created for usage, leading to notable enhancements in mechanical, barrier, and thermal properties. The multitude of applications made possible by the fine-tuned production of different nanoparticle morphologies, including quantum dots, spheres, particles, and rods, can possibly lead to an endless chance for technological growth (Huston et al., 2021).

The technology and science underpinning downsizing have created new opportunities for addressing the production, characterization, and use of nanomaterials in society. The fact that NPs work as a bridge to control the distance between bulk materials and atomic or molecular assemblies has led to the subsequent increase in the scientific attention to nanoparticles. At the nanoscale, a number of thoroughly described bulk materials exhibit intriguing characteristics. In contrast to most materials, NPs have a high aspect ratio that allows for better reactivity and efficacy (Samuel et al., 2022). Material scientists have found carbon-based materials and mineral elemental blends that have prospective optoelectronic and dimensional properties that are higher than most of their complements in the last several decades (Christian Hazael Pérez-Beltrán et al., 2021).

The green production of nanoparticles has made commendable progress in the nanotechnology field. The objective of this synthesis is to raise the environmental friendliness of these vital particles by utilizing techniques like regulation, control, cleanup, and remediation. There will be fewer hazardous and unsustainable goods made by minimizing detrimental by-products produced during the current nanoparticle creation method. The utilization of natural resources, such as organic systems, will help achieve the objective of establishing a greener and more sustainable economy; nevertheless, the acceptance of these alternative ways will need industrial support. By decreasing the detrimental by-products of current nanoparticle manufacturing, less hazardous nonrenewable goods will be generated. The utilization of expected resources, such as organic systems, will help achieve the objective of establishing a greener and more sustainable economy; nevertheless, the acceptance of these alternative ways will need industrial support (Singh et al., 2018). The plan of this chapter is given in Fig. 1.

2 Green Chemistry

The creation of chemical products and procedures that lessen the usage or manufacture of hazardous compounds is known as "green chemistry." Green chemistry encompasses every stage of the creation, use, and disposal of a chemical product (Ahluwalia, 2013; Bryan et al., 2018). Chemists have long been fascinated by the synthesis of chemicals ranging from structurally simple to structurally complex substances. These scientific endeavors resulted in the creation of a large array of chemicals of enormous value in the fields of medicine and material research.

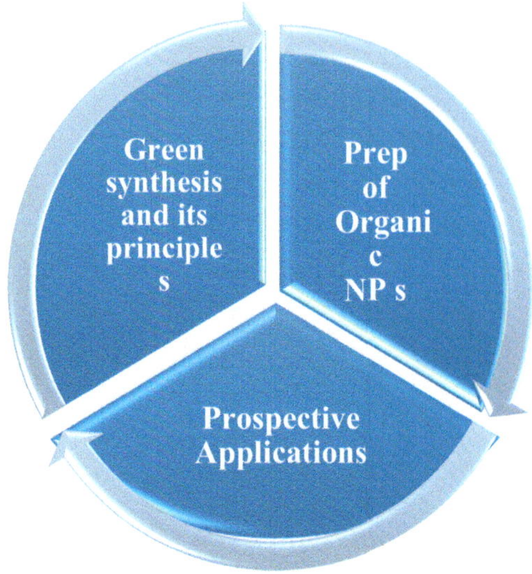

Fig. 1 Segmented cycle depicting the plan of work

2.1 Green Chemistry Objectives (Rowena, 2019)

- Green chemistry aims to avoid contamination at the molecular level.
- It is a philosophy that is applicable to all aspects of chemistry rather than just one.
- It aims to apply novel scientific solutions to real-world environmental issues.
- Source reduction occurs as a result of the prevention of pollutant creation.
- Part of the objective of green chemistry is to decrease the adverse effects of chemical methods and goods on human well-being and the atmosphere.
- It also aims to reduce and eradicate threats from prevailing processes and products.
- Finally, it resolves to develop chemical goods and techniques to decrease inherent dangers.

2.2 Green Chemistry Framework

Three main points about the green chemistry framework can be schematically represented in Fig. 2.

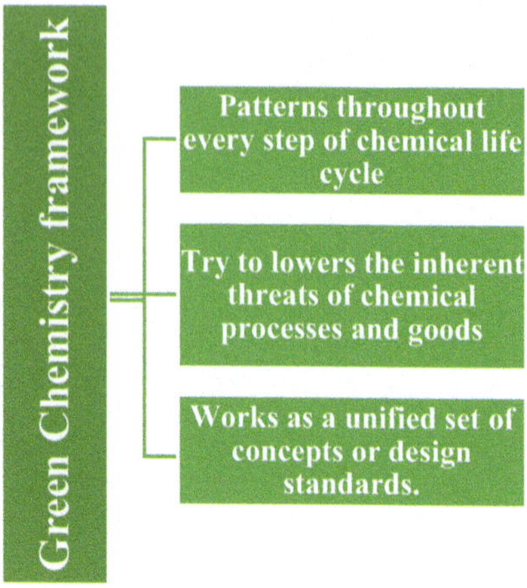

Fig. 2 Green chemistry framework

2.3 *Principles of Green Chemistry*

It is based on 12 principles, which were initially articulated by John Warner and Paul Anastas in 1998. These principles serve as a guide for creating and developing environmentally friendly and sustainable chemical processes and products and processes. The 12 green chemistry principles (Anastas & Warner, 2000; Saleh & Koller, 2018) are listed in Table 1.

This description of green chemistry principles (GCPs) in nanomaterial production resulted in efficient nanomaterials with low waste. The benefits of nanomaterial production from environmentally friendly methods such as agricultural or plant waste, sugars, and vitamins waste can be relished as an environmental reducing and capping agent. Raveendran et al. (Raveendran et al., 2003) were the first to propose the notion of producing green nanoparticles (NPs) by employing starch as a capping agent and D-glucose as a reducing agent to produce silver nanoparticles (AgNPs).

Natural polysaccharides such as chitosan, heparin (Kemp et al., 2009), and gum Arabic (Gils et al., 2010) were also used in addition to creating starch AgNPs. During the production of nanoparticles by a green approach solvent, the stabilizing agent and reducing agent must also be considered (Madani et al., 2022). The importance of green chemistry is closely tied to general rising trends in legislation, regulations, and incentives; industrial activities; as well as scientific and technological advancements involved in reimagining how materials are used. Understanding the difficulty and potential effects of green chemistry requires some experience with the environment in which it is adopted and used (Warner et al., 2004).

Table 1 Green chemistry principles

S. no.	Principle	Description	References
1	Prevention	It is better to avoid waste than to clean or treat it and clean it up after it has occurred	Ayalew and Jeevan (2022)
2	Atom economy	Artificial methods should be devised to optimize the assimilation of all components employed in the process into the finished product	Saleh et al. (2014)
3	Less hazardous chemical synthesis	Synthetic processes should be developed to employ and manufacture less harmful compounds wherever possible	Folino et al. (2020)
4	Designing safer chemicals	Chemicals that are less harmful. Chemical product design should prioritize efficacy while minimizing toxicity	Ncube et al. (2023) and Bayoumi et al. (2013)
5	Safer solvents and auxiliaries	Auxiliary compounds (e.g., solvents, separation agents, etc.) should be avoided wherever feasible and used sparingly	Murugan et al. (2016)
6	Design for energy efficiency	The environmental and economic effects of energy needs for chemical processes should be assessed, and then the needed energy input should be optimized. Wherever feasible, chemical synthesis should be carried out under moderate process conditions, i.e., at room temperature and pressure	Wale et al. (2023) and Shahzad et al. (2013)
7	Use of the renewable feedstock principle	When technologically and economically viable, synthetic processes should use renewable rather than restricted raw resources and feedstocks	Kurniawan et al. (2021) and Narodoslawsky (2010)
8	Reduction of derivatives	The temporary alteration of physical/chemical processes necessitates the use of extra reagents and frequently adds to the development of additional waste	Saleh and Koller (2018)
9	Catalysis	Stoichiometric reagents are not favored to catalytic reagents by nature; these catalysts should be as selective as feasible	Rathner et al. (2017)
10	Design for degradation	Chemical products must be designed in such a manner that when they reach the end of their useful life, they turn into harmless degradation products rather than resist in the biosphere	Cucciniello and Anastas (2021)
11	Real-time analysis for pollution prevention	Advanced analytical approaches that enable control, real-time, and in-line process monitoring, far before detrimental chemicals are formed, must be developed	Tri Wahyuni and Sri Kunarti (2022)
12	Accident prevention by inherently safer chemistry	Compounds and compound formulas used in a chemical method should be chosen in such a way that the danger of chemical mishaps, such as chemical leakage, detonations, or fire development, is minimized	Gao et al. (2021)

3 Nanomaterials

Nanomaterials have evolved as an astonishing class of materials that include a wide variety of samples with a minimum of one dimension ranging from 1 to 100 nm. On a nanometric scale, nanomaterials are typically categorized as inorganic, carbon-based, and organic particles when compared to larger quantities of respective materials with exceptional characteristics.

3.1 Classification of Nanomaterials

Organic and inorganic nanomaterials are the two primary kinds of nanomaterials depending on their carbon content. Carbon-based nanomaterials are regarded as a unique type of nanomaterial having a broad spectrum due to their various uses and extensive research. At present, nanomaterials can be classified into three material-based groups.

Carbon-Based Nanomaterials

Carbon can form covalent bonds with other carbons in different hybridization states, like Sp, Sp2, and Sp3, to generate a variety of structures, ranging from tiny molecules to longer chains, because of its remarkable catenation properties. Nanomaterials made of carbon can be shaped as spheres, ellipses, or tubes that are hollow. Graphene (Gr), carbon onions, fullerenes (C60), carbon nanotubes (CNTs), carbon nanofibers, and carbon black are the different types of carbon-based nanomaterials (Riley & Narayan, 2021).

Inorganic-Based Nanomaterials

These nanomaterials include metal-based nanoparticles, transition metal chalcogenide (TMC) nanoparticles, and metal oxide/hydroxide nanoparticles. These nanomaterials may be transformed into metal nanoparticles like gold (Au), iron (Fe), and silver (Ag) as well as metal oxides like titanium dioxide (TiO_2), ceric oxide (CeO_2), black iron oxide (Fe_3O_4), and zinc oxide (ZnO) (Bamrungsap et al., 2012).

Organic-based nanomaterials:

These nanoscale materials, with the exception of carbon-based or inorganic-based ones, are primarily formed of biological matter. Organic nanomaterials are transformed into useful structures such as dendrimers, compact polymers, ferritins, liposome nanoparticles, and micelles with the help of noncovalent interactions used for self-accumulation and molecular design. These nanoparticles are environmentally advantageous materials since they are frequently nontoxic and biodegradable (Datta et al., 2022).

Hybrid/Organic Nanoparticles

Carbon- and inorganic-based nanoparticles are collectively used to produce hybrid nanoparticles. They are manufactured to obtain numerous better drug transfer systems and carriers, such as mesoporous silica hybrid nanoparticles or lipid polymers with a lipid shell and an internal polymeric core, which are employed for a more challenging drug carrier system (Wang et al., 2015). Furthermore, the physiochemical characteristics of hybrid nanoparticles can be used in therapeutic applications such as dental implants, gene editing, and biomedical imaging (Park et al., 2020).

3.2 Synthesis of Nanoparticles

Nanoparticles produced through different methods are classified as top-down or bottom-up approaches.

1. Top-down approach: The destructive method or top-down approach is the decrease of majority materials to nanometric scale particles. A variety of procedures, including lithographic techniques, are used to prepare nanoparticles and materials. The lithographic interferometric method is the most often used top-down strategy for nanomaterial fabrication (Agarwal et al., 2017). This method includes the self-assembly of nanoparticles made of tiny atomic components. It is a rather inexpensive procedure. It is based on a process called kinetic and thermodynamic balance.

   Some of the most extensively used nanoparticle synthesis methods are thermal decomposition, laser ablation, nanolithography, mechanical milling, and sputtering.
2. Bottom-up approach: The constructive or bottom-up method involves the formation of materials ranging from atoms to clusters to nanoparticles. Bottom-up methods for producing nanoparticles include pyrolysis, chemical vapor deposition (CVD), sol-gel, spinning, and biosynthesis. This approach makes use of techniques to create nanoparticles from simpler molecules (Singh et al., 2018). These two approaches are given in Fig. 3.

   These approaches create an extensive range of nanoparticles with several uses. Some methods provide widespread scalability (Wegner et al., 2011) and regulate the morphology of nanoparticles (Amendola & Meneghetti, 2013) and applications in targeted disease therapy (Walter et al., 2010), battery conduction, energy conservation/storage (Leng et al., 2019), and electrical applications (Kubota et al., 2018).

However, it cannot be disputed that using these conventional methods has a significant negative impact. Due to the extensive use of organic solvents in the synthesis of these nanomaterials, there is a significant risk to reproductive and neurobehavioral health (Akinyemi et al., 2019). Moreover, using heat conditions and high pressure could add up to dangerous working conditions.

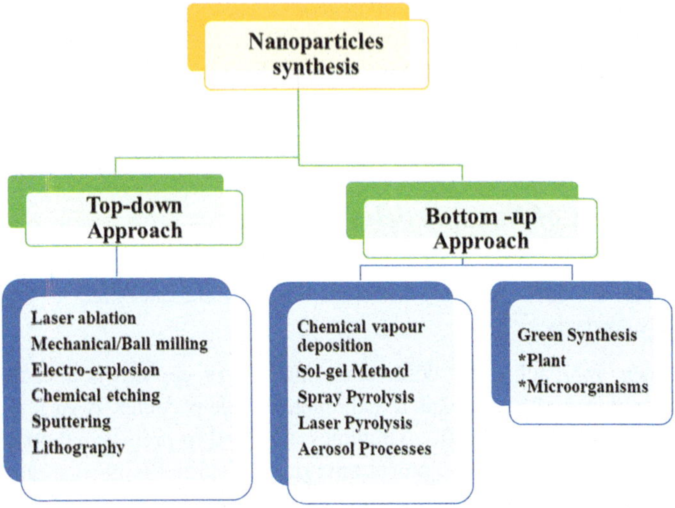

Fig. 3 Nanoparticle synthesis approaches

3.3 *Green Synthesis of Nanoparticles*

The synthesis of assorted nanoparticles using bioactive agents such as microbes and plant materials as well as various biowastes such as agricultural waste, vegetable waste, eggshells, fruit skin waste, and so on is referred to as the green synthesis of nanomaterials. The rising demand for "green" and cost-effective metal nanoparticle manufacturing techniques has motivated researchers to investigate the use of microbes, plant extracts, and other biomaterials (Kumari et al., 2021). Green nanoparticle synthesis and its uses provide insight into the creation of nanoparticles utilizing natural means. Numerous industries, including the nanotechnology and material sector, have been impacted by the demand for a more environmentally friendly and sustainable economy. The production of nanoparticles from plant extracts occurs in three phases, as given in Fig. 4.

In order to manufacture nanomaterials that are sustainable, dependable, and environmentally benign, the "green" synthesis of materials has been developed as a new field of study. Presently, many chemical and physical techniques are available for the production of nanomaterials in the literature. These techniques offer a higher production rate and well-controlled shape and size of nanomaterials, but they are not promising due to higher energy and financial losses, the use of dangerous chemicals, and the production of a substantial amount of biowaste. These crucial elements have an impact on the scale-up of nanomaterials to the commercial level, both ecologically and economically.

Additionally, due to problems with stability, biocompatibility, and toxicity, the clinical application of nanomaterials made through chemical means has been restricted. These factors increase the need for recyclable, less expensive, and

Fig. 4 Mechanism involved in the formation of green nanoparticles

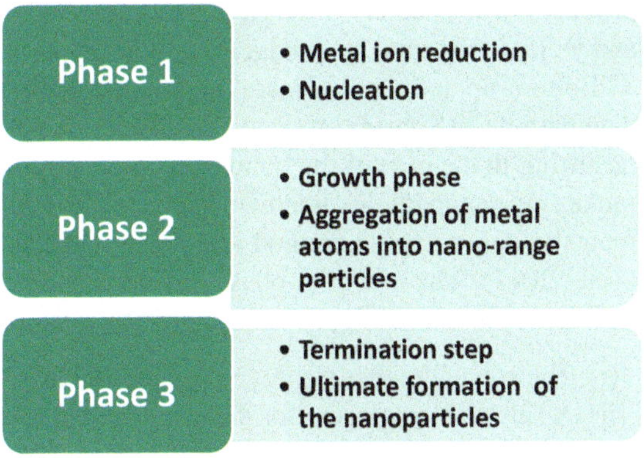

Fig. 5 Properties of nanoparticles

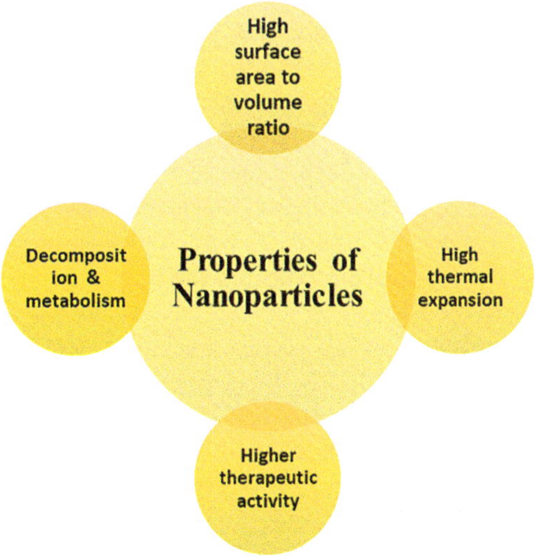

biocompatible ways of fabricating nanomaterials. The greener way for the synthesis of NPs offers cost-effective, eco-friendly, and nontoxic alternatives to traditional physical and chemical processes (Fariq et al., 2017). Green synthesis techniques are the most promising of the many synthesis methods since they do not generate any harmful chemicals or leave a tiny quantity of waste products and use environmentally friendly ingredients for the production of nanoparticles.

The worldwide vast potential of green nanoparticles and their biocompatibility for applications, such as the generation of biofuels, catalysts, and antimicrobials; biofuel cancer/gene treatment; and sensorization, have piqued the public's curiosity due to the intrinsic properties of nanomaterials, such as composition, specific form, bigger volume/superficial area, and the purity and shape of single components. The potential properties of nanoparticles are given in Fig. 5. The production method of nanoparticles differentiates them. There are various ways available for the production of nanoparticles. Both physical and chemical synthesis are expensive and yield detrimental by-products (Mukherjee et al., 2021).

Bottom-up and top-down are the two methods suggested for the biosynthesis of nanoparticles (Kalpana & Devi Rajeswari, 2018). The most important reaction occurring in the case of the bottom-up approach is oxidation/reduction. The manufacture of nanoparticles is now an important field of study that seeks an environmentally approachable method and green materials for today's reality (Rajakumar et al., 2017). The essential phases in the synthesis of nanoparticles that must be considered in terms of green chemistry are given below:

(i) The solvent media employed in the synthesis.
(ii) A safe reducing agent for the environment.
(iii) The nonhazardous material for the nanoparticle stabilization.

The bulk of the chemical and physical approaches discussed thus far are heavily reliant on organic solvents. This is mostly because the capping agents utilized are hydrophobic. Bio-organism synthesis is consistent with green chemistry concepts such as the capping agent in the reaction, the ecological approach, and the reducing agent used. Because of its peculiar features (optical, chemical, electrical, etc.), the creation of inorganic metal oxide nanoparticles utilizing biological materials has garnered a lot of interest (Dhandapani et al., 2012).

Extensive synthetic routes have been used for the synthesis of nanoparticles, with chemical, biosynthetic, and physical routes being the most common. Chemical procedures are generally excessively costly and include the use of harmful and toxic chemicals that are accountable for a variety of eco-friendly concerns (Nath & Banerjee, 2013). The biosynthetic pathway is a risk-free, biocompatible method. The biosynthetic pathway is a green, safe, and biocompatible way to produce nanoparticles for biomedical purposes utilizing microorganisms and plants (Razavi et al., 2015). This synthesis may be carried out using fungus, algae, bacteria, and plants, among other things. Because of the presence of phytochemicals in its extract, several plant components, such as roots, seeds, leaves, fruits, and stems, have been employed for the production of diverse nanoparticles (Narayanan & Sakthivel, 2011). Actually, these phytochemicals act like reducing and stabilization agents.

3.4 Factors Influencing the Green Synthesis of Various Nanoparticles

A variety of variables influence the production, characterization, and use of nanoparticles. Many research studies have documented that the nature of nanoparticles produced changes depending on the kind of adsorbate and the activity of the catalysts utilized in the manufacturing process (Somorjai & Park, 2008). Other key aspects influencing nanoparticle production consist of the temperature, the pH of the solution, the concentration of the extracts utilized, the size, the raw materials used, and, most importantly, the practices employed throughout the production process (Baker et al., 2013).

pH

The pH of the reaction plays a vital role in the structure of nanoparticles. Temperature and pH, in particular, influence the development of nucleation sites. When the pH level rises, the number of nucleation sites expands, which is critical for the creation of metal nanoparticles. It has been established that pH has an important role in determining the size and structural morphology of nanoparticles (Singh et al., 2020).

Temperature

The most essential factor influencing the dimension and shape of nanoparticles, as well as their amount of synthesis, is temperature. Physical techniques necessitate the maximum temperature (greater than 350 °C), while chemical methods demand a temperature less than 350 °C. In most situations, green technology is used to synthesize nanoparticles at temperatures lower than 100 °C or at room temperature. The type of nanoparticle generated is determined by the temperature of the reaction medium (Rai et al., 2006). During the green synthesis of nanoparticles, reaction time is a crucial component that utmost influences the size, yield, and shape of produced nanoparticles (Roy et al., 2022).

Pressure

The importance of pressure in the synthesis of nanoparticles cannot be overstated. The shape and size of the produced nanoparticles are affected by the pressure applied to the reaction media. When we make use of biological agents, the rate of metal ion reduction was discovered to be expressively quicker under ambient pressure settings (Tran et al., 2013).

Reaction Time

The kind and quality of nanoparticles produced using green technology are significantly influenced by the length of time the reaction media is incubated (Darroudi, 2011). Similar to this, the storage conditions, synthesis process, exposure to light, and other factors all had an impact on the characteristics of the newly generated nanoparticles over time (Kuchibhatla et al., 2012). Particles may aggregate as a consequence of long-term preservation, they may shrink or expand as a result of long-term storage, they may have a shelf life, and so forth. All of these variations in time have an impact on the long-term viability of the particles.

Pore Size

The degree of porosity of the generated nanoparticles has a significant impact on the quality and utilization of nanoparticles. The immobilization of biomolecules onto nanoparticles has been accomplished in order to broaden their utility in the delivery of drugs and biological applications (Ruckenstein & Kong, 1999).

Particle Shape and Size

The features of nanoparticles are significantly influenced by their size. For instance, it has been shown that when nanoparticles go closer to the nanoscale, their melting point drops (Akbari et al., 2011). The sort of energy frequently employed during nanoparticle investigation causes the nanoparticle's structure to vary. The chemical characteristics of manufactured nanoparticles are highly influenced by their dynamic nature and form (Patra & Baek, 2014).

Environment

The nature of the generated nanoparticles is significantly influenced by the environment outside. When a single nanoparticle absorbs substances or interacts with other environmental elements via oxidation or corrosion, it frequently swiftly splits into core-shell nanoparticles (Sarathy et al., 2008). In a biological system, the produced nanoparticles develop a covering that thickens and expands them (Lynch et al., 2007). Furthermore, the environment influences the physical structure and chemistry of the produced nanoparticles.

Other Factors

Secondary metabolites are abundant in various biological systems, including plants and function as stabilizing and reducing agents in the creation of nanoparticles. However, the content of these metabolites differs depending on the kind of plant, plant component, and extraction method utilized (Park et al., 2011). Different bacteria produce varied amounts of internal and extracellular enzymes that influence nanoparticle production (Haverkamp et al., 2007).

Additionally, both the quantity and quality of the created nanoparticles may be impacted by the method used to purify them.

4 Green Synthesis of Organic Nanoparticles

Organic nanoparticles are two-dimensional materials with sizes ranging from 1 to 100 nm. Nanoparticles have exceptional size-dependent physical and chemical characteristics, such as magnetic, catalytic, optical, electrochemical, and thermodynamic properties. A nanoparticle's shape and chemical composition can have an impact on its unique characteristics. Liposomes, solid-lipid nanoparticles, polymeric nanoparticles, nanoemulsions, dendrimers, and carbon nanoparticles are the types of organic nanoparticles, as given in Fig. 6.

4.1 Nanoliposomes

Liposomes are bilayered lipid nanocarriers consisting of a core that is aqueous, and it is encircled by synthetic or natural phospholipids (Fathi et al., 2012). These are used in lipophilic vitamin, mineral, and antibacterial formulations. Furthermore, the liposomes-lipid vesicles may be commanded to transport and release their content in specific locations, both inside and outside of the body (Yu et al., 2018). Thin-layer coating using a mix of cholesterol (30%) and soybean phosphatidylcholine (70%) was used to prepare a liposomal system containing essential oils. At 45 °C, the contents were dissolved in chloroform and dried under vacuum. By adding sterilized distilled water, hydration for 1 h time was carried at 55 °C. The nanoliposomes were subsequently decreased in size for 40 min using an ultrasonic bath sonicator. To homogenize the component particles, size reduction was implemented using 0.45 and 0.20 m filters (Mirzaei et al., 2020).

4.2 Solid Lipid Nanoparticles (SLNs)

SLNs are colloidal nanocarriers first produced in the year 1991 as a substitute for emulsion, nanoparticle-based systems, and liposomes. They are circular in shape, with a diameter varying from 50 to 500 nm, and are stable at body and room temperatures. Solid lipid nanoparticles consist of solid lipids with biocompatible and bioactive components as the core constituents surrounded by an emulsifier that stabilizes the particles. Because of the solid matrix, these SLNs feature significant flexibility in regulating the release profile, excellent encapsulation competence, no organic solvents, and a slower degradation rate during particle formation. All these features make SLNs more advantageous compared to liposomes and nanoemulsions (Musielak et al., 2022).

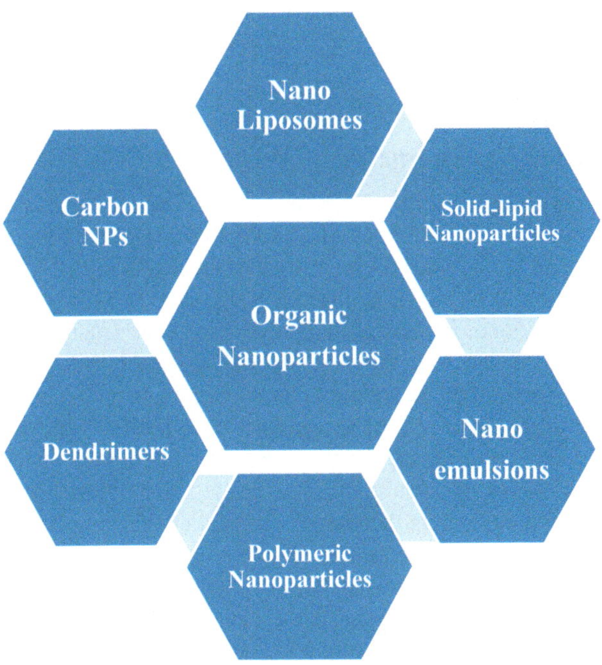

Fig. 6 Types of organic nanoparticles

4.3 Nanoemulsions

Nanoemulsions are nanoscale lipid-based droplets formed by the dispersal of one liquid in another immiscible liquid. Based on their morphology, they can be classified as either water-based or oil-in-water-based emulsions, with oil as the dispersed phase and water as the continuous phase (Singh et al., 2017). According to nanotechnology's definition, the ideal size of nanoemulsion should be less than 100 nm; however, the size of the droplet can be considerably lower than the visible wavelength, resulting in most nanoemulsions appearing to be crystal clear. These nanoemulsions have the potential to increase bioactive solubilization, such as those of poorly water-soluble dietary constituents (lipophilic vitamins and fish oil), by changing the penetrability of the gastrointestinal system through surfactants.

Green coffee nanoemulsions were made utilizing the impulsive emulsification technique by using various quantities of green coffee oil, green coffee extract, hydrophilic tensoactive poloxamer, and soy lecithin (Buzanello et al., 2020). Micelles are nanoemulsion amphiphilic surfactant molecules made up of amphiphilic compounds and lipids. Micelles instinctively combine and self-accumulate into sphere-shaped vesicles with a hydrophilic external monolayer and a hydrophobic core in the aqueous environment and may thus be utilized to integrate hydrophobic medicinal substances (Sim & Wong, 2021). Because of the unique features of micelles, hydrophobic medicines can be made more soluble, enhancing bioavailability. Micelles have diameters ranging from 10 to 100 nm. Micelles are used in various applications, including imaging agents, drug delivery agents, therapeutic agents, and contrast agents.

4.4 Polymeric Nanoparticles

Polymeric nanoparticles are made up of a matrix that consists of polymers like albumin, chitosan, and alginic acid enclosed by surfactants like lecithin. They are of a size between 20 and 1000 nm. The bioactive component is often confined inside the polymer matrix's core or in between the polymer and surfactant layers. The bioactive trapping type is determined by the polymer and by bioactive characteristics. The polymer utilized might be natural or synthetic; the latter can be substituted by natural and biocompatible polymeric nanoparticles, such as polysaccharides like chitosan, pullulan, carrageenan, alginic acid, xanthan gum, etc. and proteins like casein, β-lactoglobulin, and zein (Sabliov & Astete, 2015). However, due to the difficulties of properly regulating the various intricate chemical or thermal treatments employed in the process, the approach for generating natural polymer-based nanocapsules does not have the potential to be fully scaled up. Several aspects, including their physicochemical features (molar mass, size, and structure), and external influences such as temperature and pH might impact the breakdown rate of biodegradable polymeric nanoparticles (Su & Kang, 2020).

4.5 Dendrimers

The term "dendrimer" is a combination of two Greek words, "dendron" and "meros," which respectively mean "tree" and "parts," which explains their branching form (Carmo et al., 2013). Dendrimers are made up of three basic parts: a core, an inner shell, and an outer shell. A dendrimer may be synthesized with varying functionality in each of these sections to regulate traits such as heat stability, compound attachment, and solubility for specific purposes. Polyethylene glycol (PEG) surface modification has been used on dendrimers of several sorts to enable biological application (Kojima et al., 2000). Indeed, increasing the hydrophobicity of the dendrimer backbone has been demonstrated to significantly improve dendrimer encapsulation ability. Drug-dendrimer interaction is advantageous because it promotes solubility, which improves the absorption and bioavailability of the drug molecule or its cytotoxicity (Farooq et al., 2019).

4.6 Carbon-Based Nanomaterials

Carbon nanotubes are tubular structures comprised of carbon that are stacked in graphene sheets, which are rolled to create cylinders. Carbon nanotubes (CNTs) are divided into two types: single-walled nanotubes (SWNTs) and multiwalled nanotubes (MWNTs). Single-walled nanotubes are composed of a single graphene sheet rolled up to hollow cylinders, whereas multiwalled nanotubes are composed of several concentric cylindrical shells of graphene sheets surrounding a central hollow core (Kierkowicz et al., 2018). The addition of wood sawdust to spray pyrolysis produced CNPs of high quantity and quality as well as carbon fibers and nanotubes. Hydrogen was obtained as a by-product, together with a small amount of poisonous gases (CO, NO_2) (Ghosh et al., 2007). CNTs made from turpentine oil showed fewer flaws as well as a significantly greater current density capability and field emission strength (Ghosh et al., 2008). Rice husk (RH) has been employed as a carbon precursor as well as a catalyst source in the microwave treatment of CNTs (Asnawi et al., 2018).

4.7 Prospective Applications of Green-Synthesized Organic Nanoparticles

Recent advancements in green chemistry have led to the creation of straightforward, quick, and cost-effective techniques for producing nanoparticles, sometimes referred to as green synthesis (Selvakesavan & Franklin, 2021). Organic nanoparticles at the nanoscale can be employed to encapsulate agrochemicals such as fertilizers, insecticides, herbicides, and plant growth regulators. This enables the precise administration of plant protection inputs by facilitating gradual absorption,using active compounds, and decreasing agrochemical usage through input and waste reduction (Lorenzo et al., 2021). By working in concert with one another, a wide range of substances derived from plants or microbes, such as terpenes, polyphenols, alkaloids, carbohydrates, proteins, and genetic elements, contribute significantly to the production of nanoparticles (Álvarez-Chimal & Arenas-Alatorre, 2023). The potential applications of green-synthesized organic nanoparticles are summarized in Table 2.

5 Conclusions

Green nanomaterial synthesis methods do not harm the environment or the people involved in their fabrication. They are as effective as traditional synthesis methods. This type of synthesis method has enormous potential; still, it encounters barriers in raw material selection, reaction conditions, product quality control, and application. These issues restrict the development of green nanomaterials for production and their widespread application. There is no need for high pressure, energy, temperature, or harmful chemicals in the green synthesis approach. As a result, many researchers are increasingly deviating from utilizing synthetic approaches. Plants create more stable nanoparticles than other methods, and scaling up is quite simple. Contamination is also less likely.

Researchers studied a variety of locally available plants and determined that they are promising materials for manufacturing green NPs. These studies indicate that it is feasible to totally rely on native plants, yet significant global nanomaterial production is problematic. The use of nanoparticles in the medical, food, pharmaceutical, and agricultural industries has sparked a lot of interest, with a focus on developing more convenient methods for producing eco-friendly, nontoxic, and environmentally benign nanoparticles using green biotechnology tools. Many factors impact the quality and quantity of produced nanoparticles for prospective usage in a variety of applications.

Green nanomaterial production technologies and procedures are currently available, with more on the way in the future. In today's culture, innovative answers to problems that include automobile and stationary technologies are becoming increasingly important. As a result, the first component of green nanotechnology is to employ more environmentally friendly or energy-efficient production methods. The ability to continuously modify and improve the physical and chemical properties of nanomaterials that are beneficial to the greater scientific community is made possible by the identification of active molecules in the environmentally friendly synthesis of nanomaterials.

As the globe continues to adapt to climate change, further advancement in the green synthesis of nanomaterials is critical for the preservation of environmental health, energy, and scientific research ethics. More study is needed to evaluate and appreciate the true green synthesis of organic nanoparticles. This study is intended

Table 2 Prospective applications of organic nanoparticles

S. No.	Organic nanoparticles	Applications	References
1	Chitosan nanofibers, nanopowder, and nanoparticles	Dental field—nerve revival treatment Probable skin substitute Anticancer treatment—controlled drug delivery and active targeting	Ibrahim et al. (2023), Razavi et al. (2015), Lin et al. (2015), and Mansur et al. (2016)
2	Nanoliposomes	Cancer therapy—hormonal and targeted therapy Food industry—controlling flavor release and boosting dietary component bioavailability	Taléns-Visconti et al. (2022), Misra et al. (2010), and Chaudhry et al. (2008)
3	Solid lipid nanoparticles	Colon tumor—potential for colon cancer treatment Prostate cancer—inhibits prostate cancer cells	Rajpoot and Jain (2017) and Akanda et al. (2015)
4	Nanoemulsions	Food industry—design of smart food Pharmaceutical industry—drug release in bioactive form Nanomicelles—intravenous drug delivery	Mushtaq et al. (2023), Gupta et al. (2016), Wilson et al. (2021), and Bae et al. (2005)
5	Polymeric nanoparticles	Cancer diagnosis and imaging Breast cancer—nanoparticle-mediated drug delivery Liver cancer—to increase the efficiency of anticancer medications	Thakor and Gambhir (2013), Khanna et al. (2019), and Zhu et al. (2016)
6	Dendrimers	Drug delivery carriers Efficient immune stimulating compounds—increase the efficiency of vaccines	Noriega-Luna et al. (2014), Heegaard et al. (2010)
7	Carbon-based nanomaterials	Drug delivery, single-cell imaging, biosensing, and protein purification Drug formulation reagents	Qasim et al. (2023) and O'Driscoll and Griffin (2008)

to serve as a comprehensive resource for academics and scientists working in the field, as well as to stimulate future investigation and invention in this rapidly emerging area of nanoscience and green chemistry.

References

Agarwal, H., Venkat Kumar, S., & Rajeshkumar, S. (2017). A review on green synthesis of zinc oxide nanoparticles – An eco-friendly approach. *Resource-Efficient Technologies, 3*(4), 406–413. https://doi.org/10.1016/j.reffit.2017.03.002

Ahluwalia, V. K. (2013). *Green chemistry: A textbook*. Google Books. Alpha Science International. Available at: https://books.google.ae/books/about/Green_Chemistry.html?id=1KJuMAEACAAJ&redir_esc=y. Accessed 30 Apr 2023

Akanda, M. H., Rai, R., Slipper, I. J., Chowdhry, B. Z., Lamprou, D., Getti, G., & Douroumis, D. (2015). Delivery of retinoic acid to LNCap human prostate cancer cells using solid lipid nanoparticles. *International Journal of Pharmaceutics, 493*(1–2), 161–171. https://doi.org/10.1016/j.ijpharm.2015.07.042

Akbari, B., Pirhadi Tavandashti, M., & Zandrahimi, M. (2011). Particle size characterization of nanoparticles – A practicalapproach. *Iranian Journal of Materials Science and Engineering, 8*(2), 48–56. Available at: http://ijmse.iust.ac.ir/browse.php?a_code=A-10-3-149&sid=1&slc_lang=en. Accessed 7 May 2023.

Akinyemi, P. A., Adegbenro, C. A., Ojo, T. O., & Elugbaju, O. (2019). Neurobehavioral effects of organic solvents exposure among wood furniture makers in Ile-Ife, Osun state, southwestern Nigeria. *Journal of Health and Pollution, 9*(22). https://doi.org/10.5696/2156-9614-9.22.190604

Álvarez-Chimal, R., & Arenas-Alatorre, J. (2023). *Green synthesis of nanoparticles. A biological approach*. IntechOpen eBooks. https://doi.org/10.5772/intechopen.1002203

Amendola, V., & Meneghetti, M. (2013). What controls the composition and the structure of nanomaterials generated by laser ablation in liquid solution? *Physical Chemistry Chemical Physics, 15*(9), 3027–3046. https://doi.org/10.1039/c2cp42895d

Anastas, P. T., & Warner, J. C. (2000). *Green chemistry: Theory and practice*. Google Books. Oxford University Press. Available at: https://books.google.ae/books/about/Green_Chemistry.html?id=_iMORRU42isC&redir_esc=y. Accessed 30 Apr 2023.

Asnawi, M., Azhari, S., Hamidon, M. N., Ismail, I., & Helina, I. (2018). Synthesis of carbon Nanomaterials from Rice husk via microwave oven. *Journal of Nanomaterials, 2018*, 1–5. https://doi.org/10.1155/2018/2898326

Ayalew, M. E., & Jeevan, T. S. M. A. (2022). The Technology of Green Chemistry and its Function in redox response and: Environmentally friendly Technology for Sustainable Development: Assessment of recent findings. *Advances in Chemical Engineering and Science, 12*(03), 131–144. https://doi.org/10.4236/aces.2022.123010

Bae, Y., Nishiyama, N., Fukushima, S., Koyama, H., Yasuhiro, M. and Kataoka, K. (2005). Preparation and biological characterization of polymeric micelle drug carriers with intracellular pH-triggered drug release property: Tumor permeability, controlled subcellular drug distribution, and enhanced in vivo antitumor efficacy. Bioconjugate Chemistry, 16(1), pp. 122–130. https://doi.org/10.1021/bc0498166.

Baker, S., Rakshith, D., Kavitha, K. S., Santosh, P., Kavitha, H. U., Rao, Y., & Satish, S. (2013). Plants: Emerging as Nanofactories towards facile route in synthesis of nanoparticles. *BioImpacts: BI, 3*(3), 111–117. https://doi.org/10.5681/bi.2013.012

Bamrungsap, S., Zhao, Z., Chen, T., Wang, L., Li, C., Fu, T., & Tan, W. (2012). Nanotechnology in therapeutics: A focus on nanoparticles as a drug delivery system. *Nanomedicine, 7*(8), 1253–1271. https://doi.org/10.2217/nnm.12.87

Bayoumi, T. A., Saleh, H. M., & Eskander, S. B. (2013). Solidification of hot real radioactive liquid scintillator waste using cement–clay composite. *Monatshefte Fur Chemie, 144*(12), 1751–1758. https://doi.org/10.1007/s00706-013-1065-9

Bhattacharya, T., Soares, G. A. B. E., Chopra, H., Rahman, M. M., Hasan, Z., Swain, S. S., & Cavalu, S. (2022). Applications of Phyto-nanotechnology for the treatment of neurodegenerative disorders. *Materials, 15*(3), 804. https://doi.org/10.3390/ma15030804

Bryan, M. C., Dunn, P. J., Entwistle, D., Gallou, F., Koenig, S. G., Hayler, J. D., Hickey, M. R., Hughes, S., Kopach, M. E., Moine, G., Richardson, P., Roschangar, F., Steven, A., & Weiberth, F. J. (2018). Key green chemistry research areas from a pharmaceutical manufacturers' perspective revisited. *Green Chemistry, 20*(22), 5082–5103. https://doi.org/10.1039/c8gc01276h

Buzanello, E. B., Pinheiro Machado, G. T. B., Kuhnen, S., Mazzarino, L., & Maraschin, M. (2020). Nanoemulsions containing oil and aqueous extract of green coffee beans with antioxidant and antimicrobial activities. *Nano Express, 1*(1), 010058. https://doi.org/10.1088/2632-959x/ab9c47

Carmo, D. (2013). Synthesis and a preliminary characterization of poly (Propylene) imine hexadecylamine dendrimer (DAB-Am-16) modified with methyl acrylate. *American Chemical Science Journal, 3*(3), pp.314–324. https://doi.org/10.9734/acsj/2013/3818

Chaudhry, Q., Scotter, M., Blackburn, J., Ross, B., Boxall, A., Castle, L., Aitken, R., & Watkins, R. (2008). Applications and implications of nanotechnologies for the food sector. *Food Additives & Contaminants: Part A, 25*(3), 241–258. https://doi.org/10.1080/02652030701744538

Cucciniello, R. and Anastas, P.T. (2021). Design for degradation or recycling for reuse? Current Opinion in Green and Sustainable Chemistry, 31, 100528. https://doi.org/10.1016/j.cogsc.2021.100528.

Darroudi, M. (2011). Time-dependent effect in green synthesis of silver nanoparticles. *International Journal of Nanomedicine, 6*, 677. https://doi.org/10.2147/ijn.s17669

Datta, D., Deepak, K. S., & Das, B. K. (2022). *Membranes with FNMs for sustainable development* (pp. 355–387). Elsevier eBooks. https://doi.org/10.1016/b978-0-323-85946-2.00006-0

Dhandapani, P., Maruthamuthu, S., & Rajagopal, G. (2012). Bio-mediated synthesis of TiO_2 nanoparticles and its photocatalytic effect on aquatic biofilm. *Journal of Photochemistry and Photobiology B: Biology, 110*, 43–49. https://doi.org/10.1016/j.jphotobiol.2012.03.003

Fariq, A., Khan, T., & Yasmin, A. (2017). Microbial synthesis of nanoparticles and their potential applications in biomedicine. *Journal of Applied Biomedicine, 15*(4), 241–248. https://doi.org/10.1016/j.jab.2017.03.004

Farooq, M. A., Aquib, M., Farooq, A., Haleem Khan, D., Joelle Maviah, M. B., Sied Filli, M., Kesse, S., Boakye-Yiadom, K. O., Mavlyanova, R., Parveen, A., & Wang, B. (2019). Recent progress in nanotechnology-based novel drug delivery systems in designing of cisplatin for cancer therapy: An overview. *Artificial Cells, Nanomedicine, and Biotechnology, 47*(1), 1674–1692. https://doi.org/10.1080/21691401.2019.1604535

Fathi, M., Mozafari, M. R., & Mohebbi, M. (2012). Nanoencapsulation of food ingredients using lipid based delivery systems. *Trends in Food Science & Technology, 23*(1), 13–27. https://doi.org/10.1016/j.tifs.2011.08.003

Folino, A., Karageorgiou, A., Calabrò, P. S., & Komilis, D. (2020). Biodegradation of wasted bioplastics in natural and industrial environments: A review. *Sustainability, 12*(15), 6030. https://doi.org/10.3390/su12156030

Furxhi, I., Murphy, F., Mullins, M., Arvanitis, A., & Poland, C. A. (2020). Practices and trends of machine learning application in Nanotoxicology. *Nanomaterials, 10*(1), 116. https://doi.org/10.3390/nano10010116

Gao, X., Abdul Raman, A. A., Hizaddin, H. F., Bello, M. M., & Buthiyappan, A. (2021). Review on the inherently safer design for chemical processes: Past, present and future. *Journal of Cleaner Production, 305*, 127154. https://doi.org/10.1016/j.jclepro.2021.127154

Ghosh, P., Afre, R. A., Soga, T., & Jimbo, T. (2007). A simple method of producing single-walled carbon nanotubes from a natural precursor: Eucalyptus oil. *Materials Letters, 61*(17), 3768–3770. https://doi.org/10.1016/j.matlet.2006.12.030

Ghosh, P., Soga, T., Tanemura, M., Zamri, M., Jimbo, T., Katoh, R., & Sumiyama, K. (2008). Vertically aligned carbon nanotubes from natural precursors by spray pyrolysis method and their field electron emission properties. *Applied Physics A, 94*(1), 51–56. https://doi.org/10.1007/s00339-008-4856-9

Gils, P. S., Ray, D., & Sahoo, P. K. (2010). Designing of silver nanoparticles in gum arabic based semi-IPN hydrogel. *International Journal of Biological Macromolecules, 46*(2), 237–244. https://doi.org/10.1016/j.ijbiomac.2009.12.014

Gupta, A., Eral, H. B., Hatton, T. A., & Doyle, P. S. (2016). Nanoemulsions: Formation, properties and applications. *Soft Matter, 12*(11), 2826–2841. https://doi.org/10.1039/c5sm02958a

Haverkamp, R. G., Marshall, A. T., & van Agterveld, D. (2007). Pick your carats: Nanoparticles of gold–silver–copper alloy produced in vivo. *Journal of Nanoparticle Research, 9*(4), 697–700. https://doi.org/10.1007/s11051-006-9198-y

Heegaard, P. M. H., Boas, U., & Sorensen, N. S. (2010). Dendrimers for vaccine and Immunostimulatory uses. A review. *Bioconjugate Chemistry, 21*(3), 405–418. https://doi.org/10.1021/bc900290d

Huston, M., DeBella, M., DiBella, M., & Gupta, A. (2021). Green synthesis of nanomaterials. *Nanomaterials, 11*(8), 2130. https://doi.org/10.3390/nano11082130

Ibrahim, M. A., Alhalafi, M. H., Emam, E.-A. M., Ibrahim, H. M., & Mosaad, R. M. (2023). A review of chitosan and chitosan nanofiber: Preparation, characterization, and its potential applications. *Polymers, 15*(13), 2820–2820. https://doi.org/10.3390/polym15132820

Kalpana, V. N., & Devi Rajeswari, V. (2018). A review on green synthesis, biomedical applications, and toxicity studies of ZnO NPs. *Bioinorganic Chemistry and Applications, 2018*, 1–12. https://doi.org/10.1155/2018/3569758

Kemp, M. M., Kumar, A., Mousa, S., Dyskin, E., Yalcin, M., Ajayan, P., Linhardt, R. J., & Mousa, S. A. (2009). Gold and silver nanoparticles conjugated with heparin derivative possess anti-angiogenesis properties. *Nanotechnology, 20*(45), 455104. https://doi.org/10.1088/0957-4484/20/45/455104

Kessler, R. (2011). Engineered nanoparticles in consumer products: Understanding a NewIngredient. *Environmental Health Perspectives, 119*(3), a120. https://doi.org/10.1289/ehp.119-a120

Khanna, V., Kalscheuer, S., Kirtane, A., Zhang, W., & Panyam, J. (2019). Perlecan-targeted nanoparticles for drug delivery to triple-negative breast cancer. *Future Drug Discovery, 1*(1), FDD8. https://doi.org/10.4155/fdd-2019-0005

Kierkowicz, M., Pach, E., Santidrián, A., Sandoval, S., Gonçalves, G., Tobías-Rossell, E., Kalbáč, M., Ballesteros, B., & Tobias, G. (2018). Comparative study of shortening and cutting strategies of single-walled and multi-walled carbon nanotubes assessed by scanning electron microscopy. *Carbon, 139*, 922–932. https://doi.org/10.1016/j.carbon.2018.06.021

Kojima, C., Kono, K., Maruyama, K., & Takagishi, T. (2000). Synthesis of polyamidoamine dendrimers having poly(ethylene glycol) grafts and their ability to encapsulate anticancer drugs. *Bioconjugate Chemistry, 11*(6), 910–917. https://doi.org/10.1021/bc0000583

Kubota, K., Dahbi, M., Hosaka, T., Kumakura, S., & Komaba, S. (2018). Towards K-Ion and Na-ion batteries as 'Beyond Li-Ion'. *The Chemical Record, 18*(4), 459–479. https://doi.org/10.1002/tcr.201700057

Kuchibhatla, S. V. N. T., Karakoti, A. S., Baer, D. R., Samudrala, S., Engelhard, M. H., Amonette, J. E., Thevuthasan, S., & Seal, S. (2012). Influence of aging and environment on nanoparticle chemistry: Implication to confinement effects in Nanoceria. *The Journal of Physical Chemistry C, 116*(26), 14108–14114. https://doi.org/10.1021/jp300725s

Kumari, S. C., Dhand, V., & Padma, P. N. (2021). Green synthesis of metallic nanoparticles: A review. *Nanomaterials*, 259–281. https://doi.org/10.1016/b978-0-12-822401-4.00022-2

Kurniawan, Y. S., Priyangga, K. T. A., Krisbiantoro, P. A., & Imawan, A. C. (2021). *Green chemistry influences in organic synthesis: A review*. Social Science Research Network. Available at: https://ssrn.com/abstract=3847465. Accessed 25 Sep 2023

Leng, J., Wang, Z., Wang, J., Wu, H.-H., Yan, G., Li, X., Guo, H., Liu, Y., Zhang, Q., & Guo, Z. (2019). Advances in nanostructures fabricated via spray pyrolysis and their applications in energy storage and conversion. *Chemical Society Reviews, 48*(11), 3015–3072. https://doi.org/10.1039/c8cs00904j

Lin, H.-Y., Chen, S.-H., Chang, S.-H., & Huang, S.-T. (2015). Tri-layered chitosan scaffold as a potential skin substitute. *Journal of Biomaterials Science, Polymer Edition, 26*(13), 855–867. https://doi.org/10.1080/09205063.2015.1061350

Lorenzo, V. D., Schiavi, D., Petrucci, C., & Balestra, G. M. (2021). Organic nanoparticles from lignocellulosic agricultural residues for plant protection applications. *NanoInnovation, 2020*. https://doi.org/10.1063/5.0069309

Lynch, I., Cedervall, T., Lundqvist, M., Cabaleiro-Lago, C., Linse, S., & Dawson, K. A. (2007). The nanoparticle–protein complex as a biological entity; a complex fluids and surface science challenge for the 21st century. *Advances in Colloid and Interface Science, 134-135*, 167–174. https://doi.org/10.1016/j.cis.2007.04.021

Madani, M., Hosny, S., Alshangiti, D. M., Nady, N., Alkhursani, S. A., Alkhaldi, H., Al-Gahtany, S. A., Ghobashy, M. M., & Gaber, G. A. (2022). Green synthesis of nanoparticles for varied applications: Green renewable resources and energy-efficient synthetic routes. *Nanotechnology Reviews, 11*(1), 731–759. https://doi.org/10.1515/ntrev-2022-0034

Mansur, A. A. P., de Carvalho, S. M., & Mansur, H. S. (2016). Bioengineered quantum dot/chitosan-tripeptide nanoconjugates for targeting the receptors of cancer cells. *International Journal of Biological Macromolecules, 82*, 780–789. https://doi.org/10.1016/j.ijbiomac.2015.10.047

Mirzaei, H., & Darroudi, M. (2017). Zinc oxide nanoparticles: Biological synthesis and biomedical applications. *Ceramics International, 43*(1), 907–914. https://doi.org/10.1016/j.ceramint.2016.10.051

Mirzaei, F., Norouzi, R., Siyadatpanah, A., Haghirosadat, B. F., Rezaei, F., Nissapatorn, V., Mitsuwan, W., Nawaz, M., Pereira, M. L., Hosseini, S. A., Montazeri, M., & Majdizadeh, M. (2020). Green synthesis of nano-liposomes containing Bunium persicum and Trachyspermum ammi essential oils against Trichomonas vaginalis. *Journal of Microbiology, Immunology and Infection*. https://doi.org/10.21203/rs.3.rs-32930/v5

Misra, R., Acharya, S., & Sahoo, S. K. (2010). Cancer nanotechnology: Application of nanotechnology in cancer therapy. *Drug Discovery Today, 15*(19–20), 842–850. https://doi.org/10.1016/j.drudis.2010.08.006

Mukherjee, A., Sarkar, D., & Sasmal, S. (2021). A review of green synthesis of metal nanoparticles using algae. *Frontiers in Microbiology, 12*. https://doi.org/10.3389/fmicb.2021.693899

Murugan, P., Han, L., Gan, C.-Y., Maurer, F. H. J., & Sudesh, K. (2016). A new biological recovery approach for PHA using mealworm, Tenebrio molitor. *Journal of Biotechnology, 239*, 98–105. https://doi.org/10.1016/j.jbiotec.2016.10.012

Mushtaq, A., Wani, S. M., Malik, A. R., Gull, A., Ramniwas, S., Nayik, G. A., Ercisli, S., Marc, R. A., Ullah, R., & Bari, A. (2023). Recent insights into Nanoemulsions: Their preparation, properties and applications. *Food Chemistry: X, 18*, 100684–100684. https://doi.org/10.1016/j.fochx.2023.100684

Musielak, E., Feliczak-Guzik, A., & Nowak, I. (2022). Synthesis and potential applications of lipid nanoparticles in medicine. *Materials, 15*(2), 682. https://doi.org/10.3390/ma15020682

Narayanan, K. B., & Sakthivel, N. (2011). Green synthesis of biogenic metal nanoparticles by terrestrial and aquatic phototrophic and heterotrophic eukaryotes and biocompatible agents. *Advances in Colloid and Interface Science, 169*(2), 59–79. https://doi.org/10.1016/j.cis.2011.08.004

Narodoslawsky, M. (2010). Structural prospects and challenges for bio commodity processes. *Food Technology and Biotechnology, 48*(3), 270–275. Available at: https://hrcak.srce.hr/57558. Accessed 27 Apr 2023.

Nath, D., & Banerjee, P. (2013). Green nanotechnology—A new hope for medical biology. *Environmental Toxicology and Pharmacology, 36*(3), 997–1014. https://doi.org/10.1016/j.etap.2013.09.002

Ncube, A., Mtetwa, S., Bukhari, M., Fiorentino, G., & Passaro, R. (2023). Circular economy and green chemistry: The need for radical innovative approaches in the Design for new Products. *Energies, 16*(4), 1752. https://doi.org/10.3390/en16041752

Noriega-Luna, B., Godínez, L. A., Rodríguez, F. J., Rodríguez, A., Zaldívar-Lelo de Larrea, G., Sosa-Ferreyra, C. F., Mercado-Curiel, R. F., Manríquez, J., & Bustos, E. (2014). Applications of dendrimers in drug delivery agents, diagnosis, therapy, and detection. *Journal of Nanomaterials*. Available at: https://www.hindawi.com/journals/jnm/2014/507273/

O'Driscoll, C. M., & Griffin, B. T. (2008). Biopharmaceutical challenges associated with drugs with low aqueous solubility—The potential impact of lipid-based formulations. *Advanced Drug Delivery Reviews, 60*(6), 617–624. https://doi.org/10.1016/j.addr.2007.10.012

Park, Y., Hong, Y. N., Weyers, A., Kim, Y. S., & Linhardt, R. J. (2011). Polysaccharides and phytochemicals: A natural reservoir for the green synthesis of gold and silver nanoparticles. *IET Nanobiotechnology, 5*(3), 69. https://doi.org/10.1049/iet-nbt.2010.0033

Park, W., Shin, H., Choi, B., Rhim, W.-K., Na, K., & Keun Han, D. (2020). Advanced hybrid nanomaterials for biomedical applications. *Progress in Materials Science, 114*, 100686. https://doi.org/10.1016/j.pmatsci.2020.100686

Patra, J. K., & Baek, K.-H. (2014). Green Nanobiotechnology: Factors affecting synthesis and characterization techniques. *Journal of Nanomaterials, 2014*, 1–12. https://doi.org/10.1155/2014/417305

Pérez-Beltrán, C. H., García-Guzmán, J. J., Ferreira, B. I., Estévez-Hernández, O., López-Iglesias, D., Cubillana-Aguilera, L., Dallavalle, S., Stanica, N., Rosa, A. M., & Palacios-Santander, J. M. (2021). One-minute and green synthesis of magnetic iron oxide nanoparticles assisted by design of experiments and high energy ultrasound: Application to biosensing and immunoprecipitation. *Materials Science and Engineering: C, 123*, 112023–112023. https://doi.org/10.1016/j.msec.2021.112023

Qasim, M., Clarkson, A. N., & Hinkley, S. F. R. (2023). Green synthesis of carbon nanoparticles (CNPs) from biomass for biomedical applications. *International Journal of Molecular Sciences, 24*(2), 1023. https://doi.org/10.3390/ijms24021023

Rai, A., Singh, A., Ahmad, A., & Sastry, M. (2006). Role of halide ions and temperature on the morphology of biologically synthesized gold Nanotriangles. *Langmuir, 22*(2), 736–741. https://doi.org/10.1021/la052055q

Rajakumar, G., Gomathi, T., Thiruvengadam, M., Devi Rajeswari, V., Kalpana, V. N., & Chung, I.-M. (2017). Evaluation of anti-cholinesterase, antibacterial and cytotoxic activities of green synthesized silver nanoparticles using from Millettia pinnata flower extract. *Microbial Pathogenesis, 103*, 123–128. https://doi.org/10.1016/j.micpath.2016.12.019

Rajpoot, K., & Jain, S. K. (2017). Colorectal cancer-targeted delivery of oxaliplatin via folic acid-grafted solid lipid nanoparticles: Preparation, optimization, and in vitro evaluation. *Artificial Cells, Nanomedicine, and Biotechnology, 46*(6), 1236–1247. https://doi.org/10.1080/21691401.2017.1366338

Rathner, R., Petz, S., Tasnádi, G., Koller, M., & Ribitsch, V. (2017). Monitoring the kinetics of biocatalytic removal of the endocrine disrupting compound 17α-ethinylestradiol from differently polluted wastewater bodies. *Journal of Environmental Chemical Engineering, 5*(2), 1920–1926. https://doi.org/10.1016/j.jece.2017.03.034

Raveendran, P., Fu, J., & Wallen, S. L. (2003). Completely 'green' synthesis and stabilization of metal nanoparticles. *Journal of the American Chemical Society, 125*(46), 13940–13941. https://doi.org/10.1021/ja029267j

Razavi, M., Salahinejad, E., Fahmy, M., Yazdimamaghani, M., Vashaee, D., & Tayebi, L. (2015). Green chemical and biological synthesis of nanoparticles and their biomedical applications. In *Green processes for nanotechnology* (pp. 207–235). https://doi.org/10.1007/978-3-319-15461-9_7

Riley, P. R., & Narayan, R. J. (2021). Recent advances in carbon nanomaterials for biomedical applications: A review. *Current Opinion in Biomedical Engineering, 17*, 100262. https://doi.org/10.1016/j.cobme.2021.100262

Rowena. (2019). *Introduction to green chemistry*. The Catalyst Group. Available at: https://www.catalyst-hsegroup.com/introduction-green-chemistry/. Accessed 26 Apr 2023.

Roy, A., Pandit, C., Gacem, A., Alqahtani, M. S., Bilal, M., Islam, S., Hossain, M. J., & Jameel, M. (2022). Biologically derived gold nanoparticles and their applications. *Bioinorganic Chemistry and Applications, 2022*, 8184217. https://doi.org/10.1155/2022/8184217

Ruckenstein, E., & Kong, X. Z. (1999). Control of pore generation and pore size in nanoparticles of poly(styrene-methyl methacrylate-acrylic acid). *Journal of Applied Polymer*

Science, 72(3), 419–426. https://doi.org/10.1002/(sici)1097-4628(19990418)72:3%3C419::aid-app11%3E3.0.co;2-8

Sabliov, C. M., & Astete, C. E. (2015). Polymeric nanoparticles for food applications. In *Nanotechnology and functional foods* (pp. 272–296). https://doi.org/10.1002/9781118462157.ch17

Saleh, H. E.-D. M., & Koller, M. (2018). Introductory chapter: Principles of green chemistry. In *Green chemistry*. https://doi.org/10.5772/intechopen.71191

Saleh, H. M., Eskander, S. B., & Fahmy, H. M. (2014). Mortar composite based on wet oxidative degraded cellulosic spinney waste fibers. *International journal of Environmental Science and Technology, 11*, 1297–1304. https://doi.org/10.1007/s13762-013-0319-8

Samuel, M. S., Ravikumar, M., John, J. A., Selvarajan, E., Patel, H., Chander, P. S., Soundarya, J., Vuppala, S., Balaji, R., & Chandrasekar, N. (2022). A review on green synthesis of nanoparticles and their diverse biomedical and environmental applications. *Catalysts, 12*(5), 459. https://doi.org/10.3390/catal12050459

Sarathy, V., Tratnyek, P. G., Nurmi, J. T., Baer, D. R., Amonette, J. E., Chun, C. L., Penn, R. L., & Reardon, E. J. (2008). Aging of iron nanoparticles in aqueous solution: Effects on structure and reactivity. *The Journal of Physical Chemistry C, 112*(7), 2286–2293. https://doi.org/10.1021/jp0777418

Selvakesavan, R. K., & Franklin, G. (2021). Prospective application of nanoparticles green synthesized using medicinal plant extracts as novel Nanomedicines. *Nanotechnology, Science and Applications, 14*, 179–195. https://doi.org/10.2147/nsa.s333467

Shahzad, K., Kettl, K.-H., Titz, M., Koller, M., Schnitzer, H., & Narodoslawsky, M. (2013). Comparison of ecological footprint for biobased PHA production from animal residues utilizing different energy resources. *Clean Technologies and Environmental Policy, 15*(3), 525–536. https://doi.org/10.1007/s10098-013-0608-4

Sim, S., & Wong, N. (2021). Nanotechnology and its use in imaging and drug delivery (Review). *Biomedical Reports, 14*(5), 42. https://doi.org/10.3892/br.2021.1418

Singh, Y., Meher, J. G., Raval, K., Khan, F. A., Chaurasia, M., Jain, N. K., & Chourasia, M. K. (2017). Nanoemulsion: Concepts, development and applications in drug delivery. *Journal of Controlled Release, 252*, 28–49. https://doi.org/10.1016/j.jconrel.2017.03.008

Singh, J., Dutta, T., Kim, K.-H., Rawat, M., Samddar, P., & Kumar, P. (2018). 'Green' synthesis of metals and their oxide nanoparticles: Applications for environmental remediation. *Journal of Nanobiotechnology, 16*(1), 84. https://doi.org/10.1186/s12951-018-0408-4

Singh, A., Gautam, P. K., Verma, A., Singh, V., Shivapriya, P. M., Shivalkar, S., Sahoo, A. K., & Samanta, S. K. (2020). Green synthesis of metallic nanoparticles as effective alternatives to treat antibiotics resistant bacterial infections: A review. *Biotechnology Reports, 25*, e00427. https://doi.org/10.1016/j.btre.2020.e00427

Somorjai, G. A., & Park, J. Y. (2008). Colloid science of metal nanoparticle catalysts in 2D and 3D structures. Challenges of nucleation, growth, composition, particle shape, size control and their influence on activity and selectivity. *Topics in Catalysis, 49*(3–4), 126–135. https://doi.org/10.1007/s11244-008-9077-0

Su, S., & Kang, P. M. (2020). Systemic review of biodegradable nanomaterials in Nanomedicine. *Nanomaterials, 10*(4), 656. https://doi.org/10.3390/nano10040656

Sun, L., Wang, G., Zhang, C., Jin, Q., & Song, Y. (2021). On the rheological properties of multi-walled carbon nano-polyvinylpyrrolidone/silicon-based shear thickening fluid. *Nanotechnology Reviews, 10*(1), 1339–1348. https://doi.org/10.1515/ntrev-2021-0087

Taléns-Visconti, R., Díez-Sales, O., de Julián-Ortiz, J. V., & Nácher, A. (2022). Nanoliposomes in cancer therapy: Marketed products and current clinical trials. *International Journal of Molecular Sciences, 23*(8), 4249. https://doi.org/10.3390/ijms23084249

Thakor, A. S., & Gambhir, S. S. (2013). Nanooncology: The future of cancer diagnosis and therapy. *CA: A Cancer Journal for Clinicians, 63*(6), 395–418. https://doi.org/10.3322/caac.21199

Tran, Q. H., Nguyen, V. Q., & Le, A.-T. (2013). Silver nanoparticles: Synthesis, properties, toxicology, applications and perspectives. *Advances in Natural Sciences: Nanoscience and Nanotechnology, 4*(3), 033001. https://doi.org/10.1088/2043-6262/4/3/033001

Tri Wahyuni, E., & Sri Kunarti, E. (2022). Green methods of chemical analysis and pollutant removal. In *Green chemistry – New perspectives*. https://doi.org/10.5772/intechopen.104829

Wale, T., Suryavanshi, S., Wavare, V., Phalke, P., & Sharmale, M. (2023). Review on green chemistry. *Journal of Drug Delivery and Therapeutics, 13*(7), 190–193. https://doi.org/10.22270/jddt.v13i7.5919

Walter, J. G., Petersen, S., Stahl, F., Scheper, T., & Barcikowski, S. (2010). Laser ablation-based one-step generation and bio-functionalization of gold nanoparticles conjugated with aptamers. *Journal of Nanobiotechnology, 8*(1). https://doi.org/10.1186/1477-3155-8-21

Wang, Y., Zhao, Q., Han, N., Bai, L., Li, J., Liu, J., Che, E., Hu, L., Zhang, Q., Jiang, T., & Wang, S. (2015). Mesoporous silica nanoparticles in drug delivery and biomedical applications. *Nanomedicine: Nanotechnology, Biology, and Medicine, 11*(2), 313–327. https://doi.org/10.1016/j.nano.2014.09.014

Warner, J. C., Cannon, A. S., & Dye, K. M. (2004). Green chemistry. *Environmental Impact Assessment Review, 24*(7), 775–799. https://doi.org/10.1016/j.eiar.2004.06.006

Wegner, K., Schimmöller, B., Thiebaut, B., Fernandez, C., & Rao, T. N. (2011). Pilot plants for industrial nanoparticle production by flame spray pyrolysis. *Kona Powder and Particle Journal, 29*, 251–265. https://doi.org/10.14356/kona.2011025

Wilson, R. J., Li, Y., Yang, G., & Zhao, C.-X. (2021). Nanoemulsions for drug delivery. *Particuology., 64*, 85. https://doi.org/10.1016/j.partic.2021.05.009

Yap, Y. H., Azmi, A. A., Mohd, N. K., Yong, F. S. J., Kan, S.-Y., Thirmizir, M. Z. A., & Chia, P. W. (2020). Green synthesis of silver nanoparticle using water extract of onion peel and application in the acetylation reaction. *Arabian Journal for Science and Engineering, 45*(6), 4797–4807. https://doi.org/10.1007/s13369-020-04595-3

Yu, H., Park, J.-Y., Kwon, C. W., Hong, S.-C., Park, K.-M., & Chang, P.-S. (2018). An overview of nanotechnology in food science: Preparative methods, practical applications, and safety. *Journal of Chemistry, 2018*, 1–10. https://doi.org/10.1155/2018/5427978

Zhu, D., Tao, W., Zhang, H., Liu, G., Wang, T., Zhang, L., Zeng, X. and Mei, L. (2016). Docetaxel (DTX)-loaded polydopamine-modified TPGS-PLA nanoparticles as a targeted drug delivery system for the treatment of liver cancer. Acta Biomaterialia 30, pp. 144–154. https://doi.org/10.1016/j.actbio.2015.11.031.

Biogenic Synthesis of Nanomaterials and Their Therapeutic Application on Fishes

Birendra Thakur, Abhishek Kumar Bhardwaj, Prem Kumar Tripathi, Rishi Mittal, and Dayalanand Roy

1 Introduction

Today, malnutrition, which has become a challenge for humanity, can be resolved by the use of aquatic animal food products rich in essential nutrients. Around 20% of average per capita intake of animal protein is received from all captured and farmed aquatic animal products by more than 3 billion people in 2009 according to the food and agriculture organization (FAO). Besides having high nutritional and medicinal value, there is a sharp decline in some specific fish species such as *Heteropneustes fossilis* and *Labeo rohita.* in the past two decades because they are easily prone to epizootic ulcerative syndrome (EUS) diseases, high fishing pressure, and habitat loss (Haniffa et al., 2008). In aquaponics proteobacteria are the most abundant bacterial phylum (Wang et al., 2022) whereas in traditional farming Lactobacillaceae are rich mainly because of water pollution. In the Asian Pacific region, EUS is the most destructive disease among brackish as well as fresh water fish species (Das, 1997). A diverse group of microbiotas such as viruses, bacteria, and cutaneous ectoparasites are responsible, which may initiate skin lesions, subsequently colonized by *Aphanomyces invadans* and finally leading to EUS (Lilley et al., 1992). Various pathogenic organisms such as fungus (Chinabut et al., 1995), bacteria (Chakrabarty & Dastidar, 1991), and virus (Frerichs, 1995) have been reported in naturally infected fish. An EUS outbreak in more than 100 fish species

B. Thakur · A. K. Bhardwaj (✉) · P. K. Tripathi
Department of Environmental Sciences, Amity School of Life Sciences, Amity University, Gwalior, Madhya Pradesh, India

R. Mittal
Department of Environmental Science and Engineering, Guru Jambheshwar University of Science and Technology, Hisar, Haryana, India

D. Roy
Department of Zoology, Shri Murli Manohar Town P.G. College, Ballia, Uttar Pradesh, India

A. K. Bhardwaj et al. (eds.), *Biogenic Wastes-Enabled Nanomaterial Synthesis*,
https://doi.org/10.1007/978-3-031-59083-2_12

especially in air breathing have been reported (Roberts et al., 1994). The study of infected fishes of different countries has shown a wide range of pathogenic bacteria including *Aeromonas hydrophila* and the fungus *A. invadans* being the most prevalent (Das, 1997).

1.1 Bacterial Infection

Catfish show large diversity and approximately 4000 species are found around the world. The catfish species important for aquaculture include walking catfish (*Clarias batrachus*), African catfish (*Clarias gariepinus*), Thai catfish (*Clarias macrobrachium*), shark catfish (*Pangasius bocourti*), channel catfish, and blue catfish (*I. furcatus*). The hybrid catfish developed by mating a channel female catfish with a blue male catfish are most important in North America. In particular, three bacterial diseases including enteric septicemia of catfish (ESC), columnaris disease, and motile *Aeromonas* septicemia (MAS) are the primary disease concerns (Plumb & Hansons, 2010).

a. *Enteric septicemia of catfish*

One of the most fatal, the first catfish disease caused by *Edwardsiella ictaluri* was reported in 1976 (Hawke et al., 1981). It affects all stages of catfish with a high mortality rate. It infects channel catfishes entering via intestinal epithelia and olfactory mucosa lines. The intestinal epithelial cells uptake *E. ictaluri* by actin polymerization and receptor-mediated endocytosis (Skirpstunas & Baldwin, 2002).

b. *Columnaris disease*

Columnaris, a dreadful disease, was reported in 1922 from freshwater fish in the Mississippi river (Davis, 1922). *Flavobacterium columnare*, ubiquitous in the fresh water environment, can cause catastrophic mortalities in both wild as well as domesticated species (Morley & Lewis, 2010). The disease initiates via external infection through the body surface, which ultimately proceeds to the fins or gills developing a yellow-orange lesions on dorsal midline, known as "saddleback" (Bullock, 1986). The columnaris infection causes 90% mortality of catfish in extreme conditions. *F. columnare* secretes different extracellular enzymes that can degrade polysaccharides, destructing gills, muscles, and skin of fishes (Bernardet & Bowman, 2006).

c. *Motile* Aeromonas *septicemia*

Motile *Aeromonas* septicemia is an emerging disease threatening the catfish industry, caused by the Gram-negative bacterium *A. hydrophila* (Pang et al., 2015). *A. hydrophila* infected various fish species including eel (*Anguilla anguilla*) (Esteve et al., 1994), goldfish (*Carassius auratus*) (Harikrishnan et al., 2009), Nile tilapia (*Oreochromis niloticus*) (Abd-El-Rhman, 2009), and common carp (*Cyprinus carpio*) (Robinson et al., 2012). MAS has seriously affected the cyprinid fish industry of central China since 1989 (Zhang et al., 2014). The MAS was first reported in the catfish industry of the southern USA during the summer of 2009, which caused the loss of millions of pounds of food-sized channel catfish (Hemstreet 2010). The symptoms of *A. hydrophila* infection include dropsy, necrosis, sores on the skin, ulceration, and hemorrhagic septicemia (Pridgeon & Klesius, 2011).

Fungal Infection
The fungal infection can be dangerous for fishes which can deteriorates the fish health. It was studied that *Aspergillus*, *Alternaria*, *Penicillium*, *Blastomyces* spp. and *Rhizopus* were fungal infections in four species of carps including silver carp; *Hypophthalmichthys molitrix*, goldfish; *Carassius auratus*, rohu; *Labeo rohita* and grass carp; *Ctenopharyngodon idella* (Iqbal & Asghar, 2012). He observed the clinical symptoms in infected fish like fungal growth on skin, eyes, fins, eroded fins and scales, abdominal distension and hemorrhages on body surface. The infection rate was highest (44.4%) in *C. auratus*. The fungal species isolated were *Aspergillus* (highest 33.3%), *Alternaria*, *Penicillium*, *Blastomyces* spp. and *Rhizopus* (lowest 5%) were isolated. The fish posterior part (62.5%) is more infected comparing to anterior (31.25%) (Iqbal & Asghar, 2012). The above fish microbial problems can be resolved by the application of biosynthesized zero toxicity nanodrugs on diseased fish. The nanoparticle (NP) ranges between 1 and 100 nanometers (nm) and can be originated naturally or artificially (Khulbe & Matsuura, 2021). Nanomaterials (NMs) can be present throughout the earth's spheres (atmosphere, hydrosphere, lithosphere, and biosphere) including the troposphere, hydrosphere such as oceans, rivers, lakes, groundwater, and hydrothermal vents. Microscopic size and properties different from macro materials make them significantly important. Large surface area provides large opportunities for interaction with other molecules which can change optical, magnetic and catalytic properties. Nanomaterials show high solubility, volume ratio, thermal, electrical conductivity. Some NMs show antiviral, antibacterial, and antifungal properties. It also pour ray on surface chemistry like charge, composition, surface tension, physical structure, reactive sites, photocatalytic properties, zeta potential. Besides harmful effects, they also have some beneficial impacts (Alagawany et al., 2021).

This chapter entails the biogenic synthesis of NMs which recently being employed by the different researchers for the treatment of several fishes from various microbial diseases. Number of such reputed publications extracted to fulfillment the objectives. Here we tabulated the recent observation of silver, copper, zinc, gold and selenium NPs. Such antimicrobial substances to control fish diseases. However, this study will be helpful for researchers to develop an effective and sustainable restoration of aquatic animals using green synthesized NMs to maintain the aquatic fish in the ex-situ condition.

2 Synthesis of Nanoparticles

Nanoparticles are prepared by two main two approaches, i.e., the top-down approach and the bottom-up approach. Top-down approaches of NP synthesis have been reported to be physical vapor deposition, chemical vapor deposition, ion

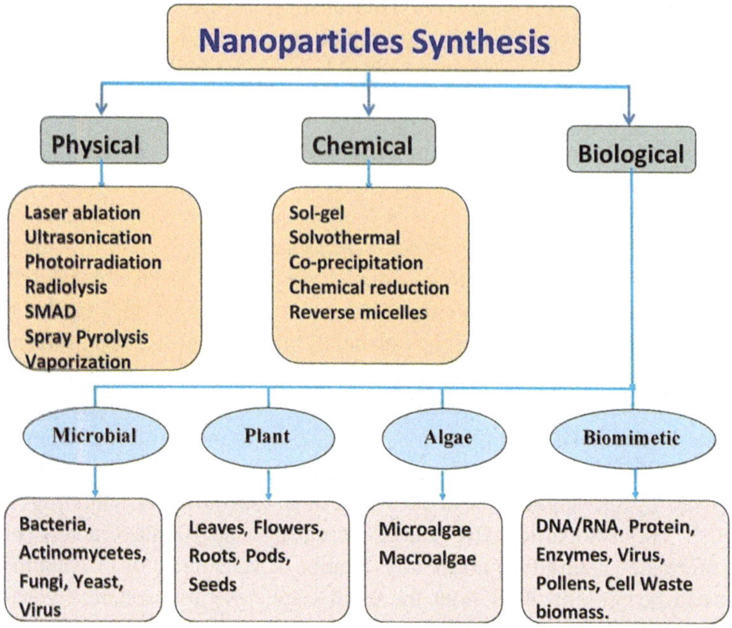

Fig. 1 Sketch diagram of the synthesis of metallic nanoparticles

implantation, electron beam lithography, X-ray lithography, etc. (Koch et al., 1997) to get the desired size and shape of NPs. On the other hand, the bottom-up approach of NPs includes Sol-gel synthesis, colloidal precipitation, hydrothermal synthesis, organometallic chemical route, and electro-deposition (Bhardwaj et al., 2021). The different shape and size of green synthesis of silver NPs are optimized (Bhardwaj & Naraian, 2021; Bhardwaj et al., 2018, 2019; Maurya et al., 2016). All the above methods of synthesis of material can be collectively organized as physical, chemical, and biological, as shown in Fig. 1.

3 Therapeutic Application of Metal Nanoparticles in Fishes

Several works are available on the studies of microbiota of different fishes to understand their health. Such studies play an important role in maintaining the health of fishes, including metabolic status, possible infection, energy consumption, and enterocyte proliferation. The disease-causing microbes/organisms can be controlled by different metallic NPs.

3.1 Silver Nanoparticles

The unique properties of silver nanoparticles (Ag NPs) can be employed in pisciculture to treat severe microbial infections (Table 1). It has been found that the green synthesized Ag NPs measuring 29.1 and 32.5 nm showed antibacterial activity when applied in fish with 1.64 μg/mL minimum bactericidal concentration (MBC)/minimum inhibitory concentration (MIC) (Adhikari et al., 2023). Antibacterial activity against *Staphylococcus aureus, Bacillus subtilis, Escherichia. coli,* and *Pseudomonas aeruginosa* increased when Ag NPs measuring 32 nm were biosynthesized using fish scale charcoal (Alagumuthu et al., 2023). Treatment using Ag NPs in *Oreochromis niloticus* (Nile tilapia) (Elgendy et al., 2022) and Mi-Ag NPs (*Mangifera indica*) (Raza et al., 2023) in *Cirrhinus mrigala* increased the fish survival rate by improving hematological, antioxidant, and immunological activities. It also optimizes kidney and liver function. Nano composites of silver and *Terminalia catappa* at 4.33 mg/L reduces 98.70% of fish body infection and increased survival rate 100% (Meneses et al., 2023). In *Oreochromis mossambicus* the spherical, 88-nm Ag NPs showed antibacterial and antioxidant activities (Thanigaivel et al., 2022) whereas in Nile tilapia and sea bass, Ag NPs of irregular shape (measuring 100 nm) with zone of inhibition (ZI) 23.7–31.3 mm at 10 μL/disk showed antibacterial activity and with ZI 11–18 mm at 10 μL/disk showed antifungal activity (Ghetas et al., 2022). Chitosan-coated spherical Ag NPs measuring 12–26 nm and Ag NPs synthesized with leaves of *Persicaria odorata* increased the wound healing rate, inhibiting wound pathogens such as *Staphylococcus epidermidis* and methicillin-resistant *Staphylococcus aureus* (MRSA) (Joughi et al., 2022). The Ag NPs also inhibited fish disease, causing *P. fluorescens* and *A. hydrophila* microbes (Essawy et al., 2021), whereas Ag NPs measuring 23 nm decreased mortality 45% in *C. carpio*, inhibiting *F. johnsoniae* growth (Shaalan et al., 2020). In the case of shrimp Ag NPs showed broad spectrum antibacterial activity (Camacho et al., 2020). The spherical Ag NPs and *A. sobria* Ag NPs (420–480 nm) in *Carassius auratus* (goldfish) inhibited *A. hydrophila, Enterobacter cloacae,* and *E. coli* respectively (Erdem et al., 2019). In *Cirrhinus mrigala* the treatment of fingerlings with 74%, spherical (35.4 nm) Ag NPs increased survival rate by modulating the fish immune system (Rather et al., 2017). Starch-stabilized Ag NPs in *Carassius auratus* were found to be effective against fungal and protozoan growth (Daniel et al., 2016). In salmonid fish and cod chitosan Ag NPs spherical in shape and measuring 281 nm showed bacteriostatic and bactericidal activity with MIC and MBC of 50 and 100 μg/mL respectively (Dananjaya et al., 2016). (Akram et al., 2016) found that Ag NPs-coupled azithromycin or clarithromycin showed highest antimicrobial activity against MRSA.

Table 1 Therapeutic role of silver nanoparticles (Ag NPs) in different stages of different fish

NPs	Shape/ Size (nm)	Specimen	Result	References
Ag NPs Aloe vera	29.1 and 32.5	Fish (not specific)	(MBC/MIC) 50 μg/ml concentration of Ag NPs showed antibacterial activity	Adhikari et al. (2023)
Ag NPs		Fish (not specific)	Antibacterial activity increased against *Staphylococcus aureus, Bacillus subtilis, Escherichia coli*, and *Pseudomonas aeruginosa*	Alagumuthu et al. (2023)
Ag NPs (Fish scale charcoal)	32	Fish (not specific)	Worked against bacteria such as *Escherichia coli, Staphylococcus aureus*, and *Pseudomonas aeruginosa*	Zhang and He (2023)
Mi-Ag NPs *Mangifera indica*	62 ± 13	*Cirrhinus mrigala*	Improved several hematological and histological parameters	Raza et al. (2023)
Ag NPs (*Terminalia catappa)*		Fish (not specific)	The nano composite Ag-*T. catappa* at 4.33 mg/L reduced the infected area on the fish body by 98.70% and 100% survival achievement	Meneses et al. (2023)
Ag NPs		*O. niloticus*	The treatment with Ag NPs increased the fish survival rate by improving immunological, hematological, and antioxidant activities. It also optimized function of the liver and kidney	Elgendy et al. (2022)
Ag NPs	Spherical (88)	*O. mossambicus*	Antibacterial and antioxidant activity shown	Thanigaivel et al. (2022)
Ag NPs	Irregular (100)	Nile tilapia and sea bass	The green Ag NPs showed antibacterial activity with ZI 23.7–31.3 mm at 10 μL/disk and antifungal activity with ZI 11–18 mm at 10 μL/disk	Ghetas et al. (2022)
Chitosan-coated Ag NPs	Spherical 12–26		Chitosan-coated Ag NPs enhanced wound healing owing to their antibacterial activities	Joughi et al. (2022)
Ag NPs *Persicaria odorata* leaves	11–13		The (PO-Ag NPs) worked against wound-causing agents such as *Staphylococcus epidermis* and methicillin-resistant *Staphylococcus aureus*	Lubis et al. (2022)

(continued)

Table 1 (continued)

Ag NPs		Fish (not specific)	It showed antimicrobial activity against *A. hydrophila* and *P. fluorescens*	Ehab Essawy et al. (2021)
Ag NPs	Spherical (10 nm)	Fish (not specific)	Antimicrobial activity	El-Adawy et al. (2021)
Ag NPs	Spherical (70–75)	Fish (not specific)	Antibacterial activity with ZI of 16–28	Fatima et al. (2020)
Ag NPs	23	*Cyprinus carpio*	Antibacterial activity against *F. johnsoniae*. Fish mortality decreases up to 45%	Shaalan et al. (2020)
A. sobria Ag NPs	420–480	Fish (not specific)	*A. hydrophila*, *E. cloacae*, and *E. coli* is strongly inhibited	Erdem et al. (2019)
Ag NPs	17 μg/mL	Fish (not specific)	Inhibits growth of *A. salmonicida* and *A. invadans*	Shaalan et al. (2017)
Ag NPs		MRSA	The triple therapy (Ag NPs, blue light, and antibiotics such as azithromycin or clarithromycin) inhibits MRSA	Akram et al. (2016)
Starch-stabilized Ag NPs		*Carassius auratus*	Antiprotozoal and antifungal activity	Daniel et al. (2016)
Chitosan Ag NPs	Spherical, 281	Salmonid fish and cod	Chitosan Ag NPs showed bacteriostatic and bactericidal activity with MIC and MBC of 50 μg/mL	Dananjaya et al. (2016)
Ag_3PO_4-loaded hydroxyapatite nano wires		Water treatment	*S. aureus* and *E. coli* are completely inhibited	Li et al. (2015)
Ag NPs with papaya leaf extract	25–45	Aquatic animals	Ag NPs synthesized with papaya leaf extract showed maximum antimicrobial activity at a concentration of 153.6 μg/mL against *A. hydrophila*	Mahanty et al. (2013)

MBC minimum bactericidal concentration, *MIC* minimum inhibitory concentration, *ZI*, *MRSA* methicillin-resistant *Staphylococcus aureus*, *MDR* multidrug resistant

3.2 Copper Nanoparticles

Copper nanoparticles (CuO NPs/Cu NPs) play a complex role in various cells, such as wound healing by modulating the action mechanism of several cytokines and growth factors (Kornblatt et al., 2016) (Table 2). The MU-CuO NPs synthesized using Mithun urine enhanced antibacterial activity against *Aeromonas hydrophila* and *Aeromonas veronii* at a concentration of 100 μg/mL (Pelesinuo et al., 2023). MS-CuO NPs (*Moringa oleifera*) at a concentration of 0.1 mg/L in African catfish improved its maximum parameters affected because of *S. aureus* (Rahman et al., 2023). In an experiment on fish (Chao et al., 2022), it was found that multiple enzymatic activities of CuP NPs and the action of the reactive oxygen species produced by them facilitate in their inherent activities for glutathione depletion and the lipid peroxidation. Spherical Cu(II) nanoflowers with a ZI of 30.66–33.66 mm at 0.5 μg/mL in fish showed antibacterial activity (Demirbas 2021) whereas *Aerva javanica* leaf extract inhibited the growth of fungi such as *C. albicans, C. krusei,* and *C. tropicalis* with 160 mg/L minimum fungicidal concentration (MFC) and MIC and growth of bacteria such as *E. coli, P. aeruginosa, S. aureus, Acinetobacter baumannii*. On the basis of their cytotoxic activity against Neuro2A cells, CuO NPs are safe below 60 μg/mL (Amin et al., 2021). In *C. carpio* infected with *Saprolegnia* spp., when treated with mint Cu NPs measuring 11–34 nm, positive improvements in fish blood parameters were found (Alwash et al., 2021). The dietary Cu NPs in *C. carpio* increased the growth performance and reduced the false discovery rate with a linear and quadratic model ($P < 0.05$) (Dawood et al., 2020). The Cu NPs synthesized with *M. oleifera* leaf extract bear antioxidant, antibacterial, and antifungal properties (Das et al., 2020a, b). In *Danio rerio* CuS NPs inhibited Gram-negative and Gram-positive bacteria growth by generating toxic hydroxyl radicals that damage cell membranes (Ahmed & Anbazhagan, 2017; Sistemática et al., 2016; Kalatehjari et al., 2015). Cu NP is a good alternative for carcinogenic malachite green used for its antifungal properties, with an MIC of 10 ppm. Cu NPs also have high potency for bacterial cell filamentation and cell killing (Chatterjee et al., 2014).

3.3 Zinc Nanoparticles

The zinc nanoparticles (ZnO NPs) of various sizes have antimicrobial, antioxidant, antidiabetic and anti-inflammatory effects (Ramana et al., 2023). Anandhi and Rajeshkumar (2023) biosynthesized Zn NPs with rod-shaped *Acacia modesta* measuring 70 nm, inhibiting MRSA, whereas Zn NPs coupled with azithromycin proved to be more effective against *S. aureus* and *E. coli* (Saddik et al., 2022). In *Oreochromis niloticus* Zn NPs in a concentration of 60 μg/mL boosted the immune system, which increased resistivity against *Aeromonas hydrophila* (Sherif et al., 2022). *Labeo rohita* fingerlings fed with Zn NPs (spherical, 22.61 nm size) showed increased growth and improved biochemical and hematological parameters (Thangapandiyan

Table 2 Therapeutic role of copper nanoparticles (Cu NPs) at different stages of different fish

NPs	Size/form	Specimen	Result	References
MU-CuO NPs		Fish (Not specific)	MU-CuO NPs enhanced antibacterial activity against *A. hydrophila* and *A. veronii*	Pelesinuo et al. (2023)
MS-CuO NPs *M.olefera*	0.1 mg/L	African catfish	MS-CuO NPs treatment for catfish improved the maximum parameters affected because of infection of *S. aureus*	Rahman et al. (2023)
Cu_3P NPs		Fish (Not specific)	The Cu_3P NPs enzymatic activities caused ROS production causing glutathione depletion and lipid peroxidation	Chao et al. (2022)
Cu(II) nanoflower	Spherical	Fish (Not specific)	Antibacterial activity with a ZI of 30.66– 33.66 mm at 0.5 µg/mL	Demirbas (2021)
Cu NP with leaf extract			CuO NPs MFC and MIC for fungus were 160 mg/mL. CuO NPs were safe for Neuro 2 A cells below 60 µL/mL	Amin et al. (2021)
Cu NPs	11–34 nm	*Cyprinus carpio*	Mint-Cu NPs showed positive improvement in blood parameters of *Cyprinus carpio* in *Saprolegnia* spp., infections	Alwash et al. (2021)
Cu NP with leaf extract			Cu NPs with *Aerva javanica* leaf extract showed antimicrobial activities against fungus such as *C. albicans, C. krusei*, and *C. tropicalis* and bacteria such as *E. coli, P. aeruginosa, S. aureus*	Amin et al. (2021)
Cu NPs		*Cyprinus carpio*	In *C. carpio* Cu NPs in the diet increased their growth and reduced FCR linearly.	Dawood et al. (2020)
Cu NP with leaf extract			Cu NPs synthesized with *M. oleifera* leaf extract showed antioxidant, antibacterial and antifungal activity	Das et al. (2020a, b)
CuS NPs		*Danio rerio*	Antibacterial activity	Ahmed and Anbazhagan (2017)
CuS NPs			Hydroxyl radicals developed by copper destroyed the bacterial cell membrane	Sistemática et al. (2016)
Cu NPs		Fish (*Rutilus frisii kutum*) eggs	Cu NPs proved to be a good alternative to carcinogenic malachite green and showed antifungal activity at an MIC of 10 ppm.	Kalatehjari et al. (2015)
Cu NPs			Cu NPs caused bacterial cell inflammation and cell death	Chatterjee et al. (2014)

ROS reactive oxygen species, *MFC* minimum fungicidal concentration, *MIC* minimum inhibitory concentration, *FCR* false discovery rate

& Monika, 2020). 1 mg/L ZnO NPs inhibited *A. veronii* in *Xiphophorus hellerii* (Das et al., 2020a, b). ZnO NPs along with β 1,3 glucan binding protein exhibits antibacterial activity against *A. hydrophila* infection in *Oreochromis mossambicus*. ZnO NPs in 15.75, 31.5, and 3.15 μg/mL in fish inhibited the growth of *A. salmonicida, Y. ruckeri*, and *A. invadans* (Shaalan et al., 2017).

3.4 Gold Nanoparticles

Besides being highly used in jewelry, gold is also being used in the field of medicine. Gold NPs (Au NPs) of cubic, rectangular shape, and 560 nm in size, biosynthesized with *Cydonia oblonga* seeds, and have microbial growth inhibitory properties up to 12 mm. In a murine model it showed 99% wound closure within 5 days (Batool et al., 2022). In Longfin yellowtail, spherical (24 nm) Au NPs show antimicrobial activity, antioxidant activity, and immune modulation (Reyes-Becerril et al., 2021). The Au NPs synthesized using seaweed (*Halymenia dilatata*) at a concentration of 100 μg/mL in *D. rerio* disease inhibited *A. hydrophila, Bacillus subtilis, S. aureus, Salmonella typhi, P. aeruginosa, Micrococcus luteus*, and *Enterobacter aerogenes* (Vinosha et al., 2019). Spherical and triangular Au NPs with a ZI of 23.2 mm at 100 μg/mL in *Oreochromis mossambicus* showed antibacterial and antibiofilm activity (Vijaykumar et al., 2017).

3.5 Selenium NPs

Algae-coated selenium nanoparticles (Se NPs) in human and shrimp showed high performance against *Vibrio harveyi* (Mansouri-Tehrani et al., 2021). Se NPs in *Oreochromis niloticus* improved growth, immunity and antioxidant power (Abu-Elala et al., 2021). Root extracts (*B. axillaris*) synthesized Se NPs inhibited aquatic pathogens (Jyoti et al., 2021). Cremonini et al., 2016 also found antimicrobial and antifungal activities of Se NPs in fish.

4 Reported Nanomaterials in Fish Disease Control

After referring to different results of articles that studied fish disease control using magnetic nanoparticles (MNPs), it was observed that more than 60% of articles discuss curing different fish diseases by using green synthesized Ag NPs followed by Cu and Zn NPs, which contribute around 15%. However, Au and Se NPs-based treatment had the lowest percentage reported of around 5% each (Fig. 2).

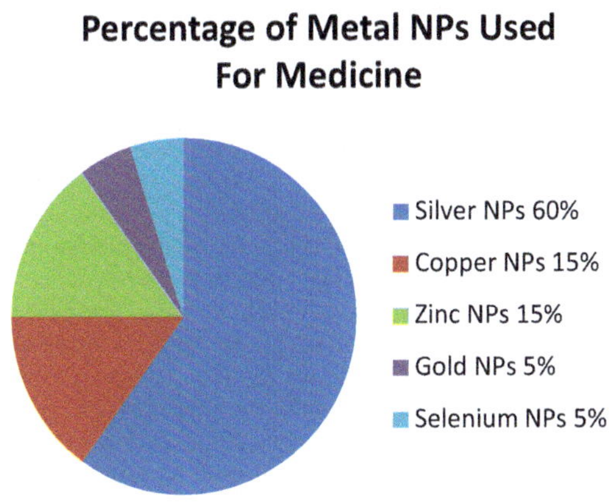

Fig. 2 Percentage of different magnetic nanoparticles reported to control fish disease

5 Conclusion

It is believed that various biologically active compounds found in different plants are able to improve the health of animals. This chapter encapsulates the recent research that prepared functional NMs using active biogenic compounds. Several kinds of MNPs, including silver, copper, zinc, gold, and selenium, have remarkable antibacterial, antioxidant, and anti-inflammatory properties. This chapter provides a dataset to various researchers working in the respective fields. The analysis of different research articles regarding the use of different metal NPs to cure various fish diseases concludes that fishes can be protected from their different diseases more efficiently by the use of biosynthesized metal NP drugs than by any other drugs. Green synthesized nano drugs are safer, with fewer side effects than any others. After reviewing hundreds of results it was observed that Ag and Cu NPs have been widely reported as an antimicrobial substance to control fish diseases. However, other metals, including zinc, gold and selenium were also reported to be strong candidates for drug design in order to maintain aquatic fish in ex situ conditions. This study will be helpful for researchers to develop an effective and sustainable restoration of aquatic animals using green synthesized NMs. Furthermore, such an approach will be helpful in keeping healthier fishes and reducing the malnutrition problem across the world.

References

Abd-El-Rhman, A. M. (2009). Antagonism of *Aeromonas hydrophila* by propolis and its effect on the performance of Nile tilapia, *Oreochromis niloticus*. *Fish & Shellfish Immunology, 27*(3), 454–459.

Abu-Elala, N. M., Shaalan, M., Ali, S. E., & Younis, N. A. (2021). Immune responses and protective efficacy of diet supplementation with selenium nanoparticles against cadmium toxicity in Oreochromis niloticus. *Aquaculture Research, 52*(8), 3677–3686.

Adhikari, S. P., Paudel, A., Sharma, A., Thapa, B., Khanal, N., Shastri, N., et al. (2023). Development of decellularized fish skin scaffold decorated with biosynthesized silver nanoparticles for accelerated burn wound healing. *International Journal of Biomaterials*.

Ahmed, K. B. A., & Anbazhagan, V. (2017). Synthesis of copper sulfide nanoparticles and evaluation of in vitro antibacterial activity and in vivo therapeutic effect in bacteria- infected zebrafish. *RSC advances, 7*(58), 36644–36652.

Akram, F. E., El-Tayeb, T., Abou-Aisha, K., & El-Azizi, M. (2016). A combination of silver nanoparticles and visible blue light enhances the antibacterial efficacy of ineffective antibiotics against methicillin-resistant Staphylococcus aureus (MRSA). *Annals of Clinical Microbiology and Antimicrobials, 15*, 1–13.

Alagawany, M., Farag, M. R., Abdelnour, S. A., Dawood, M. A., Elnesr, S. S., & Dhama, K. (2021). Curcumin and its different forms: A review on fish nutrition. *Aquaculture, 532*, 736030.

Alagumuthu, T. S., Samidurai, S., Khambhaty, Y., & James, K. (2023). Evaluation of antimicrobial property of bacteriocin capped silver nanoparticles prepared from cell free supernatant of Lactobacillus brevis isolated from marine fishes. *International Journal of Peptide Research and Therapeutics, 29*(6), 1–9.

Alwash, S. W., Alfaragi, J. K., & AlSaadi, T. M. (2021). Assessment of the mint-copper nanoparticles treatment to control the infection of saprolegniasis in Cyprinus carpio L. *Annals of the Romanian Society for Cell Biology*, 16010–16019.

Amin, F., Khattak, B., Alotaibi, A., Qasim, M., Ahmad, I., Ullah, R., et al. (2021). Green synthesis of copper oxide nanoparticles using Aerva javanica leaf extract and their characterization and investigation of in vitro antimicrobial potential and cytotoxic activities. *Evidence-Based Complementary and Alternative Medicine*.

Anandhi, P., & Rajeshkumar, S. (2023). Copper nanoparticles in wound healing: A review. *Journal of Survey in Fisheries Sciences, 10*(1S), 49–63.

Batool, Z., Muhammad, G., Iqbal, M. M., Aslam, M. S., Raza, M. A., Sajjad, N., et al. (2022). Hydrogel assisted synthesis of gold nanoparticles with enhanced microbicidal and in vivo wound healing potential. *Scientific Reports, 12*(1), 6575.

Bernardet, J. F., & Bowman, J. P. (2006). The genus flavobacterium. *The prokaryotes, 7*, 481–531.

Bhardwaj, A. K., & Naraian, R. (2021). Cyanobacteria as biochemical energy source for the synthesis of inorganic nanoparticles, mechanism and potential applications: a review. *3Biotech, 11*(10), 445.

Bhardwaj, A. K., Shukla, A., Maurya, S., Singh, S. C., Uttam, K. N., Sundaram, S., Singh, M. P., & Gopal, R. (2018). Direct sunlight enabled photo-biochemical synthesis of silver nanoparticles and their bactericidal efficacy: Photon energy as key for size and distribution control. *Journal of Photochemistry and Photobiology B: Biology, 188*, 42–49.

Bhardwaj, A. K., Gupta, M. K., & Naraian, R. (2019). Myco-nanotechnological approach for improved degradation of lignocellulosic waste: Its future aspect. In *Mycodegradation of lignocelluloses* (pp. 227–245).

Bhardwaj, A. K., Sundaram, S., Yadav, K. K., & Srivastav, A. L. (2021). An overview of silver nano-particles as promising materials for water disinfection. *Environmental Technology & Innovation, 23*, 101721.

Bullock, G. L. (1986). *Columnaris disease of fishes* (Vol. 72). US Department of the Interior, Fish and Wildlife Service, Division of Fisheries and Wetlands Research.

Camacho-Jiménez, L., Álvarez-Sánchez, A. R., & Mejía-Ruíz, C. H. (2020). Silver nanoparticles (AgNPs) as antimicrobials in marine shrimp farming: A review. *Aquaculture Reports, 18*, 100512.

Chakrabarty, A. N., & Dastidar, S. G. (1991). Repeated isolation of chemoautotrophic nocardioform bacteria from fish epizootic ulcerative syndrome. *Indian Journal of Experimental Biology, 29*(7), 623–627.

Chao, D., Dong, Q., Chen, J., Yu, Z., Wu, W., Fang, Y., et al. (2022). Highly efficient disinfection based on multiple enzyme-like activities of Cu_3P nanoparticles: A catalytic approach to impede antibiotic resistance. *Applied Catalysis B: Environmental, 304*, 121017.

Chatterjee, A. K., Chakraborty, R., & Basu, T. (2014). Mechanism of antibacterial activity of copper nanoparticles. *Nanotechnology, 25*(13), 135101.

Chinabut, S., Roberts, R. J., Willoughby, G. R., & Pearson, M. D. (1995). Histopathology of snakehead, Channa striatus (Bloch), experimentally infected with the specific Aphanomyces fungus associated with epizootic ulcerative syndrome (EUS) at different temperatures. *Journal of Fish Diseases, 18*(1), 41–47.

Cremonini, E., Zonaro, E., Donini, M., Lampis, S., Boaretti, M., Dusi, S., et al. (2016). Biogenic selenium nanoparticles: characterization, antimicrobial activity and effects on human dendritic cells and fibroblasts. *Microbial Biotechnology, 9*(6), 758–771.

Dananjaya, S. H. S., Godahewa, G. I., Jayasooriya, R. G. P. T., Lee, J., & De Zoysa, M. (2016). Antimicrobial effects of chitosan silver nano composites (CAgNCs) on fish pathogenic Aliivibrio (Vibrio) salmonicida. *Aquaculture, 450*, 422–430.

Daniel, S. C. G. K., Sironmani, T. A., & Dinakaran, S. (2016). Nano formulations as curative and protective agent for fish diseases: studies on red spot and white spot diseases of ornamental gold fish Carassius auratus. *International Journal of Fisheries and Aquatic Studies, 4*, 255–261.

Das, M. K. (1997). Epizootic ulcerative syndrome in fishes: its present status in India.

Das, P. E., Abu-Yousef, I. A., Majdalawieh, A. F., Narasimhan, S., & Poltronieri, P. (2020a). Green synthesis of encapsulated copper nanoparticles using a hydroalcoholic extract of Moringa oleifera leaves and assessment of their antioxidant and antimicrobial activities. *Molecules, 25*(3), 555.

Das, S., Aswani, R., Midhun, S. J., Radhakrishnan, E. K., & Mathew, J. (2020b). Advantage of zinc oxide nanoparticles over silver nanoparticles for the management of Aeromonas veronii infection in *Xiphophorus hellerii. Microbial Pathogenesis, 147*, 104348.

Davis, H. S. (1922). *A new bacterial disease of fresh-water fishes*. US Government Printing Office.

Dawood, M. A., Eweedah, N. M., Moustafa, E. M., El-Sharawy, M. E., Soliman, A. A., Amer, A. A., & Atia, M. H. (2020). Copper nanoparticles mitigate the growth, immunity, and oxidation resistance in common carp (Cyprinus carpio). *Biological Trace Element Research, 198*, 283–292.

Demirbas, A. (2021). Comparison study of synthesized red (or Blood) orange peels and juice extract-nanoflowers and their antimicrobial properties on fish pathogen (*Yersinia ruckeri*). *Indian Journal of Microbiology, 61*(3), 324–330.

El-Adawy, M. M., Eissa, A. E., Shaalan, M., Ahmed, A. A., Younis, N. A., Ismail, M. M., & Abdelsalam, M. (2021). Green synthesis and physical properties of Gum Arabic-silver nanoparticles and its antibacterial efficacy against fish bacterial pathogens. *Aquaculture Research, 52*(3), 1247–1254.

Elgendy, M. Y., Shaalan, M., Abdelsalam, M., Eissa, A. E., El-Adawy, M. M., & Seida, A. A. (2022). Antibacterial activity of silver nanoparticles against antibiotic-resistant Aeromonas veronii infections in Nile tilapia, Oreochromis niloticus (L.), in vitro and in vivo assay. *Aquaculture Research, 53*(3), 901–920.

Erdem, B., Dayangaç, A., Kıray, E., & Duygu, D. (2019). Biosynthesis of silver nanoparticles from *Aeromonas sobria* and antibacterial activity against fish pathogens. *International Journal of Environmental Science and Technology, 16*(9), 5125–5130.

Essawy, E., Abdelfattah, M. S., El-Matbouli, M., & Saleh, M. (2021). Synergistic effect of biosynthesized silver nanoparticles and natural phenolic compounds against drug- resistant fish pathogens and their cytotoxicity: an in vitro study. *Marine Drugs, 19*(1), 22.

Esteve, C., Amaro, C., & Toranzo, A. E. (1994). O-serogrouping and surface components of Aeromonas hydrophila and Aeromonas jandaei pathogenic for eels. *FEMS Microbiology Letters, 117*(1), 85–90.

Fatima, R., Priya, M., Indurthi, L., Radhakrishnan, V., & Sudhakaran, R. (2020). Biosynthesis of silver nanoparticles using red algae *Portieria hornemannii* and its antibacterial activity against fish pathogens. *Microbial Pathogenesis, 138*, 103780.

Frerichs, G. N. (1995). Viruses associated with the epizootic ulcerative syndrome (EUS) of fish in South-East Asia. *Veterinary Research, 26*(5-6), 449–454.

Ghetas, H. A., Abdel-Razek, N., Shakweer, M. S., Abotaleb, M. M., Paray, B. A., Ali, S., et al. (2022). Antimicrobial activity of chemically and biologically synthesized silver nanoparticles against some fish pathogens. *Saudi Journal of Biological Sciences, 29*(3), 1298–1305.

Haniffa, M. A., Dhanaraj, M., Muthu Ramakrishnan, C., Sethuramalingam, T. A., Arun Singh, S. V., Ananth Kumar, Y., & Arthi Manju, R. (2008). Threatened fishes of the world: Heteropneustes fossilis (Bloch, 1794) (Cypriniformes: Saccobranchidae). *Environmental Biology of Fishes, 82*(2), 203–204.

Harikrishnan, R., Balasundaram, C., & Heo, M. S. (2009). Effect of chemotherapy, vaccines and immunostimulants on innate immunity of goldfish infected with *Aeromonas hydrophila*. *Diseases of Aquatic Organisms, 88*(1), 45–54.

Hawke, J. P., Mcwhorter, A. C., Steigerwalt, A. G., & Brenner, D. J. (1981). Edwardsiella ictaluri sp. nov., the causative agent of enteric septicemia of catfish. *International Journal of Systematic and Evolutionary Microbiology, 31*(4), 396–400.

Hemstreet, B. (2010). An update on *Aeromonas hydrophila* from a fish health specialist for summer 2010. *Catfish J, 24*(4).

Iqbal, Z., & Asghar, M. (2012). Saprolegniasis in two commercially important carps. *Pakistan Journal of Zoology, 44*(2).

Joughi, N. F. G., Farahpour, M. R., Mohammadi, M., Jafarirad, S., & Mahmazi, S. (2022). Investigation on the antibacterial properties and rapid infected wound healing activity of silver/laterite/chitosan nanocomposites. *Journal of Industrial and Engineering Chemistry, 111*, 64–75.

Jyoti, K., Pattnaik, P., & Singh, T. (2021). Green synthesis of silver nanoparticles using sustainable resources and their use as antibacterial agents: a review. *Current Materials Science: Formerly: Recent Patents on Materials Science, 14*(1), 40–52.

Kalatehjari, P., Yousefian, M., & Khalilzadeh, M. A. (2015). Assessment of antifungal effects of copper nanoparticles on the growth of the fungus Saprolegnia sp. on white fish (Rutilus frisii kutum) eggs. *The. Egyptian Journal of Aquatic Research, 41*(4), 303–306.

Khulbe, K. C., & Matsuura, T. (2021). *Nanotechnology in membrane processes* (pp. 199–343). Springer International Publishing.).

Koch, P. L., Tuross, N., & Fogel, M. L. (1997). The effects of sample treatment and diagenesis on the isotopic integrity of carbonate in biogenic hydroxylapatite. *Journal of Archaeological Science, 24*(5), 417–429.

Kornblatt, A. P., Nicoletti, V. G., & Travaglia, A. (2016). The neglected role of copper ions in wound healing. *Journal of Inorganic Biochemistry, 161*, 1–8.

Li, H., Xin, H., & Li, S. F. Y. (2015). Multiplex PMA–qPCR Assay with internal amplification control for simultaneous detection of viable Legionella pneumophila, Salmonella typhimurium, and Staphylococcus aureus in environmental waters. *Environmental Science & Technology, 49*(24), 14249–14256.

Lilley, J. H., Phillips, M. J., & Tonguthai, K. (1992). A review of epizootic ulcerative syndrome (EUS) in Asia.

Lubis, F. A., Malek, N. A. N. N., Sani, N. S., & Jemon, K. (2022). Biogenic synthesis of silver nanoparticles using Persicaria odorata leaf extract: Antibacterial, cytocompatibility, and in vitro wound healing evaluation. *Particuology, 70*, 10–19.

Mahanty, A., Mishra, S., Bosu, R., Maurya, U. K., Netam, S. P., & Sarkar, B. (2013). Phytoextracts-synthesized silver nanoparticles inhibit bacterial fish pathogen Aeromonas hydrophila. *Indian Journal of Microbiology, 53*, 438–446.

Mansouri-Tehrani, H. A., Keyhanfar, M., Behbahani, M., & Dini, G. (2021). Synthesis and characterization of algae-coated selenium nanoparticles as a novel antibacterial agent against Vibrio harveyi, a Penaeus vannamei pathogen. *Aquaculture, 534*, 736260.

Maurya, S., Bhardwaj, A. K., Gupta, K. K., Agarwal, S., Kushwaha, A., Vk, C., Pathak, R. K., Gopal, R., Uttam, K. N., Singh, A. K., & Verma, V. (2016). Green synthesis of silver nanoparticles using Pleurotus and its bactericidal activity. *Cellular and Molecular Biology, 62*(3), 1000131.

Meneses, J. O., Dias, J. A. R., dos Santos Cunha, F., dos Santos, H. L., Santos, T. B. R., Santos, C. C. M., et al. (2023). Prophylactic and therapeutic effects of a nanocomposite (silver nanoparticle plus Terminalia catappa) against Saprolegnia parasitica in tambaqui. *Aquaculture, 574*, 739695.

Morley, N. J., & Lewis, J. W. (2010). Consequences of an outbreak of columnaris disease (Flavobacterium columnare) to the helminth fauna of perch (Perca fluviatilis) in the Queen Mary reservoir, south-east England. *Journal of Helminthology, 84*(2), 186–192.

Pang, M., Jiang, J., Xie, X., Wu, Y., Dong, Y., Kwok, A. H., et al. (2015). Novel insights into the pathogenicity of epidemic Aeromonas hydrophila ST251 clones from comparative genomics. *Scientific Reports, 5*(1), 9833.

Pelesinuo, K. B., Sattanathan, G., Haque, N., Al-Ghanim, K. A., Nicoletti, M., Sachivkina, N., & Govindarajan, M. (2023). Synthesis and characterization of Mithun (Bos frontalis) urine-based antibacterial copper oxide nanoparticles. *Biomedicines, 11*(6), 1690.

Plumb, J. A., & Hanson, L. A. (2010). *Health maintenance and principal microbial diseases of cultured fishes*. Wiley.

Pridgeon, J. W., & Klesius, P. H. (2011). Virulence of Aeromonas hydrophila to channel catfish Ictaluras punctatus fingerlings in the presence and absence of bacterial extracellular products. *Diseases of Aquatic Organisms, 95*(3), 209–215.

Rahman, A. N. A., Abdelwarith, A. A., Younis, E. M., Rhouma, N. R., Zaki, H. T., Khalil, S. S., & Ibrahim, R. E. (2023). The alleviative effects of green synthesized copper oxide nanoparticles against oxidative stress, hepato-renal alterations, and immune suppression induced by Staphylococcus aureus infection in Clarias gariepinus. *Aquaculture Reports, 32*, 101746.

Ramana, V., Rajeshkumar, S., & Jagadeesh, K. (2023). Review of the environmentally friendly production of zinc oxide nanoparticles and its anti-oxidant, anti-hyperlipidemic, and anti-diabetic properties. *Journal of Survey in Fisheries Sciences, 10*(1S), 117–127.

Rather, M. A., Bhat, I. A., Sharma, N., Gora, A., Ganie, P. A., & Sharma, R. (2017). Synthesis and characterization of Azadirachta indica constructed silver nanoparticles and their immunomodulatory activity in fish. *Aquaculture Research, 48*(7), 3742–3754.

Raza, M. A., Kanwal, Z., Riaz, S., Amjad, M., Rasool, S., Naseem, S., et al. (2023). In-vivo bactericidal potential of Mangifera indica mediated silver nanoparticles against Aeromonas hydrophila in Cirrhinus mrigala. *Biomedicines, 11*(8), 2272.

Reyes-Becerril, M., Ruvalcaba, F., Sanchez, V., López, M. G., Silva-Jara, J., Hernandez-Adame, L., & Angulo, C. (2021). Green synthesis of gold nanoparticles using Turnera diffusa Willd enhanced antimicrobial properties and immune response in Longfin yellowtail leukocytes. *Aquaculture Research, 52*(7), 3391–3402.

Roberts, R. J., Frerichs, G. N., Tonguthai, K., & Chinabut, S. (1994). Epizootic ulcerative syndrome of farmed and wild fishes. *Recent Advances in Aquaculture, 5*.

Robinson, N., Sahoo, P. K., Baranski, M., Mahapatra, K. D., Saha, J. N., Das, S., et al. (2012). Expressed sequences and polymorphisms in rohu carp (Labeo rohita, Hamilton) revealed by mRNA-seq. *Marine Biotechnology, 14*, 620–633.

Saddik, M. S., Elsayed, M. M., El-Mokhtar, M. A., Sedky, H., Abdel-Aleem, J. A., Abu-Dief, A. M., et al. (2022). Tailoring of novel azithromycin-loaded zinc oxide nanoparticles for wound healing. *Pharmaceutics, 14*(1), 111.

Shaalan, M. I., El-Mahdy, M. M., Theiner, S., El-Matbouli, M., & Saleh, M. (2017). In vitro assessment of the antimicrobial activity of silver and zinc oxide nanoparticles against fish pathogens. *Acta Veterinaria Scandinavica, 59*(1), 1–11.

Shaalan, M., Sellyei, B., El-Matbouli, M., & Székely, C. (2020). Efficacy of silver nanoparticles to control flavobacteriosis caused by Flavobacterium johnsoniae in common carp Cyprinus carpio. *Diseases of Aquatic Organisms, 137*(3), 175–183.

Sherif, A. H., Abdelsalam, M., Ali, N. G., & Mahrous, K. F. (2022). Zinc oxide nanoparticles boost the immune responses in Oreochromis niloticus and improve disease resistance to Aeromonas hydrophila infection. *Biological Trace Element Research*, 1–10.

Sistemática, D., Gabriela, S. S., Daniela, F. R., & Helia, B. T. (2016). Copper nanoparticles as potential antimicrobial agent in disinfecting root canals. A systematic review. *International Journal of Odontostomatology, 10*(3), 547–554.
Skirpstunas, R. T., & Baldwin, T. J. (2002). Edwardsiella ictaluri invasion of IEC-6, Henle 407, fathead minnow and channel catfish enteric epithelial cells. *Diseases of Aquatic Organisms, 51*(3), 161–167.
Thangapandiyan, S., & Monika, S. (2020). Green synthesized zinc oxide nanoparticles as feed additives to improve growth, biochemical, and hematological parameters in freshwater fish Labeo rohita. *Biological Trace Element Research, 195*, 636–647.
Thanigaivel, S., Vickram, S., Saranya, V., Ali, H., Alarifi, S., Modigunta, J. K. R., et al. (2022). Seaweed polysaccharide mediated synthesis of silver nanoparticles and its enhanced disease resistance in Oreochromis mossambicus. *Journal of King Saud University-Science, 34*(2), 101771.
Vijayakumar, S., Vaseeharan, B., Malaikozhundan, B., Gobi, N., Ravichandran, S., Karthi, S., et al. (2017). A novel antimicrobial therapy for the control of Aeromonas hydrophila infection in aquaculture using marine polysaccharide coated gold nanoparticle. *Microbial Pathogenesis, 110*, 140–151.
Vinosha, M., Palanisamy, S., Muthukrishnan, R., Selvam, S., Kannapiran, E., You, S., & Prabhu, N. M. (2019). Biogenic synthesis of gold nanoparticles from Halymenia dilatata for pharmaceutical applications: Antioxidant, anti-cancer and antibacterial activities. *Process Biochemistry, 85*, 219–229.
Wang, S. X., Zhang, J. Y., Du, X. K., Liu, D. J., Liu, L. X., & Shen, X. H. (2022). Comparative analysis of the intestinal microbiota in goldfish and crucian carps between different aquaponics and traditional farming. *Aquaculture Reports, 25*, 101240.
Zhang, Z., & He, Y. (2023). Synthesis and characteristics of a fish scale-based biochar–nanosilver antibacterial material. *Processes, 11*(7), 1992.
Zhang, J., Yao, J., Wang, R., Zhang, Y., Liu, S., Sun, L., et al. (2014). The cytochrome P450 genes of channel catfish: their involvement in disease defense responses as revealed by meta-analysis of RNA-Seq data sets. *Biochimica et Biophysica Acta (BBA)-General Subjects, 1840*(9).

Nanomaterial Synthesis Using Tyre and Plastic

Shikha Saxena and Shikha Baghel Chauhan

1 Introduction

The population and urbanization are increasing tremendously in today's scenario. The set-up for industry and entrepreneurship has created the space for lots of industrial waste. As per the World Bank Project, it is predicted that by the year 2050, the amount of waste will double the amount of population increase. Globally, data shows that 2.01 billion tons of municipal wastes are collected annually, and this figure will increase to 3.40 billion by the year 2025. Recycling this waste into high-end nanomaterials is creating buzz in the solid waste management industry.

Nanomaterials can be defined as a group of materials in which at least one dimension is less than approximately 100 nm (Bhardwaj et al., 2017). Based upon their scale of existence, many extraordinary optical, magnetic, electrical and other relevant properties can be expected. (Alagarasi, 2011). These features have a significant role in electronic, medical and other fields. Based on shaping, these can be 0D, ID, 2D or 3D (Kolahalam et al., 2019). Engineered Nanomaterials exploit the small size and novel properties associated with nanoparticles and thus open the gates for their commercial utilization (Khan et al., 2019). The waste generated from the nanomaterials can be categorized as biogenic (containing biomolecules), like from dead leaves, waste food materials, etc., and non-biogenic (devoid of biomolecules), like from plastic waste or tyre waste (Singh et al., 2021). In the quest to find the solution, nanomaterials played a beautiful role by not only providing the sustainable source but also decreasing the product cost.

Plastics are inexpensive, lightweight, strong, durable and corrosion-resistant materials with high thermal and electrical insulation properties. Its usage keeps on

S. Saxena (✉) · S. B. Chauhan
Amity Institute of Pharmacy, Amity University, Noida, India
e-mail: skataria@amity.edu

A. K. Bhardwaj et al. (eds.), *Biogenic Wastes-Enabled Nanomaterial Synthesis*,
https://doi.org/10.1007/978-3-031-59083-2_13

proliferating and is there in packaging, building, construction, transportation and electronic components (Kumar, 2010; Kumar et al., 2020a, b).

Plastics have major categories as follows:

(1) PE (polyethylene)
(2) PP (Polypropylene)
(3) PS (Polystyrene)
(4) PAN (Polyacrylonitrile)
(5) PET (Polyethylene terephthalate)

Different types of plastic items include plastic cloths, pellets, bottles, personal stuff and polyethene. Despite many varieties, as a whole, there are many issues (Thompson et al., 2009) associated with plastics. These are:

(1) Plastic is non-renewable source of energy.
(2) It takes millions of years to decompose naturally.
(3) Conversion of raw plastic to useful material is not easy.
(4) It is difficult to recycle.

On the similar grounds, waste produced from tyres in the form of rubber attracts the attention of environmental as well as medical scientists. The bulk of tyre tread is made of a variety of rubbers, including material rubber copolymers, butadiene rubber, styrene butadiene rubber, neoprene rubber and polysulfide rubber. In general, tyres are made up of 45–47% material rubber, 21–22% carbon black, 12–25% metals, 5–10% textile, 6–7% additives and 1–25 zinc, ~1% silver. Based on application, the basic composition of various tyres differs; all tyres are made of carbon black, steel, natural rubber and synthetic rubber. They have high calorific fuel; initially, they were used as energy sources of power stations and cement kilns.

Disposal of waste is one of the biggest challenges the environment is facing in current world. Waste can be broadly categorized into industrial waste and consumer-generated waste. Tyres usually come under the roof of industrial waste, and plastics come under consumer waste. Nowadays, tyres are burnt or discarded, and thus, legally, they are banned in many countries (Moghaddasi et al., 2013). Waste tyre stock piles are also significant fire risks, as they are long burning and emit hazardous fumes. In landfills, waste tyres are not readily biodegradable and risk leading toxic chemicals into the surrounding environment. Yet, despite doing research and employing many technological developments, applicable processes for cost effective recycling of waste tyres do not exist (Giere et al., 2004).

Motor vehicles on the road are increasing annually. Disposing of scrap tyres causes serious threats to the atmosphere. The waster rubber tyres are a thermoset polymer that cannot be reprocessed like thermoplastic polymers. Based on worldwide data, there is a generation of 1000 million scrap tyres annually. Developing as well as developed countries have banned the disposal of their tyres due to the involvement of open burning method, which aids in pollution and environmental degradation. Management of their scrap tyres is the area of interest of many scientists so that environmentally friendly methods can be launched.

The field of material sciences focuses on sustainable processes and explores new methods for the production of nanomaterials. This opens the gate to green synthesis methods. Low-income countries are leading in dumping or burning waste every year. According to the Environment Program of the United Nations, 11.2 billion tonnes of solid waste are formed. Countries with incapable waste management facilities create the need for the development, implementation and strengthening of different policy strategies like banning plastics, disposing rubber waste in ocean bodies, etc. (Wu et al., 2019).

This is accompanied by many technical, economical and resource-focused challenges; for example, the purification and extraction of plastics utilize hazardous materials, leading to health risks and safety issues (Newman & Meneley, 2006). Hence, clean manufacturing or green manufacturing is the need of the hour. Parallelly, engineered nanomaterials have promising results in terms of energy transformation, intelligent packaging, precise crop management and food and drug industry (UNEP, 2018).

Talking about the worldwide scenario for plastics, in 2012 alone, 280 million tonnes of plastic were produced. Out of this, 130 million tonnes plastic were land filled or recycled. Rest 150 million tonnes enter the daily lives of human. Some amount enters as litters in oceans and water bodies. As per UNEP (United Nations Environment program), the assessment of the negative externalities produced by mismanaging plastic is in the range of 40 billion USD/year, and it is expected that by 2050, there will be 12 billion plastic waste despites all efforts. In 2015, an agreement was signed by 195 countries called the Paris Agreement of climate change, which focused on the establishment of sustainable development goals (SDG) in order to have safe and sustainable society. This changed the overall view for plastics and its waste, and systematic approach was considered for their disposal (Kumar et al., 2020a, b). As the global market is concerned, it is valued at 10,22.45 kilometric ton in 2020 and is expected to reach 15,256.19 kiloton (Kumar et al., 2021). In 2020, the global nanomaterial market was reported to be USD 19.928.96 million. The expectation is that it will reach USD 57.608.26 million by 2026.

Potential hazards associated with plastic and rubber waste are obstruction in ground water movement, heavy metals in plastic mixes up with rain water and cause soil pollution and finally, impact on viability, development life cycle, movement and fertility of living organisms. The interfering mechanisms associated with the hazards of plastic and tyre waste include impaired fatty acid metabolism, irritation, oxidative damage, digestion impairment, tissue deposition, changes in gut mucosal communities and molecular damage.

Meeting the demands of society and the alarming rise in waste generation calls for the development of cost effective and stable methods for the energy conversion and waste recycling methods.

Sources of migration of this plastic and tyre waste into various levels of the food chain are summarized in Fig. 1.

Different types of nanomaterials that can be obtained from plastic and tyre waste are categorized into the following:

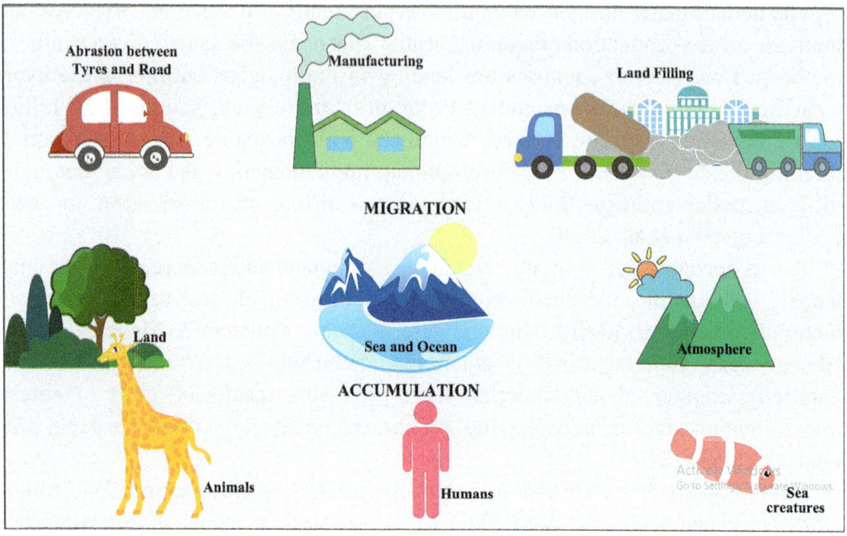

Fig. 1 Sources of micronanoparticles

1. Carbon-Based Nanoparticles.
   (a) Carbon Nanotubes (CNTs): These are specially designed rolled sheets with single-layer sp^2 hybridized carbon arranged in hexagons. Further, they can be classified into single-walled, doubled-walled and multi-walled based on the number of rolled graphene sheets (Sajid, 2022). Single-walled has characteristic one-atom-thick graphite layer, but double-walled has two single-walled nanotubes where nesting of one in the other takes place. In multi-walled, multiple sheets of graphite are being used. This rolling method can cause generation of various patterns like chiral, armchair and zigzag (Balasubramanian & Burghard, 2005). One of the major properties of nanotubes is very high tensile strength. These are some of the smartest candidates in nanomaterials for electronic devices such as rectifying diodes, single-electron transistors, and field-effect transistors due to their unique and brilliant electrochemical properties.
   (b) Fullerenes: They are allotropes of carbon and possess symmetrical cages of sp^2 hybridized carbon atoms. Because of their specificity in carbon numbers, they are different from other carbon allotropes. Fullerenes range in different sizes based on the number of carbon atoms, such as C60, C70, C72, C76, C84, and C100. Nowadays, chemical modification of fullerenes is capturing the market due to their better applicability. Exohederal fullerenes are also advantageous, as modification for outer surface is easy. Fullerenes are vital members of the carbon-based nanomaterial family, and they certainly possess exceptional properties.

(c) Graphene: It is the wonder nanomaterial with the property of being the thinnest known material with outstanding mechanical strength (Sajid, 2022). Graphene has two-dimensional sp2-hybridized carbon atom planar sheets that are tightly packed into honeycomb-like lattices. It is also used in electrochemical sensors. They have the ability to generate transparent electrodes with very good optical transmittance (Li et al., 2016).
(d) Nanodiamonds: Nanodiamonds have sp3-hybridized carbon, and they are unique carbon-based nanomaterials with extraordinary optical properties, mechanical properties, high specific surface areas, and rich surface structures. Several functional groups, such as amide, aldehyde, ketone, carboxylic acid, alkene, hydroperoxide, nitroso, carbonate ester and alcohol groups, are present on nanodiamond surfaces.
(e) Quantum Dots: These are accidentally discovered particles with less than 10 nm sizing. Carbon quantum dots have exclusive features that include tunable photoluminescence properties, good multi-photon excitation, biocompatibility and low toxicity; solubility in water due to rich oxygen-containing groups; cost effectiveness; and environmental friendliness. These are also considered nano lights due to their powerful luminescence properties. They are major players in the bioimaging and nanomedicine markets (Li et al., 2012).
(f) Nano Cones: The other name for nano corns is nano horns. They are sp2 hybridized carbon sheets having single wall and tubular structure. It has structural similarities with nano tubes and has three types: dahlia like, bud type and seed like. Their synthesis involves use of no catalyst, and they have excellent storage options. Other properties include electrochemiluminescence, adsorption, electrochemical sensing and drug delivery system.

2. *Metal Nanoparticles:* The synthesis is based upon chemical reduction of related salts. The applicability of metal nanoparticles is vast, and they can be further categorized as metal nanoparticles, metal oxide nanoparticles, bi-metallic nanoparticles and tri-metallic nanoparticles. These have incredible optical properties, as seen in optical sensors, such as gold and silver nanoparticles. Their usage is very much decided by their shape and size. They also exhibit antimicrobial properties and are therefore inclined to be used in drug industry.

   Metal oxide nanoparticles have a small size and high density, making them fit for sensing, catalysis, energy storage, antibacterial, biomedical and optoelectronic applications. Examples include Fe_2O_3, Al_2O_3, TiO_2, SiO_2, CuO, SnO_2 and WO_3. The preparation is based on the sol-gel method.

   Bimetallic nanoparticles involve two metals and have exclusive optical, electrical, thermal, and catalytic properties. Bimetallic architecture involves charge transfer and enhancement of catalytic properties (Mody, 2010; Bhardwaj et al., 2020).

   Trimetallic nanoparticles have technically three metals involved and are better performers in comparison to monometallic compounds. But their surface area is unstable, which causes the use of stabilizers like block copolymers, organic ligands, surfactants and dendrimers. Different techniques for synthesis are microwave, selective catalytic reduction, microemulsion, coprecipitation and hydrothermal technique (Jana et al., 2001).

3. *Ceramic Nanoparticles:* The composition of ceramic nanoparticles is based on oxides, carbides, phosphates and carbonates of metals. The degradation and degranulation are rare phenomena in ceramic nanoparticles due to their porosity. They involve nano scaffolds and nano clay also (Singh et al., 2016). The size is controllable and can be customized; they are heat resistant and chemically inert and hence can be easily used in biomedical domain. Their tolerance to pH change is noteworthy and can be fabricated to hydrophobic and hydrophilic system. Parallelly their usage in biomedical field becomes questionable sometimes due to their potential toxicity (Thomas et al., 2015).
4. *Semiconductor Nanoparticles:* They are made up of semiconductor materials reduced to nanoscale, thus leading to altered physical, chemical, optical and magnetic properties. These are referred to as II-VI, III-V or IV-VI based on groups in periodic table. The applications involve optoelectronic devices, single electron devices, resonant tunnelling devices, memory devices, magnetic sensors and catalysis (Bhardwaj et al., 2019; Sahu, 2019).
5. *Polymeric Nanoparticles:* It involves nanoparticles with particle size of 1–1000 nm at least in one dimension. They have the ability to enhance surface area-to-volume ratio. Their superior optical, thermal and electrical properties make them unique in nature. Different types include natural polymer-based, bio-synthetic polymer-based and chemosynthetic polymer-based nanoparticles (Han et al., 2018). They are implicated to be used in the field of drug delivery, separation, catalysis, nanoreactors, and surface coatings. They have the capacity to encapsulate drug molecules and are therefore used in drug delivery applications (Fu & Fu, 2011).
6. *Biomolecule-Based Nanoparticles:* Many biomolecules are involved in the synthesis of sophisticated nanoparticles like proteins, lipids, nucleic acids, etc. The current scenario is giving a lot of attention to this type of nanoparticles due to its biocompatibility and biodegradability. They can be very well employed in the synthesis of hybrid materials, thus enhancing the capability and applicability of nanomaterials like nucleic acid-conjugated nanomaterial (Ding et al., 2021).

Different approaches to synthesize nanomaterials using plastic and tyre wastes are broadly classified under two domains: (Fig. 2).

1. *Top-Down Approach*

It involves the breakage of bulk materials to fine particles with the help of various lithographic techniques (Baig et al., 2021) (Table 1):

2. *Bottom-Up Approach:* Self assembling of atoms into new nuclei is the key to this approach. This further involves enhancement of particles having nanoscopic dimensions (Pal et al., 2019) (Table 2).

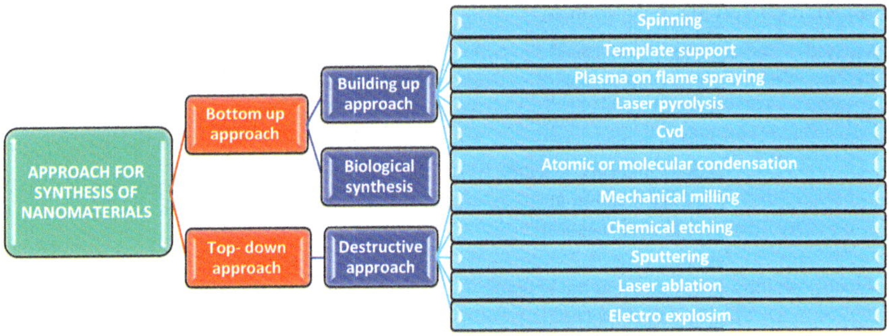

Fig. 2 Synthesis criteria for nanomaterials

2 Uniqueness of Synthesized Nanomaterials from Waste

Nanomaterials obtained from plastics and tyre wastes exhibit unique properties like other nanomaterials (Baig et al., 2021).

(a) Surface area: The surface area is high in nanomaterials compared to bulk materials.
(b) Magnetism: It is the property that can be seen at nanoscale. Even non-magnetic compound can show magnetic characteristics.
(c) Quantum nature: Nanomaterials have property to show strong quantum effect.
(d) Thermal and electrical nature: Brilliant thermal and electric conductivity occurs with nanomaterials.
(e) Mechanical properties: One of the major changes from bulk to nanomaterial is generation of excellent mechanical properties.
(f) Antimicrobial nature: Few nanomaterials exhibit antiviral, antibacterials and antifungal properties and have an exceptional power to deal with pathogenic diseases.

3 Applications of Synthesized Nanomaterials

The processed nanomaterials from the waste have vast applications, which are summarized in Fig. 3.

1. *Drugs and Medication Industry:* Designing of new nanodevices for physical, chemical, biological, pharmaceutical application. For example, the generation of transdermal patches using carbon nanotubes. In this, when voltage is applied, the pure medication is either put onto the carbon nanotube membrane or covered in it, forcing it past the stratum corneum's thick outer layer. Since carbon nanotubes have been added, the conductance capacity of carbon nanotube hydrogels has increased, forcing the medication to penetrate the skin and reach the intended

Table 1 Different top-down methods

Method	Process involved	Applications	
Mechanical milling	Series of events: Elastic deformation, plastic deformation, shear deformations, fracture, amorphization and chemical reactions	Synthesis of oxide- and carbide-strengthened aluminum alloys, wear-resistant spray coatings, aluminum/nickel/magnesium/copper-based nanoalloys, useful for the synthesis of nanocomposites	Zhuang et al. (2016)
Electrospinning	Spinneret is made up of two coaxial capillaries, two viscous liquids or combination of viscous-non viscous liquids in electric field	Useful for the synthesis of nanofibers, core shells.	Ostermann et al. (2011)
Lithography	In this beam of light is focused or electron are focused on the materials	Development of nanoarchitectures	Pimpin and Srituravanich (2012)
Sputtering	Bombardment of solid surfaces is done with high energy particles like plasma or gas.	Film of nanomaterials is being generated.	Ayyub et al. (2001)
Arc discharged method	Formation of precursor, growth and agglomeration	Formation of fullerenes, carbon nano horns (CNHs), carbon nanotubes, few-layer graphene (FLG), and amorphous spherical carbon nanoparticles	Zhang et al. (2019)
Laser ablation	Powerful laser beam hits the target material leading to vaporization of precursor moiety and finally production of nanomaterials	Metal nanoparticles, carbon nanomaterials, and oxide composites	Ismail et al. (2020)
Thermal decomposition method	Endothermic process where breaking of chemical bonds is done using heat process. It involves breakage of metal to generate nanoparticles involving chemical reaction too.	Generation of metal nanoparticles	

site of action. The drug release profile is controlled using carbon nanotube-microneedles, which have an excess mechanical strength to pierce skin and are entirely reliant on the kinetics and mechanism of polymer breakdown (Zhou et al., 2021). In another example, generation of implants can be seen. Due to an increase in surface roughness brought about using CNTs as a coating or in a coating matrix (such as hydroxyapatite), osteoblast adhesion to the implant surface has also been improved.

Table 2 Different methods of bottom-up approach

Method	Process Involved	Applications	
Chemical vapor deposition	Process a thin layer is being formed chemically on the target surface with suitable precursors	Synthesis of carbon nanotubes	Jones & Hitchman (2009)
Solvothermal and hydrothermal methods	a heterogenous reaction is being carried out in the water based medium at high pressure and temperature in the sealed compartment	Nano geometric materials, like nanowires, nanorods, nanosheets, and nanospheres	Cao et al. (2017)
Sol-gel method	Liquid based precursor is converted to sol and that is transformed to network like gel.	High-quality metal -oxide based nanoparticles	Parashar et al. (2020)
Template method	Nano casting involves use of solid materials where pores are filled with precursor molecules	Nanostructured materials, polymeric carbonaceous nanospheres, single crystal nanorods	Baig et al. (2021)
Reverse micelle method	Synthesis takes place at core acting as reactor, where size can be customized based on water to surfactant ratio	Nanomaterials of desired shape and size.	Shiri et al. (2019)
Spray pyrolysis method	Using a spray precursor into high temperature like furnace atmosphere causing the composition for production of nanoparticles.	Metal, ceramic nanoparticles of different size and shape.	Workie et al. (2023)

CNTs are employed as a medication delivery mechanism, have the potential to be a high-performance biomaterial and have great material qualities. They can also increase the mechanical strength of composite materials. Formulation of nanodrugs is another crucial application. When functionalized with anti-cancer drugs, biocompatibility-improving molecules and specific cell-targeting biomolecules, graphene, carbon nanotubes, and their derivatives can be used as photothermal agents and have strong near-infrared absorption. Hence, these materials are used as anti-cancer nanodrugs (Bhardwaj et al., 2022).

2. *Manufacturing and Material Industry:* Nanocrystalline material, consumer products, fitness products, food processing and packaging, for example, the generation of fuel cells. Carbon nanotubes could store hydrogen, allow electrons to move freely or boost catalytic activity. To increase the effectiveness of the catalytic reactions in the fuel cell, carbon nanotubes have been introduced to the platinum/carbon catalyst mixture at the anode in fuel cell research.

Another road to success is scaffolding for the stem cell research. A crucial component of tissue engineering and regenerative medicine is the use of stem cells. To develop and renew, they need a favourable microenvironment, though. As a result of their strong optical absorbance in the near-infrared range, CNTs can also be employed for stem cell labelling. Additionally, CNTs have been

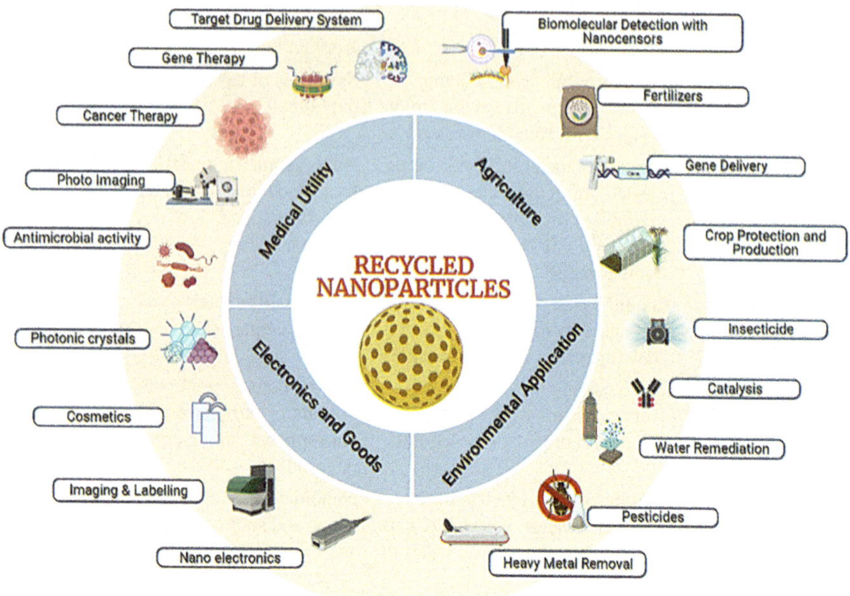

Fig. 3 Applications of recycled nanomaterials

added to frequently used scaffolds to provide useful scaffolds for controlling stem cell destiny.

3. *Electronic Industry:* Printed electronics and organic electronic molecules, for example, the generation of mechanical sensors. The active substance graphene is used to detect physical signals such as pressure and tension. Graphene materials are frequently employed as a conducting layer or electrode for graphene-based strain and pressure sensors due to their excellent electrical conductivity.

   Photonics sensors, optoelectronics, laser materials are other applications of these recycled nanomaterials. In comparison to traditional semiconductor optoelectronic devices made of silicon and GaAs, graphene exhibits extraordinarily high carrier mobility and a wide variety of photoelectric response properties.
4. *Energy Harvesting:* Nanomaterials as renewable energy resources, water splitting, electro-chemical CO_2 reduction, solar cells and piezoelectric generator and nanogenerators opened the gates of energy harvesting mode.
5. *Mechanical Industry:* Coating, lubrication and adhesive application lead to the generation of alumina, Titania and carbon-based nanomaterials having better tensile strength (Abdelbasir et al., 2020).

\

4 Case Studies and Examples of Nanomaterials from Rubber Tyres

(a) Zinc Nanoparticles from Used Tyres

In this milling approach, is used where in step one, metal wires are removed from scraped tyres using a shredder. This is followed by grinding, rolling mills and sieving, followed by ball milling. Waste rubber of tyres is put in ball mill for 5–10 h with silicon waste at the speed of 500 rpm alone or sometimes with liquid nitrogen. As a sub sample, rubber waste is given a temperature of 550 degrees to ash for 12 h. The size of resultant nanomaterials is checked using techniques like SEM and XRD. Aggregation of rubber particles takes place during the milling method, thus causing flexibility and plasticity. Silicon has compressive properties and nitrogen has cooling properties. As the nutrient medium is considered, Zn source is used for cucumber grown in nutrient solution. The benefits of NP include greater shoot and fruit yield. Now the incorporation of nitrogen leads to larger sized particles instead of reducing them; however, with the silicon waste <100 nm size can be achieved. The resulting nanomaterials are considered good source of Zn in cucumber plants. It is to mention that the size of the rubber ash particles was less than 500 nm before milling and decreased to less than 50 nm after milling (Moghaddasi et al., 2013).

(b) Carbon Nanoparticles

CNPs gained popularity by showing superior electrochemical performance in LIBs and super capacitors. The enchanting properties of high surface area and porosity are key factors in determining capability. Different techniques involved are pyrolysis technique, microwave plasma, chemical vapour predisposition using iron and nickel as catalyst, electrolysis in molten alkali halides and microemulsion polymerization at low temperatures. CNPs are readily formed by innovative high-temperature approach using waste rubber tyre primarily made of carbon 81.2%. Maroufi et al. reported a technique for quick transformation of tyre waste at very high temperature (1550 °C) with shorter span of time. The evolution of toxic gases was minimized during synthesis. The process involves a series of steps including shredding, separating wires, granulation and pulverization. At high temperatures, waste is put into furnace, and the resultant char material was milled and cold zone treatment is given to the prepared pellets. Characterization is carried out using Raman spectroscopy (Maroufi et al., 2017).

(c) Carbon Spheres

Eighty percent of waste tyres, which can't be used in the world, can be used for clustered and hierarchical carbon spheres. For this, waste tyre scraps are treated with chemical degradation and autoclaving. The steps involved are the swelling of scraps in H_2SO_4 and CH_3OH-NaOH solutions for 15 min at RT followed by degradation process. After this production takes place at 300 °C under 5 MPa for 3 h. Characterization parameters involves FT-IR, SEM, XRD, TEM and Raman spectroscopy. Mostly traditional methods involve energy intensive condition leading to expensive cost of production. Nano structures obtained from waste tyres found to be long and entangled braid like cylinders with lengths of ~40 μm and diameter of about 100 nm. In a very crude manner, nanotechnology has a capability to produce next generation raw material (Balbay & Acıkgoz, 2022).

(d) Carbon Black Nanoparticles (CB)

Carbon black is generated with the help of partial combustion of petroleum feedstock. Commercially available technique that was used previously involved series of complex reactions and purification; however, nowadays, CB can be synthesized using rubber waste under optimal conditions with the result of thermally stable and good-quality nanomaterials. These small pieces of crumbled rubber are obtained from waste and cleaned with water, soup and acetone followed by drying. This crumbled tyre waste is further ground without damaging the metallic wire and then poured into a lab-made cylindrical stainless-steel reactor. This reactor is placed into furnace kept at high temperature. Of 1000 degrees for 4 h. After the cooling-off period, it is opened and the resultant black solid was washed, filtered and dried and finally characterization is performed. The resultant nanomaterial are structurally close to those obtained from hydrocarbon pyrolysis. Therefore, carbon black, which is a high constituent of waste tyres, can be very well converted to nanomaterials using thermal transformation. Their usage can be implicated in the development of batteries, flexible electronic devices, mechanical reinforcement additive, sensors and fluorescent displays.

(e) Carbon Nanotubes

Carbon nanotubes are one of the excellent nanomaterials of carbon origin and have versatile applications. The diameter ranges from 1 nm–100 nm with the length of 10 nm to a few centimeters. As compared to conventional methods of catalytic decomposition, tyre waste acts as a cheaper feedstock. Thermal degradation of tyre waste leads to pyrolysis of oil in the atmosphere, which is devoid of oxygen.

(f) Graphene Nanoplatelets

Graphene nanoplatelets can be another form of nanomaterial that is synthesized from waste tyres. Pyrolysis allows the burning of tyres, thus forming carbon black, which further leads to the formation of nanoplatelets. These have vast applications in the field of locomotives (Mustafa et al., 2021).

Different merits connected with plastic-derived nanomaterials include catering to the issue of global environmental crisis, which can be seen with plastic waste piling up. Conversion of plastics to nanomaterial can solve the problem of impractical traditional disposal methods like landfills, incineration, and energy production. Also, the novelty of constructing nanomaterials from waste plastic plays a role (Derraik, 2002). Current plastic disposal options include landfills leading to waste of space, the generation of energy products that are costly, incineration having toxic by-products and physical recycling making low-grade products.

The generation of plastic-derived nanomaterials seems environmentally friendly and a lucrative approach to the surrounding, but there are some problems associated with plastic derived nanomaterials. These are:

(1) Boosting the nanomaterials, synthesis using plastics seems easy but has its share of challenges, like the cleaning of plastic, which is expensive process, and many times low grade products is formed, leading to financially unviable situation.
(2) Lack of scalable methods.
(3) Many countries are stopping the export of used plastics, like Australia, which is banning the polymer plastic export from July 1, 2022.
(4) Polyethylene and other polyolefins are difficult to destruct using catalytic process.
(5) Flipside is difficulty in segregating mixed plastics and removing residues.

Conversion to nanomaterials attracts lots of attention due to their high porosity, lightweight and small diameters. Also, nanomaterials can be employed in many domains, like bioengineering, electronic devices, energy storage, etc. Methods involved in the preparation of nanomaterials are, namely, drawing, phase separation, centrifugal spinning and electrospinning.

5 Impact of COVID and Generation of Plastic Waste

In 2020, the outbreak of Covid-19 generated millions of plastic because of the increased consumption of plastic as medical devices such as masks and syringes. Keeping the traditional disposal of plastics and its generation in mind, waste has to be recycled into useful products, including nanomaterials, which have applications in biomedical research, diagnostics, communication, etc.

6 Case Studies and Examples of Nanomaterials from Plastic Wastes

Different types of plastics lead to different types of nanomaterials, and, hence, different fields of applicability. The data is summarized in Table 3.

(a) Nanocomposites

Nanocomposites come under the category of hard compounds with multiple phases of which at least one is less than 100 nm. They are matrix-type materials with nanofillers like fibrils, pieces and particles encircled. As it involves integration of two materials, it clearly enhances the properties of toughness, strength and thermal conductivity. Further these are categorized into three types, i.e.,

Table 3 Types of plastics and their applicability

S. No.	Waste plastic	Polymer	Nanomaterial	Application
1	Plastic water bottles	PET	Fullerenes	Dermatology and cosmetics
2	PVC powder	PVC	Nanoporous carbon	Supercapacitance
3	Waste plastic	PP	Graphene flakes	Conductive inks, solar cells, smart windows, transistor
4	Plastic foam	PS	Nanofoams	Degradation of dyes
5	Waste PS foam	PS	Superabsorbent nanocomposite	
6	Mixed plastic		Carbon nanosheets	Degradation of dyes
		PP	Nano-clay/ nanocomposites	Improved mechanical properties of the polymer after recycling
7	Woven bags, foam sheets, bottkes, vessels, sewage pipes	PE, PET, PP, PVC	Carbon nanosheets	Supercapacitance
		PTFE	Carbon nanospheres	Energy storage
8.	Bottles, vessels	PP, PE, PS	Carbon nanotubes	Photocatalyst, medical devices
9.	Plastic bag waste	PMMA	SiC platelet	Biosensors, hydrogen storage
10.	Plastic bag	PMMA	SiC nanowires	Supercapacitance
11.	Solid plastic waste	PMMA	Carbon nanobud	Radiation source, X-ray tubes.
12.	PET waste bottles	PET	Fluorocent nanoparticle	Biosensors

nano-filamentary composites (consist of specific model nested with nanosized filaments), nanolayered composites (consist of changeable coverings of nanosized dimensions) and nanoparticulate composites (consist of specific form in which nanosized particles are placed). The method utilizes pyrolysis at 600 °C and autoclaving in the presence of ferrocene and NaN_3 for 12 h followed by cooling and getting black precipitates. The application associated with these are marvellous; they are employed in the field of magnetic resonance imaging (MRI), integrated diagnosis and therapeutics, drug delivery, ultra-high-density magnetic recording media, and electromagnetic (EM) wave absorption (Mustafa et al., 2021; Hunter, 2018).

Wang et al. reported one-step process for nanocomposite generation where waste plastic is fed into autoclave with metallic Li and Ta2O5 at 700 °C for 10 h. Removal of residual impurities is done by a washing process using hydrochloric acid, distilled water and absolute alcohol.

(b) 2D Graphene-Based Materials

Graphene can be defined as a single two-dimensional sheet of where sp2 hybridized carbon is arranged in hexagonal pattern. The two major bonds that is σ and π bonds are responsible for durability and delocalized electrons respectively. Graphene is reported to have electrical and thermal conductivity with electron mobility 100-fold higher than silicon and better heat conducting performance in comparison to diamond, also its mechanical strength is 100-fold more than steel. Conversion of plastic waste to 2D graphene-based materials has extensive application in research, electronics, sensors technology, etc. It is synthesized based on top-to-down and bottom-up approach. Plastic waste is high in carbon content; therefore, they act as feedstock for 2D graphene material (Balbay & Acıkgoz, 2022).

Technically, graphene acts as semiconductor with zero-bandgap and zero band energy. It has ability to possess strong bipolar electric field effect. Leading to these remarkable properties, it is one of the excellent nanomaterials with astounding applicability. The material appears to be translucent due to absorptivity of 2.3% in visible range.

Graphene foil is one of the useful materials processed from plastic waste primarily plastic bottles from mineral water, plastic wrappers, lunch boxes, etc. As far as integrity is concerned they are made up of PS, PVC, PP, PET, polymethyl methacrylate (PMMA) and PE. In this, CVD (chemical vapor deposition) is employed where heat treatment is given to chopped plastic at 1050 °C for the 120 min. Previously treated and cleaned Ni foil is used to generate graphene foil with the conductance power of 3824 S cm^{-1} (Cui et al., 2017).

Graphene Nanostructures: PET waste bottles can generate moulted salt, which is beautifully designed to have graphene nanostructures. Small plastic pieces are cut, and pyrolysis was performed in an aluminium crucible using sodium chloride in the furnace heated to 1300 °C. Nanostructures thus obtained have a thickness of less than 10 nm and super high conductivity. The resulting nanomaterials are characterized using TEM, XRD, Raman spectroscopy and differential scanning calorimetry (DSC).

Graphene Nanosheets: Designing of graphene is like converting the waste into value-added material. It is formed using upcycle process in the presence of bentonite nano clay. Using this two-step decomposition of plastic waste takes place where pyrolysis happens at 450 °C and 945 °C in an inert atmosphere of N2. Resultant nanosheets can be confirmed using Raman and XRD methods. Another method includes MgCNi3 nanoparticles, can be fixed as nanosheets from waste plastic. In the process, nanoparticles of metal, sodium, MgCl2 and waste plastic were kept in stainless steel autoclave at 800 °C for 10 h. The cooling of autoclave was done at room temperature and black product was obtained (Pandey et al., 2021).

(c) Carbon Nanotubes

CNTs are coaxial tubular graphene sheets having lucrative props like 100 times more conducting power than copper, high tensile strength and high thermal stability with inert natural. One pot synthesis allows mixing of solid polymer with catalyst. Pyrolysis is allowed to occur and decomposed material serves as source of carbon for growth of CNT. The other option is to convert plastic waste to CNTs stepwise with relevant conditions. This waste plastic act as cheap feedstocks for generation of nanomaterials (Magrez et al., 2010).

Another aspect of the generation carbon nanotubes is the heating of plastic waste, like from PS, PE and PP at very high temperature conditions approximately 700–900 °C under controlled nitrogen purging, followed by the burning of runoffs in air. The so-called burning causes the generation of carbonaceous sources for carbon nanotubes. The method followed here is chemical vapour deposition with the help of stainless steel (Mustafa et al., 2021). One more method reported the use of batch pyrolysis reactor at 400 °C under vacuum conditions. The integrity of carbon nanotubes was confirmed using SEM and TEM analysis.

Plastic is so hazardous to the environment that once leached into soil, water, plants and animals, it can cause cancer and other ailments, aquatic life problems and environmental pollution.

Organic pollutants from plastic include: PCBs (polychlorinated biphenyls) PAHs (polycyclic aromatic hydrocarbons), nonylphenol, PBDEs (polybrominated biphenyl ethers), DDTs (dichlorodiphenyl trichloroethane) and BPA (bisphenol A).

On similar grounds, car bumper plastic waste can be utilized to form multiwalled carbon tubes in a two-step procedure. The method has step one as pyrolysis of waste in a screw kiln over montmorillonites and zeolites at a temperature 750 °C. Step two involves decomposition in a moving bed kiln with catalyst in the form of Ni. As the temperature increases, the yield of multi-walled carbon nanotubes also increases (Mustafa et al., 2021).

Nanomaterial include many materials like nanoparticles, nanotubes, quantum dots, nanosheets, etc. However, when the cost of fabrication is taken into account, nanomaterial synthesis is costly and expensive process. The use of waste allows us to reduce product costs. This waste acts as carbonaceous feed for production and, at the same time, reduces pollution process.

(d) Florescent Carbon Dots

Carbon dots are newly formed nanomaterials with wide range of applicability and properties like biocompatibility, unique optical properties, low cost, eco-friendliness and abundant functional groups. These can be synthesized from plastic waste. Water-soluble nitrogen-doped carbon dots can be formed using a single-step solvothermal process using polystyrene derived from plastic waste. This carbon dot has a small photoluminescence lifetime, low cytotoxicity, better photostability and stable luminescence quantum yield. These dots can be used as fluorescence probes for Au3+ ion detection. They can be used in bio-imaging, metallic ion detection, optical imaging, etc. (Song et al., 2019).

(e) Electro Spun Micro/Nanofibers

Plastics like PET (Polyethylene terephthalate polystyrene) or expanded polystyrene (PS) and polyamide (PA) gained commercial popularity due to their lightweight, resistance to corrosion and processability. But their disposal and coming to food chain, impacting humans and animals, raises the alarming situation.

Synthesis method from waste includes two major steps:

(a) *Waste Sample Preparation:* This is the first step that needs to be taken care of. New methods keep hovering over the researchers in order to decrease production costs and maximize waste from production. Composition and waste type are key factors to decide fate of sample preparation. It involves the use of three methods: physical, chemical and combined method (Samaddar et al., 2018).

1. *Physical Pretreatment:* This method is devoid of chemical use and mainly involves grinding and milling processes. Grinding is the use of mortar and pestle and providing mechanical energy to the system. Milling is concerned with ball mills, where utmost care is taken related to energy input and time for reproducible results. Both the processes have size reduction as an outcome. Ball milling process is one of the reliable approaches as a part of physical treatment of waste. Example, use of ball mill in the production of zinc nanoparticles from the tyre waste, which is collected as scrap form rubber industry. Reduction of waste is achieved up to less than 10 cm after removing metal wires from raw waste. Shredding method further causes lowering of size less than 6 cm and finally particles are sieved for segregation of smaller particles followed by ball milling process.
2. *Chemical Pretreatment:* The impurities present in the waste are chemically treated, heated and solubilized to remove them quickly. It involves strong acid hydrolysis, where concentrated strong acids (e.g., HNO3, HCl, H2SO4) are reacted with waste, but the limitation of this hydrolysis is the corrosive nature of strong acids. Secondly, these products have to be neutralized after treatment with acids, which increases the production cost. Example is the synthesis of γ- and α-Al nanopowder from waste nanofoils (Marcoux et al., 2013; Singh et al., 2012).

   Thermal degradation is another method that works on many factors like heating reaction medium, pressure and reactor geometry, hence making it complex technique. Further high viscosity of plastics interferes with the process, therefore making it more complicated. The issue can be resolved with the addition of a suitable solvent or sometimes heavy oil, like in the case of polystyrene.
3. *Combined Pretreatment:* It is the process involving both chemical and physical methods and therefore seems more logical and convenient. For example, the preparation of nanomaterials with waste elm samaras, bagasse, sugarcane waste, etc. (Samaddar et al., 2018).

(b) *Processing of Pretreated Material.*

The processing methods (Samaddar et al., 2018). are computed in Table 4.

7 Characterization of Nanomaterials from Plastic and Tyre Waste

Nanomaterials have gained huge popularity and captured many markets. Therefore, their characterization becomes equally important as many are used in drug and biomedical fields. Several methods have been used to characterize working mainly on

Table 4 Processing methods of waste management

Method	Process	Outcome
Chemical-thermal methodology	Involves mixing of waste with chemical reagent in a glass container	Chemical impregnation on waste material
Electric arc discharge	Hollow graphite rod is used as anode in the rotating cathode are discharged lamp. N2 is purged throughout	Packing of waste is done at anode and finally production of nanoparticles
Vacuum evaporation	Deposition of solid occurs on heated medium chemically from vapor phase	Vaporization takes place with vacuum levels of 75000.6 to 750.062 torr
Sodium borohydride reduction	Reduction process is carried out after adding Na borohydride solution rapidly to waste	Metallic nanoparticles
Solvent thermal method	Under critical temperature condition, chemical reduction is carried out in closed vessels.	High end nanoparticles

size, crystal structure, elemental composition and other physical properties of nanoparticles. Each technique has its own share of merits and demerits; therefore, combination of methods are always reliable (Mourdikoudis et al., 2018).

7.1 Structural Characterization

7.1.1 FT-IR (Fourier Transform Infrared Spectroscopy)

The method works on the absorption of electromagnetic radiation in IR region, i.e. 400–4000 cm^{-1}. This absorption leads to change in dipole moment in sample, and hence recorded spectrum gives the position of bands related to the strength and nature of bonds and specific functional groups and provides information concerning molecular structures and interactions to interpret various modes of precursors and products. It also helps to determine the impurities in nanomaterials (Dobson, 2016).

7.1.2 X Ray Techniques

XRD (X-Ray Diffraction): It is one of the most extensively used technique. The information gathered from this includes crystalline structure, nature of the phase, lattice parameters and crystalline grain size. The plus point of XRD is that it can be directly performed on the powder form after drying, and finally, particle composition can be derived from the comparison of peaks with the reference patterns available from the International Centre for Diffraction Data (ICDD). But the problem is, it cannot be used for amorphous particles or particles with size below 3 nm.

SAXS: It is the special type of X ray technique called as small angle X-ray scattering, exclusively designed for nano and meso scale particles. SAXS leads to quantification of the mass or concentration of NPs as well as their size simultaneously, as a function of time. For example, the evolution of size, size distribution, amount, and total volume of Rh and Pd NPs were quantitatively determined by SAXS (Li et al., 2013).

XPS (X-ray photoelectron spectroscopy): It works on the principle of photoelectric effect and very well employed for the nanomaterial characterization. It can very well depict electronic structure, elemental composition and oxidation states of elements in a nanomaterial. Further, it can be useful to determine ligand exchange interactions and surface functionalization of NPs (Rambau et al., 2018).

7.1.3 NMR (Nuclear Magnetic Resonance) Spectroscopy

It is the technique that is used for the characterization of nanomaterials and is based on the magnetic property showed by nuclei having non-zero spin. This nucleus, when positioned in a strong magnetic field, leads to small energy difference between the 'spin-up' and 'spin-down' states. Transitions between these states can be investigated by electromagnetic radiation in the radio wave range. The technique is employed to study the interactions or coordination between the ligand and the surface of diamagnetic or antiferromagnetic NPs. The problem is that it is not suitable for ferri- or ferromagnetic materials.

7.1.4 BET (Brunauer–Emmett–Teller) Technique

The method is based on physical adsorption of a gas on a solid surface. It allows to check the surface area of nanomaterials. For example, BET analysis of Au/poly (ethylene glycol dimethacrylate-co-acrylonitrile) composite microspheres helps to determine the microsphere porosity.

7.1.5 TGA (Thermogravimetric Analysis)

It gives data related to the mass and composition of the stabilizers. In this, heat is given to nanomaterial, and mass variation is recorded based on decomposition and vaporization. It is one of the simplest and direct technique.

7.1.6 LEIS (Low Energy Ion Scattering)

The technique is modern and rapid, leading to quick characterization of thickenings, for example, in case of gold nanoparticles. In this low-energy gas ionic molecules impinged on sample and elemental composition id based on loss and scattering of energy.

7.1.7 UV (Ultra Violet) Spectroscopy

It is the low-cost technique that works on the measurement of intensity of reflected light form nanomaterial in terms of lambda-max. As NMs are having optical properties, this technique is useful to check size, shape, concentration, agglomeration state and refractive index.

7.1.8 MS (Mass Spectroscopy)

It is one of the most sophisticated techniques and thus most trustworthy in terms of data derived. It works on the elemental level and provides the molecular information on composition, structure and chemical state of nanomaterial.

7.1.9 ICP-MS (Inductively Coupled Plasma Mass Spectrometry)

It is a sophisticated analytical method for identification and quantification of the elemental composition of nanomaterials, including metals and select nonmetals with atomic masses ranging from 7 to 250. The detection happens in ppt, i.e. parts per trillion. The sample gets aerosolized in nebulizer with high temperature range to obtain plasma like state (Peurifoy et al., 2018).

7.2 *Morphological Characterization*

When nanomaterials are concerned, particle size plays a very important role. Therefore, shape and size are major criteria and can be depicted using some microscopic methods. Internal dispersion of nanosized particles is characterized by SEM (Scanning electronic microscopy). Distribution patterns of particle is determined by EDS (Energy dispersive X Ray spectroscopy). Pre size of structured Nanomaterial can be withdrawn using TEM (Transmission electron microscopy). Particle determination at various conditions can be carried out using DLS (Dynamic Light scattering).

7.2.1 TEM (Transmission Electron Microscopy)

The method is high magnification microscopic process and based on the communication between electron beam of energy 60–150 keV and nanomaterial sample. Samples are prepared for imaging by drying nanoparticles on a copper grid that is coated with a thin layer of carbon. On impinging the sample, some part of beam of electrons gets transmitted leaving remaining scattered. The power of this coupling is based on size, sample density and composition of nanomaterial. Finally, the image is reflected on digital screen. Here, the direct imaging of nanomaterial is obtained with accurate depiction of its homogeneity. But sometimes, the results are misguiding due to orientation effect when large number of particles are involved (Mourdikoudis et al., 2018; Jiang et al., 2022).

HRTEM (High-Resolution Transmission Electron Microscopy) It is based on phase-contrast imaging and gives more accurate information on the structure of nanomaterials. It takes transmitted and scattered light into account simultaneously. It can also differentiate between single and multi-crystal components (Meneses et al., 2022).

Cryo TEM (Cryo-electron Microscopy)
It involves the depiction of near-unaltered samples in their frozen-native environment by vitrifying them at cryogenic temperatures. The method helps to learn complex growth mechanisms, like the aggregative growth of zeolite crystals. It also works to check kinetic manipulation of copolymer nanomaterials.

7.2.2 Electron Diffraction (ED)

The technique works on specific area and thus determines the structure of crystal. In this process, the acceleration of electrons is achieved, and then final interaction with nanomaterial sample takes place. The final diffraction pattern of molecule depicts its structural nature, especially its lattice planes.

7.2.3 Scanning Transmission Electron Microscopy (STEM)

The technique is based on the principle of focusing an electron beam on fine spot and then performing a scan in a raster. This allows Z-contrast annular dark-field imaging and spectroscopic mapping, and finally, the atomic features of the sample can be revealed. STEM electron tomography is also useful to check embedded nanomaterials. The only problem here is time consumption to check full tomograph (Mourdikoudis et al., 2018).

7.2.4 Dynamic Light Scattering (DLS)

The method is helpful for nonmaterial characterization of colloidal particles. The analysis works on the diffusive motion of particles in solution (Brownian motion), where larger particles will move more slowly and scatter more light than smaller particles.

7.2.5 Safety Considerations with Nanomaterials (Hidden Truth)

The indulgence of nanomaterials originated from waste in the food industry, water purification and commercialized products cannot be ignored. Exposure of the human race to these customized materials is again a matter of concern due to the fact that engineered products are being tested in vitro rather than in vivo impact. Till date, the toxicity issues have not been properly addressed, and there is a lack of proper dose regimen to test the material. As usage of these nanomaterials is not restricted to the commercial level but is becoming as a major contributor in the drug delivery domain, their ingestion and connection with the human race raise many eyebrows related to their clearance and distribution in the body. Decreased size leads to high reactivity, especially more invasive power in plant and animal tissues. The shape and charge related to nanomaterials also escalate the translocation process in tissues. The safety of workers handling nanomaterials is equally important; therefore, proper ventilated labs should be established to ensure safe environment. Regulatory requirements are the need of the hour as far as nanomaterials are concerned (Baig et al., 2021).

8 Conclusion and Future Prospects

Plastics and tyres are one of the key materials in the consumer market, and despite many hazards and challenges associated with their use, they cannot be avoided. They are the biggest players in the case of material sciences, but unfortunately the earth is slowly becoming a big dump yard when the waste associated with the tyres and plastics is considered. The situation is alarming as the waste is entering not only human food chain, but also the atmosphere, oceans and lands. This creates the need to convert this waste into wealth by using the scrap in the production of nanomaterials. One man's food is another man's poison—when we deeply think about the famous quote of Lucretius, we can say that plastics and rubber tyre parts are poison to the ocean body but can be very well utilized in the field of nanomaterials that generate fruitful products for the human race. On looking at the bigger picture, the utilization of this waste globally is still at a very naïve stage.

But, the best part of this situation is that the nanomaterials derived from this waste have very good mechanical, optical, magnetic, antibacterial and sometimes florescent properties. Therefore, they can be very well employed in the biomedical field, thermal energy, food packaging, agriculture, construction and electronic industries. In this chapter, plastic and tyre waste as non-biogenic resources are taken into consideration by looking at all the aspects involved, right from the production to activity.

The future lies in the hands of green and cleaner synthesis of nanomaterials to have dual benefit of better products while at the same time decreasing the environmental load. Parallelly there is no hard-and-fast rule for disposal; hence, proper regulatory guidelines need to be fabricated for the disposal of waste. Also properly established schemes need to be implicated to address possible toxic issues with the nanomaterials. Hence, the field of nanomaterials science can boom with the sharing of environmental pollutants.

References

Abdelbasir, S. M., McCourt, K. M., Lee, C. M., & Vanegas, D. C. (2020). Waste-derived nanoparticles: Synthesis approaches, environmental applications, and sustainability considerations. *Frontiers in Chemistry, 8*, 782.

Alagarasi, A. (2011). *Chapter—Introduction to nanomaterials.*

Ayyub, P., Chandra, R., Taneja, P., Sharma, A. K., & Pinto, R. (2001). Synthesis of nanocrystalline material by sputtering and laser ablation at low temperatures. *Applied Physics A, 73*, 67–73.

Baig, N., Kammakakam, I., & Falath, W. (2021). Nanomaterials: A review of synthesis methods, properties, recent progress, and challenges. *Materials Advances, 2*(6), 1821–1871.

Balasubramanian, K., & Burghard, M. (2005). Chemically functionalized carbon nanotubes. *Small, 1*(2), 180–192.

Balbay, S., & Acıkgoz, C. (2022). New method for producing carbon sphere from waste tyre (NEWCSWT). *Waste and Biomass Valorization, 13*(12), 4951–4962.

Bhardwaj, A. K., & Naraian, R. (2020). Green synthesis and characterization of silver NPs using oyster mushroom extract for antibacterial efficacy. *Journal of Chemistry, Environmental Sciences and its Applications, 7*(1), 13–18.

Bhardwaj, A. K., Shukla, A., Mishra, R. K., Singh, S. C., Mishra, V., Uttam, K. N., Singh, M. P., Sharma, S., & Gopal, R. (2017). Power and time dependent microwave assisted fabrication of silver nanoparticles decorated cotton (SNDC) fibers for bacterial decontamination. *Frontiers in Microbiology, 8*, 330.

Bhardwaj, A. K., Kumar, V., Pandey, V., Naraian, R., & Gopal, R. (2019). Bacterial killing efficacy of synthesized rod-shaped cuprous oxide nanoparticles using laser ablation technique. *SN Applied Sciences, 1*, 1–8.

Bhardwaj, A. K., Naraian, R., Sundaram, S., & Kaur, R. (2022). Biogenic and non-biogenic waste for the synthesis of nanoparticles and their applications. In *Bioremediation: Green approaches for a clean and sustainable environment* (pp. 207–218). CRC Press.

Cao, J., Qin, C., Wang, Y., Zhang, H., Zhang, B., Gong, Y., Wang, X., Sun, G., Bala, H., & Zhang, Z. (2017). Synthesis of gC_3N_4 nanosheet modified SnO_2 composites with improved performance for ethanol gas sensing. *RSC Advances, 7*(41), 25504–25511.

Cui, L., Wang, X., Chen, N., Ji, B., & Qu, L. (2017). Trash to treasure: Converting plastic waste into a useful graphene foil. *Nanoscale, 9*(26), 9089–9094.

Derraik, J. G. (2002). The pollution of the marine environment by plastic debris: A review. *Marine Pollution Bulletin, 44*(9), 842–852.

Ding, S., Zhang, N., Lyu, Z., Zhu, W., Chang, Y. C., Hu, X., Du, D., & Lin, Y. (2021). Protein-based nanomaterials and nanosystems for biomedical applications: A review. *Materials Today, 43*, 166–184.

Dobson, P. (2016). Talk on 'NPs: What do we need to know and can we measure everything we need to?' at 'nanoparticle characterisation—challenges for the Community' event—IOP.

Fu, R., & Fu, G. D. (2011). Polymeric nanomaterials from combined click chemistry and controlled radical polymerization. *Polymer Chemistry, 2*(3), 465–475.

Gieré, R., LaFree, S. T., Carleton, L. E., & Tishmack, J. K. (2004). Environmental impact of energy recovery from waste tyres. *Geological Society, London, Special Publications, 236*(1), 475–498.

Han, J., Zhao, D., Li, D., Wang, X., Jin, Z., & Zhao, K. (2018). Polymer-based nanomaterials and applications for vaccines and drugs. *Polymers, 10*(1), 31.

Hunter, P. (2018). Nucleic acid-based nanotechnology: The ability of DNA and RNA to fold into precise and complex shapes can be exploited for applications both in biology and electronics. *EMBO Reports, 19*(1), 13–17.

Ismail, R. A., Mohsin, M. H., Ali, A. K., Hassoon, K. I., & Erten-Ela, S. (2020). Preparation and characterization of carbon nanotubes by pulsed laser ablation in water for optoelectronic application. *Physica E: Low-dimensional Systems and Nanostructures, 119*, 113997.

Jana, N. R., Gearheart, L., & Murphy, C. J. (2001). Wet chemical synthesis of high aspect ratio cylindrical gold nanorods. *The Journal of Physical Chemistry B, 105*(19), 4065–4067.

Jiang, Z., Li, L., Huang, H., He, W., & Ming, W. (2022). Progress in laser ablation and biological synthesis processes:"Top-Down" and "Bottom-Up" approaches for the green synthesis of Au/Ag nanoparticles. *International Journal of Molecular Sciences, 23*(23), 14658.

Jones, A. C., & Hitchman, M. L. (Eds.). (2009). *Chemical vapour deposition: Precursors, processes and applications*. Royal Society of Chemistry.

Khan, I., Saeed, K., & Khan, I. (2019). Nanoparticles: Properties, applications and toxicities. *Arabian Journal of Chemistry, 12*(7), 908–931.

Kolahalam, L. A., Viswanath, I. K., Diwakar, B. S., Govindh, B., Reddy, V., & Murthy, Y. L. N. (2019). Review on nanomaterials: Synthesis and applications. *Materials Today: Proceedings, 18*, 2182–2190.

Kumar, C. S. (Ed.). (2010). *Semiconductor nanomaterials*. John Wiley & Sons.

Kumar, B., Pundir, A., Mehta, V., Singh, B., & Solanki, R. (2020a). A review paper on plastic, its variety, current scenario and its waste management. *Plant Arch, 20*, 53–56.

Kumar, M., Xiong, X., He, M., Tsang, D. C., Gupta, J., Khan, E., Harrad, S., Hou, D., Ok, Y. S., & Bolan, N. S. (2020b). Microplastics as pollutants in agricultural soils. *Environmental Pollution, 265*, 114980.

Kumar, M., Chen, H., Sarsaiya, S., Qin, S., Liu, H., Awasthi, M. K., Kumar, S., Singh, L., Zhang, Z., Bolan, N. S., & Pandey, A. (2021). Current research trends on micro-and nano-plastics as an emerging threat to global environment: A review. *Journal of Hazardous Materials, 409*, 124967.

Li, H., Kang, Z., Liu, Y., & Lee, S. T. (2012). Carbon nanodots: Synthesis, properties and applications. *Journal of Materials Chemistry, 22*(46), 24230–24253.

Li, W., Zamani, R., Rivera Gil, P., Pelaz, B., Ibáñez, M., Cadavid, D., Shavel, A., Alvarez-Puebla, R. A., Parak, W. J., Arbiol, J., & Cabot, A. (2013). CuTe nanocrystals: Shape and size control, plasmonic properties, and use as SERS probes and photothermal agents. *Journal of the American Chemical Society, 135*(19), 7098–7101.

Li, X., Yu, J., Wageh, S., Al-Ghamdi, A. A., & Xie, J. (2016). Graphene in photocatalysis: A review. *Small, 12*(48), 6640–6696.

Magrez, A., Seo, J. W., Smajda, R., Mionić, M., & Forró, L. (2010). Catalytic CVD synthesis of carbon nanotubes: Towards high yield and low temperature growth. *Materials, 3*(11), 4871–4891.

Marcoux, M. A., Matias, M., Olivier, F., & Keck, G. (2013). Review and prospect of emerging contaminants in waste–Key issues and challenges linked to their presence in waste treatment schemes: General aspects and focus on nanoparticles. *Waste Management, 33*(11), 2147–2156.

Maroufi, S., Mayyas, M., & Sahajwalla, V. (2017). Nano-carbons from waste tyre rubber: An insight into structure and morphology. *Waste Management, 69*, 110–116.

Meneses, R. A. M., Cabrera-Papamija, G., Machuca-Martínez, F., Rodríguez, L. A., Diosa, J. E., & Mosquera-Vargas, E. (2022). *Plastic recycling and their use as raw material for the synthesis of carbonaceous materials*. Heliyon.

Mody, V. V., Siwale, R., Singh, A., & Mody, H. R. (2010). Introduction to metallic nanoparticles. *Journal of Pharmacy and Bioallied Sciences, 2*(4), 282.

Moghaddasi, S., Khoshgoftarmanesh, A. H., Karimzadeh, F., & Chaney, R. L. (2013). Preparation of nano-particles from waste tire rubber and evaluation of their effectiveness as zinc source for cucumber in nutrient solution culture. *Scientia Horticulturae, 160*, 398–403.

Mourdikoudis, S., Pallares, R. M., & Thanh, N. T. (2018). Characterization techniques for nanoparticles: Comparison and complementarity upon studying nanoparticle properties. *Nanoscale, 10*(27), 12871–12934.

Mustafa, K., Kanwal, J., & Musaddiq, S. (2021). Waste plastic-based nanomaterials and their applications. *Waste Recycling Technologies for Nanomaterials Manufacturing*, 781–803.

Newman, S. E., & Meneley, J. C. (2006). *Adaptation of waste tire rubber for green house media and zinc fertilizer* (pp. 80401–83852). Final Report, Colorado School of Mines, Golden, Co.

Ostermann, R., Cravillon, J., Weidmann, C., Wiebcke, M., & Smarsly, B. M. (2011). Metal–organic framework nanofibers via electrospinning. *Chemical Communications, 47*(1), 442–444.

Pal, G., Rai, P., & Pandey, A. (2019). Green synthesis of nanoparticles: A greener approach for a cleaner future. In *Green synthesis, characterization and applications of nanoparticles* (pp. 1–26). Elsevier.

Pandey, S., Karakoti, M., Surana, K., Dhapola, P. S., SanthiBhushan, B., Ganguly, S., Singh, P. K., Abbas, A., Srivastava, A., & Sahoo, N. G. (2021). Graphene nanosheets derived from plastic waste for the application of DSSCs and supercapacitors. *Scientific Reports, 11*(1), 3916.

Parashar, M., Shukla, V. K., & Singh, R. (2020). Metal oxides nanoparticles via sol–gel method: A review on synthesis, characterization and applications. *Journal of Materials Science: Materials in Electronics, 31*, 3729–3749.

Peurifoy, S. R., Castro, E., Liu, F., Zhu, X. Y., Ng, F., Jockusch, S., Steigerwald, M. L., Echegoyen, L., Nuckolls, C., & Sisto, T. J. (2018). Three-dimensional graphene nanostructures. *Journal of the American Chemical Society, 140*(30), 9341–9345.

Pimpin, A., & Srituravanich, W. (2012). Review on micro-and nanolithography techniques and their applications. *Engineering Journal, 16*(1), 37–56.

Rambau, K. M., Musyoka, N. M., Manyala, N., Ren, J., Langmi, H. W., & Mathe, M. K. (2018). Preparation of carbon nanofibers/tubes using waste tyres pyrolysis oil and coal fly ash derived catalyst. *Journal of Environmental Science and Health, Part A, 53*(12), 1115–1122.

Sahu, M. K. (2019). Semiconductor nanoparticles theory and applications. *International Journal of Applied Engineering Research, 14*(2), 491–494.

Sajid, M. (2022). Nanomaterials: Types, properties, recent advances, and toxicity concerns. *Current Opinion in Environmental Science & Health, 25*, 100319.

Samaddar, P., Ok, Y. S., Kim, K. H., Kwon, E. E., & Tsang, D. C. (2018). Synthesis of nanomaterials from various wastes and their new age applications. *Journal of Cleaner Production, 197*, 1190–1209.

Shiri, M. S. Z., Henderson, W., & Mucalo, M. R. (2019). A review of the lesser-studied microemulsion-based synthesis methodologies used for preparing nanoparticle systems of the noble metals, Os, Re, Ir and Rh. *Materials, 12*(12), 1896.

Singh, S. C., Zeng, H. B., Guo, C., & Cai, W. (Eds.). (2012). *Nanomaterials: Processing and characterization with lasers*. John Wiley & Sons.

Singh, D., Singh, S., Sahu, J., Srivastava, S., & Singh, M. R. (2016). Ceramic nanoparticles: Recompense, cellular uptake and toxicity concerns. *Artificial Cells, Nanomedicine, and Biotechnology, 44*(1), 401–409.

Singh, M. P., Bhardwaj, A. K., Bharati, K., Singh, R. P., Chaurasia, S. K., Kumar, S., Singh, R. P., Shukla, A., Naraian, R., & Vikram, K. (2021). Biogenic and non-biogenic waste utilization in the synthesis of 2D materials (graphene, h-BN, g-C2N) and their applications. *Frontiers in Nanotechnology, 3*, 685427.

Song, H., Liu, X., Wang, B., Tang, Z., & Lu, S. (2019). High production-yield solid-state carbon dots with tunable photoluminescence for white/multi-color light-emitting diodes. *Science Bulletin, 64*(23), 1788–1794.

Thomas, S. C., Kumar Mishra, P., & Talegaonkar, S. (2015). Ceramic nanoparticles: Fabrication methods and applications in drug delivery. *Current Pharmaceutical Design, 21*(42), 6165–6188.

Thompson, R. C., Moore, C. J., Vom Saal, F. S., & Swan, S. H. (2009). Plastics, the environment and human health: Current consensus and future trends. *Philosophical Transactions of the Royal Society B: Biological Sciences, 364*(1526), 2153–2166.

UNEP. (2018). *Single-use plastics: A roadmap for sustainability*. Legal Limits on Single-Use Plastics and Microplastics: A Global Review of National Laws and Regulations. https://www.unenvironment.org/resources/publication/legal-limits-single-use-plastics-and-microplastics-global-review-national

Workie, A. B., Ningsih, H. S., & Shih, S. J. (2023). An comprehensive review on the spray pyrolysis technique: Historical context, operational factors, classifications, and product applications. *Journal of Analytical and Applied Pyrolysis*, 105915.

Wu, P., Huang, J., Zheng, Y., Yang, Y., Zhang, Y., He, F., Chen, H., Quan, G., Yan, J., Li, T., & Gao, B. (2019). Environmental occurrences, fate, and impacts of microplastics. *Ecotoxicology and Environmental Safety, 184*, 109612.

Zhang, D., Ye, K., Yao, Y., Liang, F., Qu, T., Ma, W., Yang, B., Dai, Y., & Watanabe, T. (2019). Controllable synthesis of carbon nanomaterials by direct current arc discharge from the inner wall of the chamber. *Carbon, 142*, 278–284.

Zhou, H., Luo, D., Chen, D., Tan, X., Bai, X., Liu, Z., Yang, X., & Liu, W. (2021). *Current advances of nanocarrier technology-based active cosmetic ingredients for beauty applications* (pp. 867–887). Clinical, Cosmetic and Investigational Dermatology.

Zhuang, S., Lee, E. S., Lei, L., Nunna, B. B., Kuang, L., & Zhang, W. (2016). Synthesis of nitrogen-doped graphene catalyst by high-energy wet ball milling for electrochemical systems. *International Journal of Energy Research, 40*(15), 2136–2149.

Current Status of Smart Nanoparticles in Drug Delivery

Shikha Baghel Chauhan

1 Introduction

In the present era, smart drug delivery systems (SDDSs) possess the remarkable capability to autonomously transmit signals, respond, dispense medication, and cease distribution. The precise objective of all smart drug delivery methods revolves around delivering the medication in the desired quantity, at the expected time, and to the appropriate location. Control signals for SDDSs encompass external triggers, such as varying wavelengths of light, magnetic fields, electric fields, and ultrasound, as well as internal cues such as redox potential, pH levels, concentrations of specific biomolecules, and enzyme activity. Various carriers for smart nanoparticles (NPs) exist, including polymers, hydrogels, liposomes, nanosheets, and micelles. The realm of pharmaceutical particles encompasses a wide spectrum of sizes and forms, ranging from traditional tablets and granules to microparticles and nanoparticles. The development of innovative methods for crafting nano formulations (nanocarriers) holds immense potential for a multitude of biotechnological applications, effectively facilitating the transportation of therapeutic molecules.

One novel approach, which harnesses nanomaterials for therapeutic management of diverse diseases, falls under the domain of nanotechnology. Maintaining medication stability against oxidation or metabolic processes within the circulatory system is of paramount importance before their release. Nanoparticles (NPs) can significantly enhance therapeutic efficacy by bolstering drug stability during blood transport and promoting enhanced cellular absorption upon entry. Smart NPs play a pivotal role in synergy and mitigation by elevating drug concentrations within target tissues or cells while simultaneously curtailing drug release in healthy tissues or cells. They offer an effective means of mitigating the common issues associated like

S. B. Chauhan (✉)
Department of Pharmaceutics, Amity Institute of Pharmacy, Amity University, Noida, UP, India

A. K. Bhardwaj et al. (eds.), *Biogenic Wastes-Enabled Nanomaterial Synthesis*,
https://doi.org/10.1007/978-3-031-59083-2_14

nonuniform distribution of drugs, fast elimination, unpredictable drug distribution, and limited bioavailability. Furthermore, smart nanostructured materials can diminish side effects tied to conventional drugs by enabling precise drug delivery to the target site, ultimately resulting in reduced dosing frequencies and controlled spatial/temporal release. A significant reduction in toxicity and/or negative reactions is the overall result. Despite the impressive advancements of recent techniques, the majority of nanocarriers' actions are linked to undesirable adverse effects that impact their effectiveness (Lee et al., 2015).

While the nanoparticle itself may not possess any unique characteristics, nanoparticle formulations can introduce novel qualities that have the potential to enhance drug delivery. Small molecules or free drugs that exist independently of nanoparticle systems remain distinct. The primary objective in formulating and assessing nanoparticles is to establish stable nanoparticles, which can be further used for the treatment of a wide array of conditions (Bhardwaj et al., 2017; Bhardwaj & Naraian, 2020). Therefore, it becomes imperative to consider the distinctive attributes of nanoparticles while resolving safety and toxicity issues. If nanoparticles can deliver large dose of drugs to the site of action as compared to the control, it should theoretically lead to a reduction in the required pharmaceutical doses, thereby minimizing the potential for toxic side effects. Additionally, nanoparticles possess the advantageous ability to enhance the aqueous solubility of poorly soluble medications, a trait particularly valuable in drugs like paclitaxel, which exhibit limited water solubility. Converting these drugs into nanoparticle formulations can significantly increase their water solubility. This transformation might not necessarily enhance efficacy but improved their water solubility without the inclusion of toxic additives. It also enhances safety, thereby providing a compelling rationale for its utilization in various applications (Lombardo et al., 2019).

2 Types of Smart Nanoparticles (NPs)

Contemporary intelligent nanostructured systems can be categorized into two main types: organic and inorganic nanocarriers. The manipulation of their physicochemical features is achievable through modifications in composition, encompassing organic, inorganic, or hybrid elements, as well as alterations in size, shape, and surface attributes. These modifications may involve adjusting dimensions, such as small or large sizes, and adopting diverse structures such as spheres, rods, hyperbranched configurations, multilamellar architectures, or multilayered forms. Moreover, surface attributes such as functional groups, coating procedures, PEGylation, integration of targeting elements, or surface charge can be customized to fulfill specific needs (Bhardwaj et al., 2019; Campora & Ghersi, 2022).

3 Polymer and Organic-Based Nanocarriers

Carbon-based nanomaterials are characterized by enhanced drug-loading capacity and great biocompatibility. Since these nanomaterials have large size and better stability, they are able to provide sustained controlled drug delivery for various drugs (Cheon et al., 2019) Fig. 1.

3.1 Lipid-Based NPs

Liposomes consist of a bilayer of phospholipids and take on a spherical structure, encompassing an aqueous core with a diameter ranging from 50 to 1000 nanometers. There are various subtypes of bilayers, which are distinguished by the number of layers they possess. These include small unilamellar vesicles (SUVs) with dimensions ranging from 20 to 100 nanometers, large unilamellar vesicles (LUVs) exceeding 100 nanometers, multivesicular vesicles (MVs), and multilamellar vesicles (MLVs) with sizes above 500 nanometers. Due to its advantages over other NP types, such as their efficiency in production, biocompatibility, and ability to carry molecules of various types, nanoparticles derived from lipids are the most frequently used nanoparticles to receive FDA approval. Liposomes are primarily made of cholesterol and a variety of phospholipids, whose chemical composition ultimately determines the liposomal characteristics. Lipid nanoparticles (LNPs) have proven to be more favorable than polymeric nanoparticles in the creation of delivery systems are primarily used in drug delivery (Arno et al., 2020). These carriers, often labeled as "nano safe," are constructed from physiological and/or biodegradable lipids, enhancing their safety profile. Due to their ability to conjugate with plasma membrane and release drugs intracellularly, solid lipid nanoparticles (SLNs) are extensively employed in drug delivery systems (DDS). Noteworthy distinctions between SLNs, liposomes, and polymeric nanoparticles encompass the practicality of preparation processes and scaling-up procedures, the generally recognized as safe (GRAS) status of all formulation ingredients, and the absence of organic

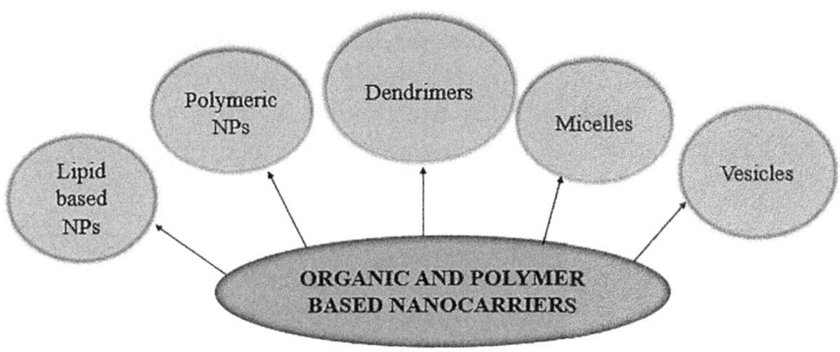

Fig. 1 Types of organic and polymers-based nanocarriers

solvent usage. Particles smaller than 100 nm face increased opsonization and clearance by the reticuloendothelial system. Additionally, the quantity of drug encapsulated in liposomes is influenced by the size and number of lipid bilayers.

There are four main types of lipid nanocarriers: liposomes, niosomes, SLNs, and nanostructured lipid carriers (NLCs). SLNs and NLCs overcome drawbacks associated with colloidal carrier systems. The incorporation of lipid nanoparticles enhances drug solubility, bioavailability, intestinal absorption, pharmacokinetic parameters, skin permeation, and ocular residence time while minimizing side effects. The lipid particle drug delivery system can be classified into SLNs and NLCs. According to findings, SLNs demonstrate a slower drug release with lower drug loading in comparison to NLCs. Nevertheless, at elevated drug loading levels, there is no notable distinction in drug release between SLNs and NLCs. NLCs exhibit superior stability than SLNs at a temperature of 25 °C (Xu et al., 2022).

4 Applications of Lipid-Based NPs

4.1 In Cosmeceuticals

A cosmetic category called as cosmeceuticals serves as a link between cosmetic and pharmaceuticals. This is a fast-growing segment in the cosmetic industry, especially in the light of advanced formulations and production methods such as lipid nanoparticles (LNPs).

LNPs-based cosmeceutical products provide a number of benefits, including enhanced visual appeal, stability of the finished products, and increased bioavailability of the cosmeceutical active ingredients (CAIs). Among the suggested technologies for developing unique, secure, potent, and elegant cosmeceuticals, cosmetic nanotechnology is emerging as a standout. Nanoparticles are absorbed onto the skin and supply the active ingredient to mend the skin's damaged areas, increasing the effectiveness of the treatment. Lipid nanoparticles adhere well to the skin when they come into contact with it. This property is due to the adsorption of these particles on the skin, which reduces the free surface energy, which directly reduces the thermodynamic driving force of the resulting system. Liposomes have been demonstrated to exhibit good adhesive effects. Liposomes are known to have excellent skin permeation properties. Different bleaching agents have issues with poor water solubility, poor stability, or inadequate skin penetration. To address the aforementioned issues, various research teams have created whitening agents enclosed in lipidic nanoparticles (Pottoo et al., 2020).

- Different issues with topical application affect the many chemicals that are utilized as antioxidants or antiaging. They might have weak stability, poor skin penetration, or low water solubility. The above difficulties might be overcome by encapsulating these compounds in various kinds of lipid nanoparticles.

- Isotretinoin is the most commonly used agent for the treatment of acne, but its topical application can lead to various problems such as erythema, skin peeling, and skin irritation. Employing the formulation by design (FbD) method, solid nanoparticles containing isotretinoin have been formulated, which overcame the side effects associated with conventional method of drug delivery.
- Use of sunscreen is recommended by many health organizations to protect the skin from damage and sunburn. Various types of chemical UV filters are used as sunscreens, including octyl methoxycinnamate (OMC), zinc oxide, and titanium dioxide. They might, however, result in some skin allergies or rashes. Therefore, these compounds could have less adverse effects and less photodegradation if they were encapsulated in lipid nanoparticles.

4.2 *In CNS Disorders*

Numerous nanotechnological techniques are used to aid drug transport to the brain. Blood-brain barrier hinders entry of pharmacological compounds to the brain. Among these methods, NLC stands out as a promising carrier system that, in part, because of its reduced size, compatible lipid structure, along with better drug-loading capacity. It also increases the drug concentration in the brain. Additionally, it promotes drug uptake across the blood-brain barrier using various modes of transport.

Alzheimer's disease is a chronic neurological disorder, a reason for dementia. It is commonly seen in older age usually after 65 years of age. In literature review, it was found that various unique approaches for efficient targeting of drugs and bioactives to the brain for the control and treatment of Alzeheimer's disease exists, including nanoparticle-based carrier systems, and new carriers through nose-to-brain administration. Pioglitazone-loaded NLC was developed for nose-to-brain drug delivery. Compared to conventional drug solutions, the prepared NLV showed superior drug penetration and sustained drug delivery. Nasal epithelial toxicity was surprisingly low. The entire study provides successful preclinical evidence supporting that pioglitazone is highly effective in treating Alzheimer's disease (Agrawal et al., 2020).

4.3 *Targeted Drug Delivery*

Drug-loaded LNPs have been widely used in the control and management of many diseases. To combat tuberculosis, one of the main causes of death worldwide, rifampicin drug-loaded LNPs with chitosan was formulated. The rifampicin drug-loaded LNPs with chitosan had improved drug uptake in cells and better mucoadhesive strength establishing enhanced alternative drug delivery strategy methods for targeted approach of drugs to the lungs.

4.4 Ocular Drug Delivery

The intracameral injection entails the direct administration of medication to the front chamber of the eye. However, its usage is occasionally limited by the necessity for general anesthesia and the potential risk of intraocular damage. Lipid nanoparticles present numerous advantages, including altered release, improved absorption, heightened stability, minimal degradation, compatibility in vivo, and versatility in delivery methods. In comparison to polymeric nanoparticles, lipid nanoparticles exhibit benefits in terms of altered release, increased stability, reduced lipid degradation, in vivo tolerance, and adaptability to various delivery systems. Solid lipid nanoparticles (SLNs) offer several advantages over alternative colloidal carriers, encompassing customized drug release, targeted delivery, prolonged stability, high entrapment efficiency, biocompatibility, bactericidal properties, a self-administrable eye-drop configuration, and a straightforward scalable production process. These nanocarriers exhibit promising potential for administration through various routes, including rectal, ophthalmic, pulmonary, peroral, parenteral, transdermal, and nasal (Jacob et al., 2022).

4.5 Polymeric NPs

Polymeric nanoparticles are compact colloidal particles, typically falling within the size range of 1–100 nm, and are primarily composed of polymers sourced from synthetic/semisynthetic origins and natural origins. These can further be biodegradable or nonbiodegradable. The regular use of drugs with brief half-lives, challenges in patient adherence, pronounced peaks, conventional peak-valley plasma concentration-time patterns, and limitations associated with traditional dosage forms hinder the attainment of consistent drug concentrations and precise targeting. To address these issues, there has been a shift toward the creation of modified and targeted dosage forms. Polymeric nanoparticles play a key role in delivering drugs, proteins, or DNA material to specific organs or cells for targeted therapeutic effects. The usage of polymeric nanoparticles as a smart polymer, as well as their target specificity and diminished side effects, has increased. Because they have longer clearance times than conventional nanoparticles, they can increase therapeutic efficacy and reduce toxicity with minimal drug use. It is also easy to manage and set up. The ability to load drugs is both high and simple (without any chemical reaction) (Begines et al., 2020).

5 Applications

5.1 *Vaginal Drug Delivery*

The primary approach for achieving a local or systemic therapeutic effect is via vaginal administration, which also bypasses hepatic first-pass metabolism and can be employed for the treatment of sexually transmitted infections or diseases. Nevertheless, the vaginal route of drug administration presents challenges and drawbacks. The significant production of vaginal mucus poses a major obstacle to conventional dosage forms, limiting their ability to sustain prolonged and targeted activity. Commonly employed in the creation of nanoparticles for delivering drugs vaginally, natural polymers like chitosan play a significant role. Chitosan, serving as a mucoadhesive in vaginal drug delivery systems, improves both drug retention and effectiveness (Sur et al., 2019).

5.2 *Cancer Treatment*

The outlook for individuals diagnosed with malignant gliomas, one of the deadliest types of brain cancer, is still poor, with an average life of only 15 months, despite advances in treatment. Due to inadequate blood-brain barrier (BBB) penetration, particularly by hydrophilic medicines, traditional chemotherapy does not deliver the necessary drug dose to the tumor location. Low molecular weight medications also have a short circulation half-life and are unable to specifically accumulate in malignant tissues. For the purpose of delivering chemotherapy medications to the brain, a variety of distinct nanoparticle compositions have been investigated. Most of these formulations made use of polymers that complied with the stringent standards necessary to be approved for biological purposes. The use of nanoparticles to treat CNS illnesses has drawn a lot of attention recently. Using nanoparticles, drugs can be delivered to the brain without any physical contact. They must be tuned in terms of their size, release kinetics, and chemical composition, and altering their surfaces may make them more able to traverse the BBB while also shielding the medication from the biological environment and increasing solubility (Deirram et al., 2019). The different applications of smart nanoparticles are mentioned in Table 1.

5.3 *Dendrimers*

Dendrimers, classified as hyperbranched organic molecules, possess three-dimensional structures containing surface functional groups. Modifying these terminal groups to improve dendrimer performance can lead to the development of numerous biocompatible products. The term “dendrimer” is derived from the Greek

Table 1 Applications of smart nanoparticles

Applications of smart nanoparticles	Material	Description	References
Smart drug delivery	Plasmonic nanoparticles, composed of inert metals, such as gold, titanium, and iron oxide, exhibit distinctive optical characteristics attributed to their size and shape	Drug release occurs due to the electrospun nanofibrous material. Plasmonomic nanoparticles further potentiate the drug delivery	("Plasmonics—nanoComposix,") Lombardo et al. (2019)
Ocular drug delivery	Hydrogels consist of polymers containing ionic groups, providing them with their responsive characteristics Polymeric materials include chitosan, hyaluronic acid, and collagen Gold, iron, and silica are among the inorganic materials employed in the creation of nanostructured materials for applications in drug delivery and imaging	Lipid-based material, hydrogels, and polymeric material have been explored as potential tool for ocular drug targeting	Wang et al. (2023), Allyn et al. (2021), Borchard et al. (2020), Lynch et al. (2020) and Thang et al. (2023)
Sensors and detectors		Nanoparticles serve as sensors to identify molecules, pathogens, or alterations in the environment, finding applications in healthcare, environmental surveillance, and security	Shoukat & Khan (2021)
Cosmetics and skincare	Nanoparticles employed in cosmetics and pharmaceutical delivery can be crafted from diverse materials, encompassing metals, metal oxides, silica, and carbon. In the cosmetic industry, nanoparticles are frequently applied as UV filters or preservatives. Silica nanoparticles and inorganic substances such as titanium dioxide, zinc oxide, and carbon are commonly utilized in this context	Cosmetic and skincare products utilize nanoparticles to enhance product stability, promote skin penetration, and provide UV protection	Pottoo et al. (2020)

(continued)

Table 1 (continued)

Applications of smart nanoparticles	Material	Description	References
Drug delivery		Smart nanoparticles are employed to improve drug delivery systems, ensuring accurate targeting, regulated release, and minimized side effects in the treatment of conditions such as cancer, diabetes, and various diseases	Suciu et al. (2020) and Agrawal et al. (2020)
Diagnostics and cancer therapy	When subjected to a magnetic field, nanoparticles disrupt the field, leading to an accelerated relaxation of water protons. This allows for detection in MRI. Contrast agents employed in MRI are commonly derived from iron oxide nanoparticles as well as complexes involving lanthanide metals	Intelligent nanoparticles function as contrast agents in imaging modalities such as MRI, CT scans, and fluorescence imaging, contributing to the early detection and monitoring of diseases	Deirram et al. (2019), Dulińska-Litewka et al. (2019), Blasiak et al. (2013) and Estelrich et al. (2015)
Regenerative medicine	Widely used in regenerative medicine, scaffolds, stem cells, and bioreactor culture systems are employed to cultivate grafts in vitro for the repair of bone and cartilage. The materials utilized in regenerative medicine encompass biopolyesters such as polylactic acid (PLA) and polyhydroxyalkanoates (PHAs), polysaccharides such as hyaluronic acid (HA), polypeptides and proteins such as collagen, gelatin, and fibroin, as well as polymers including polyglycolic acid (PGA) and polyurethane (PU)	Nanoparticles are applied to stimulate tissue regeneration and healing, particularly in scenarios such as neurodegenerative diseases and organ transplantation	Vasantharaj et al. (2019), Han et al. (2023), Liu et al. (2023), Yazdanian et al. (2022) and Cheng et al. (2023)

words "dendron," meaning "tree," and "meros," meaning "part." The typical dendrimer structure consists of three primary elements: a central core, an inner layer composed of repeated branching units, and terminal functional groups on the outer surface that determine the fundamental properties of the dendrimer. Further classification of dendrimers is based on factors such as shape, terminal functional groups, and internal cavities (Nikzamir et al., 2021).

5.4 Nano Drug

Dendrimers find application as nano-drugs, as demonstrated by their effectiveness as antiviral agents combating the herpes simplex virus. Consequently, dendrimers hold promise in preventing or reducing the HIV transmission and other STDs. Additionally, it has been observed that dendrimers, conjugated with tertiary alkyl-ammonium groups, demonstrated potent antibacterial effects against both Gram-negative and Gram-positive bacteria. Furthermore, chitosan-dendrimer hybrids have been explored for their antibacterial properties.

5.5 Oral Drug Delivery

The oral route is typically recommended for anticancer drugs because it lowers administration expenses and makes it easier to use longer-term treatment plans. However, the drug's weak solubility and low penetration across cell membranes prevent widespread use. The use of effective oral medication delivery methods is crucial to overcoming this problem and ensuring good oral absorption. Drug preservation from degradation should be a feature of an effective oral delivery system.

5.6 Transdermal Drug Delivery

The transdermal drug delivery approach involves administering medication through the skin to maintain consistent drug blood concentrations. This method simplifies dosing schedules and mitigates issues related to gastrointestinal tract degradation. Dendrimers, owing to their water solubility and biocompatibility, offer advantages such as enhanced drug solubility, prolonged plasma circulation time, and controlled drug release. Permeability through the skin is categorized based on various physical and chemical characteristics, including composition, generation size, concentration, molecular weight, and surface charge (Biomedical and 2019, n.d.).

5.7 Micelles and Vesicles

In liquid media surpassing the critical micellar concentration (CMC) value, surfactant molecules undergo self-assembly to form micelles, which are aggregated structures. The amphipathic nature of these molecules, featuring both hydrophilic (polar head groups) and hydrophobic segments (long hydrophobic chains), drives micelle formation. Consequently, surfactants spontaneously organize to shield the portions of their molecules that do not match the dispersion medium. The inner region of these micelles, termed the "hydrophobic core," in a water medium, can accommodate a lipophilic medication. This accommodation enhances the drug's solubility in water, improves bioavailability, and reduces degradation and loss. Reports indicate that the use of micelles has effectively increased the water solubility of natural products.

One of the most researched colloidal delivery systems is liposomal vesicles, which were initially created for medication delivery in the 1970s. They are composed of self-assembled vesicular structures that surround an aqueous core with one or more phospholipid bilayers. These structures can range in size from being quite small (about 100 nm) to being quite massive (exceeding 1 mm). They are able to transport both hydrophilic and lipophilic molecules since they have an aqueous compartment and lipid membrane. The utilization of nanocarriers has facilitated the inclusion of several plant-based medications in clinical trials, significantly enhancing the effectiveness, safety, and adherence of natural products. Indeed, micelles and vesicles are particularly appealing nano vectors because they can eliminate the major drawbacks of instability or limited water solubility. Vesicles exhibit remarkable adaptability by accommodating both hydrophilic and hydrophobic compounds. Encapsulating natural chemicals within vesicles can enhance drug solubility and stability, facilitate systemic drug administration, improve bioavailability, and offer control over absorption and distribution profiles (Bilia et al., 2016).

5.8 Inorganic (and Metallic) Nanocarriers

Inorganic nanocarriers typically consist of two components: a core containing an inorganic element, such as quantum dots, iron oxide, and silica and an outer shell primarily composed of an organic polymer or metal shown Fig. 2. The shell serves as a favorable platform for conjugation and shields crucial areas from unfavorable physicochemical interactions with the surrounding biological microenvironment.

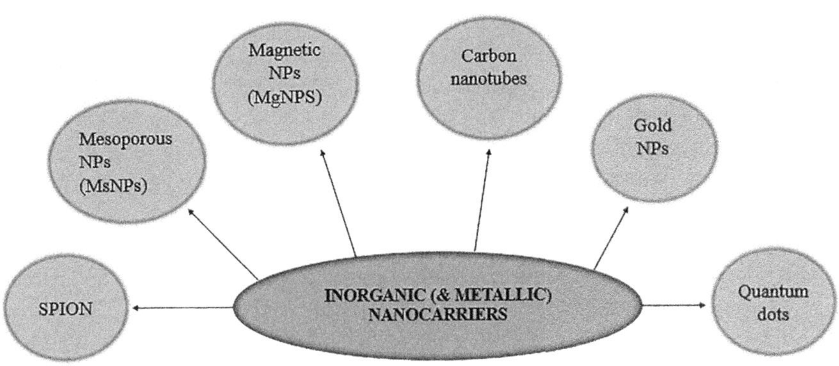

Fig. 2 Types of inorganic and metallic nanocarriers

The unique magnetic and plasmonic properties of inorganic nanomaterials allow them to generate image contrast in techniques such as positron emission tomography (PET), magnetic resonance (MR), or computed tomography (CT). This characteristic is utilized for visualizing the diseased area. Despite these benefits, inorganic nanoparticles have had relatively limited success treating disease tissues because of major problems with the amount of active drug they can carry and their high level of toxicity (Bhardwaj et al., 2022; Sanchez-Moreno et al., 2018).

6 Mesoporous Nanoparticles (MsNPs)

Mesoporous nanoparticles possess a substantial framework, an intricate porous structure, and a considerable surface area, allowing the incorporation of diverse functional groups to guide the drug moiety to a specific location. Numerous biomedical applications, including targeted drug delivery, antibacterial activity, bioimaging, and disease diagnosis and treatment, make extensive use of MNPs (Castillo et al., 2020).

7 Applications

- They can also function as imaging agents for various techniques such as ultrasound, PET, computed tomography (CT), X-ray, MRI, surface-enhanced Raman imaging (SERS), and optical imaging.
- Gold nanoparticles (AuNPs) find widespread usage due to their excellent X-ray attenuation, ease of production, and biocompatibility. Recently, there has been growing interest in utilizing AuNPs for CT imaging. Depending on the synthesis method, they exhibit unique absorption and scattering characteristics, such as surface plasmon resonance, which can be manipulated by controlling specific shape and size parameters. These nanoparticles are also applied in studies related to tumors. The reticuloendothelial system (RES) in the bloodstream can process AuNPs, distinguishing them from plasma proteins through opsonization. To prevent this, AuNPs can be functionalized by incorporating PEG (PEGylation), extending their blood circulation time.
- Mesoporous silica nanoparticles (MsNPs) have the potential to serve as carriers for antibacterial proteins. Lysozyme, the primary component of Gram-positive bacterial walls, played a significant role in facilitating the delivery of antibacterial proteins using MSNs.
- Mesoporous silica nanoparticles (MSNs) are used for phototherapy applications, including photodynamic therapy (PDT), photothermal therapy (PTT), or a combination of both. PDT and PTT, prominent classes of phototherapy, have garnered significant attention for treating diverse cancer types. PDT involves damaging and eradicating cancer cells using photosensitizers that generate

cytotoxic reactive oxygen species (ROS) or singlet oxygen ($^1O_2$) upon exposure to light. On the other hand, PTT eliminates cancer cells by inducing hyperthermia through light-activated photothermal agents.

- DNA, small-interfering RNA (siRNA), microRNA (miRNA), and antisense oligonucleotide (ASO) represent gene-based therapies with potential applications in cancer treatment. These medications have the capacity to selectively modify the expression of crucial genes. However, the negative charge of nucleic acids and their susceptibility to degradation in biofluids pose challenges for efficient delivery to their intracellular targets. There is a pressing need for effective carriers capable of encapsulating and protecting nucleic acids. Mesoporous silica nanoparticles (MSNs) are commonly employed in nucleic acid delivery and nucleic acid-guided therapeutic approaches due to their expansive surface area for functionalization, adaptable pore size for accommodating various payloads, and relatively favorable biocompatibility (Jafari et al., 2018).

7.1 Quantum Dots

A crucial element connecting the convergence of nanotechnology and drug testing is the nanoscale semiconductor known as quantum dots (QDs). Their unique photoluminescence and electronic characteristics, which include continuous absorption spectra, emission spectra near visible to infrared wavelengths, high brightness, and enduring exposure to light, render them exceptionally well-suited for immunosensing platforms (Badıllı et al., 2020).

8 Applications

8.1 Antimicrobial Drug Activity

In contrast to other nanostructures characterized by limited light responsiveness, high expenses, and toxicity, quantum dots can be easily manufactured, undergo straightforward surface modification with various antimicrobial agents, and address the growing issue of antibiotic resistance in bacterial infections. Photodynamic inactivation is one method that can be used to stop the spread of these illnesses. QDs with photosensitizer properties create microbial reactive oxygen species from molecular oxygen to brighten visible light (ROS).

Overall, the application of QDs has yielded numerous results in in vitro systems, particularly in the understanding of cellular pathways. In any case, even if CT is isolated from heavy metals, there are still some limitations to its conversion to the in vivo system due to the activity and side effects of the RES system. Surface modifications such as addition of PEG can reduce the potential toxicity of QDs.

8.2 *Antiviral Drug Delivery*

Antiviral medications are a potential option for treating infectious viral diseases since they help to reduce the pandemic mayhem brought on by specific viruses, particularly the coronavirus. Porcine epidemic diarrhea virus (PEDV), chosen as a model for coronaviruses, proved to be an effective test subject for evaluating the antiviral properties of Cur-NCDs. Through a hydrothermal process, uniform cationic CDs were synthesized and examined for their ability to inhibit viral propagation by inducing the synthesis of proinflammatory cytokines and interferon-stimulating genes (ISGs). Studies have shown that this approach can successfully utilize NCDs' antiviral potential, which can reduce the severity of this infectious, life-threatening disease (Nair et al., 2020).

9 SPION (Super Magnetic Iron Oxide Nanoparticles)

To enhance the effective utilization of superparamagnetic iron oxide nanoparticles (SPIONs), they are commonly functionalized or coated with organic surfactants. This practice helps prevent agglomeration resulting from magnetic dipole-dipole attractions. The majority of SPIONs consist of magnetite (Fe_3O_4), ferrites, and maghemite (Fe_2O_3), typically exhibiting a crystalline nature. Magnetite has gradually gained significance in research, while ferrite is also applied in hyperthermia treatments. Since it is difficult to purify magnetite from bacteria and use it in most cases, chemical co-precipitation is the best option because production is easy and fast. SPIONs with extremely small sizes, typically between 3 and 15 nm, have found use in a number of medical research applications. Hematite, magnetite, and various types of iron oxides are found as the iron ore itself. According to reports, natural magnetite performs more catalytically while degrading organic pollutants like p-nitrophenol (p-NP). One of the earliest pigments used by humans in the past was natural hematite. Recent discoveries have involved the utilization of natural hematite in the advancement of microbial fuel cells. Their responsiveness to an external magnetic field allows for controlled actions, making SPIONs particularly valuable in biomedicine for showcasing regulated behavior at specific targeted sites (Samrot et al., 2021).

10 Applications

10.1 In Imaging

Leveraging the support of superparamagnetic iron oxide nanoparticles (SPIONs), magnetic resonance imaging (MRI) has achieved successful outcomes, marking a significant advancement in diagnostics. Due to their superparamagnetic properties and lower toxicity, these SPIONs find extensive applications in various medical fields. They serve as contrast agents in MRI, facilitating the monitoring of drug delivery. Smaller particles are employed for MRI. Recent progress has made it possible to monitor stem cells using SPIONs, facilitating the evaluation of treatment outcomes and offering a more detailed understanding of the structural composition of organs (Dulińska-Litewka et al., 2019).

10.2 In Magnetic Delivery of Drugs

For the enhancement of curcumin's uptake and biocompatibility by breast cancer cells, it has been conjugated within magnetic chitosan nanoparticles (Song et al., 2018). Approximately 67.5% of curcumin was successfully encapsulated in these nanoparticles, and cancer cells absorbed the coated curcumin 3–6 times more efficiently than its equivalent, i.e., free.

10.3 Used in Hyperthermia Detection

The heating mechanisms in superparamagnetic nanoparticles (SPIONs) during hyperthermia vary based on their size: Néel relaxation for SPIONs smaller than 10 nm, a combination of Néel and Brownian relaxation for SPIONs between 10 and 13 nm, and hysteresis loss for larger SPIONs (ferromagnetic nanoparticle mechanism). The efficiency and hyperthermia capabilities of SPIONs are influenced by factors such as their composition, size, volume, magnetic anisotropy, concentration, and their ability to disperse rather than aggregate at the targeted site (Samrot et al., 2021).

10.4 In Vitro Bioseparation

The application of superparamagnetic iron oxide nanoparticles (SPIONs) is positioned to replace traditional separation methods in molecular biology, offering enhanced throughput, improved performance, quality, and specificity for the separation of biomolecules such as DNA, proteins, and antibodies.

10.5 In Cancer Therapy

In contemporary applications, magnetic nanoparticles, particularly those composed of superparamagnetic iron oxide, find widespread use in chemotherapy. One approach involves intravenous administration of SPIONs, followed by inducing a temperature increase to locally eliminate cancer cells, as seen in hyperthermia applications. Another technique involves targeted drug delivery, utilizing an external magnetic field to accurately guide the medication to the intended site of action. Iron oxide nanoparticles serve as an exemplary controlled drug delivery, overcoming the limitations of conventional chemotherapy by delivering medication directly to the site of action. This leads to a high uptake of drug absorption, enabling a more precise observation of the substance's pharmacological and pharmacodynamic effects.

10.6 In Alzheimer's Disease

Even though SPIONs' capacity to enter the brain carries some potential hazards, it also places them among the most promising treatments for disorders of the brain tissue, such as those caused by the build-up of peptide amyloids. One of the few NPs, SPIONs, does not have a counterproductive effect (such as enhanced amyloid aggregation following TiO_2 administration). Undoubtedly, there are still a lot of unsolved mysteries in this area, and additional research is required to elucidate the precise molecular pathways underlying the NPs' anti-amyloidogenic properties (Suciu et al., 2020).

10.7 Carbon Nanotubes

The unique features and potential applications of carbon nanofibers, along with carbon nanotubes, have attracted attention in the field of nanotechnology. The grouping of nanotubes is determined by the number of walls present in each tube.

11 Applications

11.1 Medicinal

The most remarkable uses of CNTs were in the biomedical domains. They are introduced in the various medical disciplines, and the outcomes are promising. They have several applications, including the transport of drugs, that are quite beneficial.

11.2 Nanoporous Membrane System

Membranes play a vital role in wastewater treatment, medical delivery systems, and chemical separation. Nanoporous membranes, especially those incorporating carbon nanotubes, prove highly effective in these applications. The integration of carbon nanotubes into the membrane occurs through their inclusion in a thin polymer film, resulting in the formation of nanoporous membranes. In this configuration, the substrate inside the nanotubes aligns perpendicularly to membrane transport. Various types of nanotubes can be oriented perpendicular to the conveyance. These membranes are employed in microfluidic channels, seeking to replicate functionalities observed in cell membranes.

11.3 Nanoprobes and Sensors

CNTs are used in nanolithography, drug administration, numerous types of sensors, high-resolution imaging, and probing the surfaces of diverse semiconducting materials. The AFM may be utilized with tips as small as 10 nm due to modern fabrication techniques. The polished silicon surface of the tip presents a problem when utilizing this probe AFM tip. The image resolution degrades, and this tip quickly wears out. The usage of VACNTs eliminates this issue. These tubes are incredibly efficient and produce images with a very high resolution due to their diameter, which is in the range of 1–10 nm (Shoukat & Khan, 2021).

12 Other Applications and Current Status of Organic and Inorganic Nanoparticles (NPs)

12.1 In Biophotolysis

Photoautotrophic organisms, such as microalgae and cyanobacteria, split water molecules through a process called biophotolysis in order to produce hydrogen and oxygen. Because just the photosystems of algae and light energy are used, biophotolysis is generally regarded as an environmentally acceptable method for producing hydrogen. In the process of photosynthesis, chlorophylls within photosystems I convert absorbed light energy into an electron, which is then transported across the membrane to produce ATP. These nanoparticles commonly play a role in utilizing feedstock to enhance hydrogen generation (Vasantharaj et al., 2019).

12.2 In Nucleic Acid Drug Delivery

For nucleic acid release, gene delivery systems must be able to form stable complexes with nucleic acids with low cytotoxicity. Functional groups capable of targeting specific tissues or cell types, stabilizing vectors in the extracellular environment, or maintaining vectors within the site of delivery are used to control the interaction with the environment. Bioactive groups may be included to aid in endosome escape or nuclear localization. By including receptor ligands in the NP, one can enhance cellular internalization, increase binding to cell-surface receptors, and target particular cell types (Khalid et al., 2020).

12.3 Enhanced Biohydrogen Production

Since hydrogen is currently the most expensive biofuel, producing it by photobiological or dark fermentation methods is seen as a sustainable and environmentally good option. However, the generation of biological hydrogen experiences some of the same problems like increased manufacturing costs and poor output. As a result, a number of tactics have been improved to increase biohydrogen generation. Among these various techniques, the technique of organic or inorganic NPs as process elements to create photobioreactors with better lighting and temperature efficiency has shown to increase the rate and output of biohydrogen (Patel et al., 2018).

12.4 Magnetic Resonance Imagining

Numerous studies have investigated the utilization of metal-organic framework nanoparticles in magnetic resonance imaging (MRI), with a specific focus on MOFs incorporating inorganic building units like gadolinium, manganese (25Mn), and iron. In the realm of MRI, starch-coated superparamagnetic nanoparticles with an average size of 6 nm were employed, and they were coated with superparamagnetic iron oxide nanoparticles. The particle size was influenced by H_2O_2 used to break glycosidic linkages in the starch, resulting in a reduction in NP diameter (Chiozzi & Rossi, 2020).

12.5 Fluorescence Imaging

The most understandable information is provided to eyesight through optical FL imaging, which is noninvasive and does not cause any negative effects from nonionizing radiation. The main challenges facing FL imaging modality are its shallow penetration depth and low spatial resolution. Fluorescent nanospheres are frequently functionalized with PVP and PEG for imaging applications to increase biocompatibility for in vivo imaging. Additionally, for focused imaging, targeting ligands like RGD could be further coupled with PEGylated carbon dots. Since the majority of high-quality semiconductor QDs are created using the high-temperature organic solution approach, the hydrophobic QDs that are produced must go through ligand exchange before being used in biomedical applications. The incorporation of water-soluble or amphiphilic ligands could be accomplished by metal-sulphur bonds, chemical reactions, or hydrophobic interactions (Zhao et al., 2019).

12.6 In Tissue Regeneration

Copper ions have the potential to trigger various processes, such as the proliferation and movement of endothelial cells, along with the release of growth factors that promote angiogenesis by cells. These ions sustain HIF-1, inducing an artificial state of hypoxia in cells, which can enhance angiogenesis. Utilizing copper ions, the nanoparticles (NPs) facilitate angiogenesis and enhance the in vivo survival of mesenchymal stem cells (MSC). Consequently, skin injuries caused by CuS nanoparticles lead to increased production of VEGF, promoting accelerated wound healing and skin regeneration.

13 Conclusion

Nanotechnology represents a groundbreaking paradigm shift with innovative approaches across various fields of contemporary science and medicine, placing a specific emphasis on nanomedicine. Thanks to their unique properties rooted in the intricate realm between classical and quantum physics, nanoparticles (NPs) possess extraordinary attributes. The NP project assumes paramount importance in this context, as it necessitates careful selection and optimization of materials, size, shape, and functionalization to achieve specific objectives. The main goal is to optimize the necessary drug dosage to achieve the desired therapeutic effect, aiming to reduce costs and minimize adverse effects. Within this domain, two primary classes of nanostructured materials, organic and inorganic, are extensively utilized in drug delivery procedures, each presenting complementary benefits. Among these, organic soft nanocarriers emerge as excellent examples of biocompatible nanostructures, demonstrating exceptional adaptability to the physicochemical conditions present in both healthy tissues.

References

Agrawal, M., Saraf, S., Saraf, S., Dubey, S. K., Puri, A., Patel, R. J., Ajazuddin, R. V., Murty, U. S., & Alexander, A. (2020). Recent strategies and advances in the fabrication of nano lipid carriers and their application towards brain targeting. *Journal of Controlled Release, 321*, 372–415.

Allyn, M. M., Luo, R. H., Hellwarth, E. B., & Swindle-Reilly, K. E. (2021). Considerations for polymers used in ocular drug delivery. *Front Med (Lausanne), 8*, 787644.

Arno, M. C., Inam, M., Weems, A. C., Li, Z., Binch, A. L. A., Platt, C. I., Richardson, S. M., Hoyland, J. A., Dove, A. P., & O'Reilly, R. K. (2020). Exploiting the role of nanoparticle shape in enhancing hydrogel adhesive and mechanical properties. *Nature Communications, 11*.

Badıllı, U., Mollarasouli, F., Bakirhan, N. K., Ozkan, Y., & Ozkan, S. A. (2020). Role of quantum dots in pharmaceutical and biomedical analysis, and its application in drug delivery. *TrAC Trends in Analytical Chemistry, 131*, 116013.

Begines, B., Ortiz, T., Pérez-Aranda, M., Martínez, G., Merinero, M., Argüelles-Arias, F., & Alcudia, A. (2020). Polymeric nanoparticles for drug delivery: Recent developments and future prospects. *Nanomaterials, 10*, 1403.

Bhardwaj, A. K., & Naraian, R. (2020). Green synthesis and characterization of silver NPs using oyster mushroom extract for antibacterial efficacy journal of chemistry. *Environmental Sciences and Its Applications, 7*(1), 13–18.

Bhardwaj, A. K., Shukla, A., Mishra, R. K., Singh, S. C., Mishra, V., Uttam, K. N., Singh, M. P., Sharma, S., & Gopal, R. (2017). Power and time dependent microwave assisted fabrication of silver nanoparticles decorated cotton (SNDC) fibers for bacterial decontamination. *Frontiers in Microbiology, 8*, 330.

Bhardwaj, A. K., Kumar, V., Pandey, V., Naraian, R., & Gopal, R. (2019). Bacterial killing efficacy of synthesized rod-shaped cuprous oxide nanoparticles using laser ablation technique. *SN Applied Sciences, 1*, 1–8.

Bhardwaj, A. K., Naraian, R., Sundaram, S., & Kaur, R. (2022). Biogenic and non-biogenic waste for the synthesis of nanoparticles and their applications. In *Bioremediation: Green approaches for a clean and sustainable environment* (pp. 207–218). CRC Press.

Bilia, A. R., Bergonzi, M. C., Guccione, C., Manconi, M., Fadda, A. M., & Sinico, C. (2016). Vesicles and micelles: Two versatile vectors for the delivery of natural products. *Journal of Drug Delivery Science and Technology, 32*, 241–255.

Biomedical, M.S.-N.P.C. for, 2019, undefined. (n.d.). *Biomedical applications of nanostructured polymeric materials*. Elsevier.

Blasiak, B., Van Veggel, F. C. J. M., & Tomanek, B. (2013). Applications of nanoparticles for MRI cancer diagnosis and therapy. *Journal of Nanomaterials, 2013*, 12.

Borchard, G., William Tibbitt, M., Zürich, E., Jianxun Ding, S., Lynch, C. R. D., Kondiah, P. P., Choonara, Y. E., du Toit, L. C., Ally, N., & Pillay, V. (2020). Hydrogel biomaterials for application in ocular drug delivery. *Frontiers in Bioengineering and Biotechnology, 8*.

Campora, S., & Ghersi, G. (2022). Recent developments and applications of smart nanoparticles in biomedicine. *Nanotechnology Reviews, 11*, 2595–2631.

Castillo, R. R., Lozano, D., & Vallet-Regí, M. (2020). Mesoporous silica nanoparticles as carriers for therapeutic biomolecules. *Pharmaceutics, 12*, 432.

Cheng, W. Y., Yang, M. Y., Yeh, C. A., Yang, Y. C., Chang, K. B., Chen, K. Y., Liu, S. Y., Tang, C. L., Shen, C. C., & Hung, H. S. (2023). Therapeutic applications of mesenchymal stem cell loaded with gold nanoparticles for regenerative medicine. *Pharmaceutics, 15*, 1385.

Cheon, J. Y., Kim, S. J., Rhee, Y. H., Kwon, O. H., & Park, W. H. (2019). Shape-dependent antimicrobial activities of silver nanoparticles. *International Journal of Nanomedicine, 14*, 2773.

Chiozzi, V., & Rossi, F. (2020). Inorganic–organic core/shell nanoparticles: Progress and applications. *Nanoscale Advances, 2*, 5090–5105.

Deirram, N., Zhang, C., Kermaniyan, S. S., Johnston, A. P. R., & Such, G. K. (2019). pH-responsive polymer nanoparticles for drug delivery. *Macromolecular Rapid Communications, 40*.

Dulińska-Litewka, J., Łazarczyk, A., Hałubiec, P., Szafrański, O., Karnas, K., & Karewicz, A. (2019). Superparamagnetic iron oxide nanoparticles-current and prospective medical applications. *Materials (Basel), 12*.

Estelrich, J., Jesús Sánchez-Martín, M., & Busquets, M. A. (2015). Nanoparticles in magnetic resonance imaging: From simple to dual contrast agents. *International Journal of Nanomedicine Dovepress, 10*, 1727–1741.

Han, F., Meng, Q., Xie, E., Li, K., Hu, J., Chen, Q., Li, J., & Han, F. (2023). Engineered biomimetic micro/nano-materials for tissue regeneration. *Frontiers in Bioengineering and Biotechnology, 11*, 1205792.

Jacob, S., Nair, A. B., Shah, J., Gupta, S., Boddu, S. H. S., Sreeharsha, N., Joseph, A., Shinu, P., & Morsy, M. A. (2022). Lipid nanoparticles as a promising drug delivery carrier for topical ocular therapy—An overview on recent advances. *Pharmaceutics, 14*.

Jafari, S., Derakhshankhah, H., Alaei, L., Fattahi, A., Varnamkhasti, B. S., & Saboury, A. A. (2018). Mesoporous silica nanoparticles for therapeutic/diagnostic applications. *Biomedicine & Pharmacotherapy, 109*, 1100–1111.

Khalid, K., Tan, X., Mohd Zaid, H. F., Tao, Y., Lye Chew, C., Chu, D. T., Lam, M. K., Ho, Y. C., Lim, J. W., & Chin Wei, L. (2020). Advanced in developmental organic and inorganic nanomaterial: A review. *Bioengineered, 11*, 328–355.

Lee, B. K., Yun, Y. H., & Park, K. (2015). Smart nanoparticles for drug delivery: Boundaries and opportunities. *Chemical Engineering Science, 125*, 158–164.

Liu, S., Yu, J. M., Gan, Y. C., Qiu, X. Z., Gao, Z. C., Wang, H., Chen, S. X., Xiong, Y., Liu, G. H., Lin, S. E., McCarthy, A., John, J. V., Wei, D. X., & Hou, H. H. (2023). Biomimetic natural biomaterials for tissue engineering and regenerative medicine: New biosynthesis methods, recent advances, and emerging applications. *Military Medical Research, 10*.

Lombardo, D., Kiselev, M. A., & Caccamo, M. T. (2019). Smart nanoparticles for drug delivery application: Development of versatile Nanocarrier platforms in biotechnology and nanomedicine. *Journal of Nanomaterials, 2019*, 1.

Lynch, C. R., Kondiah, P. P. D., Choonara, Y. E., du Toit, L. C., Ally, N., & Pillay, V. (2020). Hydrogel biomaterials for application in ocular drug delivery. *Frontiers in Bioengineering and Biotechnology, 8*, 506479.

Nair, A., Haponiuk, J. T., Thomas, S., & Gopi, S. (2020). Natural carbon-based quantum dots and their applications in drug delivery: A review. *Biomedicine & Pharmacotherapy, 132*, 110834.

Nikzamir, M., Hanifehpour, Y., Akbarzadeh, A., & Panahi, Y. (2021). Applications of dendrimers in nanomedicine and drug delivery: A review. *Journal of Inorganic and Organometallic Polymers and Materials, 31*, 2246–2261.

Patel, S. K. S., Lee, J. K., & Kalia, V. C. (2018). Nanoparticles in biological hydrogen production: An overview. *Indian Journal of Microbiology, 58*, 8–18.

Plasmonics—Nanocomposix. https://nanocomposix.com/pages/plasmonics. Accessed 26 Sept 2023.

Pottoo, F. H., Sharma, S., Javed, M. N., Barkat, M. A., Harshita, Alam, M. S., Naim, M. J., Alam, O., Ansari, M. A., Barreto, G. E., & Ashraf, G. M. (2020). Lipid-based nanoformulations in the treatment of neurological disorders. *Drug Metabolism Reviews, 52*, 185–204.

Samrot, A. V., Sahithya, C. S., Selvarani, A. J., Purayil, S. K., & Ponnaiah, P. (2021). A review on synthesis, characterization and potential biological applications of superparamagnetic iron oxide nanoparticles. *Current Research in Green and Sustainable Chemistry, 4*, 100042.

Sanchez-Moreno, P., Ortega-Vinuesa, J. L., Peula-Garcia, J. M., Marchal, J. A., & Boulaiz, H. (2018). Smart drug-delivery systems for cancer nanotherapy. *Current Drug Targets, 19*, 339–359.

Shoukat, R., & Khan, M. I. (2021). Carbon nanotubes: A review on properties, synthesis methods and applications in micro and nanotechnology. *Microsystem Technologies, 27*, 4183–4192.

Song, W., Su, X., Gregory, D. A., Li, W., Cai, Z., & Zhao, X. (2018). Magnetic alginate/chitosan nanoparticles for targeted delivery of curcumin into human breast cancer cells. *Nanomaterials, 8*(11), 907.

Suciu, M., Ionescu, C. M., Ciorita, A., Tripon, S. C., Nica, D., Al-Salami, H., & Barbu-Tudoran, L. (2020). Applications of superparamagnetic iron oxide nanoparticles in drug and therapeutic delivery, and biotechnological advancements. *Beilstein Journal of Nanotechnology, 11*, 1092.

Sur, S., Rathore, A., Dave, V., Reddy, K. R., Chouhan, R. S., & Sadhu, V. (2019). Recent developments in functionalized polymer nanoparticles for efficient drug delivery system. *Nano-Structures & Nano-Objects, 20*, 20.

Thang, N. H., Chien, T. B., & Cuong, D. X. (2023). Polymer-Based Hydrogels Applied in Drug Delivery: An Overview. *Gels, 9*, 523.

Vasantharaj, S., Sathiyavimal, S., Senthilkumar, P., Lewis Oscar, F., & Pugazhendhi, A. (2019). Biosynthesis of iron oxide nanoparticles using leaf extract of Ruellia tuberosa: Antimicrobial properties and their applications in photocatalytic degradation. *Journal of Photochemistry and Photobiology. B, 192*, 74–82.

Wang, C., Liu, Y., Jia, Z., Zhao, W., & Wu, G. (2023). Multicomponent nanoparticles synergistic one-dimensional nanofibers as heterostructure absorbers for tunable and efficient microwave absorption. *Nano-Micro Letters, 15*(1), 13.

Xu, L., Wang, X., Liu, Y., Yang, G., Falconer, R. J., & Zhao, C.-X. (2022). Lipid nanoparticles for drug delivery. *Advanced NanoBiomed Research, 2*, 2100109.

Yazdanian, M., Alam, M., Abbasi, K., Rahbar, M., Farjood, A., Tahmasebi, E., Tebyaniyan, H., Ranjbar, R., & Hesam Arefi, A. (2022). Synthetic materials in craniofacial regenerative medicine: A comprehensive overview. *Frontiers in Bioengineering and Biotechnology, 10*, 987195.

Zhao, N., Yan, L., Zhao, X., Chen, X., Li, A., Zheng, D., Zhou, X., Dai, X., & Xu, F. J. (2019). Versatile types of organic/inorganic nanohybrids: From strategic design to Biomedical applications. *Chemical Reviews, 119*, 1666–1762.

Advances of Plant-Assisted Synthesized Metal Nanoparticles and Their Environmental Application

Sonal Singh, Abhishek Kumar Bhardwaj, and Kuldip Dwivedi

Abbreviations

AFM	Atomic Force Microscopy
CNT	Carbon Nanotubes
DPPH	2,2-diphenyl-β-picrylhydrazyl
EDTA	Ethylenediaminetetraacetic acid
FE-SEM	Field emission scanning electron microscopy
FTIR	Fourier Transmission Infrared Spectrophotometer
HR-TEM	High-Resolution Transmission Electron Microscopy
NPs	Nanoparticles
nZVI	Nano Zero Valent Iron
SAED	Selected area (electron) diffraction
SEM	Scanning Electron Microscopy
SWCNTs	Single-walled Carbon Nanotubes
TEM	Transmission Electron Microscopy
TGA	Thermogravimetric analysis
XRD	X-ray Diffraction

1 Introduction

Nanoparticles (NPs) are extremely small particles measuring between 1 and 100 nm that have begun to behave more in accordance with the laws of quantum physics than with those of classical physics (Bhardwaj et al., 2018). The same material has different optical/electrical properties on the nanoscale compared with the macroscale. Nanoscience is the study of materials and/or particles that, despite their small size, exhibit remarkable properties, processes, and phenomena.

S. Singh · A. K. Bhardwaj · K. Dwivedi (✉)
Department of Environmental Science, Amity School of Life Science, Amity University Madhya Pradesh, Gwalior, India
e-mail: kdwivedi@gwa.amity.edu

A. K. Bhardwaj et al. (eds.), *Biogenic Wastes-Enabled Nanomaterial Synthesis*,
https://doi.org/10.1007/978-3-031-59083-2_15

"Nanotechnology" is the industrial and commercial application of nanoscience to create NPs that could enhance human health. The development of nanotechnology has a very lengthy history. NPs can be produced by physical, chemical, or biological processes. Figures 1 and 2 provide a schematic illustration of various techniques for creating NPs and their uses. In contrast to the chemical and biological procedures, which employ a bottom-up strategy, the physical method produces NPs in a top-down manner. According to Maurya et al. (2016), evaporation-condensation, electrolysis, diffusion, laser ablation, sputter deposition, pyrolysis, plasma arcing, and high energy ball milling are a few of the most widely used physical methods for the synthesis of NPs. The primary disadvantages of chemical and physical approaches are their high energy consumption, high operating costs, and low output rates. As biological organisms have a variety of chemical compositions, NPs can be created by a variety of biological or biogenetic processes; previously, only physical and chemical processes could get more attention for the synthesis of nanoparticles. However, the traditional and most popular means of creating metallic NPs use chemical synthesis techniques such as chemical reduction, micro-emulsion/colloidal, electrochemical, and thermal degradation. Green chemistry has been growing over the last few decades to follow sustainable goals.

Nanoparticle biogenesis became necessary because physical and chemical processes were too expensive. Reduction and oxidation are the main processes involved in the bottom-up biogenesis of NPs. However, plant enzymes or phytochemicals have unique reducing properties that have the capability to frequently reduce metal compounds into the corresponding NPs. Microorganisms such as bacteria, yeast,

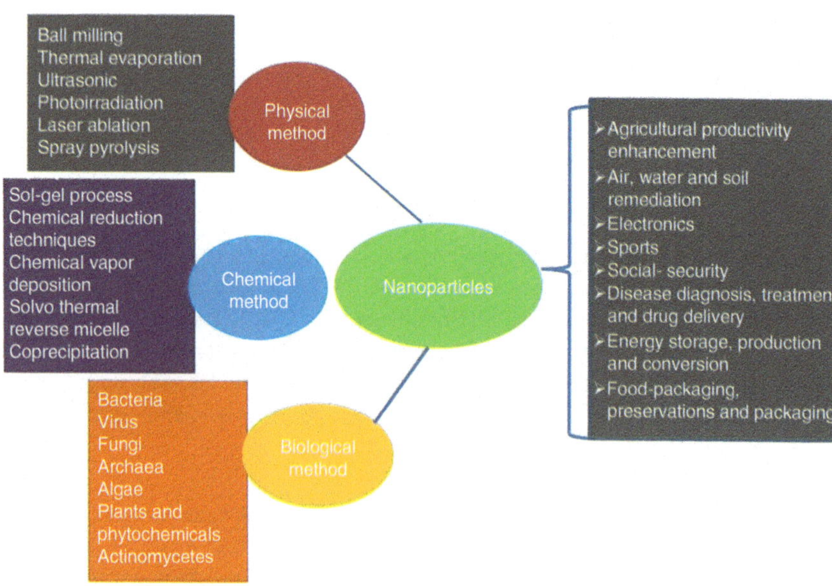

Fig. 1 Various techniques for the synthesis of NPs and their uses

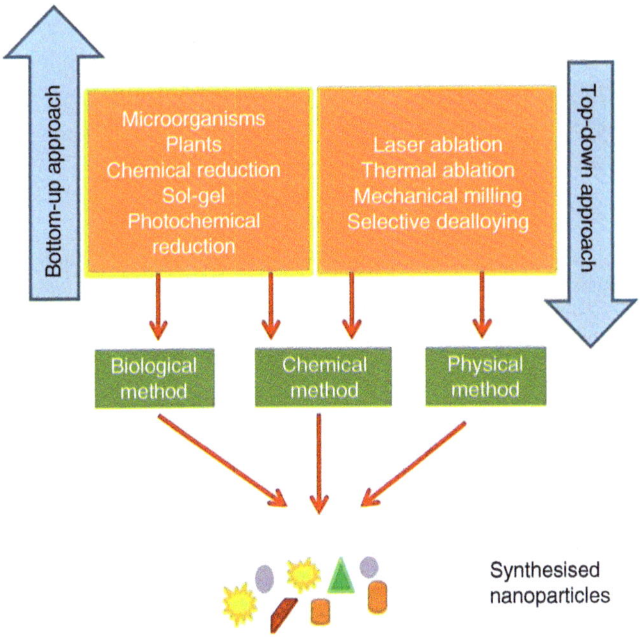

Fig. 2 Top-down and bottom-up methods for synthesizing NPs

algae, fungi, and actinomycetes can also be employed to create NPs in addition to vascular plants (Bhardwaj et al., 2019; Bhardwaj & Naraian, 2021; Maurya et al., 2016). In this chapter an attempt has been made to discuss the recently published research papers related to synthesis of metal NPs using plant extracts, mediated metal NPs and their properties, size, shape, characteristics, synthesis mechanism and application.

2 Synthesis of Nanoparticles Using Plants

The environmentally benign method of NP synthesis does not harm other living things (Samuel et al., 2022). Metal NPs have been produced from a variety of plant materials, including fruits, flowers, roots, tubers, leaves, and stems. Figure 3 illustrates the novel process of "green synthesis," which blends biology and nanotechnology to create NPs. This technology effectively addresses the limitations of physical and chemical procedures by producing NPs from biological resources (Grammatikopoulos et al., 2016). Cu NPs from seeds of *Myristica fragrans* (nutmeg) and *Illicium verum* (star anise), Ag NPs from agricultural waste, and extracts of red currant and bilberry waste are a few examples (Vijaykumar et al., 2021; Wolny-Koładka et al., 2022). This method is eco-friendly because it does not involve

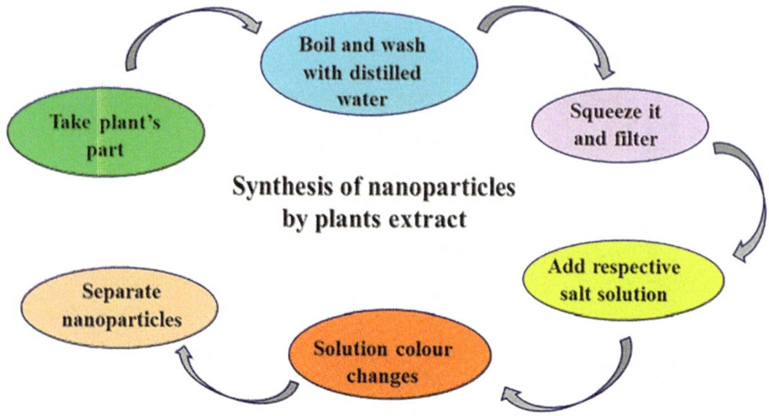

Fig. 3 Steps of green synthesis of NPs from plant parts

harmful chemicals or harsh weather. It is affordable and might be suitable for massive production (Bhardwaj et al., 2021). The benefits of the green synthesis approach include its simplicity, reproducibility, biocompatibility, environmental friendliness, and affordability. Every biological agent recognizes metal ions differently in order to produce its own NPs.

The biomolecules of the biological system retain the properties of NPs, giving the utilization of biological resources over nonbiological ones major advantages, and eliminating the need for stabilizing or capping agents (Ovais et al., 2018). Plant components are reduced in size and dried during extraction, increasing the surface area (Vincent et al., 2022). Plant extracts must be cleaned, heated, and boiled in order to be used in green synthesis (Fig. 3). After that, the plant extract is squeezed and filtered. The solution's color changes and NPs are created when the proper salt solution is added at the very end. Powdered *Punica granatum* (pomegranate) peels were boiled in distilled water for 30 min at 55 °C, 2 h at 90 °C, 2 h at 85 °C, and 20 min at 60 °C to produce copper oxide (CuO) NPs (Siddiqui et al., 2022), pumpkin seeds produced TiO2 NPs (Abisharani et al., 2019), and *Chromolaena odorata* roots produced smart nanocomposites (Nnadozie et al., 2020). To create Ag NPs, *Nepeta leucophylla* roots were cooked in methanol for 8 h (Singh, et al., 2019). Other techniques have also been used, including sonication, maceration, and autoclaving. For the green synthesis of CuO, Ag, (Fe/Si/Cu-Ag), and ZnO NPs respectively, ground *Caesalpinia bonducella* kernels were sonicated for 30 min (Sukumar et al., 2020), fruits of *Solanum mammosum* (Pilaquinga et al., 2019) and *Crateagus pentagyna* were macerated (Ebrahimzadeh et al., 2019), and roots of *Scutellaria baicalensis* were autoclaved at 100 °C for 30 min (Chen et al., 2019).

The NPs created are verified and further investigated using various characterization techniques. NP production is confirmed by UV–Vis spectroscopy. It provides details on the size, composition, aggregation, and stabilization of the NPs produced. Through the use of techniques such as X-ray photoelectron spectroscopy), HR-TEM,

FE-SEM, SAED, AFM, and DLS, their morphology and size are identified. Similarly, zeta potential measurements, thermal gravimetric analysis (TGA), and FT-IR spectroscopy provide details regarding the surface properties of the generated NPs. The zeta potential can be used to determine the surface charge of the NPs, any surface coatings, and any materials inside its cavity. An overall idea of the crystallinity of the particles can be obtained by using the X-ray diffraction method.

3 Classification of Nanoparticles

The classification of NPs is based on dimension and chemical nature. Hett (2004) divided the NPs into groups according to their various dimensional configurations shown in Fig. 4. One-dimensional NPs are thin films or engineered surfaces utilized in solar cells, catalysis, and biological sensors. Carbon nanotubes (CNTs) measuring 1 nm and 100 nm long are among the two-dimensional NPs. Single-walled CNTs and multi-walled CNTs are both possible. The utilization of CNTs is based on their metallic and semiconductor characteristics, which have better electrical performance, making them work as semiconductors. Dendrimers, quantum dots (QDs), and fullerenes (carbon 60) are the three-dimensional NPs. Materials called fullerenes are formed of carbon 60 (C60) and come in the shape of hollow spheres that resemble soccer balls. Regarding high-pressure subjection and returning to their natural shape after pressure release, these show special characteristics. Owing to the fact that fullerenes are hollow structures, their function may be based on filling them with various materials; as a result, they are utilized in medical

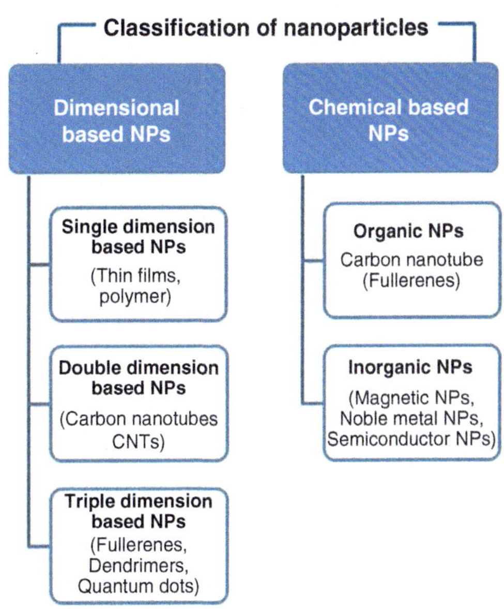

Fig. 4 Classification of NPs

applications. Dendrimers are polymers with a diameter of 10–100 nm and include many functional molecules, indicating a possible use for them as drug delivery systems. QDs are nanocrystals with a diameter of 2–10 nm that may have one electron or thousands of them. Color-coded QDs are utilized for quick sensing applications, furthermore, it can be employed for optical and optoelectronic devices, quantum computing, and information storage (Pal et al., 2011).

In addition, two kinds of NPs can be distinguished by their chemical composition: organic and inorganic. A CNT is an example of organic NPs. Inorganic NPs consist of three types: magnetic, noble metal, and semiconductor NPs. Nanoformulations such as nanomembranes, nanotubes, nano polymers, graphene-based NPs, and metal-, carbon-, and silicon-based NPs have been extensively used in the remediation of environmental toxic elements (Hussain et al., 2022).

4 Application of Plant-Assisted Synthesized NPs

Plant-mediated synthesized NPs have unique properties owing to the existence of specific reducing and coating agents. These unique properties have different application as mentioned in Fig. 5. Several studies are summarized in Table 1. According to Mittal et al. (2013), leaves of mango (*Medicago sativa*) are employed for the manufacture of gold (Au) NPs, which are then used in optoelectronics and as antioxidants. According to Sarathi Kannan et al. (2021), silver NPs have antibacterial capabilities against a variety of bacterial infections, including *Staphylococcus aureus* and *Escherichia coli*. Additionally, they observed the action taken against harmful polluting dyes such as rhodamine B, methylene blue, alizarin red, and Congo red. According to Padmavathi et al. (2022), *Corchorus aestuans* silver NPs exhibit potent antibacterial action against Gram-negative bacteria, which are crucial in the medical industry for treating a variety of illnesses. According to Zhang et al.

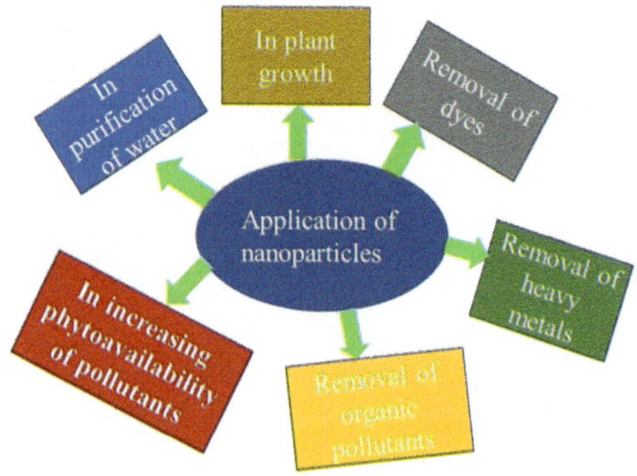

Fig. 5 Applications of plant-based NPs

Table 1 Synthesis of NPs by plant with their shape and size and their pharmaceutical uses

NPs	Shape and size(nm)	Synthesized by plant	Applications	References
Gold NP	NR, images presented (02–40) Decahedral, Icosahedral (20–40) Spherical, triangular, Hexagonal (NR) Triangular (50–350) Spherical, (10–40) Triangular, pentagonal (5–300) Spherical (20–30) Crystalline (0–35)	*Medicago sativa* (leaf) *Pelargonium graveolens* *Azadirachta indica* *Aloe vera* *Magnolia kobus and Diopyros kaki* *Capsicum annuum* *Curculigo orchioides*	Antioxidant and optoelectronic Antimicrobial and anticancerous Antimicrobial Optical properties Catalytic and antibacterial Antiangiogenic Antioxidant activity against free radicals such as DPPH and suppressor of breast cancer cells	Mittal et al. (2013) Das et al. (2013) Kora et al. (2012) Firdhouse & Jannathul (2013) Lukman et al. (2011) Ghosh et al. (2012) Patil, T.P. et al. (2023)
Silver NP	Icosahedral (2–20, 2–4) Spherical (<100)	*Medicago sativa* *Brassica juncea* *Aloe vera* *Calendula officinalis* *Coccinia indica* *Ludwigia octovalvis* *Croton macrostachyus*	Antioxidant In arthritis Optical properties Analgesic and antiseptic Antibacterial Antimicrobial against bacterial pathogens such as *E. coli*	Mittal et al. (2013) Narayanan & Park, (2014) Firdhouse & Jannathulet (2013) Chandra Paul et al. (2020) Chinni et al. (2021) Sarathi Kannan et al. (2021) Padmavathi et al. (2022) Logambal et al. (2022)
	Spherical (15) Spherical (50–60) Spherical (8–48) Spherical (28–50) Spherical			
	Crystallite (0–18) Crystalline with a face-centered cubic (0–32)	*Corchorus aestuans* *Couroupita guianensis*	Antibacterial Antibacterial against Gram-negative bacteria Antibacterial	

(continued)

Table 1 (continued)

NPs	Shape and size(nm)	Synthesized by plant	Applications	References
Platinum NPs	Spherical	*Diopyros kaki*	Antibacterial	Song et al. (2010) Soundarrajan et al. (2012) Zhang et al. (2022)
	Triangular	*Ocimum sanctum*	Antimicrobial	
	Spherical (200–729)	*Nymphaea tetragona*	Antiaging used for skin protection Antioxidant	
Palladium NPs	Spherical (15–20)	*Cinnamomum zeylanicum* Blume	Antibacterial	Smitha et al. (2009) Satyavani et al. (2011) Vivekanandhan et al. (2009) Vivek et al. (2011)
	Spherical (3.2–6.4)	*Cinnamomum camphora* L.	Antioxidant and also used in jewelry, dental fillings, and electronic components	
	Rock like appearance	*Gardenia jasminoides* Ellis	Antibacterial	
	Cubic hexagonal crystalline (35)	Soybean (*Glycine max*)	Antilarvicidal and antifungal	
Silver NPs	Face-centered cubic (4–10)	*Phyllanthus amarus*	Antibacterial against *Pseudomonas aeruginosa* clinical isolates that are multidrug resistant	Singh et al. (2014)
Iron NPs	Spheres (30–38)	*Platanus orientalis*	Antioxidant	Mittal et al. (2015)
Copper NPs	Spherical (3–7)	*Mukia maderaspatana*	Antibacterial	Vijayakumar et al. (2017)
Nickel NPs	Spherical (5–15)	*Musa balbisiana*	Antimicrobial	Amooaghaie et al. (2015)
Magnesium oxide	Cubic hexagonal crystalline	*Cucurbita maxima*	Antimicrobial	Shameli et al. (2012)
Zinc oxide	Spherical (5–50)	*Aloe barbadensis* *Tephrosia purpurea* *Heliotropium indicum*	Antimicrobial Anti-inflammatory, antioxidants Antiulcer	Wijesinghe et al. (2023)
Cobalt tetraoxide	Hexagonal (50–100)	*Solanum tuberosum*	Gram-positive bacteria are susceptible to antibacterial activity. *Staphylococcus aureus*	Bekele et al. (2022)
Titanium dioxide	Crystalline (50)	*Citrus sinensis*	Antimicrobial and also acts as an inhibitor of methylene blue dye	H. O. Mulugeta et al. (2022)

DPPH 2,2-diphenyl-β-picrylhydrazyl, *NPs* nanoparticles

(2022), in vitro tests of synthesized platinum NPs revealed strong antioxidant and antityrosinase activity.

Vivek et al. (2011), reported that the antifungal activity of Ag NPs was evaluated against the following human pathogenic strains: *Humicola insolens* (MTCC 4520), *Fusarium dimerum* (MTCC 6583), *Mucor indicus* (MTCC 3318), and *Trichoderma reesei* (MTCC 3929). Synthetic titanium dioxide NPs were studied as a green alternative photocatalyst for the breakdown of methylene blue dye under visible light irradiation, according to Mulugeta et al. (2022). According to Singh et al. (2014), *Pseudomonas aeruginosa* infections are a major source of morbidity and death in patients with burns and weakened immune systems. He also discussed how *Phyllanthus amarus* aqueous extract was used to create silver NPs. The produced NPs were investigated as a promising source of nanomedicine against *P. aeruginosa* multi drug resistant (MDR) burn isolates.

4.1 Nanobioremediation

The use of NPs in bioremediation, which is the process of removing toxicants or pollutants from polluted sites utilizing NPs and nanomaterials made from microorganisms such as bacteria, fungi, and plants. Nanobioremediation is a brand-new technique for eliminating pollutants in order to clean up the environment. Nowadays, contaminated areas are cleaned using chemical, physical, incineration, and bioremediation methods. Thanks to recent advancements, bioremediation can now remove toxins from contaminated areas in an economical and environmentally friendly manner. Microbes, plants, and enzymatic remediation are the three basic methods of bioremediation. Nanotechnology improves the effectiveness of phytoremediation; additionally, the cleanup of heavy metal-contaminated water, soils, and organic and inorganic contaminants can be done using NPs. Recent research has demonstrated that nanosized zerovalent ions can destroy organic pollutants such atrazine, molinate, and chlorpyrifos (Zhang, 2003; Ghormade et al., 2011). Phytoremediation can be employed in conjunction with NPs in enzyme-based bioremediation. For instance, a number of complex organic substances, such as organochlorines and long-chain hydrocarbons, are extremely resistant to microbial and plant breakdown. This constraint could be overcome by combining nanotechnology and biotechnology. Nano-encapsulated enzymes would break down complex organic materials into simpler chemicals, which would then be swiftly broken down by microbes and plants working together.

4.2 Nanomaterials in Agriculture

When selecting plant species for phytoremediation, plant biomass and growth rate are two crucial factors to take into account. Many applied plants are often unsatisfactory owing to poor plant biomass and a slow development rate, which are caused by inadequate resistance to pollutants and unfavorable soil conditions for plant growth. As a result, certain tactics are employed in phytoremediation processes to encourage plant growth, such as employing transgenic plants, plant growth regulators, and inoculating plant growth-promoting rhizobacteria (PGPR). Some nanomaterials, including graphene QDs, CNTs, Ag NPs, ZnO NPs, nano zero valent iron (nZVI) particles, and upconversion NPs, have been proven in studies on nanomaterials and plants to promote plant growth (Mokarram-Kashtiban et al., 2019). NPs increase crop production by stimulating seed germination and quality of soil (Fig. 6). NPs are used as nanofertilizers. With the use of a foliar spray, a cutting-edge technique, liquid fertilizers are directly applied to the leaves or foliage of plants, facilitating the quick uptake of nutrients through the leaf surface. The technique uses the application of the nanofertilizer to the leaf surface to provide precise, optimal, fast, and targeted plant uptake. One potential way of giving plants critical materials, including fungicides, herbicides, preservatives, and fertilizers, is by applying NPs topically.

In soil polluted with Cd, the application of TiO2 NPs and PGPR increased the development of *Trifolium repens* and improved the plant's absorption and accumulation of Cd. There are several potential applications for the intelligent fusion of plants, nanomaterials, and PGPR in soil remediation. Aluminum oxide NPs affect root elongation in hydroponic trials (Yang & Watts, 2005). According to several studies by Zheng et al., 2005, spinach can develop more quickly when given

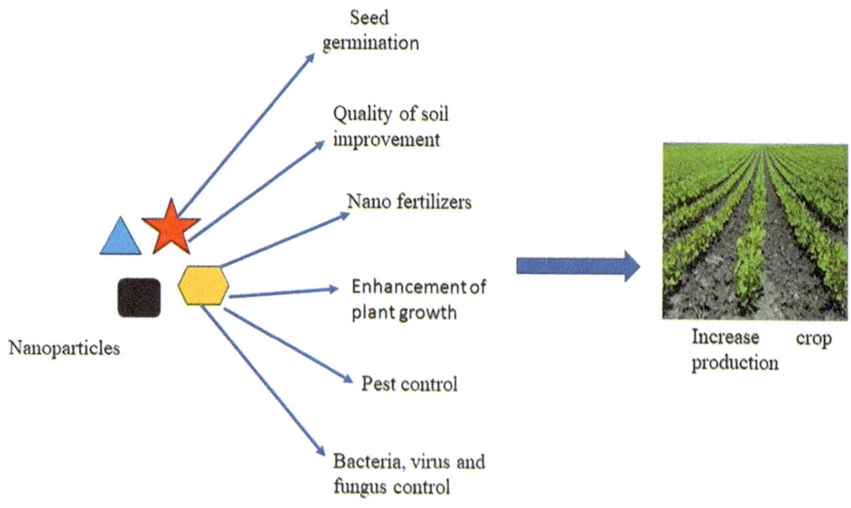

Fig. 6 Role of nanoparticles in promoting plant growth

nano-sized TiO_2 by spraying it onto the leaves. It was shown that nano-TiO_2 promoted nitrate adsorption, accelerated the transformation of inorganic nitrogen into organic nitrogen, and boosted the activity of several enzymes.

4.3 Nanomaterials in Contaminant Removal

The use of NPs is site specific, and in addition to the size of the particles, other physical and chemical characteristics of the soil, such as soil aggregates, texture, etc., also affect its efficiency. Pollutant types, concentrations, sizes, charge/non-charged materials, and organic content are among the factors that affect this. NPs are applied to the soil using a variety of techniques as part of the restoration process.

4.3.1 Nanoparticles Used in Heavy Metal Removal

The ecosystem is greatly threatened by the build-up of poisonous heavy metals in various environmental compartments, and the gravity of the issue is worsened by the bioaccumulation of these toxic substances at higher trophic levels through a process known as biomagnification (Ali et al., 2013). The bulk of metals and metalloids do not degrade through chemical or microbiological processes like organic pollutants do, which results in a continuous increase in the concentrations of these deadly metals and long-term soil ecotoxicity (Adriano et al., 2004). Pandey et al. (2015) reported that the persistent enrichment of these toxic elements in aquatic and soil ecosystems is a major global concern, and the pursuit of novel technological advancements has made the contamination even worse, raising concerns about the viability of the current biotic communities (Ebbs et al., 2016). There are currently a few methods of removing these dangerous metals/metalloids from ecosystems, with the use of nanomaterials being one of the most promising. Adsorption is a very effective and widely used chemical technology for this purpose owing to its ease of use, low cost, and treatment stability (Mohan & Pittman, 2007). Eliminating or removing heavy metals from contaminated soils and streams is one of the key approaches to relieving heavy metal stress in plants. nZVI) has been successfully used to reduce the concentrations of arsenic (As-III) and chromium (Cr-VI) from water, among other different nanomaterials tested for remediation purposes (Poguberović et al., 2016). Decades back there was a significant increase in the use of nZVI NPs as a reducing agent for environmental remediation reasons. These nZVI NPs high surface energy and reaction activity enable the quick detoxification of a variety of aquatic contaminants, including lethal heavy metals.

4.3.2 Reduction of Organic Pollutants by Nanoparticles

The majority of organic contaminants found in soil come from human activity. Organochlorine pesticides, polycyclic aromatic hydrocarbons, chlorinated hydrocarbons, polychlorinated biphenyls, phenols, and their derivatives are common organic pollutants in soil, and it is crucial to remove these pollutants owing to their toxicity, persistence, and bioaccumulation (Jones & de Voogt, 1999). Phytoremediation has been an effective approach for remedying soil contaminated with organic pollutants, especially pesticides, polycyclic aromatic hydrocarbons, petroleum, and explosives (Kang, 2014). It has been observed that applying nanomaterials enhanced the phytoremediation of soil contaminated with trichloroethylene, endosulfan, and trinitrotoluene (Pillai & Kottekottil, 2016). Ma and Wang (2010) suggest that fullerene NPs might improve the phytoremediation system's capacity of eastern cottonwood to absorb trichloroethylene. The use of fullerene NPs increased the absorption of trichloroethylene by 26% and 82% respectively, at concentrations of 2 mg/L and 15 mg/L. Additionally, fullerene NPs were introduced, although the plants showed no immediate negative consequences from them. Jiamjitrpanich et al. (2012) added nZVI particles of various concentrations (100, 500, and 1000 mg/kg) to the soil and measured the residual trinitrotoluene concentrations in soil during a remediation period of 120 days in the study of removing trinitrotoluene from contaminated soil by *Panicum maximum*. The results showed that the efficiency of phytoremediation may be greatly enhanced by the addition of nZVI particles. A trinitrotoluene–nZVI ratio of 1/10 was shown to have the best removal performance in their trials, and when the initial trinitrotoluene concentration was 100 mg/kg, the phytoremediation process could be completed in 60 days. Pillai and Kottekottil (2016) employed ZVI particles to aid in the phytoremediation of soil polluted with endosulfan. In the experiments, three plant species – *Alpinia calcarata, Ocimum sanctum*, and *Cymbopogon citratus* – were employed, in either the presence or the absence of nZVI particles. The removal rates of endosulfan from soil were raised for *Alpinia calcarata, Ocimum sanctum,* and *Cymbopogon citratus* by nZVI particles from 81.2% to 100%, 20.76% to 76.28%, and 65.08% to 86.16% respectively.

4.3.3 Nanomaterials in Dye Degradation

Since the middle of the twentieth century, there have been many studies on the photocatalytic destruction of dyes and other contaminants. Products including furniture, clothing, plastics, and many more, frequently use dyes. A little under 12% of the colors used in dyeing are wasted, and 20% of that waste seeps into the environment and poses a substantial health risk to humans (Ali, 2010). Because dye particles are inhaled when using these reactive dyes, lung issues can arise (Vakili et al., 2014). One of the most recent techniques for dye degradation is phytoremediation using nanomaterials. For cleaning up environmental pollutants, nanoscale particles are gaining popularity. Owing to their enhanced metabolic activities and efficient use of native nutrients by fostering microbial activities, NPs are promoted as an effective nutrient source for biomass generation from plants (Marchiol et al., 2020). Using fewer toxic chemicals is the main issue with nanomaterials. Compared with other processes, photocatalytic degradation offers the advantages of being environmentally friendly, cost effective, and producing no secondary pollutants.

4.4 *Nanomaterials in Water Purification*

Over the past few decades, nanomaterials have been studied and developed for a variety of applications, including catalysis (Singh et al., 2015), medicine (Adeleye et al., 2016), sensing (Sen et al., 2016), and biology (Santhosh et al., 2016). Particularly when it comes to NPs being injected into the water supply, the use of NPs in the treatment of water and wastewater is a popular concern.

In Fig. 7, the ability of nanomaterials to adsorb molecules and their reactivity are considerably enhanced by the large specific surface area and compact size. Additionally, NPs that have undergone solution processing are very mobile. It has been claimed that certain nanomaterials can remove different kinds of heavy metals, organic contaminants, inorganic anions, and bacteria. Applications concerning the treatment of water and wastewater show enormous potential for nanomaterials. The concentrations of nanomaterials, such as CNTs, metal oxide NPs, zero-valent metal NPs, and nanocomposites, are now being studied primarily in the field of water and wastewater treatment.

4.5 *In Phytoremediation of Contaminants*

Phytoremediation is the process of using plants to eliminate, break down, or purify environmental elements such as soil, silt, and water. Along with plants, the accompanying rhizospheric microbial fauna is also exploited to purify these contaminated resources (Cunningham et al., 1997). Numerous plant species have been studied for their potential in the remediation of contaminated soils, including alfalfa, tomato, sunflower, willow, poplar tree, Chinese cabbage, and sunbeam. Many of these plants have been found to be effective phytoremediators (Mitton et al., 2014).

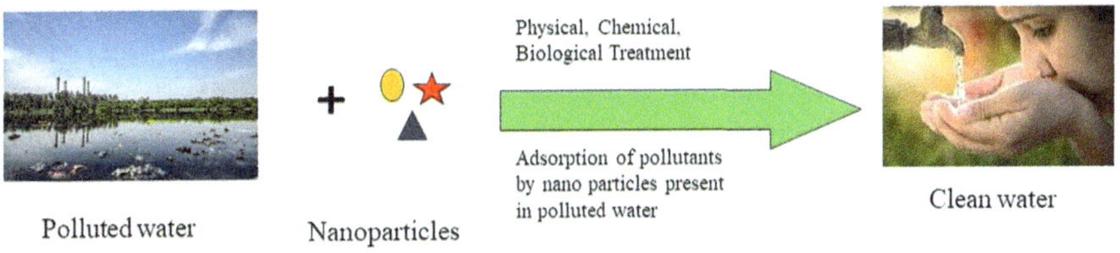

Fig. 7 Use of NPs in the purification of water

The same principles utilized by plants and microorganisms in nature for the breakdown and sequestration of organic and inorganic contaminants are applied in phytoremediation (Negri & Hinchman, 2000). Numerous phytoremediation methods, such as phytovolatilization, phytoextraction, phytodegradation, phytostabilization, and rhizodegradation, are employed to treat soil contamination (Song et al., 2019).

Owing to its fast mode of action and on-site treatment, nanotechnology is being viewed as a realistic and practical way of removing environmental toxins (Gong et al., 2018; Wang et al., 2016). It has been demonstrated that NPs may affect the destiny, behavior, and uptake of contaminants in phytoremediation systems (Abhilash et al., 2012). Furthermore, because of the expected increased breakdown capability of NPs compared with the removal of contaminants with a single technology, a significant amount of research is presently being carried out on the combined usage of NPs with soil bacteria and plants. When cleaning up polluted soil, nano-phytoremediation combines the use of nanotechnology and phytotechnology. Carbon-based nanomaterials are the ones that have been investigated the most, followed closely by metal and metal oxide NPs (Chen et al., 2017; Gong et al., 2009). Owing to their huge specific surface area, CNTs have shown excellent adsorption ability for a variety of contaminants, especially hydrophobic organic pollutants (Pan & Xing, 2008; Song et al., 2017). The fate and transport of contaminants, especially organic pollutants, could be drastically altered by these organic molecules, which are loaded with carbon NPs (Hofmann & Von der Kammer, 2009).

4.6 To Increase the Phytoavailability of Contaminants in Soil

Pollutant phytoavailability has a significant impact on the effectiveness of phytoremediation, particularly in the case of phytoextraction. Only available forms of pollutants are absorbed by plants. The chemical speciation and distribution of pollutants in the soil have a significant impact on their phytoavailability. Generally, the highest phytoavailability of metals is in exchangeable forms (dissolved in the soil solution), and then in combined forms with minerals, oxides, and organic matter, and the lowest is in the crystalline phase (Liang et al., 2017; Sheoran et al., 2016). Additionally, the physicochemical properties of soil and physiological characteristics of plants also affect the phytoavailability of pollutants (Ren et al., 2018; Sheoran et al., 2016). Low phytoavailability often limits the phytoremediation process. For example, lead usually exists in insoluble forms in soil owing to adsorption, complexation, and precipitation, which makes it difficult for phytoextraction to occur (Zaier et al., 2014). Fullerene is one kind of NP that has drawn a great deal of attention from researchers because it can increase the phytoavailability of contaminants. Ma and Wang (2010) used C_{60} fullerene NPs in the phytoremediation system with eastern cottonwood to facilitate the removal of trichloroethylene from contaminated soil. The pollutant was absorbed by all plants 30% to 65% more when fullerene was added, and this effect was particularly noticeable in the roots. Additionally, it has been demonstrated that some other nanomaterials, such as short CNTs, Fe_3O_4 NPs, TiO_2 NPs, silica NPs, and QDs, could be directly taken in by plants (Wang et al., 2016). Because of this, it is feasible to employ more nanomaterials in subsequent studies to increase the phytoavailability of contaminants in phytoremediation systems.

5 Conclusion

Plant-mediated nanomaterials have been successfully prepared by different researchers from different geographic places and have been finally employed in solving various environmental concerns. Metallic NPs, including silver, gold, copper, platinum, zinc, iron, palladium, titanium, titanium dioxide, manganese, and silica, were prepared with different configurations (shape and sizes) using a variety of plant products (leaves, root, bark, fruits, and flowers). Compared with chemical and physical methods, such synthesis may be a cheaper, more environmentally friendly, and more sustainable process. During environmentally friendly synthesis, the special enzymes and proteins found in plant extract are first used as capping and reducing agents. NPs generated from plants are used to eliminate a range of contaminants, such as dyes, heavy metals, and pollutants, both organic and inorganic. Future research will primarily focus on identifying functional compounds so that nanomaterials can be tailored to fit desired application scenarios.

References

Abhilash, P. C., Powell, J. R., Singh, H. B., & Singh, B. K. (2012). Plant–microbe interactions: Novel applications for exploitation in multipurpose remediation technologies. *Trends in Biotechnology, 30*(8), 416–420.

Abisharani, J. M., Devikala, S., Dinesh Kumar, R., Arthanareeswari, M., & Kamaraj, P. (2019). Green synthesis of TiO2 nanoparticles using Cucurbita pepo seeds extract. *Materials Today Proceedings, 14*, 302–307.

Adeleye, A. S., Conway, J. R., Garner, K., Huang, Y., Su, Y., & Keller, A. A. (2016). Engineered nanomaterials for water treatment and remediation: costs, benefits, and applicability. *Chemical Engineering Journal, 286*(2016), 640–662.

Adriano, D. C., Wenzel, W. W., Vangronsveld, J., & Bolan, N. S. (2004). Role of assisted natural remediation in environmental cleanup. *Geoderma, 122*, 121–142.

Ali, H. (2010). Biodegradation of synthetic dyes—A review. *Water, Air, and Soil Pollution, 213*(1), 251–273.

Ali, H., Khan, E., & Sajad, M. A. (2013). Phytoremediation of heavy metals-concepts and applications. *Chemosphere, 91*, 869–881.

Amooaghaie, R., Saeri, M. R., & Azizi, M. (2015). Synthesis, characterization and biocompatibility of silver nanoparticles synthesized from *Nigella sativa* leaf extract in comparison with chemical silver nanoparticles. *Ecotoxicology and Environmental Safety, 120*, 400–408.

Bekele, E. T., Ananda Murthy, H. C., Muniswamy, D., Lemenh, Y. A., Shume, S., Ayanie, G. T., Kumar, A. P., Ravikumar, C. R., Balachandra, G., & Roy, A. (2022). *Solanum tuberosum* leaf extract templated synthesis of Co_3O_4 nanoparticles for electrochemical sensor and antibacterial applications. *Bioinorganic Chemistry and Applications, 2022*, 8440756.

Bhardwaj, A. K., & Naraian, R. (2021). Cyanobacteria as biochemical energy source for the synthesis of inorganic nanoparticles, mechanism and potential applications: a review. 3. *Biotech, 11*(10), 445.

Bhardwaj, A. K., Shukla, A., Maurya, S., Singh, S. C., Uttam, K. N., Sundaram, S., Singh, M. P., & Gopal, R. (2018). Direct sunlight enabled photo-biochemical synthesis of silver nanoparticles and their Bactericidal Efficacy: Photon energy as key for size and distribution control. *Journal of Photochemistry and Photobiology B: Biology, 188*, 42–49.

Bhardwaj, A. K., Gupta, M. K., & Naraian, R. (2019). Myco-nanotechnological approach for improved degradation of lignocellulosic waste: Its future aspect. *Mycodegradation of Lignocelluloses*, 227–245.

Bhardwaj, A. K., Sundaram, S., Yadav, K. K., & Srivastav, A. L. (2021). An overview of silver nano-particles as promising materials for water disinfection. *Environmental Technology & Innovation, 23*, 101721.

Chandra Paul, S., Bhowmik, S., Rani Nath, M., Islam, M. S., Kanti Paul, S., Neazi, J., Binta, S., Monir, T., Dewanjee, S., & Abdus Salam, M. (2020). Silver nanoparticles synthesis in a green approach: size dependent catalytic degradation of cationic and anionic dyes. *Oriental Journal of Chemistry, 36*, 353–360.

Chen, M., Zeng, G., Xu, P., Zhang, Y., Jiang, D., & Zhou, S. (2017). Understanding enzymatic degradation of single-walled carbon nanotubes triggered by functionalization using molecular dynamics simulation. *Environmental Science: Nano, 4*(3), 720–727.

Chen, L., Batjikh, I., Hurh, J., Han, Y., Huo, Y., Ali, H., Li, J. F., Rupa, E. J., Ahn, J. C., Mathiyalagan, R., et al. (2019). Green synthesis of zinc oxide nanoparticles from root extract of *Scutellaria baicalensis* and its photocatalytic degradation activity using methylene blue. *Optik, 184*, 324–329.

Chinni, S. V., Gopinath, S. C. B., Anbu, P., Fuloria, N. K., Fuloria, S., Mariappan, P., Krusnamurthy, K., Reddy, L. V., Ramachawolran, G., Sreeramanan, S., & Samuggam, S. (2021). Characterization and antibacterial response of silver nanoparticles biosynthesized using an ethanolic extract of Coccinia indica leaves. *Crystals, 11*, 1–12.

Cunningham, S. D., Shann, J. R., Crowley, D. E., & Anderson, T. A. (1997). Phytoremediation of contaminated water and soil. In E. L. Kruger, T. A. Anderson, & J. R. Coats (Eds.), *Phytoremediation of soil and water contaminants. ACS symposium series 664* (pp. 2–19). American Chemical Society.

Das, S., Das, J., Samadder, A., Bhattacharyya, S. S., Das, D., & Khuda-Bukhsh, A. R. (2013). Biosynthesized silver nanoparticles by ethanolic extracts of *Phytolacca decandra, Gelsemium sempervirens, Hydrastis canadensis* and *Thuja occidentalis* induce differential cytotoxicity through G2/M arrest in A375 cells. *Colloids and Surfaces, B: Biointerfaces, 101*, 325–336.

Ebbs, S. D., Bradfield, S. J., Kumar, P., White, J. C., Musante, C., & Ma, X. (2016). Accumulation of zinc, copper, or cerium in carrot (*Daucus carota)* exposed to metal oxide nanoparticles and metal ions. *Environmental Science. Nano, 3*, 114–126.

Ebrahimzadeh, M. A., Mortazavi-Derazkola, S., & Zazouli, M. A. (2019). Eco-friendly green synthesis and characterization of novel Fe3O4/SiO2/Cu2O–Ag nanocomposites using *Crataegus pentagyna* fruit extract for photocatalytic degradation of organic contaminants. *Journal of Materials Science: Materials in Electronics, 30*, 10994–11004.

Firdhouse, M., & Jannathul, P. L. (2013). Biosynthesis of silver nanoparticles using the extract of *Alternanthera sessilis*—antiproliferative effect against prostate cancer cells. *Cancer Nanotechnology, 4*(6), 37–143.

Ghormade, V., Deshpande, M. V., & Paknikar, K. M. (2011). Perspectives for nano-biotechnology enabled protection and nutrition of plants. *Biotechnology Advances, 29*(6), 792–803.

Ghosh, S., Patil, S., Ahire, M., Kitture, R., Kale, S., Pardesi, K., & Chopade, B. A. (2012). Synthesis of silver nanoparticles using *Dioscorea bulbifera* tuber extract and evaluation of

its synergistic potential in combination with antimicrobial agents. *International Journal of Nanomedicine, 7*, 483.
Gong, J. L., Wang, B., Zeng, G. M., Yang, C. P., Niu, C. G., Niu, Q. Y., et al. (2009). Removal of cationic dyes from aqueous solution using magnetic multi-wall carbon nanotube nanocomposite as adsorbent. *Journal of Hazardous Materials, 164*(2–3), 1517–1522.
Gong, X., Huang, D., Liu, Y., Peng, Z., Zeng, G., Xu, P., et al. (2018). Remediation of contaminated soils by biotechnology with nanomaterials: Bio-behavior, applications, and perspectives. *Critical Reviews in Biotechnology, 38*(3), 455–468.
Grammatikopoulos, P., Stephan, S., Jerome, V., Singh, V., & Sowwan, M. (2016). Nanoparticle design by gas-phase synthesis. *Advances in Physics: X, 1*, 81–100.
Hett, A. (2004). *Nanotechnology: small matter, many unknowns*. Swiss Reinsurance Company.
Hofmann, T., & Von der Kammer, F. (2009). Estimating the relevance of engineered carbonaceous nanoparticle facilitated transport of hydrophobic organic contaminants in porous media. *Environmental Pollution, 157*(4), 1117–1126.
Hussain, A., Rehman, F., Rafeeq, H., Waqas, M., Asghar, A., Afsheen, N., Rahdar, A., Bilal, M., & Hafiz, M. N. (2022). In-situ, ex-situ, and nano-remediation strategies to treat polluted soil, water, and air—A review. *Chemosphere, 289*, 133252.
Jiamjitrpanich, W., Parkpian, P., Polprasert, C., & Kosanlavit, R. (2012). *Enhanced phytoremediation efficiency of TNT-contaminated soil by nanoscale zero valent iron* (Vol. 35). 2nd International Conference on Environment and Industrial Innovation.
Jones, K. C., & de Voogt, P. (1999). Persistent organic pollutants (POPs): State of the science. *Environmental Pollution, 100*(1–3), 209–221.
Kang, J. W. (2014). Removing environmental organic pollutants with bioremediation and phytoremediation. *Biotechnology Letters, 36*, 1129–1139.
Kannan, D. S., Mahboob, S., Al-Ghanim, K. A., & Venkatachalam, P. (2021). Antibacterial, antibiofilm and photocatalytic activities of biogenic silver nanoparticles from Ludwigia octovalvis. *Journal of Cluster Science, 32*, 255–264.
Kora, A. J., Sashidhar, R. B., & Arunachalam, J. (2012). Aqueous extract of gum olibanum (*Boswellia serrata*): a reductant and stabilizer for the biosynthesis of antibacterial silver nanoparticles. *Process Biochemistry, 47*(10), 1516–1520.
Liang, J., Yang, Z., Tang, L., Zeng, G., Yu, M., Li, X., & Luo, Y. (2017). Changes in heavy metal mobility and availability from contaminated wetland soil remediated with combined biochar-compost. *Chemosphere, 181*, 281–288.
Logambal S., Thilagavathi T., Chandrasekar M., Inmozhi C., Ebanda Kedi B. P., Bassyouni F. A., Uthrakumar R., Muthukumaran A., Naveenkumar S., Kaviyarasu K. (2022). Synthesis and antimicrobial activity of silver nanaoparticles; incorporated *Couroupita guianensis* flower petal extract for biomedical applications.
Lukman, A. I., Gong, B., Marjo, C. E., Roessner, U., & Harris, A. T. (2011). Facile synthesis, stabilization, and anti-bacterial performance of discrete Ag nanoparticles using *Medicago sativa* seed exudates. *Journal of Colloid and Interface Science, 353*(2), 433–444.
Ma, X., & Wang, C. (2010). Fullerene nanoparticles affect the fate and uptake of trichloroethylene in phytoremediation systems. *Environmental Engineering Science, 27*, 989–992.
Marchiol, L., Iafisco, M., Fellet, G., & Adamiano, A. (2020). Nanotechnology support the next agricultural revolution: Perspectives to enhancement of nutrient use efficiency. *Advances in Agronomy, 161*, 27–116.
Maurya, S., Bhardwaj, A. K., Gupta, K. K., Agarwal, S., Kushwaha, A., Vk, C., Pathak, R. K., Gopal, R., Uttam, K. N., Singh, A. K., & Verma, V. (2016). Green synthesis of silver nanoparticles using Pleurotus and its bactericidal activity. *Cellular and Molecular Biology, 62*(3), 1000131.
Mittal, A. K., Chisti, Y., & Banerjee, U. C. (2013). Synthesis of metallic nanoparticles using plant extracts. *Biotechnology Advances, 31*(2), 346–356.
Mittal, A. K., Tripathy, D., Choudhary, A., Aili, P. K., Chatterjee, A., Singh, I. P., & Banerjee, U. C. (2015). Bio-synthesis of silver nanoparticles using *Potentilla fulgens* And its therapeu-

tic evaluation as anticancer and antimicrobial agent. *Materials Science & Engineering. C, Materials for Biological Applications, 53*, 120–127.

Mitton, F. M., Miglioranza, K. S., Gonzalez, M., Shimabukuro, V. M., & Monserrat, J. M. (2014). Assessment of tolerance and efficiency of crop species in the phytoremediation of DDT polluted soils. *Ecological Engineering, 71*, 501–508.

Mohan, D., & Pittman, C. U., Jr. (2007). Arsenic removal from water/wastewater using adsorbents—A critical review. *Journal of Hazardous Materials, 142*, 1–53.

Mokarram-Kashtiban, S., Hosseini, S. M., Tabari Kouchaksaraei, M., & Younesi, H. (2019). The impact of nanoparticles zero-valent iron (nZVI) and rhizosphere microorganisms on the phytoremediation ability of white willow and its response. *Environmental Science and Pollution Research International, 26*, 10776–10789.

Mulugeta, H. O., Fedlu, K. S., Eneyew, T. B., & Bedasa, A. G. (2022). *Citrus sinensis* and *Musa acuminata* peel waste extract mediated synthesis of TiO2/rGO nanocomposites for photocatalytic degradation of methylene blue under visible light irradiation. *Bioinorganic Chemistry and Applications*.

Narayanan, K. B., & Park, H. H. (2014). Antifungal activity of silver nanoparticles synthesized using turnip leaf extract (*Brassica rapa L.*) against wood rotting pathogens. *European Journal of Plant Pathology, 140*, 185–192.

Negri, M. C., & Hinchman, R. R. (2000). *The use of plants for the treatment of radionuclides*. Using plants to clean up the environment. Wiley.

Nnadozie, E. C., & Ajibade, P. A. (2020). Green synthesis and characterization of magnetite (Fe3O4) nanoparticles using *Chromolaena odorata* root extract for smart nanocomposite. *Materials Letters, 263*, 127145.

Ovais, M., Khalil, A., Ayaz, M., Ahmad, I., Nethi, S., & Mukherjee, S. (2018). Biosynthesis of metal nanoparticles via microbial enzymes: A mechanistic approach. *International Journal of Molecular Sciences, 19*, 4100.

Padmavathi, P. J., Mani, M., Gokulakumar, B., Ramesh, A., Ananatharaj, A., & Kaviyarasu, K. (2022). A study on the antimicrobial activity of silver nanoparticles derived from *Corchorus aestuans* leaves and their characterization. *Chemical Physics Letters, 805*, 139952.

Pal, S. L., Jana, U., Manna, P. K., Mohanta, G. P., & Manavalan, R. (2011). *Nanoparticle: an overview of preparation and characterization*. J Appl. Pharm.

Pan, B., & Xing, B. (2008). Adsorption mechanisms of organic chemicals on carbon nanotubes. *Environmental Science & Technology, 42*(24), 9005–9013.

Pandey, V. C., Pandey, D. N., & Singh, N. (2015). Sustainable phytoremediation based on naturally colonizing and economically valuable plants. *Journal of Cleaner Production, 86*, 37–39.

Pilaquinga, F., Morejón, B., Ganchala, D., Morey, J., Piña, N., Debut, A., & Neira, M. (2019). Green synthesis of silver nanoparticles using *Solanum mammosum* L. (Solanaceae) fruit extract and their larvicidal activity against *Aedes aegypti* L. (Diptera: Culicidae). *PLoS One, 14*(10), e0224109.

Pillai, H. P., & Kottekottil, J. (2016). Nano-phytotechnological remediation of endosulfan using zero valent iron nanoparticles. *Journal of Environmental Protection, 7*, 734–744.

Poguberović, S. S., Krčmar, D. M., Maletić, S. P., Kónya, Z., Pilipović, D. D., Kerkez, D. V., & Rončević, S. D. (2016). Removal of As (III) and Cr (VI) from aqueous solutions using "green" zero-valent iron nanoparticles produced by oak, mulberry and cherry leaf extracts. *Ecological Engineering, 90*, 42–49.

Ren, X., Zeng, G., Tang, L., Wang, J., Wan, J., Liu, Y., & Deng, R. (2018). Sorption, transport and biodegradation: An insight into bioavailability of persistent organic pollutants in soil. *The Science of the Total Environment, 610–611*, 1154–1163.

Samuel, M. S., Ravikumar, M., John, J. A., Selvarajan, E., Patel, H., Chander, P. S., Soundarya, J., Vuppala, S., Balaji, R., & Chandrasekar, N. (2022). A review on green synthesis of nanoparticles and their diverse biomedical and environmental applications. *Catalysts, 12*, 459.

Santhosh, C., Velmurugan, V., Jacob, G., Jeong, S. K., Grace, A. N., & Bhatnagar, A. (2016). Role of nanomaterials in water treatment applications: a review. *Chemical Engineering Journal, 306*, 1116–1137.

Sathyavathi, R., Krishna, M. B., Rao, S. V., Saritha, R., & Rao, D. N. (2010). Biosynthesis of silver nanoparticles using *Coriandrum sativum* leaf extract and their application in nonlinear optics. *Advanced Science Letters, 3*(2), 138–143.

Satyavani, K., Ramanathan, T., & Gurudeeban, S. (2011). Plant mediated synthesis of biomedical silver nanoparticles by using leaf extract of *Citrullus colocynthis*. *Research Journal of Nanoscience and Nanotechnology, 1*(2), 95–101.

Sen, S. K., Raut, S., Bandyopadhyay, P., & Raut, S. (2016). Fungal decolouration and degradation of azo dyes: A review. *Fungal Biology Reviews, 30*(3), 112–133.

Shameli, K., Ahmad, M. B., Zamanian, A., Sangpour, P., Shabanzadeh, P., Abdollahi, Y., & Zargar, M. (2012). Green biosynthesis of silver nanoparticles using *Curcuma longa* tuber powder. *International Journal of Nanomedicine, 7*, 5603.

Sheoran, V., Sheoran, A. S., & Poonia, P. (2016). Factors affecting phytoextraction: A review. *Pedosphere, 26*, 148–166.

Siddiqui, S., Dickens, J. M., Cunningham, B. E., Hutton, S. J., Pedersen, E. I., Harper, B., Harper, S., & Brander, S. M. (2022). Internalization, reduced growth, and behavioural effects following exposure to micro and nano tire particles in two estuarine indicator species. *Chemosphere, 296*, 133934.

Singh, J., & Dhaliwal, A. S. (2019). Novel green synthesis and characterization of the antioxidant activity of silver nanoparticles prepared from *Nepeta leucophylla* root extract. *Analytical Letters, 52*, 213–230.

Singh, K., Panghal, M., Kadyan, S., Chaudhary, U., & Yadav, J. P. (2014). Green silver nanoparticles of Phyllanthus amarus: as an antibacterial agent against multi drug resistant clinical isolates of Pseudomonas aeruginosa. *Journal of Nanobiotechnology, 12*(1), 40.

Singh, R. L., Singh, P. K., & Singh, R. P. (2015). Enzymatic decolorization and degradation of azo dyes—A review. *International Biodeterioration and Biodegradation, 104*, 21–31.

Smitha, S. L., Philip, D., & Gopchandran, K. G. (2009). Green synthesis of gold nanoparticles using Cinnamomum zeylanicum leaf broth. *Spectrochimica Acta A, 74*(3), 735–739.

Song, J. Y., Kwon, E. Y., & Kim, B. S. (2010). Biological synthesis of platinum nano-particles using Diopyros kaki leaf extract. *Bioprocess and Biosystems Engineering, 33*(1), 159–164.

Song, B., Zeng, G., Gong, J., Liang, J., Xu, P., Liu, Z., et al. (2017). Evaluation methods for assessing effectiveness of in situ remediation of soil and sediment contaminated with organic pollutants and heavy metals. *Environment International, 105*, 43–55.

Song, B., Xu, P., Chen, M., Tang, W., Zeng, G., Gong, J., et al. (2019). Using nanomaterials to facilitate the phytoremediation of contaminated soil. *Critical Reviews in Environmental Science and Technology, 49*(9), 791–824.

Soundarrajan, C., Sankari, A., Dhandapani, P., Maruthamuthu, S., Ravichandran, S., Sozhan, G., & Palaniswamy, N. (2012). Rapid biological synthesis of platinum nanoparticles using Ocimum sanctum for water electrolysis applications. *Bioprocess and Biosystems Engineering, 35*, 827–833.

Sukumar, S., Rudrasenan, A., & Padmanabhan Nambiar, D. (2020). Green-synthesized rice-shaped copper oxide nanoparticles using *Caesalpinia bonducella* seed extract and their applications. *ACS Omega, 2020*(5), 1040–1051.

Vakili, M., Rafatullah, M., Salamatinia, B., Abdullah, A. Z., Ibrahim, M. H., et al. (2014). Application of chitosan and its derivatives as adsorbents for dye removal from water and wastewater: A review. *Carbohydrate Polymers, 113*, 115–130.

Vijayakumar, S., Vaseeharan, B., Malaikozhundan, B., Gopi, N., Ekambaram, P., Pachaiappan, R., & Suriyanarayanamoorthy, M. (2017). Therapeutic effects of gold nanoparticles synthesized using *Musa paradisiaca* peel extract against multiple antibiotic resistant *Enterococcus faecalis* bioflms and human lung cancer cells (A549). *Microbial Pathogenesis, 102*, 173–183.

Vijaykumar, G., Kesavan, H., Kannan, A., Arulanandam, D., Kim, J. H., Kim, K. J., Song, H. J., Kim, H. J., & Rangarajulu, S. K. (2021). Phytosynthesis of copper nanoparticles using extracts of spices and their antibacterial properties. *Processes, 2021*(9), 1341.

Vincent, J., Lau, K. S., Evyan, Y. C.-Y., Chin, S. X., Sillanpää, M., & Chia, C. H. (2022). Biogenic synthesis of copper-based nanomaterials using plant extracts and their applications: current and future directions. *Nanomaterials, 12*, 3312.

Vivek, M., Kumar, P. S., Stef, S., & Sudha, S. (2011). Biogenic silver nanoparticles by *Gelidiella acerosa* extract and their antifungal effects. *Avicenna Journal Of Medical Biotechnology, 3*(3), 143.

Vivekanandhan, S., Misra, M., & Mohanty, A. K. (2009). Biological synthesis of silver nanoparticles using Glycine max (soybean) leaf extract: An investigation on different soybean varieties. *Journal of Nanoscience and Nanotechnology, 9*(12), 6828–6833.

Wang, P., Lombi, E., Zhao, F. J., & Kopittke, P. M. (2016). Nanotechnology: A new opportunity in plant sciences. *Trends in Plant Science, 21*(8), 699–712.

Wijesinghe, U., Thiripuranathar, G., Menaa, F., & Almukhlifi, H. (2023). Eco-friendly synthesis of zinc oxide nanoparticles and assessment of their activities as efficient antioxidant agents. *Current Nanoscience, 19*(1), 132–146. (15).

Wolny-Koładka, K., Malina, D., Suder, A., Pluta, K., & Wzorek, Z. (2022). Bio-based synthesis of silver nanoparticles from waste agricultural biomass and its antimicrobial activity. *Processes, 10*, 389.

Yang, L., & Watts, D. J. (2005). Particle surface characteristics may play an important role in phytotoxicity of alumina nanoparticles. *Toxicology Letters, 164*(2), 185–187.

Zaier, H., Ghnaya, T., Ghabriche, R., Chmingui, W., Lakhdar, A., Lutts, S., & Abdelly, C. (2014). EDTA-enhanced phytoremediation of lead-contaminated soil by the halophyte *Sesuvium portulacastrum. Environmental Science and Pollution Research, 21*, 7607–7615.

Zhang, W. (2003). Nanoscale iron particles for environmental remediation—An overview. *Journal of Nanoparticle Research, 5*, 323–332.

Zhang, Y., Cheng, S., Jia, H., Zhou, J., Xi, J., Wang, J., Chen, X., & Wu, L. (2022). Green synthesis of platinum nanoparticles by *Nymphaea tetragona* flower extract and their skin lightening, antiaging effects. *Arabian Journal of Chemistry, 16*, 104391.

Zheng, L., Hong, F. S., Lu, S. P., & Liu, C. (2005). Effect of nano-TiO2 on strength of naturally and growth aged seeds of spinach. *Biological Trace Element Research, 104*, 83–91.

Index

A. K. Bhardwaj et al. (eds.), *Biogenic Wastes-Enabled Nanomaterial Synthesis*,
https://doi.org/10.1007/978-3-031-59083-2

SPRINGER NATURE

GPSR Compliance

The European Union's (EU) General Product Safety Regulation (GPSR) is a set of rules that requires consumer products to be safe and our obligations to ensure this.

If you have any concerns about our products, you can contact us on ProductSafety@springernature.com

In case Publisher is established outside the EU, the EU authorized representative is:

Springer Nature Customer Service Center GmbH
Europaplatz 3
69115 Heidelberg, Germany

The manufacturer's authorised representative in the EU is Springer Nature Customer Service Centre GmbH, Europaplatz 3, 69115 Heidelberg, Germany. If you have any concerns regarding our products, please contact ProductSafety@springernature.com

Printed and bound by CPI Group (UK) Ltd, Croydon, CR0 4YY
28/09/2026
02224077-0011